Bruce C. Glavovic • Gavin P. Smith
Editors

# Adapting to Climate Change

## Lessons from Natural Hazards Planning

Springer

*Editors*
Bruce C. Glavovic
School of People, Environment & Planning Program
Massey University
New Zealand

Gavin P. Smith
Department of City and Regional Planning
University of North Carolina at Chapel Hill
North Carolina
USA

ISBN 978-94-017-8630-0 ISBN 978-94-017-8631-7 (eBook)
DOI 10.1007/978-94-017-8631-7
Springer Dordrecht Heidelberg New York London

Library of Congress Control Number: 2014931854



Printed on acid-free paper

Springer is part of Springer Science+Business Media (www.springer.com)

*Bruce C. Glavovic:*
*For Christopher and Julia who give meaning and purpose to life.*

*Gavin P. Smith:*
*For Libby who makes all good things possible.*

# Foreword

This book is aimed at helping communities adapt to climate change by outlining a set of lessons derived from natural hazards experience around the world together with insights from contemporary scholarship. These lessons are framed in the context of building resilient and sustainable communities in this era of climate change. The book addresses several vital questions. What are the root causes of increasing climate change risk? What are the key considerations for reducing climate change impacts? What institutional barriers and opportunities do communities face as they attempt to adapt to climate change? What role should key actors, including government, the private sector, civil society and the research community, play in building adaptive capacity? What can be done to recover from the adverse impacts of natural hazards, disasters and climate change? These questions are answered in a manner that helps to expand our knowledge base, placing a particular emphasis on integrative scholarship across complementary fields of research. The lessons drawn from experience around the world will help communities take more effective actions to develop and implement integrated risk reduction and climate change adaptation strategies.

The book will be of particular interest to public and private sector professionals working on issues related to climate change adaptation, including policy analysts, land-use planners, environmental managers, engineers, public administrators and community development specialists. It provides practical guidance for key decision-makers, including elected and appointed government officials, business leaders and leaders of non-governmental and community based organizations. The book will be relevant to students and scholars in fields ranging from environmental science to geography, planning, development studies, politics, public administration, policy analysis, emergency management, and emerging interdisciplinary fields such as sustainability studies and adaptive management. The book will also be of special interest to people in communities who are searching for innovative and practical ways to manage natural hazard risks and adapt to climate change.

There is unequivocal evidence that our climate is changing due to human activity and this change is contributing to escalating disaster risk. The ability to adapt to these changes represents one of the great challenges facing humankind in the twenty-first century. The magnitude of the challenges surrounding climate change adaptation (and hence the importance of the author's endeavor) can be illustrated by recent statistics on extreme events around the world. Between 2002 and 2012, there were just over 4,100 disasters around the world, equating to about 370 per year, and almost 91 % of them were related to climate and weather-related events, including floods, storms and extreme temperatures (Centre for the Research on the Epidemiology of Disasters, CRED 2013). During this same time period (2001–2012) were the 13 warmest years on record (World Meteorological Organization, 2013).

In December 2009, most countries endorsed the Copenhagen Accord of 18 December 2009, which stated that "*We underline that climate change is one of the greatest challenges of our time. We emphasise our strong political will to urgently combat climate change in accordance with the principle of common but differentiated responsibilities and respective capabilities.*" The Accord went on to emphasize the need for deep cuts in global emissions. In the following paragraph, it was stated that "*Adaptation to the adverse effects of climate change and the potential impacts of response measures is a challenge faced by all countries.*" Climate change adaptation is also included in the United Nations Framework Convention on Climate Change which states that "*adaptation to the adverse effects of climate change is vital in order to reduce the impacts of climate change that are happening now and increase resilience to future impacts.*"

In 2005, nations of the world gathered in Kobe, Japan, for the World Conference on Disaster Reduction and declared that: "*Disasters have a tremendous detrimental impact on efforts at all levels to eradicate global poverty; the impact of disasters remains a significant challenge to sustainable development.*" The Conference also adopted the Hyogo Framework for Action (UNISDR 2005), which opened with: "*The starting point for reducing disaster risk and for promoting a culture of disaster resilience lies in the knowledge of the hazards and the physical, social, economic and environmental vulnerabilities to disasters that most societies face, and of the ways in which hazards and vulnerabilities are changing in the short and long term,….*" In 2009, the same year as the Copenhagen Accord, the Assistant Secretary-General for Disaster Risk Reduction (UNISDR, International Strategy for Disaster Reduction) and Special Representative of the U.N. Secretary-General for the implementation of the Hyogo Framework for Action (Wahlstrom 2009) stated "*Over the last two decades (1988–2007), 76 % of all disaster events were hydrological, meteorological or climatological in nature; these accounted for 45 % of the deaths and 79 % of the economic losses caused by natural hazards.*" She concluded that: "*The real tragedy is that many of these deaths can be avoided.*" The relationship between disasters, climate change and development was further emphasized by the World Bank President (Kim 2013) who stated that "*global warming imperils all of the development gains we have made.*"

While the natural hazards risk management community has identified the need to address the social and economic costs of extreme events through proactive planning and coordinated actions across broad governance networks for more than 80 years, the climate change adaptation community has more recently been moving forward along parallel tracks that remain somewhat disconnected from this important base of research findings and experiential knowledge. In 1993, Burton et al. stated that it is important to consider "*The Environment as Hazard.*" They recognized that the natural environment is essential for human development but it is also essential that humans, in order to sustain their development, take actions to reduce the growing impacts of hazards on societies. The disconnect between the science of disaster risk reduction and the implementation of actions to reduce risks led the same authors in 2001 to publish a paper entitled "*Knowing better and losing even more: the use of knowledge in hazards management*" (White et al. 2001). An overall concern for addressing climate change, including adaptation, and for disaster risk reduction, is that there has been an inadequate use of available knowledge; namely a failure to act. Working together across disaster risk reduction and climate change adaptation governance networks through the use of natural hazards planning principles and practices provides a grounded and action-oriented way to address the issues and key questions explored throughout this book.

In March, 2009, as a result of discussions between the Intergovernmental Panel on Climate Change (IPCC) and the United Nations International Strategy for Disaster Reduction (UNISDR), a scoping meeting was held in Oslo to frame the content of a report that would ultimately become the IPCC Special Report on "*Managing the Risks of Extreme Events and Disasters to Advance Climate Change Adaptation*" (IPCC 2012). As one of the co-authors of the Special Report, I recall many lengthy and enlightening if sometimes heated discussions while preparing the report. Among the report's conclusions was that as the climate changes in the future, extreme events will likely increase in numbers and costs, in victims, deaths and dollars (IPCC 2012). It is clear that there need to be stronger links between responding to climate change via adaptation and disaster risk reduction programmes, policies and plans.

This book addresses a key part of how to accomplish this aim—how to bring together the knowledge of the disaster risk reduction community with that of the climate change adaptation community. The book's introductory chapter, *Learning from Natural Hazards Experience to Adapt to Climate Change* sets the scene and outlines how the book proposes to address this challenge. Chapter 2 is focused on conceptual approaches, synergies and mismatches between climate change adaptation and disaster risk reduction. Chapters 3 through 15 present case studies that provide insights from real world hazards and disaster experiences that are directly relevant to developing and implementing integrated approaches to risk reduction and adaptation. The case studies, written by a stellar group of scholars from appropriately diverse backgrounds, elucidate the distinctive features that shape exposure and vulnerability to natural hazards and climate change; the lessons learned (and not learned) from past experience; existing barriers and opportunities; and future actions aimed at addressing identified challenges. These chapters also draw upon the theory and practice of natural hazards planning and climate change adaptation, describing disasters and how planning can help to build resilient and sustainable communities.

Case studies on coastal cities generally and specifically in South Africa and in the context of hurricanes Fran, Floyd, Camille and Katrina in the United States are an important focus, recognizing this special at-risk zone. Although tsunamis are not the result of climate change, there are also important lessons to be learned from the tsunami in 2004 in the Indian Ocean and the 2009 event affecting Samoa. Experiences with earthquakes in the Peruvian highlands as well as the cities of Berkeley and San Francisco, California (USA) also provide valuable lessons, including the value of recognizing the adaptive capacities of long-standing cultures and indigenous knowledge and the importance of understanding societal and physical dimensions of resilience.

Studies of floods in England and New Zealand are critically important given the fact that floods are the single most common and costly natural hazard worldwide and the strategies devised to reduce their impact are among the most developed. However, the challenges inherent in engineering versus land-use or spatial planning based approaches provide important insights to explore in the context of a changing climate. Similarly, framing policy options intended to address the threat of bushfires in Victoria, Australia from the perspective of emergency management or sustainability helps to inform how varied actors perceive risk and take action based on these perceptions. The Mississippi Delta with its confluence of floods, hurricanes, oil spills and climate change brings together a broad set of hazards and leads the author to recommend adopting a deliberative risk governance strategy.

The conclusions chapter, *Integrating Natural Hazards Risk Management and Climate Change Adaptation through Natural Hazards Planning* outlines four imperatives (governance, capability, planning and moral) for integrating and mainstreaming natural hazards risk management and climate change adaptation. Nested within each of these imperatives are lessons derived from the preceding chapters. Considering the diversity of cases and breadth of topics addressed by the book's authors, the emergence of a strikingly similar set of common lessons strengthens the argument that an integrative framework for risk reduction and adaptation is possible and will benefit from the rich information assembled in this book.

The global scientific community, professionals and practitioners from diverse backgrounds and community members around the world are faced with the grand challenge of adapting to a changing climate. A critically important part of this endeavour requires providing the basis for evidence-based decision-making and taking action leading to reduced losses due to extreme events, including those driven by a changing climate. This book advocates natural hazards planning as a practical means to bridge the divide between the hitherto parallel discourses on disaster risk reduction and climate change adaptation. It identifies lessons drawn from scholarship and real-world experience and as such represents a major contribution towards addressing one of the greatest challenges of the twenty-first century: reducing risk and adapting to climate change.

Gordon McBean (Western University, London, Ontario, Canada; President-elect from 2013, International Council for Science).

## References

Burton, I., Kates, R.W. and White, G.F. (1993). The Environment as Hazard. The Guilford Press: New York and London, 290 pp.

Copenhagen Accord (2009). http://unfccc.int/meetings/copenhagen_dec_2009/items/5262.php

CRED, (2013): Disaster Data: A Balanced Perspective. March 2013, Issue No. 31, http://www.cred.be

IPCC (2012). Summary for Policymakers. In: Managing the Risks of Extreme Events and Disasters to Advance Climate Change Adaptation [Field, C.B., V. Barros, T.F. Stocker, D. Qin, D.J. Dokken, K.L. Ebi, M.D. Mastrandrea, K.J. Mach, G.-K. Plattner, S.K. Allen, M. Tignor, and P.M. Midgley (eds.)]. A Special Report of Working Groups I and II of the Intergovernmental Panel on Climate Change. Cambridge University Press, Cambridge, UK, and New York, NY, USA, pp. 1–19.

Kim, J.Y. (2013). OPED – Make climate change a priority. http://www.worldbank.org/en/about/president

UNISDR (2005). Hyogo Framework for Action 2005–2015: Building the Resilience of Nations and Communities to Disasters. World Conference on Disaster Reduction http://www.unisdr.org/wcdr/intergover/official-doc/L-docs/Hyogo-framework-for-actionenglish.

Wahlstrom. M. (Assistant Secretary-General for Disaster Risk Reduction and Special Representative of the U.N. Secretary-General for the implementation of the Hyogo Framework for Action) (2009). Quoted. In: Birkmann J., Tetzlaff G., Zentel K. O., (eds). *Addressing the Challenge: Recommendations and Quality Criteria for Linking Disaster Risk Reduction and Adaptation to Climate Change. DKKV Publication Series*, 38:5.

White, G.F., Kates, R.W. and Burton, I. (2001). Knowing better and losing even more: the use of knowledge in hazards management. *Environmental Hazards, 3*(3–4), 81–92.

World Conference on Disaster Reduction (2005). World Conference on Disaster Reduction: proceedings of the conference. http://www.unisdr.org/we/inform/publications/3800

World Meteorological Organization (2013). WMO Annual Climate Statement Confirms 2012 as among the top ten warmest years. http://www.wmo.int/pages/mediacentre/press_releases/pr_972_en.html

# Preface

The genesis for this book, drawing lessons from natural hazards planning and applying them to challenges associated with adapting to a changing climate, represents a natural evolution of our ongoing working relationship. We first met in the aftermath of Hurricane Katrina as Gavin served as the Executive Director of the Mississippi Governor's Office of Recovery and Renewal, and Bruce was studying the effects of Hurricane Katrina on the City of New Orleans and the challenges of recovery in the region. Gavin's experiences with creating a way to explain the complexities of recovery to a growing staff in the Governor's Office, and a recognised gap in the academic literature, led to the writing of the book, *Planning for post-disaster recovery: A review of the United States disaster recovery assistance framework.* The book provides a critical analysis of disaster recovery as practiced in the United States and offers a series of recommendations tied to the critically important roles of planning and governance, both central components in *Adapting to climate change: Lessons from natural hazard planning.*

Shortly before the arrival of Hurricane Katrina, Bruce had been appointed to a new position at Massey University, sponsored by the New Zealand Earthquake Commission—which provides Government backed insurance for homes, contents and land, to explore the role of natural hazards planning in reducing disaster risk and building resilient, sustainable communities. He was seeking firsthand insight into the factors that predispose communities to disaster and to understand the barriers and opportunities for charting recovery pathways. Bruce has continued his research on the post-Katrina recovery experience, returning to the region regularly since 2006 to observe and learn from key recovery actors. Our combined practice and research-based experiences led us to realise that Katrina provided invaluable lessons about how to reduce risk and build more resilient communities; and that we were in a unique position to draw from and expand upon such lessons. We reflected on the root causes of disasters and the complexity of the hazard mitigation and recovery process and pondered how we might distil and share the lessons from this experience while drawing from natural hazards planning scholarship and practice to help build stronger, safer and more sustainable communities. We were struck by the need to learn from history and to look into the future and envisage how to plan for and manage natural hazard risks in an era of climate change. We were troubled by the limited attention focused on this imperative as the Gulf recovery process evolved. These conversations provided the origin for this book, and the start of our ongoing collaboration.

In the summer of 2008 Gavin gave the keynote address at the Australasian Hazards Management Conference in Wellington, New Zealand, focusing on the application of research and practice to recovery in Mississippi following Hurricane Katrina. Two years later Gavin wrote an article in a Special Issue of the *Australasian Journal of Disaster and Trauma Studies* (http://trauma.massey.ac.nz/issues/previous.shtml#2010-1), focusing on disaster recovery lessons. Bruce had initiated the Special Issue to distil insights from natural hazards planning scholarship for disaster risk reduction in Australasia, drawing, in part, from Gavin's experience in the USA. After returning from two years working in Mississippi, Gavin accepted a position as the Executive Director of the Department of Homeland Security's Coastal Hazards Center of Excellence at the University of North Carolina at Chapel Hill. Bruce was appointed Associate Director of the Massey University—GNS Joint Centre for Disaster Research. We established a Memorandum of Understanding between the two universities, which was signed in 2009, encouraging scholars to identify opportunities to collaborate on disaster research and scholarship. Bruce made a number of visits to the University of North Carolina at Chapel Hill (UNC) to share his research findings on the post-Katrina recovery experience and to work on this book with Gavin. In November, 2010, Gavin and Bruce contributed to a Theory of Recovery workshop at UNC that was convened by the Public Entity Research Institute with the support of the National Science Foundation. Participants developed a definition of disaster recovery, identified variables that influence the recovery process and outlined a research agenda to analyse these variables and ultimately inform policy-making and facilitate improved recovery outcomes. The workshop findings were published in a Special Issue of the journal *International Journal of Mass Emergencies and Disasters* (http://www.ijmed.org/issues/30/2/). The need to draw from and apply lessons from natural hazard mitigation and disaster recovery to arguably the most pressing challenge of the twenty-first century (i.e., climate change) was clear and compelling.

The co-editing of this book thus became the focus of our collaboration. We invited leading disaster and natural hazards planning scholars to contribute to the book and were encouraged by the overwhelmingly positive response. Bruce's ability to sustain focused attention editing the book was, however, impacted by the necessity to contribute to the disaster recovery process in the aftermath of a series of major earthquakes and aftershocks that struck Greater Christchurch. Seismic activity, which began on the 4th of September 2010, continued for at least 2 years, severely affecting many liquefaction prone neighbourhoods, killing 185 people and devastating the central business district on the 22nd of February 2011. This experience underscored the need to learn from and apply lessons from natural hazards planning scholarship and experience; especially given ongoing seismic risk and the exposure of some areas of Christchurch to sea-level rise. The imperative to learn from and apply natural hazards planning lessons in North Carolina, USA, were also underscored by Gavin's work with the State as he served as the Assistant Director for Hazard Mitigation during the two worst disasters in the state's history. Hurricane's Fran (1996) and Floyd (1999), which struck in close proximity to one another, provided a unique opportunity to draw lessons from one event to another and for Gavin to work closely with Governor James B. Hunt Jr. and the Director of the North Carolina Division of Emergency Management, Eric Tolbert, both of whom embraced efforts to link sustainable development, hazard mitigation and disaster recovery. Tangible examples of the state's efforts included the creation of 22 state-level programs intended to address gaps in post-disaster federal assistance, the acquisition and relocation of over 5,000 flood-prone homes in one of the largest single-state efforts of its kind in the USA, the state-led re-mapping of North Carolina's floodplains, and the creation of a state-led hazard mitigation planning effort, which influenced the federal rules promulgated under the Disaster Mitigation Act of 2000. The pivotal role of planning to reduce disaster risk and build resilience and adaptive capacity was underscored by these experiences and is increasingly recognized in the USA and New Zealand. But there is little information available that provides practical insights into real-world hazard experiences that can assist communities in these and other countries to plan for and adapt to a changing climate. This book seeks to fill this gap.

We would like to thank the many people and organisations who enabled us to complete this book. First, we would like to sincerely thank Katharine Moody for providing invaluable assistance in the technical editing of the book as well as working tirelessly with authors to ensure that each chapter was satisfactorily completed. Second, we would like to thank each of the contributing authors for sharing their knowledge, insights and experience with us. Together, we believe, these contributions provide a vital platform for natural hazards planning in this era of climate change.

Bruce would like to thank his current and previous Head of School at Massey University, Allanah Ryan and Henry Barnard, for their support over the years; and Massey University for providing the financial means to employ Katharine to assist with technical editing and undertake replacement teaching, and enable Bruce to conduct fieldwork in the Gulf Coast and work on this book. He would also like to thank the Earthquake Commission, and in particular Hugh Cowan and former CEO David Middleton, for their support and funding that enabled him to study the post-Katrina experience in the Gulf Coast and travel to UNC to work with Gavin on this book. In addition, he would like to thank David Johnston, Director: Joint Centre for Disaster Research, for providing a 2 year sub-contract from a Foundation for Research, Science and Technology grant that helped to cover fieldwork and travel costs incurred during the writing of this book.

Gavin would like to thank the Science and Technology Directorate, Office of University Programs in the U.S. Department of Homeland Security. Their support of the Coastal Hazards Center of Excellence has proven invaluable and enabled him to focus on a number of research projects, including the writing of this book, the findings of which are being used in the teaching of a DHS-supported class at the University of North Carolina at Chapel Hill titled Planning for Natural Hazards and Climate Change Adaptation.

We believe that this book provides an important source of information for scholars, students and practitioners as they frame new questions and continue to research unanswered questions, seek to gain deeper understanding of interrelated complexities, and develop robust, flexible and actionable plans that guide and better integrate natural hazards risk management and climate change adaptation efforts.

# Contents

**1 Introduction: Learning from Natural Hazards Experience to Adapt to Climate Change** ..... 1
Bruce C. Glavovic and Gavin P. Smith

**Part I Climate Change Adaptation: Theory and Practice**

**2 Climate Change Adaptation and Disaster Risk Reduction: Fundamentals, Synergies and Mismatches** ..... 41
Jörn Birkmann and Joanna Pardoe

**3 Tackling Barriers to Climate Change Adaptation in South African Coastal Cities** ..... 57
Gina Ziervogel and Susan Parnell

**Part II The Nature of Disasters and the Role of Natural Hazards Planning in Building Resilient Communities**

**4 Climate Change Adaptation and Disaster Risk Reduction in Highland Peru** ..... 77
Anthony Oliver-Smith

**5 Firm Foundations or Castles on Sand? The Shifting Sources of Flood Risk and the Implications for Flood Governance: An English Case Study** ..... 101
Iain White

**6 Planning for Resilient Coastal Communities: Emerging Practice and Future Directions** ..... 123
Timothy Beatley

**7 Adaptation to Seismic Risk and Climate Change: San Francisco and Berkeley, California, USA** ........ 145
William Siembieda

**8 Rising to the Challenge: Planning for Adaptation in the Age of Climate Change** ........ 171
Philip R. Berke

**Part III Lessons from Disaster Experience**

**9 Applying Hurricane Recovery Lessons in the United States to Climate Change Adaptation: Hurricanes Fran and Floyd in North Carolina, USA** ........ 193
Gavin P. Smith

**10 The 2004 Manawatu Floods, New Zealand: Integrating Flood Risk Reduction and Climate Change Adaptation** ........ 231
Bruce C. Glavovic

**11 Learning From Analyses of Policy Frames and Informal Institutions in the Fire Management Sector of Victoria, Australia** ....... 269
Karyn Bosomworth, John Handmer and Stephen Dovers

**12 Recovering from the 2004 Indian Ocean Tsunami: Lessons for Climate Change Response** ........ 291
Ahana Lakshmi, Purvaja R. and Ramesh R.

**13 Coastal Hazards Planning: The 2009 Tsunami and Lessons Learned for Climate Change Adaptation in Samoa** ........ 315
Namouta Poutasi, Michele Daly, Jude Kohlhase and Filomena Nelson

**14 Disaster Recovery in Coastal Mississippi, USA: Lesson Drawing from Hurricanes Camille and Katrina** ........ 339
Gavin P. Smith

**15 Waves of Adversity, Layers of Resilience: Floods, Hurricanes, Oil Spills and Climate Change in the Mississippi Delta** ........ 369
Bruce C. Glavovic

**16 Conclusions: Integrating Natural Hazards Risk Management and Climate Change Adaptation through Natural Hazards Planning** ......... 405
Gavin P. Smith and Bruce C. Glavovic

**Index** ........ 451

# Contributors

**Timothy Beatley** University of Virginia, Charlottesville, VA, USA

**Philip R. Berke** University of North Carolina at Chapel Hill, Chapel Hill, NC, USA

**Jörn Birkmann** Institute for Environment and Human Security, United Nations University, Tokyo, Japan

**Karyn Bosomworth** RMIT University, Melbourne, Australia

**Michele Daly** Risk and Society Department, GNS Science Ltd, Avalon, Lower Hutt, New Zealand

**Stephen Dovers** Australian National University (ANU), Canberra, Australia

**Bruce C. Glavovic** Massey University, Palmerston North, New Zealand

**John Handmer** RMIT University, Melbourne, Australia

**Jude Kohlhase** Planning and Urban Management Agency, Ministry of Natural Resources and Environment Apia, Samoa

**Ahana Lakshmi** National Centre for Sustainable Coastal Management, Anna University Campus, Chennai, India

**Filomena Nelson** Disaster Management Office, Ministry of Natural Resources and Environment

**Anthony Oliver-Smith** University of Florida, Gainesville, FL, USA

**Joanna Pardoe** Institute for Environment and Human Security, United Nations University, Tokyo, Japan

**Susan Parnell** Cape Town, South Africa

**Namouta Poutasi** Integrity Professionals, Auckland, New Zealand

**Purvaja R.** National Centre for Sustainable Coastal Management, Anna University Campus, Chennai, India

**Ramesh R.** National Centre for Sustainable Coastal Management, Anna University Campus, Chennai, India

**William Siembieda** California Polytechnic State University, San Luis Obispo, CA, USA

**Gavin P. Smith** University of North Carolina at Chapel Hill, Chapel Hill, North Carolina, USA

**Iain White** University of Waikato, Hamilton, New Zealand

**Gina Ziervogel** University of Cape Town, Cape Town, South Africa

# Chapter 1
# Introduction: Learning from Natural Hazards Experience to Adapt to Climate Change

**Bruce C. Glavovic and Gavin P. Smith**

**Abstract** This book explores lessons learned from the study and real-world experience of natural hazards to help communities plan for and adapt to climate change.

**Keywords** Climate change adaptation · Natural hazards risk management · Natural hazards planning · Climate resilience · Sustainable development

Notwithstanding persistent skepticism, and even deliberate obfuscation about the climate change issue (Oreskes and Conway 2010), there is unequivocal scientific evidence that the climate system is warming on a scale that is unprecedented in modern times (IPCC[1] 2007a). Climate change is arguably the most serious foreseeable threat to human development because it has the potential to undermine efforts to date (Stern 2007; UNDP 2008; World Bank 2010) and alter the physical and human geography of the planet (Stern 2009). Least developed and developing nations are most vulnerable to climate risks (IPCC 2007b; Boyd et al. 2009; McBean and Rodgers 2010). Adapting to climate change is thus imperative and poses a major challenge for nations and communities around the world (IPCC 2007b). Many will have to contend with unprecedented impacts due to rising sea levels, temperature and rainfall shifts, and geographic shifts in disease vectors. Many face the prospect of more frequent and extreme weather and climate events. Rural and urban livelihoods will be profoundly affected. Poor and marginalised groups will be especially hard hit. Increasing attention is now being focused on how to adapt to climate change. But much remains to be done to understand underlying adaptation limits, barriers and opportunities.

[1] The Intergovernmental Panel on Climate Change (IPCC) was established in 1988 by the United Nations Environment Programme (UNEP) and World Meteorological Organization (WMO). It is the leading international scientific body for assessing climate change (see http://www.ipcc.ch/organization/organization.shtml).

B. C. Glavovic (✉)
School of People, Environment & Planning Massey University,
Palmerston North, New Zealand
e-mail: b.glavovic@massey.ac.nz

G. P. Smith
UNC at Chapel Hill, North Carolina, USA
e-mail: gavin_smith@unc.edu

B. C. Glavovic, G. P. Smith (eds.), *Adapting to Climate Change,* Environmental Hazards,
DOI 10.1007/978-94-017-8631-7_1, © Springer Science+Business Media Dordrecht 2014

Moreover, limited progress has been made in taking practical steps to build community resilience and sustainability in the face of escalating disaster risk. Statistics show that climate-related hazard events have been the trigger for the increase in disasters in recent decades (Centre for Research on the Epidemiology of Disasters 2010[2]), with floods, storms and droughts accounting for more than 75% of the disasters that occurred in the first decade of the millennium (McBean and Rodgers 2010).

Much can be learned from decades of experience in natural hazards risk management to better understand adaptation and take proactive steps to reduce climate risk, build adaptive capacity and facilitate recovery from climate change-driven disasters. This book provides conceptual insights and practical guidance by drawing on diverse streams of scholarship on climate change, natural hazards and disasters, with a particular focus on natural hazards planning, and real-world case studies of natural hazard and disaster experience from a range of settings.

The case studies in this book are written by leading hazards scholars from a variety of disciplines and include analyses of historical and more contemporary disaster experiences in settings ranging from industrialized to so-called developing countries, and rural to urban localities, at scales ranging from small communities to transnational regions in North America, Europe, Australasia, Asia, Africa and Small Island Developing States. A wide range of climate related natural hazard experiences are explored, including severe storms, sea-level related hazards, droughts, heat waves, wildfires, and floods. Additional natural hazard experiences discussed in this book include earthquakes and tsunami as they offer important lessons for climate change adaptation. The concluding chapter synthesizes insights from these case studies to provide practical recommendations on planning for and adapting to climate change.

This introductory chapter sets the scene for the book. First, it outlines the relationship between climate change adaptation, risk management and local community development; and explains key concepts used throughout the book, including climate change adaptation, natural hazards, disasters, risk, vulnerability and exposure. Second, it provides an introduction to insights from natural hazards risk management scholarship and experience that are relevant to climate change adaptation; and highlights the pivotal role that natural hazards planning can play in reducing disaster risk, facilitating post-disaster recovery and adapting to climate change. Finally, it outlines the structure of the book and provides a brief overview of the main sections and associated chapters.

## 1.1 Climate Change Adaptation, Natural Hazards and Local Community Development

The United Nations Framework Convention on Climate Change (UNFCCC) came into force in 1994 and constitutes the primary international framework to address climate change. Much attention was initially focused on addressing the source of the

[2] See http://www.emdat.be/Database/Trends/trends.html.

problem by seeking agreement on curbing or mitigating greenhouse gas emissions. However, even if such efforts were immediately effective—and despite decades of effort to this end, meaningful agreement remains elusive—climate change impacts will be experienced worldwide due to lag effects (IPCC 2001, 2007a). Climate change adaptation is therefore urgent and compelling. Yet focused international attention on adaptation is relatively recent, with momentum building following the 2007 Copenhagen Climate Change Summit.

### *1.1.1 Adapting to Climate Change*

What does climate change adaptation mean? According to the IPCC (2001, 2007b), it is the adjustment in natural or human systems in response to actual or anticipated climatic stimuli or their effects which moderates harm or exploits beneficial opportunities. Adaptation can take a variety of forms. ***Anticipatory adaptation*** is proactive and takes place before climate change impacts are experienced. ***Reactive adaptation*** takes place after impacts have been experienced. ***Private adaptation*** is driven by individuals, households or private entities whereas ***public adaptation*** is initiated and undertaken by government, possibly in partnership with non-state actors, to realize preferred public outcomes. ***Autonomous adaptation*** refers to spontaneous adjustments that are not necessarily consciously or deliberately initiated in response to climate change. ***Planned adaptation*** is an intentional and deliberate choice to take a course of action to return to, maintain or attain a desired state in light of anticipated climate change. Some responses or actions can paradoxically increase exposure and vulnerability to climate change and can thus be described as being ***maladaptive***. Furthermore, the adaptive choices affect groups and sectors of society in different ways, benefiting some while harming others.

Early thinking about adaptation tended to suggest that adaptation involved a once-off choice between different options in anticipation of or response to climate change impacts. For example, in the context of sea-level rise, a distinction was made between protect, accommodate and retreat options (Dronkers et al. 1990). The **protect** option is typically based on deploying hard-engineering approaches (e.g., seawalls) and/ or soft-engineering approaches (e.g., dune rehabilitation) to protect property and infrastructure against sea-level related impacts such as coastal erosion or storms. Hard-engineering approaches may be a last resort for protecting at risk assets but often generate secondary or downstream impacts and increasing attention is being focused on the feasibility of soft-engineering approaches and alternative risk management options. The **accommodate** option typically involves occupation of the seashore but mitigating climate impacts by taking steps to reduce risk (e.g., elevating buildings; growing salt-tolerant crops). The **retreat** option involves the progressive movement of human settlements landwards from the seashore to reduce exposure to climate driven increased storminess and sea-level related impacts through, for example, land-use planning provisions, the relocation of at-risk structures and infrastructure, adjustments to insurance premiums, and/or tax or other incentives. **Avoidance** is another strategy that involves the proactive decision to avoid putting people in harm's way, i.e., not allowing development in localities exposed to sea-level related hazards. Each of these adaptation options has widely varying financial, social and environmental consequences that need to be carefully considered in choosing the preferred option, bearing in mind private and public interests in both the short and long term.

There is understandable reluctance to stop or even alter long-standing practices because of the uncertain future impacts of climate change, especially when such activities are reliant on significant historic investment in property and infrastructure or are part of established cultural norms and traditions. When substantial investment is threatened by climate change impacts, there is an understandable desire to adopt protective measures—even if the secondary impacts are significant and inequitable (e.g., if a beachfront property owner erects a seawall that leads to an adjacent property owner suffering accelerated coastal erosion; and/or the costs of such protection is subsidized by general ratepayers who do not necessarily derive direct benefits from the seawall). Additional factors to consider include the reversal of policies that incentivise maladaptive behavior such as large post-disaster funding programs that pay to rebuild vulnerable communities to their pre-event condition or insurance policies or building codes that do not account for escalating risks tied to a changing climate.

Adaptation measures such as avoidance, accommodation and retreat only become compelling when the costs of 'business as usual' and/or protection are prohibitive. But judging what is appropriate and when alternative strategies should be triggered to chart an alternative development pathway is by no means easily defined let alone agreed upon. Successful adaptation is thus a relative concept and will vary depending on who makes this judgment at a particular point in time and locality. What might be seen to be appropriate at a local community scale in the short term might not be deemed prudent from a wider regional or national perspective in the long run.

From a resilience and sustainability perspective, one can argue that normative criteria of effectiveness, efficiency, equity and legitimacy should be employed (Adger et al. 2005). But institutionalising such an 'independent' evaluation is difficult to achieve in the context of prevailing distributions of power and influence. Contemporary responses to climate change are thus typically mired in cultural practices, technological pathways and institutional practices that are unsustainable and subject to deep inertia (Burch 2011); exposing many to escalating disaster risk with short-term private interests often prioritised over long-term public safety, sustainability and resilience (Boyd et al. 2009; McBean and Rodgers 2010). Confronting the root causes and drivers of maladaptive path dependency is vital and underscores the need to frame adaptation in the context of local community development possibilities and pitfalls.

### *1.1.2 Adaptation as a Local Community Development Imperative*

Adaptation needs to be located within the context of the factors shaping community development including the underlying causes of vulnerability, exposure to extreme events and the institutional structures and processes that facilitate individual and community decision-making and access to community assets and ultimately determine livelihood outcomes.

In practice, adaptation takes place through a wide array of decisions and actions at the local community level as individuals and groups pursue alternative development pathways and livelihood strategies. Some actions involve private adaptation and others are initiated, supported, or influenced by the policies of various government entities with or without partners from the private sector and civil society, including non-government organisations (NGOs), the scientific community and in some cases donor agencies and even the United Nations. Support for adaptation initiatives thus extends from the local to region-wide and international level. At the international level, for example, the UNFCCC National Adaptation Programmes of Action[3] (NAPAs) enable Least Developed Countries to identify priority activities to address urgent and immediate needs to adapt to climate change which if otherwise delayed would result in increased vulnerability or costs. The Pilot Program for Climate Resilience[4], with USD1.3 billion pledged, aims to demonstrate how to integrate climate risk and resilience into development planning and implementation.

Adaptation is an integral part of community decision-making and is thus a key dimension of **governance**. There is no consensus definition of 'governance' and there is debate about its nature, parameters and application (see e.g., Rhodes 1997; Kooiman 2003). Governance is distinct from government. Governance involves government and other actors from civil society and the private sector whose interactions together facilitate the sharing of power, social coordination and collective action. According to Kooiman (2003, p. 4):

> Governing can be considered as the totality of interactions, in which public as well as private actors participate, aimed at solving societal problems or creating societal opportunities; attending to the institutions as contexts for those governing interactions; and establishing a normative foundation for all those activities. Governance can be seen as the totality of theoretical conceptions on governing.

Governance is thus principled, interactive and stakeholder driven. Actors involved in the governance process interact through a variety of institutional structures and processes in pursuit of their objectives. Institutions can be defined as systems of societal 'rules' and norms that shape social interactions and choices. They provide the structure, order and predictability that enable actors to manage public affairs. Institutions are embedded in social settings; they are social constructs with historical underpinnings and they are subject to change over time. Their perceived effectiveness and legitimacy varies. Therefore institutions need to be critically evaluated and, where appropriate, modified or even transformed to promote 'better' governance. This view of institutions extends beyond narrow 'rules' to include norms and cognitive attributes, and underlying social and cultural foundations. A variety of institutional mechanisms (from law to cultural norms and traditions) interact and shape the pursuit of community resilience and sustainability. Hence the need to understand the roles and responsibilities of and interactions within and between the

---

[3] See http://www.napa-pana.org/.

[4] See https://www.climateinvestmentfunds.org/cif/Pilot_Program_for_Climate_Resilience.

State, private sector and civil society. No one actor or institution on their own can reduce disaster risk or build the adaptive capacity and resilience of communities and society: innovative hybrid collaborative governance arrangements and strategies are needed to govern climate risk and disaster risk more generally.

The need to explore adaptation and risk in the context of governance is compelling because there is a persistent gap between the obvious need to reduce risk and 'business as usual' practices that continue to expose people and property to escalating levels of risk. It has long been known that extreme events do not always result in disasters (see e.g., White 1936, 1945; Burton et al. 1968, 1993). A **disaster** occurs when the 'normal' functioning of a community, region or society has been severely disrupted due to an extreme event, such as a hurricane or cyclone, interacting with vulnerable social conditions and causing extensive adverse human, material, economic and/or environmental effects that necessitate an urgent emergency response to meet vital human needs and external support may be required if local coping capacity is overwhelmed (IPCC 2012). **Disaster risk** can thus be defined as the likelihood of a disaster occurring over a specified period of time. Such a definition reinforces the long held view that risk is a function of likelihood (or probability) and consequence (or impact) (see e.g., Knight 1921). But not all risk problems can be reduced to measurable uncertainty and consideration also needs to be given to non-quantifiable aspects of uncertainty, ambiguity, ignorance and surprise (Renn 2008; Stirling 2010; Klinke and Renn 2012). As the above definition of disaster indicates, disaster risk is a function of a physical peril and the root causes and drivers of social vulnerability which are shaped by complex socio-political and economic factors (Hewitt 1983; Alabala-Bertrand 1993; Hoffman and Oliver-Smith 2002; Pelling 2003; Wisner et al. 2004; CDRSS 2006; Haque and Etkin 2007). **Natural hazard risk** refers to the likelihood and consequences of a particular natural hazard event (that does not necessarily become a disaster). Our focus is on natural hazards as distinct from technological or anthropogenic hazards but it is important to recognise that a natural hazard event can trigger and be coupled in complex ways with technological or anthropogenic hazards (e.g., the March 2011 Tōhoku earthquake and tsunami triggered the Fukushima Daiichi nuclear disaster). **Natural hazards risk management** thus refers to the design, implementation, monitoring and evaluation of strategies, policies and measures to improve understanding about natural hazard risk, foster risk reduction and transfer and promote continuous improvement in disaster preparedness, response, and recovery practices, with the explicit goal of increasing human security, well-being, quality of life, resilience and sustainable development (IPCC 2012).

International experience and scholarship demonstrates that it is imperative to complement natural hazards risk management measures with deeper institutional reforms that reduce social vulnerability and build resilience and adaptive capacity—especially in this age of climate change (Lemos et al. 2007; Kelman and Gallard 2010). For example, reforms in land tenure arrangements, income distribution, public health and education, are central to reducing vulnerability and building resilience (Brooks et al. 2005) even though they are not framed as part of conventional emergency management practice. Clearly then, both hazard- and non-hazard specific institutional structures and processes play a pivotal role in climate change adaptation and natural hazards risk management.

There is growing recognition of the need to better align and integrate natural hazards risk management and climate change adaptation efforts. Tompkins et al. (2008) explores the relationship between disaster risk reduction and climate change adaptation in the Cayman Islands and northeast Brazil and identified four critical factors that reduce long-term risk: (i) flexible, learning based responsive governance; (ii) committed, reform-minded and politically active actors; (iii) disaster risk reduction integrated into other social and economic policy processes and (iv) a long-term commitment to managing risk:

> … we argue that the adoption of good governance mechanisms (such as stakeholder participation, access to knowledge, accountability and transparency) in disaster risk reduction policy may create the policy environment that is conducive to the kind of structural reform needed to build long-term adaptive capacity to climate-driven impacts. We conclude that without a synergistic two-tiered approach that includes both disaster risk reduction and structural reform, disaster risk reduction, in the face of climate changes, will prove to be an expensive and ineffective palliative treatment of changing risks (Tompkins et al. 2008, p. 736).

Natural hazards risk management and climate change adaptation are locally nuanced and specific processes that need to be responsive and tailored to address local needs and circumstances. Adaptation and risk management thus need to be located in the context of local development imperatives, and the institutional barriers and opportunities that determine pathways of community resilience and sustainability (see Fig. 1.1).

Figure 1.1 integrates and explores insights from the IPCC's (2012) portrayal of the relationship between climate change and development, and the imperative to align natural hazards risk management and climate change adaptation, and Chambers and Conway's (1991) sustainable livelihoods framework. This figure locates the issue of climate risk in a wider local development context. It recognises that natural hazards risk management (NRM) and climate change adaptation (CCA) strategies influence exposure and vulnerability to weather and climate events that together constitute climate risk. **Exposure** refers to the presence of people and associated livelihoods, as well as the array of resources or assets available in localities that could be adversely impacted by an extreme event. **Vulnerability** is the subject of vast scholarship but typically refers to the propensity or predisposition of a population or group to suffer harm or be adversely affected by a hazard event.

> Vulnerability is related both to the differential exposure and sensitivity of communities to stimuli such as climate change and also to the particular adaptive capacities of those communities to deal with the effects or risks associated with the exposures. While exposures, sensitivities and adaptive capacities are evident at community or local levels, they reflect broader forces, drivers or determinants that shape or influence local level vulnerabilities (Smit and Wandel 2006, p. 289).

Whilst vulnerability has been framed differently and somewhat in isolation by the natural hazards risk management and climate change adaptation communities, and distinctive methodologies for vulnerability assessment have been developed and deployed, there is growing awareness of the need to bridge this gap. Romieu et al.

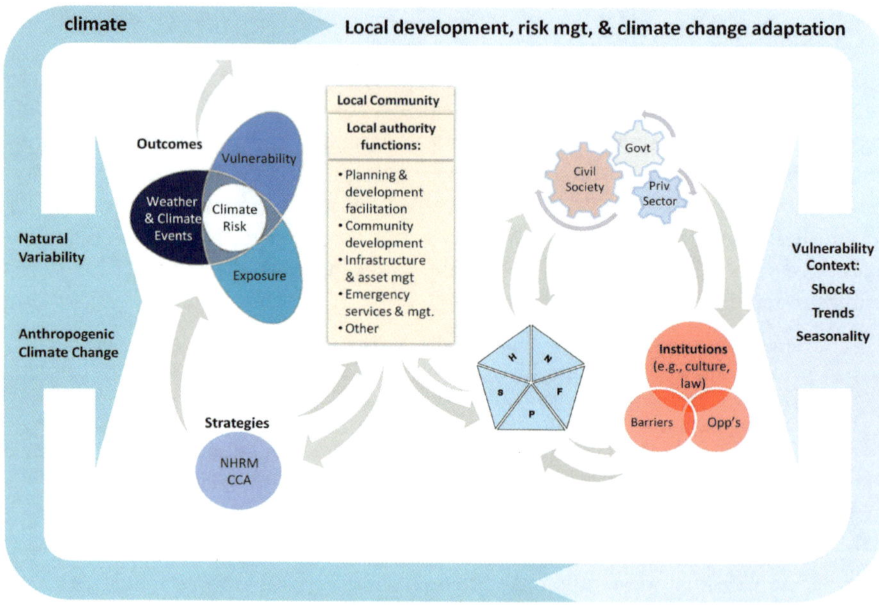


**Fig. 1.1** Local development, risk management and climate change adaptation.(Source: Bruce C. Glavovic, drawn by Paul Schneider)

(2010) point out that notwithstanding differences with respect to process (stress vs. shock), scale (spatial, functional and temporal), assessment approach (statistical vs prospective) and levels of uncertainty, the concept of vulnerability is a focal point that behooves interconnection between these two streams of scholarship and practice.

Climate risk is one component of disaster risk; with experience indicating that climate change is causing escalating disaster risk because more and more people are exposed to extreme hydro-meteorological events. The intersection of climate and local development and risk management is shaped by the adaptation limits, barriers and opportunities people face in pursuit of alternative livelihood opportunities.

Figure 1.1 shows that livelihood alternatives are located in a vulnerability context, influenced by shocks (e.g., natural hazard events, economic shocks, etc.), trends (e.g., in population growth or resource availability) and seasonal changes. This vulnerability context includes natural variability in weather and climate as well as anthropogenic climate change. Livelihood options are shaped by access to a variety of assets, including financial (F), physical (P), human (H), social (S) and natural (N) capital. Access to these assets is mediated by prevailing institutional structures and processes, and the governance choices made by key actors from government, the private sector and civil society. Depending on their access to these assets, individuals and groups adopt various livelihood strategies (e.g., subsistence agriculture, migration or formal employment) to realize desired livelihood outcomes (e.g., more income, food security or risk reduction). Inequitable access to assets creates conditions that predispose exposed groups and communities to disaster. Climate change affects livelihood options and prospects in a variety of ways, including the wider vulnerability context (e.g., more frequent and severe storms), available assets (e.g., changing temperature and rainfall affect crop viability) and the feasibility of alternative livelihood strategies and hence possible livelihood outcomes. Natural hazards risk management and climate change adaptation are two critical strategies taken by individuals and communities to manage climate risk in pursuit of preferred livelihood outcomes.

In practice, natural hazards risk management and climate change adaptation strategies are significantly influenced by actions taken by local authorities (e.g., local, district or metropolitan municipalities or councils; or traditional local authorities such as 'tribal councils'). The local level of government can play a pivotal role in enabling (or hindering) the ability of local communities to adapt to climate change; through the many functions it performs, including spatial or land use planning and development facilitation, provision of municipal infrastructure and assets, community development and emergency services and management which together enable provision of municipal services (such as rubbish collection, municipal roads and street lighting, etc.). Local authorities operate under a mandate from higher levels of government and engage in local development actions, including climate change adaptation, together with a variety of civil society and private sector actors as well as the scientific and research community. The extent to which natural hazards risk management and climate change adaptation strategies reduce climate and wider disaster risk, and enable community members and groups to realise desired livelihood outcomes, is shaped by the extent to which local authorities are able to work with other actors to overcome adaptation limits and barriers and unlock opportunities to foster resilience and sustainability.

### *1.1.3 Adaptation Limits, Barriers and Opportunities*

Increasing attention is being focused on the nature of adaptation limits and barriers and the opportunity to build adaptive capacity (Adger et al. 2007; Burch 2010; Gupta et al. 2010; Moser and Ekstrom 2010; Storbjörk 2010; Jones and Boyd 2011; Juhola and Westerhoff 2011; Lebel et al. 2011; Peñalba et al. 2012; Termeer et al. 2012). **Limits** to adaptation are absolute obstacles or thresholds beyond which social-ecological systems and/or their constituent species, ecosystems, activities, or livelihoods cannot avert a change of state. Transgressing such limits can lead to irreversible changes, such as the submergence of a low lying island due to sea-level rise or even the collapse of a society (Diamond 2005; cf. McAnany and Yoffe 2009). **Barriers** are obstacles to adaptation that can be overcome through adaptive actions. What appears to be a fundamental limit might in fact be a barrier that can be overcome through innovation and political will (Adger et al. 2009).

Limits and barriers come in many forms. First, there are natural or ecological and physical constraints such as climate change that occurs at a pace faster than prevailing ecosystems can adjust leading to potentially irreversible loss of ecosystem function and services. Second, human, resource and informational constraints are inevitable in the face of the 'super-complex' problem of climate change (Levin et al. 2012) making it difficult to understand the nature of the problem, secure the requisite resources or develop and adopt appropriate adaptive technologies. Third, social constraints include cognitive, normative and institutional dimensions rooted in divergent perceptions, interpretations, experiences, entitlements and responses to climate change. Addressing the latter is especially important in pursuit of community resilience and sustainability.

Jones and Boyd (2011) explore social barriers to adaptation and show that it is not sufficient to inform people about climate change and expect automatic behavioural change solely on the basis of such information transfer. Cognitive barriers are shaped by psychological and thought processes that people use to make sense of potential threats to their well-being, and responses range from denial to apathy, resignation, uncertainty and acceptance. Normative barriers are shaped by prevailing social values and mores that may foster adaptation in some situations but be maladaptive in others. Institutional constraints are shaped by both formal and informal 'rules' that govern behavior and frame adaptation options. Institutional structures and processes mediate access to assets and consequently the choices available to people to pursue alternative livelihood strategies and realize desired livelihood outcomes. These cognitive, normative and institutional barriers typically vary significantly within and between communities. In their study of adaptation in Nepal, Jones and Boyd (2011) demonstrate that social barriers to climate change adaptation involve complex intersecting cognitive, normative and institutional factors that are locality specific but change over time. Moreover, adaptive capacity and local action is retarded under conditions of marginalization and therefore requires that the root causes of vulnerability and social exclusion are addressed. Such challenges are, however, not limited to so-called developing countries.

In a critical review of National Adaptation Strategies in the Netherlands, United Kingdom, Finland and Sweden, Termeer et al. (2012) identified five institutional barriers: (i) limited openness to learning and variety, (ii) over-reliance on scientific experts (iii) tension between top-down policy development and bottom-up implementation, (iv) distrust in the problem-solving capacity of civil society, and (v) the profound difficulty of reserving funding for long-term action. Lebel et al. (2011) identified related 'institutional traps' in their study of climate vulnerability and flood risk management in Thailand. They found that fragmentation due to bureaucratic compartmentalization and competition leads to poor coordination, gaps in service provision and a compromised level of inter-agency capacity. Rigidity due to a fixation with certainty, control and stability fosters inflexible institutional arrangements and processes. A focus on a single scale concentrates capacity and resources at that scale without adequate attention to cross-scale interactions. Agenda capture by elites, who use their resources and influence to serve their own self-interests, marginalizes vulnerable groups. A crisis mentality prevails that is fueled by insufficient long-term strategic planning and ad hoc responses to immediate pressures and opportunities. Lebel et al. (2011) argue that escaping these institutional traps necessitates active and authentic public participation in natural hazards risk management, prioritization of risk reduction for socially vulnerable groups, building adaptive capacity at multiple scales and levels, integrating adaptation in development planning and strengthening the linkages between knowledge and practice.

In a study of 10 OECD countries, Bauer et al. (2012) found that despite increasing levels of awareness, it was not clear how many governments intended to formulate and implement adaptation policies. Attention was focused on four adaptation governance challenges, namely how to integrate adaptation policies across policy sectors; vertically across jurisdictional levels; how to address uncertainty and integrate and build knowledge into adaptation policies; and how to engage the private sector and civil society in decision-making processes. Overall, they found that adaptation governance approaches tend to rely on voluntary modalities of steering and coordination, with a national adaptation strategy being a pivot around which various adaptation governance regimes evolve.

Reflecting on adaptation in so called developing (the 'South') and developed countries (the 'North'), Wamsler and Lawson (2012) explain that, in broad terms, local coping strategies are core to climate change adaptation and disaster risk reduction in the South, not the least reason being limited access to social security and other forms of governmental assistance. In the North, however, local coping capacity is not well developed, and traditional institutional structures and processes are also not well equipped to respond to a changing climate. For instance, while wealthy nations may possess an abundance of resources to address identified problems, this is but one of many necessary elements. Additional factors include acknowledging and understanding the problem; garnering the political will to act; identifying and mobilizing vulnerable populations, assets and communities; applying knowledge to action and implementing programmes and projects (National Research Council of the National Academies 2010a, pp. 160–161). Thus, an integrated approach is needed to build both local and broader institutional adaptive capacity.

Understanding and addressing adaptation barriers requires consideration of both the proximate drivers and root causes of vulnerability. From this vantage point, unsustainable practices, such as anthropogenic climate change, reflect some of the inherent contradictions of capitalism (Pelling and Manuel-Navarette 2011). Risk reduction and adaptation therefore challenge the prevailing hegemony that drives development pathways that lead to escalating environmental degradation, poverty and inequity and disaster risk (Brooks et al. 2009; Pelling and Manuel-Navarette 2011). To what extent has progress been made in overcoming the foregoing barriers given the dramatic increase in adaptation scholarship (Berrang-Ford et al. 2011) and adaptation initiatives around the world in recent years (Tompkins et al. 2010)?

### *1.1.4 Recent Adaptation Experience*

In what may be the most comprehensive comparative analysis of adaptive actions to date, Lesnikowski et al. (2013, p. 3) studied 4,104 discrete adaptive actions taken by

117 parties to the UNFCCC and found that effective translation of good intentions into practice remains nascent and that there is a persistent "gap in understanding." There is limited evidence of effective adaptive actions in developed nations—other than in some climate sensitive sectors and municipal level institutions, often facilitated by higher levels of government; with unequal reporting across regions and sectors and little indication of adaptation with respect to vulnerable groups (Ford et al. 2011). Juholla and Westerhoff (2011) found that adaptation efforts in Finland and Italy emerged autonomously at lower levels of governance and typically take place through formal institutions as well as actor networks across spatial scales; with a lack of coordination at the national scale that limits adaptation prospects. In a study of adaptation plans in Sydney, Australia, Measham et al. (2011) found that adaptation is recognized as an important planning matter but has yet to be effectively embedded in practice. They point out that limited resources and information are obvious constraints on adaptive action at the local level, but insufficient attention has been focused on the influence of leadership, institutional context and competing planning agendas in enabling or hampering adaptation. In an evaluation of seven local climate change adaptation plans in southeast Queensland, Australia, Baker et al. (2012) found that local governments were well aware of climate change but had limited capacity to develop and implement effective locality specific action plans; highlighting a range of structural, procedural and contextual barriers.

Preston et al. (2011) reviewed 57 adaptation plans from Australia, the United Kingdom and the United States and found that adaptation plans are to a large extent under-developed with critical weaknesses because key dimensions of adaptive capacity (such as entitlements to key assets) are ignored and influential non-climatic factors were not given due consideration. They noted the opportunity to make better use of available guidance for adaptation planning and to consider more critically the governance context within which adaptation takes place. Their findings suggest serious deficiencies in climate preparedness exist in so called developed countries, including those with purportedly high adaptive capacity.

In a study of over 300 early adopters of adaptive action in the United Kingdom, Tompkins et al. (2010) found that government initiatives predominate and most of these take the form of research on climate change impacts. These initiatives nonetheless prompted a variety of actions across sectors and scales but there was little evidence of these initiatives having a meaningful impact at the local level of government. Notably, sectors with large-scale investment in climate sensitive infrastructure, such as flood defence, demonstrate a higher level of adaptation awareness and action. Top-down government driven action in the United Kingdom has yielded anticipatory action at relatively low cost in some arenas and created niche activities that may lead to diffusion of adaptation practices over time. In a study of adaption by three municipalities in British Columbia, Canada, Burch (2010) found that overcoming limited financial, human or technical resources did not necessarily require additional resources but more effective use of available capacity. Explicit high-level directives, leadership approaches that foster organizational innovation and collaboration and mainstreaming climate change responses into day-to-day business practices were found to be vital enablers of action. Doing so, however, requires overcoming unsustainable business as usual path dependent institutional structures and processes. In the United States, by 2008, 29 states had a climate change plan, and 170 local governments had become members of the Cities for Climate Change Protection Programme that among other things requires an adaptation plan. Wheeler (2008) analysed this first generation of plans and found, among other things, that few had addressed adaptation in a meaningful way. In a more recent study of 40 local climate change plans in the United States, Tang et al. (2010) found that they typically reflect a high level of awareness but moderate use of analytical capability and few had made significant progress in the implementation of tangible strategies.

The barriers to effective adaptation planning and action appear almost insurmountable. The experience in Australia exemplifies the challenge: the option of planned retreat, for example, is disappearing in southeast Queensland, Australia, according to Abel et al. (2011), for a variety of reasons, including (i) population growth is actively encouraged by the State Government; (ii) there is no apparent urgency to protect coastal ecosystems which are adversely impacted when houses are built along the seashore; (iii) liability laws tend to favour property development; (iv) planning processes ignore cumulative environmental impacts and foster unsustainable development; and (v) there is mounting political pressure to armour the coast as the value of built assets and infrastructure grows. In some cases, adaptation could be described as manipulation of the social-ecological setting rather than adaptation to changing circumstances (Thomsen et al. 2012). Adaptive actions are grounded in an ethos of respect towards the integrity of social-ecological systems and the trajectory of change is modulated by the overarching theme of sustainability; whereas manipulation tends to emphasize a short-term preoccupation that is driven by the interests of the status quo rather than the social-ecological system as a whole. In effect, such manipulation is ultimately maladaptive because it ignores system dynamics, hinders social learning, creates potentially destructive path dependencies and constrains future adaptation prospects (Thomsen et al. 2012).

In sum, despite the proliferation of adaptation scholarship and 'actions,' real progress has been limited on the ground. Much can be learned from related fields of endeavor that have a longer history than the relatively recent focused attention on climate change adaptation. The next section explores insights from decades of natural hazards risk management scholarship and experience which provide valuable insights for overcoming barriers and unlocking opportunities to build adaptive capacity, resilience and sustainability.

## 1.2 Learning from Natural Hazards Risk Management Scholarship and Experience

Viewing climate change through a natural hazards risk management lens provides valuable insights about overcoming barriers to adaptation because an extensive knowledge base, significant pre- and post-disaster experience (including the formulation and post-event evaluation of varied institutional frameworks) and specialised fields of practice exist in what has been referred to as disaster management, emergency management, sustainable hazards mitigation, sustainable disaster recovery, risk management and risk governance (White and Haas 1975; Mileti 1999; National Research Council 2006; Renn 2008; Handmer 2009; Intergovernmental Panel on Climate Change 2011; Renn et al. 2011; Intergovernmental Panel on Climate Change 2012 For the purposes of this book we refer to these fields of practice collectively as natural hazards risk management. Key concepts and terms are used somewhat differently in different parts of the world. For example, in the United States, four phases of the hazard cycle are distinguished: hazard mitigation, preparedness, response and recovery. In New Zealand, the phases are described in terms of the four "Rs"—reduction, readiness, response and recovery. In New Zealand, the term 'mitigation' is commonly used to mean reduce or alleviate hazard risk; not eliminate it. Mitigation in the United States is commonly used to mean reduce or eliminate hazard risk. In the chapters that follow, we will use the terms risk reduction and hazard mitigation interchangeably to mean reduce or eliminate natural hazard and disaster risk.

In many ways the natural hazards risk management community, like those involved in the study and practice of climate change adaptation, has adopted policies and plans, assessed risk using various analytical tools, and monetised expected impacts in relative isolation, placing a limited emphasis on the nexus between these two camps, although this has begun to change (Birkman and Von Teichman 2010; IPCC 2011; Carmin et al. 2012).[5]

Directly tied to the successful creation of hazard mitigation, disaster recovery and adaptation strategies are the challenges inherent in developing effective governance approaches that coordinate international, national, sub-national and local government actions with those of the private sector and civil society (Adger 2001; Tompkins et al. 2008; Paleo 2009; Glavovic 2010a; Birkmann et al. 2010; Tierney 2012). A growing number of studies have helped to better understand the complexities of climate change and its associated impacts, while the use of improved analytical techniques provide increasingly accurate information at more refined geospatial scales that can be used by countries, regions, cities and communities to better assess risk and develop scenario-based plans and policies to curtail greenhouse gas emissions and adapt to climate change (Bicknell et al. 2009; National Research Council 2010a; Hagemann et al. 2011; Blakely and Carbonell 2012). While there remains much work to be done, there is a need to capitalise on these advances and capabilities, including existing knowledge, experience and institutional capacity

[5] Examples of international forums that are striving to improve the connectivity between climate change adaptation and natural hazards risk management include the United Nations International Strategy for Disaster Reduction (the Working Group on Climate Change and Disaster Risk Reduction of the former Interagency Task Force for Disaster Risk Reduction), the Hyogo Framework for Action, the World Conference on Disaster Reduction, and the Climate Centre of the Red Cross/ Red Crescent. Examples of tools used to assess risk and monetize expected impacts include those created by national governments (e.g., HazardsUS or HAZUS), insurance corporations like Swiss Re and Munich Re, and the International Federation of Red Cross and Red Crescent Societies.

that already exist across prevailing governance networks. The integration of these networks presents its own set of significant, but surmountable challenges.

Communities have adopted risk reduction and disaster recovery strategies focused on natural hazards as part of informal and formal arrangements since before the beginning of recorded time. The more recent formation of nation states and modern systems that include institutions focused on this task has resulted in an array of risk reduction and post-disaster recovery strategies and practices. Given the rise of scientific, institutional and public awareness about the preconditions linking climate change and natural hazards, the sharing of practicable lessons can help to demonstrate the value of cooperation achieved through expanded and integrated adaptive governance networks (National Research Council 2010b, pp. 117–120). Understood in the context of this book, this means recognising the dual and complementary value of assessing differing climate change scenarios and adapting to the associated consequences even though these effects remain uncertain. At the same time it means drawing on established methods employed to assess natural hazard risks and taking pre-event actions through collaborative networks to reduce the damaging effects of what amounts to the worsening of hazards already prevalent in an area or the occurrence of new threats tied to climatic changes.[6] It also means instituting new and improved practices and targeted investments after disasters that recognise the larger integrative goal of sustained risk reduction in the face of a changing climate.

### *1.2.1 Disaster Risk Reduction and Post-disaster Recovery*

We focus on two phases of the natural hazards cycle: risk reduction and postdisaster recovery. While the majority of the literature that explores the links between natural hazards risk management and climate change adaptation has focused on the relationship between risk reduction and adaptation, the post-disaster recovery process also provides strategic points of intervention relevant to adaptation. Exploring risk reduction and recovery recognises the dual (i.e., proactive and reactive) nature of how societies and their communities deal with extreme events (Birkmann et al. 2008; Birkmann and von Teichman 2010, p. 176).

There are two distinguishing characteristics that need to be highlighted when considering slow-onset climate related risk relative to episodic sudden shock events such as flooding or coastal storms: the deep uncertainty and the long time frames of climate related disaster risk. The high levels of complexity and uncertainty associated with climate change, compared to what we know about many natural hazards, presents serious challenges for understanding and implementing adaptive actions

[6] The destructive nature of natural hazards has been described as the social amplification of risk due to the impacts of societal choices (Kasperson et al. 1988). While the anthropogenic alteration of our climate and the resulting increase in natural hazards and disasters expands on the concept of social amplification, it also provides a means to attenuate risk through heretofore underutilized and expanded social and institutional networks.

(Pelling et al. 2008). This condition compels us to apply the observations and lessons drawn from institutions that have dealt with disasters—including the means by which these lessons have been transferred to action—to help build an improved adaptive capacity to not only 'typical' natural hazard events, but to the expanded reality of climate change-induced or -influenced extreme events and the impacts that will unfold over coming decades and beyond.

In order to frame the discussions throughout the remainder of this book, risk reduction and disaster recovery are defined next. The United Nations International Strategy for Disaster Reduction (2009) defines **risk reduction** as the:

> …practice of reducing disaster risks through systematic efforts to analyse and manage the causal factors of disasters, including through reduced exposure to hazards, lessened vulnerability of people and property, wise management of land and the environment, and improved preparedness for adverse events.

**Disaster recovery** can be defined as the "differential process of restoring, rebuilding, and reshaping the physical, social, economic and natural environment through pre-event planning and post-event actions" (Smith and Wenger 2005, p. 237). Analyzing risk reduction and disaster recovery as practiced around the world offers a number of lessons that are potentially transferable to those who recognise that adaptation is required in both pre- and post-disaster settings if we are to promote more sustainable and resilient communities.

The evolution of natural hazards risk management is strongly influenced by the episodic nature of extreme events. Disasters are often referred to as focusing events, which can affect the creation of policy agendas among decision-makers (Birkland 1996, 1997) and influence changes in national and sub-national policy and practice (Birkland 2007; Olson et al. 1998, 1999; Rubin and Tanali 2001; Rubin 2007). Major disasters can also result in informal responses and the creation of organisations that emerge after extreme events (Birkman et al. 2008; Smith 2011). International, national, regional and local policies are often developed in response to events that draw media attention (Molotch and Lester 1974; Olson 2000); an outcry by concerned citizens; and strong criticism by technical experts, hazards scholars (Mileti 1999) and policymakers (Birkland 1997) to shortfalls in pre-event agreements, programmes, plans and policies. The media can frame stories in a way that marginalise community organizations and suggest that the event does not warrant serious attention as described in Eric Klinenberg's *Heat Wave: Social Autopsy of Disaster* (2002) or in Gregory Button's study of oil spills in Great Britain (1999). In their study of the media following Hurricane Katrina, Gawronski et al. (2006) found that the farther the media outlet was from the area impacted (where basic decisions are made surrounding development and the adoption of land use-based, pre-event hazard mitigation measures) the more likely the media was to challenge the dominant pro-growth paradigm that exacerbated high levels of exposure and vulnerability. Conversely, local media outlets that served impacted communities were less likely to challenge dominant 'business as usual' regimes. Insights about the role of the media in climate change adaptation can thus be gained from the study and experience of the role of the media in disaster situations.

Hazard mitigation is ideally practiced in advance of an event as a way to lessen future, often predictable losses. On its face, this proactive stance is well suited to both natural hazards risk management and climate change adaptation. A comprehensive risk reduction approach means taking actions to reduce risk before and after an extreme event occurs. This necessitates more effectively understanding and conveying the threat in the pre- and post-disaster setting, adopting proactive risk reduction measures, learning from destructive events and amending policies and practices as necessary to achieve complementary goals, and being prepared to implement risk reduction and adaptation measures after a disaster when those affected may be more willing to take action and the availability of post-disaster resources allows stakeholders to do so. Linking risk reduction, disaster recovery and climate change adaptation therefore requires garnering the support needed to achieve these goals by fostering a more inclusive and collaborative governance approach, and effective social learning about the close association between risk reduction, disaster recovery, resilience, sustainability and climate change adaptation.

### *1.2.2 Integrative Themes: Sustainable Development, Resilience and Risk Reduction*

Sustainable development, resilience and risk reduction, terms that have long been used to conceptualise and understand natural hazards risk management (Cuny 1983; Anderson and Woodrow 1989; Albala-Bertrand 1993; Cannon 1994; Berke 1995; May et al. 1996; Berke and Beatley 1997; Burby 1998a, b; Mileti 1999; Godschalk 2003; Pelling 2003; Campanella 2006; Glavovic 2008; Renn 2008) have been increasingly applied to issues surrounding climate change adaptation, including their use as a means to integrate both domains (Schipper 2004; Paton and Johnston 2006; Walker and Salt 2006; Lemos et al. 2007; Nelson et al. 2007; Grist 2008; Beatley 2009; Newman et al. 2009; Handmer and Dovers 2009; Wilbanks and Kates 2010; Gaillard 2010; IPCC 2012). Efforts to formulate hazard mitigation and disaster recovery policy approaches tied to these concepts have been implemented in a number of areas across the globe with limited to moderate degrees of success as will be shown in the chapters that follow. At the same time, risk reduction and climate change adaptation are increasingly being discussed in an integrative manner (O'Brien et al. 2006; Schipper and Pelling 2006; Tomalla et al. 2006; Beatley 2009) the results of which are being used to inform public policy at various spatial scales. While the effective cross-pollination of knowledge and practice between the fields of natural hazards risk management and climate change adaptation are improving, they remain insufficient and under-utilised in terms of how this knowledge is being transferred across disciplines and policy domains (Schipper and Pelling 2006; National Research Council 2009). For instance, the information associated with natural hazards risk management has been gathered at different spatial scales and applied in parallel with climate change adaptation studies and the emergence of climate change plans, policies and programmes put in place by governments, non-profits and others rather than being utilised in an integrated fashion (Schipper and Pelling 2006; Birkmann and von Teichman 2010). The failure to integrate this information and codify it in collaboratively derived and vertically integrated plans (linking community, city, region, nation and international policies) stem from a mix of perspectives and approaches to the problem, some of which differ, some of which provide opportunities for new partnerships. Drawing on these well-established constructs provides valuable insights for those striving to operationalise these important linkages (Cuny 1983; Harrell-Bond 1986; Berke and Beatley 1997; Oliver-Smith and Hoffman 1999; Nakagawa and Shaw 2004; Fordham 2006; Albala-Bertrand 2006; Smith and Wenger 2006; Gibson et al. 2009).

Natural hazards risk management scholarship in general, and natural hazards planning scholarship in particular, have been shaped by the unifying theme of **sustainable development** (Berke 1995; Geis and Kutzmark 1995; Munasinghe and Clark 1995; Beatley 1998; Burby 1998a; Godschalk et al. 1998; Mileti 1999; Schneider 2002; Mileti and Gailus 2005; Smith and Wenger 2006; Puszkin-Chevlin et al. 2006/2007; Glavovic 2008, 2010a). Making land-use decisions that expose people and property to disaster risk is obviously not sustainable; especially if alternative, less risky settlement patterns and land-use practices are feasible. The sustainable development literature underscores the imperative to reconcile ecological, economic and social goals in a manner that meets the needs of current and future generations (WCED 1987) and, when applied to natural hazards and disasters, describes how extreme events often result from long standing and interrelated issues of poverty, inequality, environmental degradation and the episodic (i.e., post-disaster) imposition of external assistance rather than undertaking an enduring effort to build institutional capacity across interacting governance regimes based on local needs and endemic resources like social capital (Burby 1988).

> In the pre-disaster period, sustainability values seek to avoid saddling future generations with sprawling, wasteful land use patterns that not only reduce the social livability and economic viability of communities, but also undermine the ability of the natural environment to absorb hazard forces and expose people to significant hazard risks. In the post-disaster period, sustainability values seek opportunities to relocate land use out of hazard areas and rebuild damaged homes and infrastructure in more resilient ways instead of replicating brittle and unsustainable development practices (Godschalk et al. 1998, p. 86).

The challenge is to translate sustainability rhetoric into reality; a challenge that necessitates confronting the root causes and underlying drivers of unsustainable practices, including social vulnerability as previously explained. Framing risk reduction and adaptation as resilience can help to address this challenge.

According to the IPCC (2001), **resilience** is the amount of change that a social-ecological system (e.g., a coastal community and associated ecosystems upon which they depend) can undergo without experiencing a change in state (e.g., rising sea level forcing abandonment of a low lying coastal locality). A resilient system can thus maintain its characteristic structure, functions and feedbacks in the face of change, including the shock of an extreme event or slow onset climate change, and plan for the future (Holling 1973; Holling and Gunderson 2002; Walker et al. 2004; Walker and Salt 2006). The nature of climate change is such that changes in state may be inevitable for some social-ecological systems and a resilient community may therefore have to navigate the transition from one state to another. Resilience thus reflects (i) that amount of change that a social-ecological system can absorb whilst retaining key structures, functions and feedbacks, (ii) the capacity of the system to reorganise and (iii) the capacity for learning and adaptation in the face of change (Holling 1973; Carpenter and Gunderson 2001; Folke 2006).

Resilience is an important but contested organizing concept for various strands of scholarship (Folke 2006; Gallopin 2006), with different definitions that reflect perspectives from engineering to ecology, psychology and more integrated and evolutionary perspectives (Folke 2006). Handmer and Dovers (1996) distinguish three different views of resilience: (i) resistance and maintenance, (ii) change at the margins and (iii) openness and adaptability. The first view indicates a controlling and defensive response to change and a resolve to maintain the status quo, even in the face of inevitable change and escalating risk. Maintaining stability may be socially desirable as long as resistance to change doesn't lead to societal 'collapse' (see e.g., Diamond 2005; cf. McAnany and Yoffee 2009). The second view reflects the typical incremental adjustments made in communities and societies to adapt to change without disrupting the status quo. Such an approach may address the symptoms of climate change but fails to address the underlying causes of climate change and unsustainable development. In short, this approach perpetuates 'business as usual'. The third view recognizes the need to address the underlying drivers and root causes of unsustainable development and thus envisages transformative or radical change. However, prospects for realizing such transformative change appear limited in the face of extant power distribution and lock-in to unsustainable development pathways. Planners must therefore confront systemic barriers as they seek to enable communities to build resilience and sustainability (Glavovic 2008).

Resilience from a disaster perspective has tended to focus on the capacity of interconnected physical, social and economic systems to rebound from an episodic shock or extreme event (Godschalk 2003; Paton and Johnston 2006; Berke and Campanella 2006). But there is also recognition of the wider societal context within which disasters occur and the imperative to confront the drivers and root causes of vulnerability that predispose communities to disaster (Wisner et al. 2004). A 'resistance and maintenance' view of resilience perpetuates the status quo and fails to confront the underlying causes of vulnerability and unsustainable development. Transformative societal change is needed to chart development pathways that reduce vulnerability and exposure to climate change impacts, and foster sustainable outcomes. Disaster experience underscores the shortcomings of a 'resistance and maintenance' view of resilience. Pre-event conditions, such as the depth and extent of vulnerability, determine the predisposition of a community or locale to disaster because an extreme event only becomes a disaster in the face of vulnerable conditions. Pre-event adoption of proactive risk reduction strategies can reduce both vulnerability and exposure, and consequently affects the speed with which individuals, organisations, communities and societies recover and return to a sense of "normalcy" after being struck by an extreme event; and transition towards a "new normal" ideally characterized by reduced vulnerability and exposure. Using resilience as an integrative theme underlines the importance of taking action in advance of predictable sudden shock and slow onset events by adopting a comprehensive suite of interconnected risk reduction measures that span natural hazard and climate change-related perils. It also means learning from past events, and enhancing one's adaptive capacity—no longer striving to return to a past condition, but rather recognising a "new normal" tied to the realities of climate change and the inherently dynamic nature of natural hazards (Beatley 2009). More fundamentally, resilience as coupled incremental and transformative change underscores the imperative to frame both natural hazards risk management and climate change adaptation as a fundamental socio-political course of action that explicitly recognises and confronts the 'business as usual' practices and inequitable power distributions that foster vulnerability and unsustainable development. Albeit contested, framing risk reduction and adaptation under the organizing principles of resilience and sustainability from this perspective recognizes the imperative for both incremental and transformative socio-political change (Tobin 1999; Nelson et al. 2007; Pelling and Manuel-Navarette 2011; Berke et al. 2011; Pelling 2011; Davoudi et al. 2013; McEvoy et al. 2013).

What then are key considerations for integrating risk management and adaptation in practice?

### *1.2.3 Key Considerations for Integrating Natural Hazards Risk Management and Climate Change Adaptation in Practice*

Drawing on the foregoing discussion, we identify key considerations for better aligning and integrating natural hazards risk management and climate change adaptation[7]:

1. **Knowledge base/intellectual underpinnings/terminology:** Climate change policies have traditionally focused on mitigation of greenhouse gas emissions and consequently placed a strong emphasis on environmental science and international mechanisms for securing agreements to address the source of the problem. The focus on adaptation at the local level is much more recent. Natural hazards risk management has focused on the local scale and drawn heavily on engineering, architecture, economics, the social sciences and land use planning, though past practices have tended to be driven by a reactive approach that, in the aftermath of an extreme event, prioritises physical and economic recovery over more complex

[7] The categorization of the natural hazards risk management and climate change adaptation communities are drawn from "Reducing hazard vulnerability: towards a common approach between disaster risk reduction and climate adaptation" (Thomalla et al. 2006, p. 40) and "Integrating Disaster Risk Reduction and Climate Change Adaptation: Key Challenges-Scales, Knowledge, and Norms" (Birkmann and von Teichman 2010, pp. 174–177). We expanded on the initial categorizations to include additional layers of specificity such as the actors involved, methods and data used by these actors and how the categorization can be broadened to include other points of possible collaboration. These factors are further discussed in Chap. 2.

social and institutional dimensions. Climate change adaptation initiatives emphasise the importance of building adaptive capacity at the local level, within a nested collaborative governance framework, which provides a clear link to the study of institutions, governance strategies and land use planning. Hindering integration is the lack of clearly defined indicators, terminology and standards that help to bridge the two fields (Birkmann et al. 2009; Schipper 2009) though progress has been made towards this end in recent years (IPCC 2012).

2. **Temporal and Spatial Aspects:** Historically, the study of climate change and natural hazards risk management-related issues, including the creation of policies addressing these phenomena, was shaped by the analysis of differing temporal (slow onset versus rapid onset) and spatial scales (global versus local). The more recent focus on adaptation at the local level aligns with the spatial focus of natural hazards risk management, and the inherently multi-scale reality of resilience and sustainability endeavours. Whilst the historical temporal and spatial bifurcation remains a problem in practice, there is growing recognition that climate change has local-global interconnections and is altering the frequency, spatial extent, duration and timing of extreme events (IPCC 2012). Major challenges remain, including the ability of stakeholders to understand these relationships in a way that results in the adoption of complementary adaptation and risk reduction strategies before extreme events occur. A closely associated subtext includes the need to improve the manner in which disaster relief is provided by members of the disaster assistance network in order to advance long-term development aims, including risk reduction and adaptation at the local level.
3. **Policy Frameworks and Initiatives:** Climate change and natural hazards risk management policy frameworks and initiatives involve a complex array of actors and institutional structures and processes that shape prevailing practices and the trajectory of adaptation, risk reduction and disaster recovery. The discrete policies and initiatives of different actors tend to reflect the nature of the dominant functional areas in each group and include perspectives grounded in disciplines, such as environmental science, climatology, engineering, the social sciences and planning that are not easily aligned. In other instances, organisations are drawn into the process unexpectedly or groups emerge following perceived climate change threats or the occurrence of an extreme event to confront specific needs that were not met by prevailing formal institutions. In order for these policy frameworks and initiatives to better coalesce, it is necessary to better integrate bottom-up and top-down policies and practices informed by real-world experiences from the local to trans-national level (Gaillard 2010).
4. **Risk Assessment, Communication and Management:** Climate change and natural hazards risk management scholars and practitioners apply varied risk assessment methods (e.g., climatological models, historical records analysis, stochastic models and scenario-based land use models), use differing datasets (e.g., post-disaster damage assessments, climatic and weather data and geologic sampling including sediment layering and ice thickness) and rely upon multiple means to communicate risk to various audiences (e.g., return periods; scientific reports; plans and policies; the media; and education, outreach and training).

These approaches are not always well aligned. One way that climate change adaptation and disaster risk reduction scholars and practitioners can bridge the limits of what amounts to a mix of emerging and well-grounded bases of knowledge is through the utilisation of scenario-based modeling and planning, both of which have the potential to guide action under conditions of uncertainty. In the realm of natural hazards risk management, the field of risk communication is relatively well-developed and the display of natural hazards data relies on a variety of mapping, and, more recently, visualisation techniques to assist in the development of risk reduction policies and plans. The ability to measure risk, including the monetised benefits (e.g., losses avoided) of implementing hazard mitigation measures is an important but underutilised tool that can be used to convey the value of hazard mitigation, and eventually climate change adaptation.[8] The development of analytical tools, like those that measure future losses avoided, are not likely to be effective if they are developed by outside "technical experts" with little or no involvement by those whose communities are being assessed or those who are responsible for conducting the assessment at the local level where the capacity to do so varies greatly. It is essential to recognise that the nature of risk problems varies considerably from measurable uncertainty to risk problems that cannot be quantified because they have high levels of socio-political ambiguity, scientific uncertainty and complexity. Risk assessment and management approaches therefore need to be carefully selected to match the nature of risk problem being addressed. The evolving conceptualization and practice of risk governance underscores the need to move beyond traditional risk assessment, communication and treatment approaches that tend to view all risk as if it were measurable uncertainty.

5. **Funding and Implementation Strategies:** Organisations involved in natural hazards risk management and climate change adaptation utilise a number of funding and implementation strategies. These include international,[9] national, and non-profit capacity building initiatives and pre- and post-disaster financial assistance in the form of grants-in-aid, loans and insurance settlements that are delivered by governments, non-profits, private sector lenders and insurance companies, and quasi-governmental organisations. International financial institutions such as the International Monetary Fund, World Bank and the Kyoto Protocol Adaptation Fund represent entities that support the funding of natural hazards risk management and climate change adaptation. Most nations have

[8] The ability to develop quantifiable losses avoided measures associated with climate change-induced events will require developing clearer expected return periods associated with these events as well as monetizing expected losses tied to slow-onset events and the benefits of differing adaptive measures.

[9] In the international arena, issues such as disaster diplomacy (using disaster aid to advance political aims) (see Kelman 2006, 2007), conditionality (establishing prescriptive pre-conditions for the receipt of aid that may preclude local needs) (see Gibson et al. 2009) and the imposition of aid without focusing enough attention on building the capacity of those receiving it to effectively accept and utilize the assistance (see Harrell-Bond 1986) play important roles in the funding of disaster relief.

developed a disaster management system to prepare for, respond to and recover from extreme events. The degree to which these institutions address disaster risk reduction and climate change adaptation in a coordinated manner varies widely, although many possess pre- and post-disaster hazard mitigation resources tied to grants, insurance and loan programmes. The ability to use climate change and natural hazards risk management funds, including those available in the pre- and post-disaster environment to address the broader and mutually reinforcing concept of risk management is beginning to be discussed and applied by a growing number of organisations. Bouwer and Aerts (2006) argue that the interrelated issues of development, risk reduction and adaptation should be part of a more "mainstreamed" financing strategy used to fund proposed development projects that are "climate proofed," thereby improving their resilience and sustainability in the face of climate change

6. **Collaborative governance:** Climate change adaptation and natural hazards risk management strategies necessitate collaborative governance (Glavovic 2010a; Glavovic et al. 2010; Quay 2010; Smith 2011). Actors include communities and civil society organisations and informal groupings; national, sub-national and local agencies and departments tied to natural hazards risk management; local and regional planning agencies; national and international nonprofit assistance and relief organisations; corporations and businesses (including insurers, contractors, design firms, etc.); and others. Each of these groups provides varied resources at differing points in time, including before and after extreme events (the degree to which these resources assist or compound the explicit climate change adaptation challenges remains less understood). The ability to coordinate overlapping governance frameworks and networks is necessary to share individually and collectively derived knowledge, experience, tools and data. In order to enhance our ability to develop integrative governance strategies we need to understand how these networks function horizontally (e.g., across locally-based groups) and vertically (e.g. across local, sub-national, national and international or global organizations).

How might these considerations be translated into practices that better align and integrate natural hazards risk management and climate change adaptation; and, in particular, what insights does the natural hazards risk management community offer? The concept and practice of risk reduction, or sustainable hazard mitigation, lies at the heart of this challenge; and spatial or land use planning plays a pivotal role in this endeavour. According to Mileti (1999, pp. 155–156), lead author of the second national assessment of natural hazards in the United States:

> No single approach to bringing sustainable hazard mitigation into existence shows more promise at this time than increased use of sound and equitable land-use management. Many political, social, and economic forces conspire to promote development and redevelopment patterns that set the stage for future catastrophes. However, by planning for and managing land use to accomplish sustainable hazards mitigation, disasters—though not wholly eliminated—can be reduced to a scale that can be borne by the governments, communities, individuals, and businesses exposed to them.

The natural hazards risk management community views risk reduction as a proactive set of actions, policies and plans that serve to reduce the loss of life and damages to property due to natural and human-caused hazards. Important conceptual components include taking a long-term view that involves a continuous re-evaluation of risk over time and modifying actions, policies and plans as needed to reflect these changing conditions. Sound hazard mitigation planning involves drawing on a range of activities including preventive or avoidance actions, property protection, natural resource protection, structural projects and public information. More specific examples of these activities include limiting or restricting development in hazardous areas, adopting building codes and standards, managing and protecting natural resources that can provide protective buffers from natural hazards like wildfire and flooding, relocating at-risk properties away from hazard-prone areas, constructing durable and resilient infrastructures and engaging and informing the public about risk and the steps they can take to reduce their exposure to natural hazards. Effective risk reduction also necessitates taking steps to address not only the proximate drivers of vulnerability but its underlying root causes. Taken together, this array of measures can be described as natural hazards planning which provides a key integrating concept and mechanism for natural hazards risk management and climate change adaptation.

### 1.2.4 Natural Hazards Planning

Disaster risk can only be eliminated if people stay out of harm's way. Natural hazards planning can help communities understand the risks they face and make informed decisions to promote community safety, resilience and sustainability. The broad field of natural hazards risk management draws upon a range of other mitigation strategies and methods, ranging from warnings and evacuation provisions to structural protection, insurance and emergency relief. Natural hazards planning can play a crucial role in coordinating and integrating the actions of actors across scales and through both formal and informal institutional structures and processes.

The promise of natural hazards planning is to reduce disaster risk by enabling communities to work together to (i) make informed collective decisions before an event occurs about where to locate physical development, infrastructure and land use activities whilst retaining the protective functions of ecosystems; (ii) design more sustainable and resilient buildings; (iii) raise levels of risk awareness, understanding and preparedness; (iv) reduce exposure and social vulnerability; (v) facilitate recovery in the aftermath of extreme events; and (vi) chart development pathways that foster resilience and sustainability (Board of Natural Disasters 1999; Burby 1998a, b; Burby et al. 1999; Godschalk et al. 1998; Godschalk et al. 1999; Mileti 1999; Godschalk 2003; Puszkin-Chevlin et al. 2006/2007; Smith 2008; Glavovic 2010a, b).

Like adaptation, the effective practice of natural hazards planning faces a number of barriers, including: (i) a reluctance among local governments to develop a robust hazard mitigation strategy grounded in land use planning tools and techniques (Burby 1998a; Mileti 1999, Handmer 2008; Smith et al. 2013; Lyles et al. 2013); (ii) the creation of national policies and public investment strategies (including aid provided after a disaster) that stimulate development in known hazard areas (Mileti 1999; Platt 1999); (iii) the ineffective use of knowledge and past experience to guide wise decision-making processes (White et al. 2001); and (iv) the failure to develop horizontally and vertically integrated, multi-party partnerships focused on the development and maintenance of a cooperatively derived comprehensive risk reduction strategy (May and Williams 1986; May and Deyle 1998; Paleo 2009; Glavovic 2010a; Smith 2011). More fundamentally, planning takes place in the context of prevailing distributions of power and influence that structure and often perpetuate social vulnerability. Writing in the context of city planning, Godschalk (2003, p. 140) argues that:

> Building a disaster resilient city goes beyond changing land use and physical facilities. It must also build the capacity of the multiple involved communities to anticipate and respond to disasters. … An important limit on the adaptability of communities is their vulnerability to disaster. … In effect, the poorest and most vulnerable communities within a city are the weakest links in its mitigation capacity. Here is an important opportunity to integrate hazard mitigation with economic development and social justice, achieving the multiple objectives needed for a resilient system.

A national study of coastal states and local governments in the United States found that states have been largely ineffective in encouraging local governments to develop hazard mitigation plans that emphasise land use planning as a central element of their risk reduction strategy (Berke et al. 2012; Lyles et al. 2013; Smith et al. 2013). This finding is closely associated with what hazard scholar Ray Burby refers to as the local government paradox in which limited incentives are put in place by states and the federal government of the United States to encourage local governments to adopt strong land use measures that guide development away from known hazardous areas (2006). Nor do local governments and individuals necessarily bear the full burden of these choices as federal and sub-national governments and other institutions, like the insurance industry, corporations, and non-profit aid organisations and foundations, provide post-disaster assistance and insurance payouts that may not require the adoption of more stringent risk reduction strategies, thereby incentivising risky development (Kunreuther 1973; Platt 1999; Kunreuther and Roths 1998; Kunreuther 2006; Smith 2011).

Further hindering the development of sound natural hazards risk management plans is the degree to which practicing land use planners, many of whom claim to adhere to the practice of sustainable development, become actively involved in hazard mitigation and disaster recovery planning (Smith 2011; Lyles et al. 2013). Low levels of involvement by planners in the United States is disconcerting as it has been shown that planners can influence the adoption of important hazard mitigation measures (Burby 1998; Olshansky and Kartez 1998; Welsh and Esnard 2009; Stevens 2010) and the development of improved disaster recovery plans (Geipel 1982; Spangle et al. 1987; Oliver-Smith 1990; Berke and Beatley 1997; Schwab et al. 1998; Smith and Wenger 2006; Smith 2011). Even though the concepts of risk reduction, resilience and sustainable development contain a number of complementary elements, including those that are applicable to climate change adaptation, many planners have deferred to emergency managers to take on hazard mitigation planning-related activities. The result has played a role in the development of local hazard mitigation plans that are weak and fail to incorporate proven land use measures (Smith 2008; Lyles 2012; Lyles et al. 2013). This is unfortunate as the adoption of risk reduction measures should be an important part of any effort to increase sustainability and resilience (Berke 1995; Burby 1998a, b; Godschalk et al. 1999; Beatley 2009).

Despite the long-standing study and practice of natural hazards planning, planning scholars have been surprisingly slow to focus attention on the role planning can play in climate change adaptation (cf. Bulkeley 2006; Davoudi et al. 2009; Wilson and Piper 2010; Davoudi 2012; Hurlimann and March 2012; Picketts et al. 2012; Porter and Davoudi 2012; Wilkinson 2012). Hurlimann and March (2012) describe six qualities of planning that makes it a powerful tool for adaptation and, by extension in the context of the argument above, reducing risk and building resilience and sustainability. Planning can enable communities to (i) address matters of common concern; (ii) resolve conflicting interests; (iii) take into account interconnections across spatial, temporal and governance scales, whilst taking into account local nuances; (iv) address uncertainty and build capacity to cope with change; (v) facilitate social learning and function as a repository, integrator and conduit of knowledge; and (vi) coordinate and integrate a wide range of actors and institutional processes to realize a shared vision for the future. Planning draws on a range of approaches and tools that can be constructively used to foster incremental as well as transformative adaptation. Hurlimann and March (2012) identify three core challenges for planning. First, planners need to develop conviction. Second, planning processes must facilitate equitable outcomes. Third, planning institutional structures and processes must be transformed from passive to proactive enablers of adaptation.

Realising the promise of natural hazards planning takes place in the context of prevailing inequitable power relationships and practice must therefore be grounded in this reality (Few et al. 2007). Building adaptive capacity therefore necessitates much more than providing information about potential climate impacts to decision-makers; it is a multi-stage reflexive learning process that requires actors to (i) recognise and reconcile different framings of the issues that arise in considering climate risk and disaster risk; (ii) understand the underlying interests, perceptions and motivations of different actors, and institutional barriers and opportunities, and use incentives and sanctions to facilitate behavioural change; (iii) formulate feasible alternatives and facilitate access to resources for individual and collective transformation; and (iv) institutionalize new norms and 'rules', including rights and responsibilities, with effective feedback to facilitate learning and adaptation to change and surprise in the long run (Tàbara et al. 2010).

Lessons learned from natural hazards planning scholarship and experience provide an important bridge between the significant, albeit shrinking gulf between the fields of risk management and adaptation, emphasizing the mutually reinforcing themes of resilience and sustainability. The case studies in this book underscore the vital bridging role of planning and provide valuable insights into lessons learned from natural hazards and disaster experiences around the world.

## 1.3 Organisation of this Book

The purpose of this book is to learn from and improve our ability to draw from what we know about natural hazard and disaster experience and apply this knowledge in practice to the challenges associated with climate change adaptation. We do this by first sharing lessons derived from the study and practice of natural hazards risk management across a global suite of case studies. The use of a case study approach allows for a more in-depth and critical review of lessons learned in diverse social-ecological settings and from different natural hazard experiences at various spatial and geo-political scales. Distilling lessons from these diverse experiences highlights key problems and opportunities for adaptation and risk management. Examples include the pitfalls of failing to proactively plan for and reduce the potentially damaging effects of natural hazards on human settlements; the limited attention placed on building a shared understanding of risk amongst policy-makers, communities and individuals in a manner that leads to proactive targeted action; and the perpetuation of vulnerability through poor recovery and reconstruction strategies including the lack of pre-event planning for post-disaster recovery. These case studies also underscore the importance of establishing broad, supportive networks that facilitate collaborative governance and reflexive learning across formal and informal institutions; fostering an inclusive dialogue that links climate change adaptation strategies and post-disaster assistance, including the ability to maximise the use of available resources following major disasters to achieve complementary aims such as risk reduction and climate change adaptation, sustainable development and disaster resilience; and the importance of addressing endemic problems like low levels of institutional capacity and commitment, poverty, environmental degradation and fragile economies that are "exposed" during disasters.

This book also promotes the sharing of lessons within and between communities, nations and regions, including so-called developing and developed nations. This is particularly relevant to the United States, for example, which has lagged behind many other nations in addressing the climate change issue at a national level. Furthermore, the United States has long undervalued lesson drawing from other nations when it comes to natural hazards risk management (Garnett and Moore 2010; Moore et al. 2009). While there are numerous lessons that can be drawn from the rich literature and practice tied to development, the addition of this important dimension to the natural hazards risk management-climate change adaptation dialogue remains understudied (Bicknell et al. 2009) and underutilized in practice due to a lack of institutional coordination and a failure to effectively communicate shared goals and lessons (Schipper and Pelling 2006).

The book is organised in the following manner. After the introductory chapter, we begin with a section titled Climate Change Adaptation: Theory and Practice. This section draws on chapters written by Jörn Birkmann and Joanna Pardoe (Climate Change Adaptation and Disaster Risk Reduction: Fundamentals, Synergies and Mismatches) who provide an overview of the linkages and gaps between climate change adaptation and disaster risk reduction concepts and associated policies and practices. Gina Zievogel and Sue Parnell's chapter, Tackling Barriers to Climate Change Adaptation in South African Coastal Cities, describes the issues facing Cape Town and eThekwini as a means to understand the challenges and opportunities of adaptation at the urban city-scale. Both chapters help to unpack important topics that are found throughout the remainder of the book. These include the identification of key barriers and opportunities to achieving adaptation, the importance of developing a broad institutional framework for action that is cognisant of the growing base of knowledge (including that which is locally informed), the need to recognise and account for varied temporal and spatial scales and the ability to develop an appropriate mix of flexible and holistic policies and practices that address underlying issues such as risk reduction and development.

The next section, titled The Nature of Disasters and the Role of Natural Hazards Planning in Building Resilient Communities, is comprised of five chapters that highlight the importance and dynamics of planning and collective action. Anthony Oliver Smith's chapter, Climate Change Adaptation and Disaster Risk Reduction in Highland Peru, describes adaptation as a long-standing cultural phenomenon closely associated with the persistence of societies over time. This perspective helps the reader to understand the complexities of what often amounts to a largely reactive process and compares this with hazard mitigation approaches which are ideally proactive in nature. In Chapter. 5, Iain White (Firm Foundations or Castles on Sand? The Shifting Sources of Flood Risk and the Implications for Flood Governance: An English Case Study) explains the challenges facing cities and regions in England as increased urbanisation and repeated flood-related disasters have led to a change in thinking from an approach driven by the adoption of flood defenses in the aftermath of an event to a more pre-emptive focus on risk management. In Planning for Resilient Coastal Communities: Emerging Practice and Future Directions, Timothy Beatley provides a vision of what it means to be a resilient coastal community, discusses principles underlying this designation, and highlights ways that communities have achieved this objective. Bill Simbieda's chapter, Adaptation to Seismic Risk and Climate Change: San Francisco and Berkeley, California, USA describes how two cities assume different approaches to achieving disaster resilience, emphasising physical and institutional methods respectively. Philip Berke, in Rising to the Challenge: Planning for Adaptation in the Age of Climate Change, argues that scenario-based planning provides a sound means to confront what still amounts to a great deal of uncertainty in our understanding of climate change-related impacts while providing the flexibility needed to account for unexpected outcomes and new information.

In the last major section, Lessons from Disaster Experience, we present a collection of case studies from across the world, including Applying Hurricane Recovery Lessons in the United States to Climate Change Adaptation: Hurricanes Fran and Floyd in North Carolina (USA) (Gavin Smith); The 2004 Manawatu Floods, New Zealand: Integrating Flood Risk Reduction and Climate Change Adaptation (Bruce Glavovic); Learning from Analyses of Policy Frames and Informal Institutions in the Fire Management Sector of Victoria, Australia (Karen Bosomworth, John Handmer and Steven Dovers); Recovering from the 2004 Indian Ocean Tsunami: Lessons for Climate Change Response (Ahana Lakshmi, R. Purvaja and R. Ramesch); Coastal Hazards Planning, the 2009 Tsunami and Lessons Learned for Climate

Change Adaptation in Samoa (Namouta Poutasi, Michele Daly, Jude Kohlhase and Filomena Nelson); Disaster Recovery in Coastal Mississippi (USA): Lesson Drawing from Hurricanes Camille and Katrina (Gavin Smith); and Waves of Adversity, Layers of Resilience: Floods, Hurricanes, Oil Spills, and Adapting to a Climate Change in the Mississippi Delta (Bruce Glavovic).

Each of the case studies addresses a number of important topical areas, including the pre- and post-event setting of the locales being discussed and how these conditions shape the policies, programmes, and plans developed and implemented in the face of natural hazards and disasters; lessons drawn from these experiences; the identification of barriers and opportunities for mainstreaming hazard mitigation and disaster recovery policies into climate change adaptation; and a set of recommendations for action. A review of the chapters show that a number of common themes emerge, including the importance of effective collaborative governance; the influence of pre-event conditions such as culture, wealth (or its absence, i.e., poverty), policy frameworks and institutions on desired outcomes; the value of establishing good vertical connectivity between national policy and local plans; adopting varied and flexible risk management strategies; and viewing disasters as focusing events, including the ability to adopt new policies and practices that reduce exposure to extreme events and confront the drivers and root causes of vulnerability.

The final chapter, Conclusions: Integrating Natural Hazards Risk Management and Climate Change Adaptation through Natural Hazards Planning consolidates the lessons learned and priority actions recommended by contributing authors and organises them under a broad vision and set of themes that are framed as imperative statements. The chapter concludes with an examination of how these imperatives can be operationalised in both existing, and where necessary, new plans, policies and collective arrangements that span diverse settings.

## References

Abel, N., Gorddard, R., Harman, B., Leitch, A., Langridge, J., Ryan, A., & Heyenga, S. (2011). Sea level rise, coastal development and planned retreat: Analytical framework, governance principles and an Australian case study. *Environmental Science and Policy, 14,* 279–288.

Adger, W. N. (2001). Scales of governance and environmental justice for adaptation and mitigation of climate change. *Journal of International Development, 13,* 921–931.

Adger, W. N., Arnell, N. W., & Tompkins, E. L. (2005). Successful adaptation to climate change across scales. *Global Environmental Change, 15,* 77–86.

Adger, W. N., Dessai, S., Goulden, M., Hulme, M., Lorenzoni, I., Nelson, D. R., Naess, L. O., Wolf, J., & Wreford, A. (2009). Are there social limits to adaptation to climate change? *Climatic Change, 93,* 335–354.

Albala-Bertrand, J. M. (1993). *The political economy of large natural disasters, with special reference to developing countries*. New York: Clarendon Press.

Albala-Bertrand, J. M. (2006). Globalization and localization: An economic approach. In H. Rodriguez, E. Quarantelli, & R. Dynes (Eds.), *Handbook of disaster research* (pp. 147–167). New York: Springer.

Anderson, M., & Woodrow P. (1989). *Rising from the ashes: Development strategies in times of disaster*. Boulder: Westview.

Baker, I., Peterson, A., Brown, G., & McAlpine, C. (2012). Local government response to the impacts of climate change: An evaluation of local climate adaptation plans. *Landscape and Urban Planning, 107,* 127–136.

Bauer, A., Feichtinger, J., & Steurer, R. (2012). The governance of climate change adaptation in 10 OECD countries: Challenges and approaches. *Journal of Environmental Policy and Planning, 14*(3), 279–304.

Beatley, T. (1998). The vision of sustainable communities. In R. J. Burby (Ed.), *Cooperating with nature: Confronting natural hazards with land-use planning for sustainable communities* (pp. 233-262). Washington, D.C.: Joseph Henry Press.

Beatley, T. (2009). *Planning for coastal resilience: Best practices for Calamitous times*. Washington, D.C.: Island Press.

Berke, P. R. (1995). Natural hazard reduction and sustainable development: A global reassessment. *Journal of Planning Literature, 9*(4), 370–382.

Berke, P. R., & Beatley, T. (1997). *After the hurricane: Linking recovery to sustainable development in the Caribbean*. Baltimore: Johns Hopkins University Press.

Berke, P. R., & Campanella, T. J. (2006). Planning for postdisaster resiliency. *The Annals of the American Academy of Political and Social Science, 604*(1), 192–207.

Berke, P. R., Kartez, J., & Wenger, D. (1993). Recovery after disasters: Achieving sustainable development, mitigation and equity. *Disasters, 17*(2), 93–109.

Berke, P. R., Cooper, J., Salvesen, D., Spurlock, D., & Rausch, C. (2011). Building capacity for disaster resiliency in six disadvantaged communities. *Sustainability, 3,* 1–20.

Berke, P. R., Smith, G., & Lyles, W. (2012). Planning for resiliency: Evaluation of state hazard mitigation plans under the disaster mitigation act. *Natural Hazards Review, 13*(2), 139–149.

Berrang-Ford, L., Ford, J. D., & Paterson, J. (2011). Are we adapting to climate change? *Global Environmental Change, 21,* 25–33.

Bicknell, J., Dodman, D., & Satterthwaite, D. (Eds.). (2009). *Adapting cities to climate change: Understanding and addressing the development challenges*. London: Earthscan.

Birkland, T. A. (1996). Natural disasters are focusing events: Policy communities and political response. *International Journal of Mass Emergencies and Disasters, 14*(2), 221–243.

Birkland, T. A. (1997). *After disaster: Agenda setting, public policy, and focusing events*. Washington, D.C.: Georgetown University Press.

Birkmann, J., Von Teichmann, K., Aldunce, P., Bach, C., Binh, N. T., Garschagen, M., Kanwar, S., Setiadi, N., & Thach, L. N. (2009). Addressing the challenge: Recommendations and quality criteria for linking disaster risk reduction and adaptation to climate change. In J. Birkmann, G. Tetzlaff, K. O. Zentel (Eds.). Bonn: KKV Publication Series 38. http://www.dkkv.org/.

Birkmann, J., & Von Teichman, K. (2010). Integrating disaster risk reduction and climate change adaptation: Key challenges-scales, knowledge, and norms. *Sustainability Science, 5,* 171–184.

Birkmann, J., Buckle, P., Jaeger, J., Pelling, M., Setiadi, N., Garschagen, M., Fernando, N., & Kropp, J. (2008). Extreme events and disasters: A window of opportunity for change? Analysis of organizational, institutional and political changes, formal responses after mega-disasters. *Natural Hazards, 55,* 637–655.

Birkmann, J., Garschagen, M., Kraas, F., & Quang N. (2010). Adaptive urban governance: New challenges for the second generation of urban adaptation strategies to climate change. *Sustainability Science, 5*(2), 185–206.

Blakely, E., & Carbonell, A. (2012). *Resilient coastal cities: Planning for climate change in the United States and Australia*. Cambridge: Lincoln Institute of Land Policy.

Board on Natural Disasters. (1999). Mitigation emerges as major strategy for reducing losses caused by natural disasters. *Science, 284,* 1943–1947.

Bouwer, L. M., Jeroen, C., & Aerts, J. H. (2006). Financing climate change adaptation. *Disasters, 30*(1), 49–63.

Boyd, E., Grist, N., Juhola, S., & Nelson, V. (2009). Exploring development futures in a changing climate: Frontiers for development policy and practice. *Development Policy Review, 27*(6), 659–674.

Brooks, N., Adger, W. N., & Kelly, P. M. (2005). The determinants of vulnerability and adaptive capacity at the national level and the implicaitons for adaptation. *Global Environmental Change, 15,* 151–163.
Brooks, N., Grist, N., & Brown, K. (2009). Development futures in the context of climate change: Challenging the present and learning from the past. *Development Policy Review, 27*(6), 741–765.
Bulkeley, H. (2006). A changing climate for spatial planning. *Planning Theory and Practice, 7*(2), 203–214.
Burby, R. J. (1998a). Natural hazards and land use: An introduction. In R. J. Burby (Ed.), *Cooperating with nature: Confronting natural hazards with land-use planning for sustainable communities* (pp. 1–26). Washington, D.C.: Joseph Henry Press.
Burby, R. J. (Ed.). (1998b). *Cooperating with nature: Confronting natural hazards with land-use planning for sustainable communities*. Washington, D.C.: Joseph Henry Press.
Burby, R. (2006). Hurricane Katrina and the paradoxes of government disaster policy: Bringing about wise government decisions for hazardous areas. *The Annals of the American Academy of Political and Social Science, 54*(May), 171–191.
Burby, R. J., Beatley, T., Berke, P. R., Deyle, R. E., French, S. P., Godschalk, D. R., Kaiser, E. J., Kartez, J. D., May, P. J., Olshansky, R., Paterson, R. G., & Platt, R. H. (1999). Unleashing the power of planning to create disaster resistant communities. *Journal of the American Planning Association, 65*(3), 247–258.
Burch, S. (2010). Transforming barriers into enablers of action on climate change: Insights from three municipal case studies in British Columbia, Canada. *Global Environmental Change, 20,* 287–297.
Burch, S. (2011). Sustainable development paths: Investigating the roots of local policy responses to climate change. *Sustainable Development, 19,* 176–188.
Burton, I., Kates, R. W., & White, G. F. (1968). *The human ecology of extreme geophysical events.* Natural Hazards Research Working Paper No.1. Department of Geography, University of Toronto, Toronto.
Burton, I., Kates, R., & White, G. F. (1993). *The environment as hazard.* New York: Guilford Press.
Button, G. V. (1999). The negation of disaster: The media response to oil spills in Great Britain. In *The angry earth: Disaster in anthropological perspective* (pp. 113–132). New York: Routledge.
Campanella, T. J. (2006). Urban resilience and the recovery of New Orleans. *Journal of the American Planning Association, 72*(2), 141–146.
Cannon, T. (1994). Vulnerability analysis and the explanation of "Natural" disasters. In A. Varley (Ed.), *Disasters, development and environment* (pp. 13–30). Chichester: John Wiley and Sons.
Carmin, J., Nadkarni, N., & Rhie, C. (2012). *Progress and challenges in urban climate adaptation planning: Results of a global survey*. Cambridge: MIT.
Carpenter, S. R., & Gunderson, L. H. (2001). Coping with collapse: Ecological and social dynamics in ecosystem management. *BioScience, 51,* 451–457.
CDRSS (Committee on Disaster Research in the Social Sciences). (2006). *Facing hazards and disasters: Understanding human dimensions.* Washington, D.C.: National Academies Press.
Chakraborty, A., Kaza, N., Gerrit-Jan, K., & Deal, B. (2011). Robust plans and contingent plans scenario planning for an uncertain world. *Journal of the American Planning Association, 77*(3), 251–266.
Chambers, R., & Conway, G. R. (1991). *Sustainable rural livelihoods: Practical concepts for the 21st century.* Institute of Development Studies DP 296, 1991. University of Sussex: Brighton.
Cohen, J. D., & Butler, B. W. (1996). *Modeling potential structure ignitions from flame radiation exposure with implications for wildland/urban interface fire management.* 13th Fire and Forest Meteorology Conference. Lorne Australia.
Colten, C. E. (2005). *An unnatural metropolis: Wresting New Orleans from nature*. Baton Rouge: Louisiana State University Press.
Cuny, F. (1983). *Disasters and development*. Oxford: Oxford University Press.

Davoudi, S. (2012). Resilience: A bridging concept or a dead end? *Planning Theory and Practice, 13*(2), 299–307.

Davoudi, S., Crawford, J., & Mehmood, A. (2009). *Planning for climate change: Strategies for mitigation and adaptation for spatial planners*. London: Earthscan.

Davoudi, S., Brooks, E., & Mehmood, A. (2013). Evolutionary resilience and strategies for climate adaptation. *Planning Practice and Research, 28*(3), 307–322.

Diamond, J. (2005). *Collapse: How societies choose to fail or succeed.* New York: Viking, Penguin Group.

Dronkers, J., Gilbert, J. T. E., Butler, L. W., Carey, J. J., Campbell, J., James, E., McKenzie, C., Misdorp, R., Quin, N., Ries, K. L., Schroder, P. C., Spradley, J. R., Titus, J. G., Vallianos, L., & von Dadelzen, J. (1990). *Strategies for adaption to sea level rise. Report of the IPCC coastal zone management subgroup: Intergovernmental panel on climate change*. Geneva: Intergovernmental Panel on Climate Change.

Few, R., Brown, K., & Tompkins, E. L. (2007). Public participation and climate change adaptation: Avoiding the illusion of inclusion. *Climate Policy, 7*(1), 46–59.

Folke, C. (2006). Resilience: The emergence of a perspective for social-ecological systems analysis. *Global Environmental Change, 16*(3), 253–267.

Ford, J. D., Berrang-Ford, L., & Paterson, J. (2011). A systematic review of observed climate change adaptation in developed nations. A letter. *Climatic Change, 106*, 327–336.

Fordham, M. (2006). Disaster and development research and practice: A necessary eclecticism? In H. Rodriguez, E. Quarantelli, & R. Dynes (Eds.), *Handbook of disaster research* (pp. 335–346). New York: Springer.

Gaillard, J. C. (2010). Vulnerability, capacity and resilience: Perspectives for climate and development policy. *Journal of International Development, 22*, 218–232.

Gallopin, G. C. (2006). Resilience, vulnerability and adaptation: A cross-cutting theme of the international human dimensions programme on global environmental change. *Global Environmental Change, 16*(3), 293–303.

Garnett, J. D., & Moore, M. (2010). Enhancing disaster recovery: Lessons from exemplary international disaster management practices. *Journal of Homeland Security and Emergency Management, 7*(1), 1–20.

Gawronski, V., Olson, R. S., & Carvalho, P. (2006). Locally influenced, distantly courageous? Exploring media constructions of Katrina's gulf coast impacts. In Learning from catastrophe: Quick response research in the wake of hurricane Katrina (pp. 443–456). Boulder: Institute for Behavioral Science, University of Colorado at Boulder.

Geipel, R. (1982). *Disaster and reconstruction. The Friuli (Italy) earthquake of 1976*. London: Allen and Unwin.

Geis, D., & Kutzmark, T. (1995). *Developing sustainable communities: The future is now. Public Management*. August, 4–13.

Gibson, C. C., Andersson, K., Ostrom, E., & Shivakumar, S. (2009). *The Samaritan's dilemma: The political economy of development aid.* Oxford: Oxford University Press.

Glavovic, B. C. (2008). Sustainable coastal communities in the age of coastal storms: Reconceptualising coastal planning and ICM as 'new' naval architecture. *Journal of Coastal Conservation, 12*(3), 125–134.

Glavovic, B. C. (2010a). The role of land-use planning in disaster risk reduction: An Australasian perspective. *Special Issue of Australasian Journal of Disaster and Trauma Studies*. See http://www.massey.ac.nz/~trauma/issues/2010-1/editorial.htm.

Glavovic, B. C. (2010b). Future challenges and opportunities for natural hazards planning in Australasia. *Special Issue of Australasian Journal of Disaster and Trauma Studies*. See http://www.massey.ac.nz/~trauma/issues/2010-1/glavovic2.htm.

Glavovic, B. C., Saunders, W., & Becker, J. (2010). Land-use planning for natural hazards in New Zealand: The setting, barriers, 'burning issues' and priority actions. *Natural Hazards, 54*(3), 679–706.

Godschalk, D. R. (2003). Urban hazard mitigation: Creating resilient cities. *Natural Hazards Review, 4*(3), 136–142.

Godschalk, D. R., Kaiser, E. J., & Berke, P. R. (1998). Integrating hazard mitigation and local land-use planning. In R. J. Burby (Ed.), *Cooperating with nature: Confronting natural hazards with land-use planning for sustainable communities* (pp. 85-118). Washington, D.C.: Joseph Henry Press.

Godschalk, D. R., Beatley, T., Berke, P. R., Brower, D. J., & Kaiser, E. J. (1999). *Natural hazard mitigation: Recasting disaster policy and planning*. Washington, D.C.: Island Press.

Grist, N. (2008). Positioning climate change in sustainable development discourse. *Journal of International Development, 20*(6), 783–803.

Gupta, J., Termeer, C., Klostermann, J., Meijerink, S., van den Brink, M., Jong, P., Nooteboom, S., & Bergsma, E. (2010). The adaptive capacity wheel: A method to assess the inherent characteristics of institutions to enable the adaptive capacity of society. *Environmental Science and Policy, 13,* 459–471.

Hagemann, M., Hendel-Blackford, S., Hohne, N., Harvey, B., Naess, L. O., & Urban, F. (2011). *Guiding climate compatible development: User-Orientated analysis of planning tools and methodologies*. Report produced by Ecofys and Institute of Development Studies for the Climate and Development Knowledge Network.

Handmer, J. (2008). Risk creation, bearing and sharing on Australian floodplains. *Water Resources Development, 24*(4), 527–540.

Handmer, J. (2009). Adaptive capacity: What does it mean in the context of natural hazards? In E. Lisa, F. Schipper, & I. Burton (Eds.), *Adaptation to climate change* (pp. 213–227). London: Earthscan.

Handmer, J., & Dovers, S. (1996). A typology of resilience: Rethinking institutions for sustainable development. *Organization and Environment, 9*(4), 482–511.

Handmer, J., & Dovers, S. (2009). A typology of resilience: Rethinking institutions for sustainable development. In E. Lisa, F. Schipper, & I. Burton (Eds.), *Adaptation to climate change* (pp. 187–210). London: Earthscan.

Haque, C. E., & Etkin, D. (2007). People and community as constituent parts of hazards: The significance of societal dimensions in hazards analysis. *Natural Hazards, 41,* 271–282.

Harrell-Bond, E. (1986). *Imposing aid: Emergency assistance to refugees*. Oxford: Oxford University Press.

Henson, R. (2008). *The rough guide to climate change*. London: Penguin Books.

Hewitt, K. (Ed.). (1983). *Interpretations of calamity from the viewpoint of human ecology*. London: Allen and Unwin.

Hoffman, S., & Oliver-Smith, A. (Eds.). (2002). *Catastrophe and culture*. Santa Fe: School of American Research Press.

Holling, C. S. (1973). Resilience and stability of ecological systems. *Annual Review of Ecology and Systematics, 4,* 1–23.

Hurlimann, A. C., & March, A. P. (2012). The role of spatial planning in adapting to climate change. *Wires Climate Change, 3*(5), 477–488.

IPCC. (2001). Climate change 2001: impacts adaptation, and vulnerability. Contribution of Working Group II to the Third Assessment Report of the Intergovernmental Panel on Climate Change. McCarthy, J. J., Canziani, O. F., Leary, N. A., D. J. Dokken, & K. S. White (Eds.). Cambridge: Cambridge University Press.

IPCC. (2007a). Climate Change 2007: The physical science basis. Contribution of Working Group I to the Fourth Assessment Report of the Intergovernmental Panel on Climate Change. In S. Solomon, D. Qin, M. Manning, M. Marquis, K. Averyt, M. M. B. Tignor, H. L. Miller Jr., & Z. Chen (Eds.). Cambridge: Cambridge University Press.

IPCC. (2007b). Climate Change 2007: Impacts, adaptation and vulnerability. Contribution of Working Group II to the Fourth Assessment Report of the Intergovernmental Panel on Climate Change. In M. L. Parry, O. F. Canziani, J. P. Palutikof, van der Linden P. J., & C. E. Hanson (Eds.). Cambridge: Cambridge University Press.

IPCC. (2011). Special report on managing the risks of extreme events and disasters to advance climate change adaptation: Summary for policymakers. In C. B. Field, V. Barros, T. F. Stocker,

D. Qin, D. Dokken, K. L. Ebi, M. D. Mastrandrea, K. J. Mach, G.-K. Plattner, S. Allen, M. Tignor, & P. M. Midgley (Eds.), Cambridge: Cambridge University Press.

IPCC. (2012). Managing the risks of extreme events and disasters to advance climate change adaptation. A Special Report of Working Groups I and II of the Intergovernmental Panel on Climate Change. In C. B. Field, V. Barros, T. F. Stocker, D. Qin, D. J. Dokken, K. L. Ebi, M. D. Mastrandrea, K. J. Mach, G.-K. Plattner, S. K. Allen, M. Tignor, & P. M. Midgley (Eds.). Cambridge: Cambridge University Press.

Jones, L., & Boyd, E. (2011). Exploring social barriers to adaptation: Insights from Western Nepal. *Global Environmental Change, 21,* 1262–1274.

Juhola, S., & Westerhoff, L. (2011). Challenges of adaptation to climate change across multiple scales: A case study of network governance in two European countries. *Environmental Science and Policy, 14,* 239–247.

Kasperson, R. E., Renn, O., Slovic, P., Brown, H. S., Emel, J., Gobel, R., Kasperson, J. X., & Ratrick, S. (1988). The social amplification of risk. *Risk Analysis, 8,* 177–187.

Kelman, I. (2006). Acting on disaster diplomacy. *Journal of International Affairs, 59*(2), 215–240.

Kelman, I. (2007). Hurricane Katrina disaster diplomacy. *Disasters, 31*(3), 288–309.

Kelman, I., & Gaillard, J. C. (2010). Embedding climate change adaptation within disaster risk reduction, Chapter 2. In R. Shaw, J. M. Pulhin, & J. J. Pereira (Eds.), *Climate change adaptation and disaster risk reduction: Issues and challenges. Community, environment and disaster risk management* (pp. 23–46). Bingley: Emerald Group Publishing Limited.

Klinenberg, E. (2002). *Heat wave: A social autopsy of disaster in Chicago*. Chicago: University of Chicago Press.

Klinke, A., & Renn, O. (2012). Adaptive and integrative governance on risk and uncertainty. *Journal of Risk Research, 15*(3), 273–292.

Knight, F. H. (1921). *Risk, uncertainty, and profit*. Boston: Hart, Schaffner and Marx/ Houghton Mifflin.

Kooiman, J. (2003). *Governing as governance*. London: Sage.

Kunreuther, H. (1973). *Recovery from natural disasters: Insurance or federal aid*? Washington, D.C.: American Enterprise Institute.

Kunreuther, H. (2006). Has the time come for comprehensive natural disaster insurance? In R. Daniels, D. Kettl, & H. Kunreuther (Eds.), *On risk and disaster: Lessons from hurricane Katrina* (pp. 175–201). Philadelphia: University of Philadelphia Press.

Kunreuther, H., & Roths, R. J. (Eds.). (1998). *Paying the price: The status and role of insurance against national disasters in the United States*. Washington, D.C.: Joseph Henry Press.

Lebel, L., Manuta, J. B., & Garden, P. (2011). Institutional traps and vulnerability to changes in climate and flood regimes in Thailand. *Regional Environmental Change, 11,* 45–58.

Lemos, M., Boyd, E., Tompkins, E., Osbahr, H., & Liverman, D. (2007). Developing adaptation and adapting development. *Ecology And Society, 12*(2), 26–29.

Lesnikowski, A. C., Ford, J. D., Berrang-Ford, L., Barrera, M., Berry, P., Henderson, J., & Heymann, S. J. (2013). National-level factors affecting planned, public adaptation to health impacts of climate change. *Global Environmental Change*, Available online 4 June 2013.

Levin, K., Cashore, B., Bernstein, S., & Auld, G. (2012). Overcoming the tragedy of super wicked problems: Constraining our future selves to ameliorate global climate change. *Policy Science, 45,* 123–152.

Lyles, L. W. (2012). *Stakeholder network influences on local-level hazard mitigation planning outputs*. The University of North Carolina at Chapel Hill, ProQuest, UMI Dissertations Publishing, 3549567.

Lyles, L. W., Berke, P. R., & Smith, G. (2013). Do planners matter? Examining factors driving incorporation of land use approaches into hazard mitigation plans. *Journal of Environmental Planning and Management*. Published online 3 May 2013.

May, P., & Deyle, R. E. (1998). Governing land use in hazardous areas with a patchwork system. In R. J. Burby (Ed.), *Cooperating with nature: Confronting natural hazards with land use planning for sustainable communities* (pp. 57–84). Washington, D.C.: Joseph Henry Press.

May, P., & Williams, W. (1986). *Disaster policy implementation: Managing programs under shared governance*. New York: Plenum Press.

May, P. J., Burby, R. J., Ericksen, N. J., Handmer, J. W., Dixon, J. E., Michaels, S., & Smith, D. I. (1996). *Environmental management and governance: Intergovernmental approaches to hazards and sustainability*. London: Routledge.

McAnany, P. A., & Yoffee, N. (2009). *Questioning collapse: Human resilience, ecological vulnerability, and the aftermath of empire*. Cambridge: Cambridge University Press.

McBean, G., & Rodgers, C. (2010). Climate hazards and disasters: The need for capacity building. *WIRES Climate Change, 1,* 871–884.

McEvoy, D., Fünfgeld, H., & Bosomworth, K. (2013). Resilience and climate change adaptation: The importance of framing. *Planning Practice and Research, 28*(3), 280–293.

Measham, T. G., Preston, B. L., Smith, T. F., Brooke, C., Gorddard, R., Withycombe, G., & Morrison, C. (2011). Adapting to climate change through local municipal planning: Barriers and challenges. *Mitigation and Adaptation Strategies for Global Change, 16*(8), 889–909.

Mileti, D. S. (Ed.). (1999). Disasters by design: A reassessment of natural hazards in the United States. Washington, D.C.: Joseph Henry Press.

Mileti, D. S., & Gailus, J. L. (2005). Sustainable development and hazards mitigation in the United States: Disasters by design revisited. *Mitigation and Adaptation Strategies for Global Change, 10,* 491–504.

Molotch, H., & Lester, M. (1974). News as purposeful behavior: On the strategic use of routine events, accidents, and scandals. *American Sociological Review, 39,* 101–112.

Moore, M., Trujillo, H. R., Stearns, B. K., Basurto-Davilia, R., & Evans, D. K. (2009). Learning from exemplary practices in international disaster management: A fresh avenue to inform U.S. policy? *Journal of Homeland Security and Emergency Management, 6*(1), Article 35.

Moser, S. C., & Ekstrom, J. A. (2010). A framework to diagnose barriers to climate change adaptation. *Proceedings of the National Academies of Sciences, 107*(51), 22026–22031.

Munasinghe, M., & Clark, C. (Eds.). (1995). *Disaster prevention for sustainable development: Economic and policy issues*. Washington, D.C.: The International Bank for Reconstruction and Development/ The World Bank.

Nakagawa, Y., & Shaw, R. (2004). Social capital: A missing link to disaster recovery. *International Journal of Mass Emergencies and Disasters, 22*(1), 5–34.

National Research Council of the National Academies. (2006). *Facing hazards and disasters: Understanding human dimensions*. Washington, D.C.: The National Academies Press.

National Research Council of the National Academies. (2009). *Informing decisions in a changing climate*. Washington, D.C.: The National Academies Press.

National Research Council of the National Academies. (2010a). *Adapting to the impacts of climate change*. Washington, D.C.: The National Academies Press.

National Research Council of the National Academies. (2010b). *Informing an effective response to climate change*. Washington, D.C.: The National Academies Press.

Nelson, D. R., Adger, W. N., & Brown, K. (2007). Adaptation to environmental change: Contributions of a resilience framework. *Annual Review of Environment and Resources, 32,* 395–419.

Newman, P., Beatley, T., & Boyer, H. (2009). *Resilient cities: Responding to peak oil and climate change*. Washington, D.C.: Island Press.

O'Brien, G., O'Keefe, P., Rose, J., & Wisner, B. (2006). Climate change and disaster management. *Disasters, 30*(1), 64–80.

Oliver-Smith, A. (1990). Post-disaster housing reconstruction and social inequality: A challenge to policy and practice. *Disasters, 14*(1), 7–19.

Oliver-Smith, A., & Hoffman, S. M. (Eds.). (1999). *The angry earth: Disaster in anthropological perspective*. New York: Routledge.

Olshansky, R. B., & Kartez, J. (1998). Managing land-use to build resilience. In R. J. Burby (Ed.), *Cooperating with nature: Confronting natural hazards with land-use planning for sustainable communities* (pp. 167–201). Washington, D.C.: Joseph Henry Press.

Olson, R. S. (2000). Toward a politics of disaster: Losses, values, agendas, and blame. *International Journal of Mass Emergencies and Disasters, 18,* 265–287.

Olson, R. S., Olson, R. A., & Gawronski, V. T. (1998). Night and day: Mitigation policymaking in Oakland California before and after the Loma Prieta earthquake. *International Journal of Mass Emergencies and Disasters, 16*(2), 145–179.

Olson, R. S., Olson, R. A., & Gawronski, V. T. (1999). *Some buildings just can't dance: Politics, life safety, and disaster*. Stamford: JAI Press.

Oreskes, N., & Conway, E. M. (2010). *Merchants of doubt: How a handful of scientists obscured the truth on issues from tobacco smoke to global warming*. New York: Bloomsbury Press.

Paleo, U. F. (2009). *Building safer communities. Risk governance, spatial planning and response to natural hazards*. Amsterdam: IOS Press.

Paton, D., & Johnston, D. (2006). *Disaster resilience: An integrated approach*. Springfield: Charles C. Thomas Publisher.

Pelling, M. (2003). *Natural disasters and development in a globalizing world*. London: Routledge.

Pelling, M. (2011). *Adaptation to climate change: From resilience to transformation*. London: Routledge.

Pelling, M., & Manuel-Navarrete, D. (2011). From resilience to transformation: The adaptive cycle in two Mexican urban centers. *Ecology and Society, 16*(2), 11. http://www.ecologyandsociety.org/vol16/iss2/art11/

Pelling, M., High, C., Dearing, J., & Smith, D. (2008). Shadow spaces for social learning: A relational understanding of adaptive capacity to climate change within organizations. *Environmental Planning A, 40*(4), 867–884.

Peñalba, L. M., Elazegui, D. D., Pulhin, J. M., & Cruz, R. V. O. (2012). Social and institutional dimensions of climate change adaptation. *International Journal of Climate Change Strategies and Management, 4*(3), 308–322.

Picketts, I. M., Curry, J., & Rapaport, E. (2012). Community adaptation to climate change: Environmental planners' knowledge and experiences in British Columbia, Canada. *Journal of Environmental Policy and Planning, 14*(2), 119–137.

Platt, R. (1999). *Disasters and democracy: The politics of extreme natural events*. Washington, D.C.: Island Press.

Porter, L., & Davoudi, S. (2012). The politics of resilience for planning: A cautionary note. *Planning Theory and Practice, 13*(2), 329–333.

Preston, B. L., Westaway, R. M., & Yuen, E. J. (2011). Climate adaptation planning in practice: An evaluation of adaptation plans from three developed nations. *Mitigation and Adaptation Strategies for Global Change, 16,* 407–438.

Puszkin-Chevlin, A., Hernandez, D., & Murley, J. (2006/2007). Land use planning and its potential to reduce hazard vulnerability: Current practices and future possibilities. *Marine Technology Society Journal, 40*(4), 7–15.

Quay, R. (2010). Anticipatory governance a tool for climate change adaptation. *Journal of the American Planning Association, 76*(4), 496–511.

Renn, O. (2008). *Risk governance: Coping with uncertainty in a complex world*. London: Earthscan.

Rhodes, R. A. W. (1997). *Understanding governance: Policy networks, governance, reflexivity and accountability*. Buckingham: Open University Press.

Romieu, E., Welle, T., Schneiderbauer, S., Pelling, M., & Vinchon, C. (2010). Vulnerability assessment within climate change and natural hazard contexts: Revealing gaps and synergies through coastal applications. *Sustainability Science, 5,* 159–170.

Rubin, C. (2007). *Emergency management: The American experience, 1900-2005.* Fairfax: Public Entity Risk Institute.

Rubin, C., & Tanali, I. R. (2001). *Disaster timeline*. Arlington: Claire Rubin and Associates. Project website http:\\www.disastertimeline.com.

Satterthwaite, D., Huq, S., Reid, H., Pelling, M., & Lankao, P. R. (2009). In J. Bicknell, D. Dodman, & D. Satterthwaite (Eds.), *Adapting cities to climate change: Understanding and addressing the development challenges* (pp. 3–47). London: Earthscan.

Schipper, E. L. F. (2004). *Exploring adaptation to climate change: A development perspective*. Ph. D. Thesis, School of Development Studies, University of East Anglia, Norwich.

Schipper, L. (2009). Meeting at the crossroads? Exploring the linkages between climate change adaptation and disaster risk reduction. *Climate Development, 1*(1), 16–30.

Schipper, L., & Pelling, M. (2006). Disaster risk, climate change and international development: Scope for, and challenges to, integration. *Disasters, 30*(1), 19–38.

Schneider, R. O. (2002). Hazard mitigation and sustainable community development. *Disaster Prevention and Management, 11*(2), 141–147.

Schwab, J., Topping, K. C., Eadie, C. C., Deyle, R. E., & Smith, R. A. (1998). *Planning for post-disaster recovery and reconstruction*. PAS Report 483–484. Chicago: American Planning Association.

Smit, B., & Wandel, J. (2006). Adaptation, adaptive capacity and vulnerability. *Global Environmental Change, 16,* 282–292.

Smith, G. (2008). Planning for sustainable and disaster resilient communities. In J. Pine (Ed.), *Natural hazards analysis: Reducing the impact of disasters* (pp. 221–247). Boca Raton: CRC Press.

Smith, G. (2011). *Planning for post-disaster recovery: A review of the United States disaster assistance framework*. Washington, D.C.: Island Press.

Smith, G., & Wenger, D. (2006). Sustainable disaster recovery: Operationalizing an existing framework. In H. Rodriguez, E. Quarantelli, & R. Dynes (Eds.), *Handbook of disaster research* (pp. 234–257). New York: Springer.

Smith, G., Ward, L. W., & Berke, P. R. (August 2013). The role of the state in building local capacity and commitment for hazard mitigation planning. *International Journal of Mass Emergencies and Disasters*.

Spangle, W., Meehan, R., Degenkolb, H., & Blair, M. (1987). *Pre-event planning for post-earthquake rebuilding*. Los Angeles: Southern California Earthquake Preparedness Project.

Stern, N. H. (2007). *The economics of climate change: The stern review*. Cambridge: Cambridge University Press.

Stern, N. (2009). *Managing climate change and overcoming poverty: facing the realities and building a global agreement*. Policy paper, Centre for Climate Change Economics and Policy, London and Leeds, UK.

Stevens, M. (2010). Implementing natural hazard mitigation provisions: Exploring the role that individual land use planners can play. *Journal of Planning Literature, 24*(4), 362–371.

Stirling, A. (2010). Keep it complex. *Nature, 468,* 1029–1031.

Storbjörk, S. (2010). 'It takes more to get a ship to change course': Barriers for organizational learning and local climate adaptation in Sweden. *Journal of Environmental Policy and Planning, 12*(3), 235–254.

Tang, Z., Brody, S. D., Quinn, C., Chang, L., & Wei, T. (2010). Moving from agenda to action: evaluating local climate change action plans. *Journal of Environmental Planning and Management, 53*(1), 41–62.

Termeer, C. R., Biesbroek, & van den Brink, M. (2012). Institutions for adaptation to climate change: Comparing national adaptation strategies in Europe. *European Political Science, 11,* 41–53.

Thomalla, F., Downing, T., Spanger-Sigfried, E., Guoyi, H., & Rockstrom, J. (2006). Reducing hazard vulnerability: Towards a common approach between disaster risk reduction and climate adaptation. *Disasters, 30*(1), 39–48.

Thomsen, D. C., & Smith, T. F., & Keys, N. (2012). Adaptation or manipulation? Unpacking climate change response strategies. *Ecology and Society, 17*(3), 20. http://dx.doi.org/10.5751/ES-04953-170320.

Tierney, K. (2012). Disaster governance: Social, political, and economic dimensions. *Annual Review of Environment and Resources, 37,* 341–363.

Tobin, G. A. (1999). Sustainability and community resilience: The holy grail of hazards planning? *Environmental Hazards, 1,* 13–25.

Tompkins, E. L., Lemos, M. C., & Boyd, E. (2008). A less disastrous disaster: Managing response to climate-driven hazards in the Cayman islands and NE Brazil. *Global Environmental Change, 18,* 736–745.

Tompkins, E. L., Adger, W. N., Boyd, E., Nicholson-Cole, S., Weatherhead, K., & Arnell, N. (2010). Observed adaptation to climate change: UK evidence of transition to a well-adapted society. *Global Environmental Change, 20,* 627–635.

UNDP. (2008). Human development report 2007/2008. Fighting climate change: Human solidarity in a divided world, UNDP.

United Nations International Strategy for Disaster Reduction. (2009). 2009: UNISDR terminology on disaster risk reduction. http://www.unisdr.org/eng/terminology/terminology-2009-eng.html.

Walker, B. H., & Salt, D. (2006). *Resilience thinking: Sustaining ecosystems and people in a changing world.* Washington, D.C.: Island Press.

Walker, B. H., Holling, C. S., Carpenter, S. C., & Kinzig, A. P. (2004). Resilience, adaptability and transformability. *Ecology and Society, 9*(2), 5. http://www.ecologyandsociety.org/vol9/iss2/art5/.

Wamsler, C., & Lawson, N. (2012). Complementing institutional with localized strategies for climate change adaptation: A South-North comparison. *Disasters, 36*(1), 28–53.

WCED (World Commission on Environment and Development). (1987). *Our common future.* Oxford: Oxford University Press.

Welsh, M. G., & Esnard, A.-M. (2009). Closing gaps in local housing recovery planning for disadvantaged displaced households. *Cityscape: A Journal of Policy Development and Research, 11*(3), 87–104.

Wheeler, S. M. (2008). State and municipal climate change plans: The first generation. *Journal of the American Planning Association, 74*(4), 481–496.

White, G. F. (1936). Notes on flood protection and land use planning. *Planners Journal, 3*(3), 57–61.

White, G. F. (1945). Human adjustment to floods. Research Paper No. 29. Chicago: Il: Department of Geography, University of Chicago.

White, G. F., & Haas, E. (1975). *Assessment of research on natural hazards.* Cambridge: MIT Press.

White, G. R., Kates, W., & Burton, I. (2001). Knowing better and losing even more: The use of knowledge in hazards management. *Environmental Hazards, 3,* 81–92.

Wilbanks, T. J., & Kates, R. W. (2010). Beyond adapting to climate change: Embedding adaptation in responses to multiple threats and stresses. *Annals of the Association of American Geographers, 100*(4), 719–728.

Wilkinson, C. (2012). Urban resilience: What does it mean in planning practice? *Planning Theory and Practice, 13*(2), 312–324.

Wilson, E., & Piper, J. (2010). *Spatial planning and climate change.* Oxon: Routledge.

Wisner, B., Blaikie, P., Cannon, T., & Davis, I. (2004). *At risk. Natural hazards, people's vulnerability and disasters.* London: Routledge.

World Bank. (2010). *World development report 2010: Development and climate change.* World Bank.

# Part I
# Climate Change Adaptation: Theory and Practice

# Chapter 2
# Climate Change Adaptation and Disaster Risk Reduction: Fundamentals, Synergies and Mismatches

**Jörn Birkmann and Joanna Pardoe**

**Abstract** The IPCC special report on Managing the Risks of Extreme Events to Advance Climate Change Adaptation (see IPCC, A Special Report of Working Groups I and II of the Intergovernmental Panel on Climate Change, 2012a, p. 582) underscores the importance of linking disaster risk reduction and climate change adaptation. However, in reality, practical approaches in adaptation and risk reduction have primarily been developed in isolation, rather than as a part of a parallel and intertwined process. This chapter examines the options and concepts that allow for the strengthening of the link between Climate Change Adaptation (CCA) and Disaster Risk Reduction (DRR). In addition, barriers and mismatches between the two communities will be addressed. The chapter also discusses how limited cooperation between different institutions and ministries has hampered effective synergies between CCA and DRR in praxis. Finally, the chapter outlines recommendations and measures that need to be adopted in order to overcome existing barriers. In this regard criteria are formulated that should be applied in order to constantly monitor and evaluate adaptation strategies designed to simultaneously meet disaster risk reduction requirements.

**Keywords** Disaster risk reduction · Climate change adaptation · Synergies · Mismatches · Concepts

---

The paper is based on key findings of the IPCC Special Report SREX and a study conducted by Birkmann and Teichmann for the DKKV which was also published in a peer-reviewed paper (see Birkmann and von Teichmann 2010).

---

J. Birkmann (✉) · J. Pardoe
Institute for Environment and Human Security, United Nations University, Tokyo, Japan
e-mail: birkmann@ehs.unu.edu

B. C. Glavovic, G. P. Smith (eds.), *Adapting to Climate Change,* Environmental Hazards,
DOI 10.1007/978-94-017-8631-7_2, 

## 2.1 Introduction

The IPCC Special Report on Managing the Risk of Extreme Events and Disasters to Advance Climate Change Adaptation (see IPCC 2012a) as well as the discourse around the special programme of the United Nations Framework for the Convention on Climate Change on Loss and Damage (UNFCCC 2012) are two prominent examples of the emerging reality that the international community has recognised the need to discuss and develop both climate change and disaster risk strategies in a more coherent manner. Although the IPCC SREX report and the programme on Loss and Damage underscores the various synergies between both schools of thought, it must also be acknowledged that there are existing challenges and gaps that hinder an effective combination of adaptation and risk reduction strategies. Various challenges have been identified by different studies at all political levels (see Few et al. 2006; Red Cross/Red Crescent Climate Centre 2007; Commission on Climate Change and Development 2008a, b, c, 2009; O'Brien et al. 2008; Moench 2009; Schipper and Burton 2009; Tearfund 2009). This chapter will present a common concept and starting point for vulnerability and adaptation research in disaster risk reduction (DRR) and climate change adaptation (CCA), followed by an overview of areas that would benefit if DRR and CCA approaches were to be applied jointly and coherently. Based on existing synergies, the chapter will also examine key challenges when linking DRR and CCA by focusing on three key areas: different spatial and temporal scales, norm systems and knowledge types and sources.

## 2.2 Linking CCA and DRR

The first IPCC assessment reports were rather limited in terms of their approach to adaptation, reflecting a concern that a stronger emphasis on adaptation might detract from mitigation goals and efforts. However, when the third assessment report of the IPCC (2001a) drew the world's attention to the unavoidable impacts of human induced climate change, the need for adaptation moved onto the international agenda (IPCC 2001a, b). At present the fifth assessment report of Working Group II, which is underway and is expected to be finalised by the middle of 2014, takes a different perspective when compared to previous reports, emphasising and promoting the importance of climate change adaptation through four separate chapters that explicitly deal with the topic (IPCC 2012b). In addition, the actual meaning and content of adaptation has been discussed during various international conferences. Furthermore, special funds, such as the Least Developed Countries Fund and the Special Climate Change Fund, have been created to provide financial support to assist with the implementation of adaptation strategies. Today, there exists an overall consensus and acknowledgement that adaptation to climate change affects various sectors of society such as agriculture, health and infrastructure in which respective measures will have to be taken to safeguard the future. DRR is another key sector affected by climate change, although the relationship between DRR and adaptation

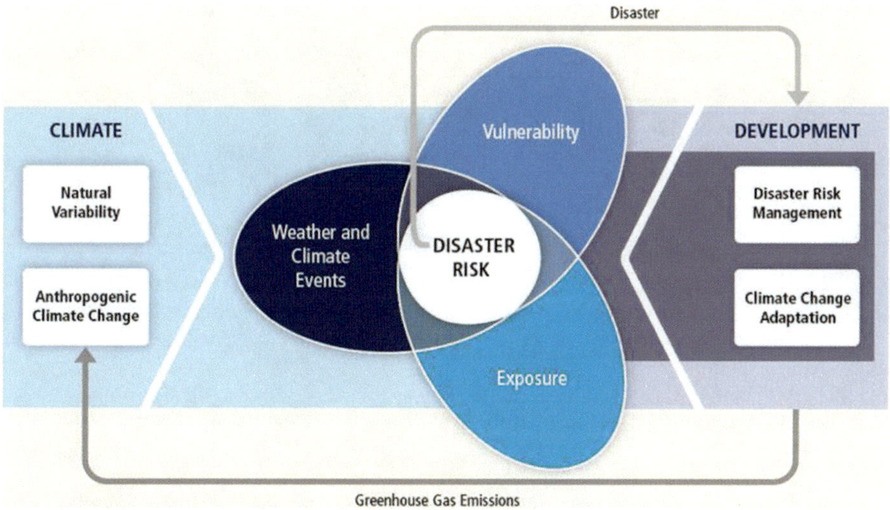


**Fig. 2.1** Determinants of disaster risk. (Source: IPCC 2012a, p. 31)

to climate change has often remained relatively opaque, particularly concerning any practical cooperation between different institutions or ministries on the ground.

### *2.2.1 Conceptual Approaches: Determinants of Risk*

The IPCC SREX framework differentiates three key factors tied to disaster risk. Disaster risk, according to the IPCC SREX is determined by physical events, such as weather and climate events on the one hand and the vulnerability and exposure of a system at risk on the other. In this regard the framework introduced in the SREX report emphasises that changes in the physical climate system due to natural variability and anthropogenic climate change need to be clearly separated from vulnerability and exposure of humans or ecosystems which is in turn influenced by development processes (see Fig. 2.1). In former approaches, the IPCC vulnerability definition encompassed issues concerning the frequency and magnitude of climate change, which clearly shifts vulnerability towards the understanding of risk in the Disaster Risk Research Community. In this regard the SREX report stresses the need to strengthen an understanding of the social construction of risk through the lens of vulnerability. Vulnerability is not a characteristic of physical phenomena; rather it is shaped by human and societal processes and patterns that are heavily influenced by different aspects of development.

In addition, Fig. 2.1 underscores that CCA needs to address vulnerability and exposure and that the respective understanding of adaptation cannot be solely based on the act of adapting to physical changes. Rather, DRR and CCA are embedded and closely linked to development processes and adaptation to climate change must, therefore,

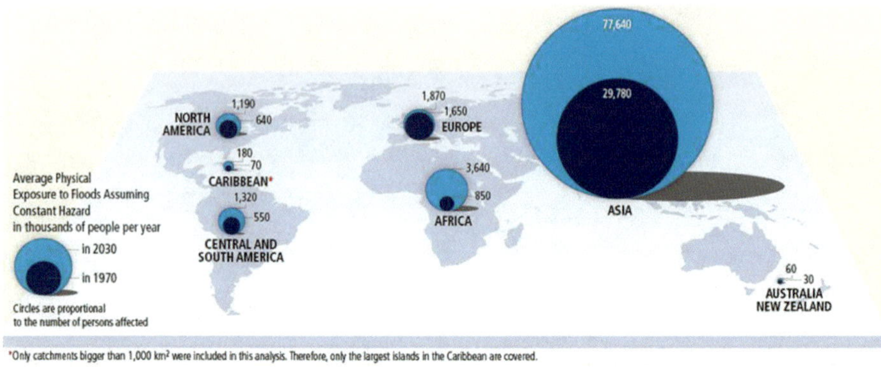


**Fig. 2.2** Past and future exposure to floods (average physical exposure to floods assuming a constant hazard). (Source: IPCC 2012a, p. 241)

account for both adaptation needs due to changes in the physical climate as well as due to societal processes. Many adaptation strategies initially focused on different climatic conditions in the future and respective challenges for adaptation, while less emphasis was given to the question of how different scenarios that encompass societal vulnerability might look and how these scenarios generate challenges for adaptation as well.

The challenges associated with climate change adaptation that are understood in the context of development processes can for example be illustrated by using the physical exposure to floods in the future. Based on work of Peduzzi et al. (2009) the IPCC SREX report underscores that major increases in the number of people exposed to floods will be seen in Asia as well as in Africa. Although the sheer number of people exposed in Africa is significantly lower than in Asia, the percentage change in exposure from 1970 to 2030 in Africa demonstrates a four-fold increase in the number of people at risk, compared to a two and a half fold increase in Asia. However, these figures must be considered carefully; the comparison of the average physical exposure to floods in 1970 compared to the 2030 scenario is based on an estimate of population increase, while the flood hazard is assumed to be constant (see Fig. 2.2). In other words, this means that the increase in exposure in Asia and Africa is primarily due to the expected increases in population growth and migration to flood prone areas. It is important to understand that even if the flood hazard does not change, increases in disaster risk are likely to materialise due to the increase in exposure linked to overall development patterns in Asia and Africa. If the increasing exposure is combined with a reduction in susceptibility, risk might remain the same or even decline. Both DRR and CCA have, so far, paid insufficient attention to the question of how macro-development trends, such as demographic changes and migration trends which have a considerable bearing on current exposure and future risk profiles, should be dealt with and can be addressed by different governance systems.

Linking CCA and DRR therefore requires an improved knowledge base describing how development trends influence disaster risk through vulnerability and exposure patterns over time (Schipper and Pelling 2006). In addition, attention needs to be paid to how DRR strategies and CCA concepts can influence development processes.

### 2.2.2 *Areas of Common Concern*

The IPCC SREX report clearly underscores that there is a wide range of complementary approaches spanning adaptation and risk reduction. Common entry points are, for example, linked to concepts and goals such as resilience building, the reduction of social vulnerability and the maintenance of healthy social-ecological systems. In this regard, both CCA strategies, such as National Adaptation Programs of Action (NAPAs), and programmes in DRR aim to improve preparedness and risk reduction initiatives and to inject adaptation to climate change into recovery and reconstruction processes following disasters. In addition, specific tools such as risk transfer mechanisms are mentioned as well as the more general goal of transformation (see Fig. 2.3).

While the IPCC SREX report and programme of the United Nations Framework for the Convention on Climate Change (UNFCCC) on Loss and Damage clearly refer to conceptual issues at the international level, several countries have adopted practical approaches to CCA and DRR at the national level. For example, the NAPAs provide a process for Least Developed Countries (LDCs) to identify areas in which urgent activities and projects are needed in order to adapt to climate change[1]. In developed countries in the north several major documents regarding national or sub-national adaptation programmes have been published. Examples include the German Adaptation Strategy (DAS) to Climate Change (2008) and the United Kingdom Climate Impacts Programme (UK-CIP) which was established in 1997 (see UK-CIP 2009). Whereas the German Adaptation Strategy, DAS, describes the effects that climate change might have on different societal sectors and suggests possible adaptation measures, the UK-CIP emphasises a cooperative effort with the scientific community to develop climate change scenarios. The UK-CIP also provides a tool for use by companies and organisations to assess their respective exposure to climate change and to derive individual adaptation and prevention measures based on the findings.

In spite of the practical approach of these national programs, DRR, as understood in the context of climate change and extreme events, often remains underdeveloped, particularly in terms of improved linkages between institutions and organizations responsible for CCA and those responsible for DRR. Although DRR was identified as an urgent problem by many of the LDCs, only 24 of the 38 LDCs that have submitted their NAPAs to the UNFCCC so far have called for immediate action in the field of disaster management and early warning. Of these 24 countries, only seven requested funding for projects that included capacity building and the development of preparedness measures (UNFCCC 2010). All other countries called for structural or technical measures (e.g. early warning systems) that primarily focus on natural hazards detection, rather than on broader policies, strategies and measures tied to DRR.

[1] The process of the development of NAPAs was initiated during the UNFCCC COP 7 conference in Marrakesh in 2001 and is funded by the least developed countries fund, which is based on voluntary contributions from developed countries and managed through the Global Environmental Facility.

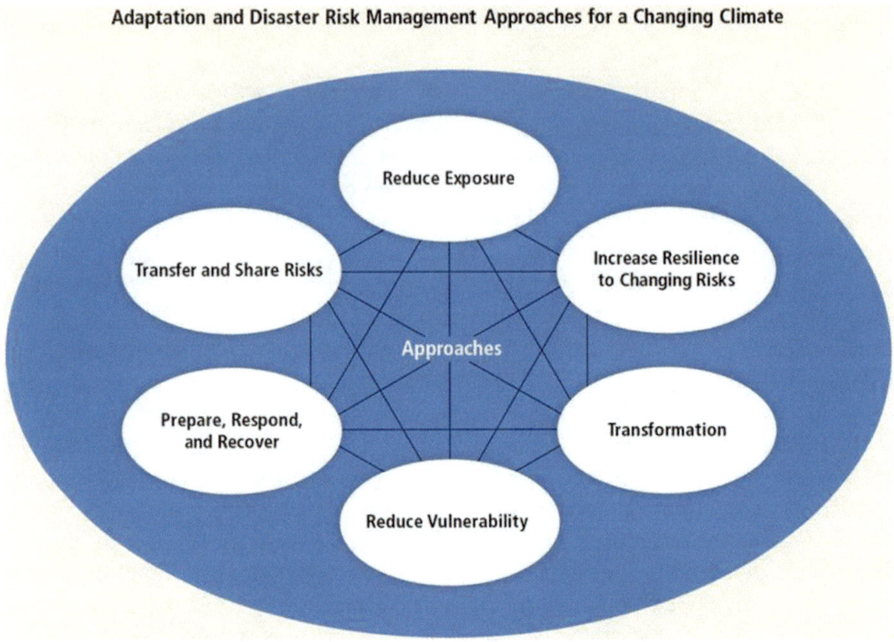


**Fig. 2.3** Adaptation and disaster risk management approaches for reducing and managing disaster risk in a changing climate. (Source: IPCC 2012a, p. 6)

In addition, Strategic National Action Plans (SNAP) have been developed and in some cases approved based on recommendations found in the Hyogo Framework. Example plans include those of the Philippines, the Maldives and Cambodia (see National Committee for Disaster Management 2009 (Cambodia); National Disaster Coordinating Council of the Philippines 2009; Office of Civil Defense and National Disaster Coordinating Council of the Philippines 2009; UN/ISDR and World Bank 2009; UN/ISDR 2009). In the case of the Philippines, the NAPA and the SNAP are considered important toolkits for dealing more effectively with disaster risk and threats related to climate change (see Birkmann and von Teichman 2010). While the Philippines plan represents an important tool, it only contains a few DRR indicators that would allow for the evaluation of the plan's implementation over time (Benson 2009, p. 45). Within the German Adaptation Strategy (DAS) DRR is mentioned as one cross cutting issue—besides spatial planning—that should support adaptation processes in terms of facilitating risk communication and developing guidelines on preventive measures for businesses, especially those responsible for critical infrastructures (CIs). Besides these conceptual approaches there are no concrete suggestions on how to create effective synergies between CCA and DRR in practice, for example with regard to joint funding mechanisms. The same is found with respect to the UK-CIP. It only refers to flood risk as a topic to be linked with CCA, but no concrete measures are proposed.

Developing countries have also launched their own initiatives and national adaptation programmes, including, for example Indonesia and Vietnam (see Republic of Indonesia, State Ministry of Environment 2007; Socialist Republic of Vietnam 2008). While a general consensus seems to exist that linking CCA and DRR would be beneficial, the challenges associated with developing effective integrative processes at the national level remain due to mismatches between CCR and DRR and different or even uncoordinated responsibilities across ministries. A workshop in Hanoi in 2012 on the occasion of the national launch of the IPCC SREX report showed, among other issues, that the two ministries responsible for DRR and CCA in Vietnam still face major challenges in communicating and cooperating with each other. The lack of cooperation between different ministries and agencies involved in DRR and CCA is often an important barrier that hinders the realisation of practical synergies between both fields in various countries.

At the local level it often appears equally difficult to effectively take advantage of synergies between both fields. For example, the opportunities that disaster recovery and reconstruction processes offer as a catalyst for change (Birkmann et al. 2009a; Birkmann and Fernando 2008), including the development of climate-proof structures in the aftermath of an extreme event, is not sufficiently taken into consideration. The reconstruction of coastal areas affected by the Indian Ocean Tsunami in Sri Lanka and Indonesia is an example of this missed opportunity. However, various local communities often view risk reduction to extreme events, CCA and resilience building as three interconnected fields that need to be addressed simultaneously in order to improve the livelihood security of communities and people at risk.

Additionally, climate change-related risks are hardly considered when designing new standards for protection systems (e.g., early warning, dyke systems, etc.) and urban redevelopment (e.g., housing standards, urban planning after a disaster). A focus on a single hazard and on experiences drawn from the past often dominates the thinking of technical experts and collective action, whereas wider aspects of climate change adaptation-including scenarios for vulnerability-are rarely addressed (see Birkmann and von Teichman 2010; Birkmann et al. 2013).

Even though the IPCC SREX report was an important contribution to an improved level of cooperation between the DRR and CCA communities, including the identification of various areas for further cooperation and synergies—as outlined in the first part of this chapter—the existing shortcomings and persisting mismatches between DRR and CCA need to be identified and dealt with in specific contexts or case studies at various levels in order to ensure a more effective and in-depth cooperation between DRR and CCA in the future. The following sections will analyze the reasons for these shortcomings in more detail.

## 2.3 Major Challenges and Gaps Between DRR and CCA

A review of the literature, the analysis of current approaches and a series of interviews conducted with recognised experts revealed a range of practical barriers to effectively link DRR and CCA (see Birkmann and von Teichman 2010)[2]. The main barriers have been categorized and are described in detail in the following section. They can be categorised across spatial, temporal and functional scales; legislative, cultural and behavioral norms; and knowledge-based mismatches (see Birkmann and von Teichman 2010).

### *2.3.1 Scale Mismatches*

When dealing with the development of appropriate strategies to reduce disaster risk, to respond to an actual disaster when it occurs and to develop appropriate adaptation strategies to climate change that are founded upon sound information, an understanding of differing spatial, temporal and functional scales is critically important.

#### 2.3.1.1 Spatial Scale Challenges

Mismatches at the spatial scale stem from the fact that climate change issues have primarily been analyzed on a global scale—even though downscaling approaches receive increasing attention—whereas disasters have been studied in the respective regions and localities where they occur (meso- or local/micro-scale). Climate scientists have mostly designed global models and predicted global trends based on universal laws, whereas the DRR community looks at local vulnerabilities and risks in specific areas, including groups of people potentially or actually affected. Local, down-scaled data on the effects of climate change or the localization of the impacts of extreme events in the future (e.g., heat waves, heavy precipitation, storms, floods etc.) is needed in order to facilitate the preparation of specific adaptation and DRR strategies, including scenario-based plans, to address one of the major concerns of risk reduction and adaptation managers. Climate impact forecasts regarding extreme events and scenarios regarding the effectiveness of adaptation strategies under different environmental and socio-economic conditions are uncommon at the local scale. Various impact studies tend to be designed for entire countries or regions (see German Adaptation Strategy to Climate Change 2008; Red Cross/Red Crescent Climate Centre 2007); but this is improving with some work being done to downscale global model outputs to the local level (Cooney 2012). Furthermore, vulnerability is also being considered on a larger scale as global

---

[2] The following sections are based on the paper of Birkmann and von Teichman 2010 and complemented with additional findings of more recent reports.

vulnerability assessments such as the World Risk Index are produced (Birkmann et al. 2011; Welle et al. 2012). Thus linking CCA and DRR more effectively requires further improvements in the exchange and combination of different spatial scales on which the two communities primarily focus and act. This also requires an improved link between local adaptation and risk reduction measures with national adaptation programs (NAPAs).

To this vertical mismatch of spatial scales the horizontal spatial scale mismatch can be added, which occurs because the sources of climate change often lie in regions and countries other than those it ultimately affects. This mismatch between countries, some of whom are more responsible for climate change (e.g. developed and rapidly developing nations) and those that carry the burden of experiencing more extreme weather events, or threats to their very existence as a sovereign state (e.g. small island nations) could lead to political conflict and thus to questions of global justice and security (Huq and Toulmin 2006). Furthermore, horizontal spatial scale mismatches become increasingly visible when examining secondary effects and second order adaptation needs introduced by risk reduction and adaptation measures (see Birkmann 2011b).

#### 2.3.1.2 Temporal Scale Challenges

DRR, particularly that delivered through humanitarian assistance agencies such as the Red Cross and the United Nations (e.g., UN/OCHA), as well as national donor programs, is often event-related and therefore tends to emphasize short-term interventions and procedures. Additionally, it is noteworthy that most of the countries requesting disaster aid, risk reduction and recovery support—especially after a disaster has occurred or in crisis situations—often issue work permits for such institutions and organizations for only a short period of time. In contrast, CCA strategies are (or should be) characterized by long-term perspectives that might also require the long-term presence of respective stakeholders in countries at high risk. However, the actors that promote vulnerability and risk reduction through the lens of CCA often face serious challenges (funding, work permits, access to conflict regions) when aiming to stay in such countries for the long-term. Thus, the establishment of a longer assistance timeframe and the development of supportive and enduring institutional structures that could effectively link DRR and CCA, for example in the aftermath of a crisis or disaster, are often not envisaged by the requesting country. In addition, temporal scale challenges between a short-term oriented strategy of dealing with the immediate consequences of climate related extreme events—such as air–conditioning to deal with the effects of heat waves in living spaces—and long-term adaptation and climate change mitigation goals has to be addressed more rigorously.

#### 2.3.1.3 Functional Scale Challenges

Functional scale mismatches refer to the differential organisation and management of crises and adaptation by actors affiliated with different institutions[3] and the related distribution of responsibilities (see the discourse in the resilience community e.g., Cumming et al. 2006). Climate change issues have been tackled in various countries by the environmental ministries and meteorological services whereas disaster risk management often lies within the responsibility of the ministry of the interior, defense or development.

Tied to the challenges of responsibilities being shared across institutions, there are further challenges relating to funding mechanisms. Existing funding schemes, which are structured according to the objectives of the issuing institution or convention, hence not allowing for the integration of measures that are inconsistent with its respective scope of responsibility, provide evidence of this incoherent search for solutions. Therefore, various governmental organizations are often discouraged from including both adaptation strategies and DRR goals in their project proposals or workplans, since this would require inter-ministerial or inter-organizational coordination and cooperation that in some cases is not seen as advantageous by the respective ministry or agency.

### *2.3.2 Mismatches Regarding Norms*

Norms—such as legislative, cultural or behavioural norms—decisively influence the functioning of human society as well as the interactive processes and dependencies between society and nature or within coupled social-ecological systems (Berkes et al. 2003; Folke 2006; Walker et al. 2006). It is not only individuals that are guided by certain rules, but also larger organisations and whole societies which follow standards that have been set by influential individuals or have evolved over time as a way to address new problems and seek agreed upon solutions. The different eras of climate change (see Huq and Toulmin 2006) provide an example of the dynamics that frame problems differently every time new developments become obvious and therefore new actors get involved in finding solutions. In the first era of climate change (1980s to 2000) climate change was seen as an environmental problem and the response emphasised the reduction of greenhouse gases. Even in this era, the discussion of climate change adaptation in the IPCC was seen, to a certain extent, as a threat to more rigorous climate change mitigation goals. The second era, beginning in 2000, was defined by the recognition that the effects of climate change are unavoidable and as such require humanity to adapt in the near term. As the negative impacts of climate change are primarily felt in poorer countries whereas their origin is attributed to industrialised countries, the issue of climate change was also linked to the question of "global justice" in what could be described as the third era.

[3] Institutions refer to rules, norms and rights as well as the organisations that enforce them.

With the development and publication of the IPCC SREX report in 2012, a fourth era can be identified that places greater emphasis on actual cooperation and synergies between DRR and CCA in international negotiations (e.g., UNFCCCs program on Loss and Damage or the integration of climate change issues in the post Hyogo Framework discussion), national programs and local activities. However, it is still important to consider the different funding scales that these approaches and communities can access.

Overall, adaptation is sometimes seen as a cross-cutting issue that needs to be mainstreamed into other development programmes. In other cases, it is seen as a separate strand of activities that should be driven by its own set of institutions and funding strategies. In other instances adaptation is seen as an additional burden that developing agencies must now, on top of all other development sectors, integrate into their growing scope of work. In addition, indicators or metrics that could help to monitor and evaluate progress toward achieving combined adaptation and risk reduction goals are still in their initial development phase. Moreover, the different spatial and time scales which currently define CCA and DRR activities are also related to the norms of the different communities, further complicating their successful integration.

One of the most important challenges, however, is the fact that after disasters, the opportunity to rebuild in an adaptive way considering future climate change is in most cases not exploited. More commonly, infrastructure is rapidly re-built back to pre-disaster conditions and standards. This relates to the prevalent view that disasters and crises due to natural hazards and climate related events are primarily seen as a threat imposed by external forces. This perspective, including the closely associated notion of stability (as opposed to the more accurate notion of dynamism), leads to a lack of awareness and acceptance of the need to promote change, including transformative change (see Nelson 2009; Pelling 2010; O'Brien 2012).

### *2.3.3 Knowledge Mismatches*

Within the general sphere of knowledge, important barriers and constraints can also be identified. One of the core challenges in this context is the competition between different types and sources of knowledge and the weak links between different types of data and work applied by climate and risk scientists and practitioners, which hinders straightforward communication, collaboration and joint programming across larger governance networks. The failure to effectively communicate scientifically acquired knowledge about climate change in a practical way and the lack of substantial guidance on how to deal with uncertainty provide major challenges for practitioners. An important aspect of knowledge (referred to as guiding knowledge) is an awareness of the limits of our knowledge and hence, the necessity to make decisions under conditions of uncertainty and possible surprise. Alternative ways of dealing with the limits of knowledge, such as scenario based planning and policy making, are not sufficiently discussed between both communities and are only just emerging as issues in both research fields. Furthermore, some important information is not yet available. For example, social and economic census data in addition to data on governance issues, especially in dynamic areas with high fluctuations of people and economic as well as political instability, would be essential in order to assess changing vulnerabilities and develop appropriate adaptation strategies. However, appropriate methodologies to detect such changes and transformations as well as the databases are not sufficiently developed yet. The development of scenarios for vulnerability at different scales might be a promising first approach to better account for potential dynamics in socio-economic conditions and in societal vulnerability.

In addition, the use of local and indigenous knowledge needs to be improved. For example, deep, locally held knowledge also reveals much about the capacities of local societies that might be difficult to assess from the outside. In other cases indigenous knowledge might also be marginalised by so-called technical experts in policy processes. Local and indigenous knowledge needs to be valued and considered in DRR and CCA. Local and indigenous knowledge is, however, often based on experiences in the past and hence may be insufficient for addressing new challenges or new hazards linked to climate change. Overall, the systematic consideration of different knowledge types is important and a pre-requisite for inclusive adaptation and risk reduction strategies.

## 2.4 Recommendations

Based on the discussion of selected key findings of the IPCC SREX report (IPCC 2012a) and the challenges identified for linking CCA and DRR along the categories of scale, norms and knowledge, a range of recommendations have been formulated and are discussed next.

While linking CCA and DRR concepts and strategies, it is important to utilise the synergies between both communities and approaches resulting in more effective disaster risk management in the context of climate change. However, as the IPCC Special Report SREX points out, this can only be achieved by an appropriate framing of the problem that takes into account the wider implications of climatic changes, particularly of climate variability and anthropogenic climate change, and their impacts on certain hazards and environmental stressors. This needs to be done from the outset.

In addition, adaptation and risk reduction strategies must be grounded on sound data tied to the vulnerability and exposure of societies, communities and social-ecological systems. In this context vulnerability and exposure also have to be viewed and understood within the broader context of development processes and interactions between DRR and CCA. Consequently, linking DRR and CCA depends on the acknowledgement of the importance of climate change and societal changes and the interactions between the two.

Compared to the conceptual linkages and the joint research reports of the DRR and CCA communities, such as the IPCC Special Report SREX, the actual cooperation between agencies and ministries responsible for DRR on the one hand and CCA on the other is often limited. Finding appropriate mechanisms to stimulate and improve the cooperation between different ministries and agencies responsible for DRR and CCA is essential. Cooperative agreements would benefit from a situation where criteria and funding for adaptation and risk reduction programmes required collaboration among DRR and CCA stakeholders and agencies. The new UNFCCC program on loss and damage, for example, could facilitate such a process if respective criteria and joint activities between different ministries or agencies were a prerequisite for funding.

Beside these points, important challenges remain with regard to spatial, temporal and functional scale mismatches. Discussions of these issues are found in DRR political science literature and agenda setting literature. This literature points towards the need for a more efficient way of combining different adaptation strategies and DRR at different spatial and temporal scales. For example, it is questionable to fund mainly national adaptation programmes, if one of the key factors that contribute to vulnerability is the failure of governance at the national level. Particularly, countries which face major challenges due to climate change and extreme events, such as Somalia or Haiti, for example, would clearly have to be targeted differently. Involving local and national stakeholders and decision-makers in such programmes will be essential in order to enhance the effectiveness and coherence of risk reduction and adaptation strategies across spatial scales. Temporal scale mismatches can be minimised if respective strategies and programmes for adaptation, risk reduction or both include different timescales and clearly define different targets for different periods. Many adaptation strategies reviewed so far, often lack a management oriented goal and strategy discussion. This means that targets are often not systematised into different time phases or time horizons. Improving the consideration of different timescale and different actors at different spatial scales requires new or modified governance approaches for adaptation and risk reduction.

In order to ensure that strategies for DRR and CCA span different timescales and spatial scales as well as recognise different types of knowledge, it is essential to also modify and re-direct adaptation and DRR funding mechanisms. For example, more flexible DRR-funding, to include the opportunity to utilise the money received for a specific disaster to implement medium- and long-term adaptation strategies, is needed. In addition, funding for adaptation strategies and measures should not be based on the individual strategy alone, but should include a procedural requirement linking different actors at different scales while considering the benefits and costs of the adaptation measures at different temporal scales. Inclusive adaptation strategies and respective funding mechanisms would also need to provide incentives to bring together different types of knowledge, such as expert and indigenous knowledge and to evaluate potential commonalities and conflicts. Various structural adaptation measures in the past and at present focus mainly on the adjustments to physical processes, without the consideration of how these measures might affect the adaptive capacity of different groups in the long-run. Consequently, the adaptiveness of adaptation strategies and measures should be a part of integrated plans.

Finally, one has to address mismatches between governmental/formal adaptation strategies and norms on the one hand and non-governmental/informal adaptation strategies and norms on the other. It would be naïve to assume that such divergences between different norms could be easily eliminated. Our recommendation would be to first identify and reveal these mismatches between different norm systems in order to create a basis from which to address them. At present many governmental adaptation strategies, such as relocation or the development of hard physical infrastructure often neglect the potential and actual conflicts of these measures with the norms of individual households and other relevant stakeholders. Developing procedures that enable these issues to be addressed in an inclusive manner would allow the coherence and coordination of different adaptation and risk reduction strategies as well as underlying norm systems to be improved.

## References

Benson, C. (2009). Mainstreaming disaster risk reduction into development: Challenges and experience in the Philippines. http://www.proventionconsortium.org/themes/default/pdfs/mainstreaming_philippines.pdf.

Berkes, F., Colding, J., & Folke, C. (Eds.). (2003). *Navigating social-ecological systems: Building resilience for complexity and change*. Cambridge: Cambridge University Press.

Birkmann, J. (2011a). Regulation and coupling of society and nature in the context of natural hazards—Different theoretical approaches and conceptual frameworks and their applicability to analyse social-ecological crises phenomena. In H. G. Brauch, U. Oswald Spring, P. Kameri-Mbote, C. Mesjasz, J. Grin, B. Chourou, P. Dunay, & J. Birkmann (Eds.), *Coping with global environmental change, disasters and security threats, challenges, vulnerabilities and risks*. Springer.

Birkmann, J. (2011b). First and second-order adaptation to natural hazards and extreme events in the context of climate change. *Natural Hazards, 58*(2), 811–840. doi:10.1007/s11069-011-9806-8.

Birkmann, J., & von Teichmann, K. (2010). Integrating disaster risk reduction and climate change adaptation: key challenges-scales, knowledge, and norms. *Sustainability Science, 5*(2), 171–184

Birkmann, J., & Fernando, N. (2008). Measuring revealed and emergent vulnerabilities on coastal communities to tsunami in Sri Lanka. *Disasters, 32*(1), 82–105.

Birkmann, J., Buckle, P., Jaeger, J., Pelling, M., Setiadi, N., Garschagen, M., Fernando, N., & Kropp, J. (2009a). Extreme events and disasters: A window of opportunity for change?—Analysis of changes, formal and informal responses after mega disasters. *Natural Hazards*. doi:10.1007/s11069-008-9319-2.

Birkmann, J., Cardona, O.D., Carreno, L., Barbat, A., Pelling, M., Schneiderbauer, S., Kienberger, S., Kelier, M., Alexander, D., Zeil, P., & Welle, T. (2013). Framing vulnerability, risk and societal responses: the MOVE framework. *Natural Hazards*, 67(2): 193–211.

Birkmann, J., Cutter, S., Rothman, D., Welle, T., Garschagen, M., Van Ruijven, B., O'Neil, B.; Preston, B., Kienberger, S., Cardona, O.D., Siagian, T., Hidayati, D., Setiadi, N., Binder, C., Hughes, B., & Pulwarty, R. (2013). Scenarios for Vulnerability-Opportunities and constraints in the context of climate change and disaster risk. *Climatic Change* (online first). doi 10.1007/s10584-013-0913-2.

Birkmann, J., Welle, T., Krause, D., Wolfertz, J., Suarez, DC., & Setiadi, N. (2011). WorldRiskIndex: Concept and Results. In Alliance Development Works (Eds.), *WorldRiskReport* (pp. 13–42). Berlin.
CCD (Commission on Climate Change and Development). (2008a). Incentives and constraints to climate change adaptation and disaster risk reduction—A local perspective. Stockholm.
CCD (Commission on Climate Change and Development). (2008b). Overview of adaptation mainstreaming initiatives. Stockholm.
CCD (Commission on Climate Change and Development). (2008c). Links between disaster risk reduction, development and climate change. Geneva Stockholm. http://www.ccdcommission.org/Filer/pdf/pb_disaster_risk_reduction.pdf. Accessed 31 May 2009.
CCD (Commission on Climate Change and Development). (2009). Closing the gaps—Disaster risk reduction and adaptation to climate change in developing countries. Stockholm http://www.ccdcommission.org/Filer/report/CCD_REPORT.pdf. Accessed 31 May 2009.
Cooney, C. M. (2012). Downscaling climate models: Sharpening the focus on local-level changes. *Environmental Health Perspectives, 120*(1), a22–a28.
Cumming, G. S., Cumming, D. H. M., & Redman, C. L. (2006). Scale mismatches in social-ecological systems: Causes, consequences, and solutions. *Ecology and Society, 11*(1), 14.
Folke, C. (2006). Resilience: The emergence of a perspective for social–ecological systems analyses. *Global Environmental Change,16,* 253–267.
Few, R., Osbahr, H., Bouwer, L. M., Viner, D., & Sperling, F. (2006). Linking climate change adaptation and disaster risk management for sustainable poverty reduction. Synthesis Report, Study carried out for the Vulnerability and Adaptation Resource Group (VARG).
German Adaptation Strategy to Climate Change. (2008). The German Government. http://www.bmu.de/files/pdfs/allgemein/application/pdf/das_gesamt_bf.pdf. Accessed 31 May 2009.
Huq, S., & Toulmin, C. (2006). Three eras of climate change. In IIED Sustainable Development Opinion. http://www.iied.org/pubs/pdfs/11060IIED.pdf. Accessed 30 June 2009.
IPCC (Intergovernmental Panel on Climate Change). (2001a). Climate change 2001. Third Assessment Report, 3 Volumes, Cambridge.
IPCC (Intergovernmental Panel on Climate Change). (2001b). Climate change 2001: Impacts, adaptation, and vulnerability, contribution of working group II to the IPCC Third Assessment Report. http://www.grida.no/publications/other/ipcc_tar/. Accessed 20 May 2009.
IPCC. (2012a). Managing the risks of extreme events and disasters to advance climate ahange adaptation. In C. B. Field, V. Barros, T. F. Stocker, D. Qin, D. J. Dokken, K. L. Ebi, M. D. Mastrandrea, K. J. Mach, G.-K. Plattner, S. K. Allen, M. Tignor, & P. M. Midgley (Eds.), *A special report of working groups I and II of the intergovernmental panel on climate change* (p. 582). Cambridge: Cambridge University Press.
IPCC. (2012b). Agreed Reference Material for the IPCC Fifth Assessment Report. Working Group I outline, Working Group II outline, Working Group III outline, at IPCC website. http://www.ipcc.ch. Accessed 1 Feb 2012.
Moench, M. (2009). Adapting to climate change and the risks associated with other natural hazards: Methods for moving from concepts to action. In E. L. F. Schipper & I. Burton (Eds.), *Adaptation to climate change—The Earthscan Reader* (pp. 249–280). London: Earthscan.
National Committee for Disaster Management. (2009). Cambodia: National progress report on the implementation of the Hyogo.
National Disaster Coordinating Council Philippines (NDCC). (2009). Strengthening disaster risk reduction in the Philippines: Strategic National Action Plan 2009–2019. European Commission on Humanitarian Aid Organisation for Disaster Preparedness, ISDR, NDCC, UNDP. http://www.adrc.asia/countryreport/PHL/2009/PHL_attachment.pdf.
Nelson, D. R. (2009). Conclusions: Transforming the world. In W. N. Adger, I. Lorenzoni, & K. O'Brien (Eds.), *Adapting to climate change: Thresholds, values, governance* (pp. 491–500). Cambridge: Cambridge University Press.
O'Brien, K. (2012). Global environmental change II: From adaptation to deliberate transformation. *Progress in Human Geography, 36*(5), 667–676

O'Brien, K., Sygna, L., Leichenko, R., Adger, W. N., Barnett, J., Mitchell, T., Schipper, L., Thanner, T., Vogel, C., & Mortreux, C. (2008). Disaster risk reduction, climate change and human security—A study for the Foreign Ministry of Norway. GECHS (Global Environmental change and human security) project. http://www.gechs.org/downloads/GECHS_Report_3_08.pdf. Accessed 31 May 2009.

Office of Civil Defence, National Disaster Coordinating Council of the Philippines. (2009). Philippines: National progress report on the implementation of the Hyogo Framework for Action. OCD and NDCC. http://www.preventionweb.net/files/7495_Philippines%5B1%5D.pdf.

Peduzzi, P., Dao, H., Herold, C., & Mouton, F. (2009). Assessing global exposure and vulnerability towards natural hazards: The disaster risk index. *Natural Hazards and the Earth System Science, 9,* 1149–1159.

Pelling, M. (2010). *Adaptation to climate change: From resilience to transformation*. London: Routledge.

Red, Cross/Red, Cresent Climate Centre. (2007). *Climate guide*. Den Hague.

Republic of Indonesia, State Ministry of Environment. (2007). National Action Plan addressing Climate Change. http://www.adaptationlearning.net/profiles/country/files/IndonesiaNationalClimateChangeActionPlan_2007_English.pdf. Accessed 28 May 2009.

Schipper, L., & Burton, I. (Eds.). (2009). *Adaptation to climate change*. London: Earthscan.

Schipper, L., & Pelling, M. (2006). Disaster risk, climate change and international development: Scope for, and challenges to, integration. *Disasters, 30*(1), 19–38.

Socialist Republic of Vietnam. (2008). Decision on Approval of the National Target Programme to Respond to Climate Change. Decision No: 158/2008/QD-TTg. Hanoi.

Tearfund. (2009). *Linking climate change adaptation and disaster risk reduction*. Teddington.

UK CIP (Climate Impacts Programme). (2009). Adapting to Climate Change in England: A framework for action. http://www.ukcip.org.uk/index.php. Accessed 4 June 2009.

UNFCCC. (2012). Approaches to address loss and damage associated with climate change impacts in developing countries that are particularly vulnerable to the adverse effects of climate change to enhance adaptive capacity. Agenda item 9. http://unfccc.int/resource/docs/2012/sbi/eng/l12.pdf.

UNFCCC (United Nations Framework Convention on Climate Change). (2010). http://unfccc.int/cooperation_support/least_developed_countries_portal/items/4751.php. Accessed 13 Jan 2010.

UN/ISDR (United Nations/International Strategy for Disaster Risk Reduction). (2009). UNISDR terminology on disaster risk reduction 2009. http://www.unisdr.org/eng/library/lib-terminology-eng-2004.htm. Accessed 31 May 2009.

United Nations International Strategy for Disaster Reduction (UN/ISDR), & World Bank. (2009). Disaster risk management programs for priority countries: Global facility for disaster reduction and recovery. UN/ISDR and World Bank. http://gfdrr.org/docs/6thCG_DRM_Programs_for_Priority_Countries.pdf.

Walker, B. H., Gunderson, L. H., Kinzig, A. P., Folke, C., Carpenter, R., & Schultz, L. (2006). A handful of heuristics and some propositions for understanding resilience in social-ecological systems. *Ecology and Society, 11*(1), 13.

Welle, T., Birkmann, J., Rhyner, J., Witting, M., & Wolfertz, J. (2012). WorldRiskIndex: Concept and Results. In Alliance Development Works (Eds.), *WorldRiskReport* (pp. 11–26). Berlin.

# Chapter 3
# Tackling Barriers to Climate Change Adaptation in South African Coastal Cities

**Gina Ziervogel and Susan Parnell**

**Abstract** Cities are starting to develop policies and plans to adapt to the impacts of climate change. As these policies and plans are implemented, the barriers and opportunities of adaptation in practice are starting to be realised. It is clear that addressing these barriers is key to achieving more systemic adaptation that is not just focused on projects but on an institutional environment that is supportive of integrating climate knowledge, acting in a timely, flexible, and holistic manner and engaging appropriate regulatory mechanisms. This chapter uses the case studies of the City of Cape Town and eThekwini municipalities along the South African coast to explore what the barriers and opportunities have been in implementing adaptation and how these can be addressed moving forward on an urban city-scale.

**Keywords** Barriers to adaptation · South Africa · Governance · Adaptation implementation · Urban

## 3.1 Introduction

Rising rates of urbanisation alongside increasing consumption and high emissions puts the spotlight on how cities can mitigate and adapt to climate change (Parnell et al. 2007; Satterthwaite et al. 2007; Wilbanks et al. 2007). Globally, cities are starting to develop policies and plans to adapt to the impacts of climate change, including the likelihood of an increase in extreme events (Corburn 2009; Birkmann et al. 2010; Horton et al. 2010). This response is in part driven by the international scientific community that is encouraging adaptation as an important and urgent way to complement on-going mitigation efforts that have formerly tended to dominate policies and finance (Pielke et al. 2007; Romero-Lankao 2008). It is also driven by bottom-up initiatives that promote the need to better plan for climate variability, including extreme events, in order to increase the resilience of cities and protect its

G. Ziervogel (✉)
University of Cape Town, Cape Town, South Africa
e-mail: gina.ziervogel@uct.ac.za

S. Parnell
Cape Town, South Africa

B. C. Glavovic, G. P. Smith (eds.), *Adapting to Climate Change,* Environmental Hazards, DOI 10.1007/978-94-017-8631-7_3, © Springer Science+Business Media Dordrecht 2014

inhabitants. The momentum around adaptation is becoming evident at the international as well as at the national and local level.

Although the focus on urban adaptation has emerged more recently than earlier adaptation research that tended to focus on sectors such as agriculture, coasts or biodiversity, cities are catching up and developing adaptation responses (Storch et al. 2009; Bloomberg et al. 2010; GLA 2010). Despite urban adaptation plans being developed fairly rapidly, many are slow to be translated into action, with the peer-reviewed literature mirroring limited evidence of adaptation actions (Granberg and Elander 2007; Heinrichs et al. 2009; Birkmann et al. 2010). Developing policy is an important step but translating policy into practice is often where the barriers to climate change adaptation start to emerge more clearly.

Using a governance lens that incorporates the management of the course of events in a social system characterised by a plurality of actors (Burris et al. 2005) is helpful to highlight the different nature of barriers that can include issues related to information, knowledge, perceptions, policy, finance, capacity and cooperation. This is particularly important in urban areas, where different actors often have different goals and social, economic and ecological responses need to be integrated within urban policy priorities. Municipalities are an important factor in the sense that they manage cities at the meso scale and play a key role in multi-level governance because of their link to the local and national scale (Ziervogel and Taylor 2008; Corfee-Morlot et al. 2009). Exploring municipalities' adaptation to climate change through a governance lens is therefore of paramount importance in order to overcome barriers.

Many cities in the global South have started to explore climate change adaptation recently. However, two cities in South Africa, namely eThekwini and the City of Cape Town, have been developing adaptation policies and plans for the last decade (Cartwright et al. 2008; Roberts 2008; Mukheibir and Ziervogel 2007; Satterthwaite et al. 2007). These policies and plans are slowly being translated into action. This chapter explores how eThekwini and the City of Cape Town have engaged in climate change adaptation, what the barriers and opportunities have been, and where future efforts need to focus. A framework for assessing the barriers and opportunities helps position support for adaptation to climate change to address specific needs such as knowledge gaps, action gaps or institutional gaps. This is important in developing lessons for other global South cities, where there are large numbers of people exposed to climate hazards and adaptation policies and plans might not have been well developed. Because climate impacts represent one of many challenges, it is necessary to carefully position adaptation within a complex political and institutional landscape (Ziervogel et al. 2010).

## 3.2 Framing Urban Adaptation Barriers and Opportunities

Within the field of adaptation to climate change, a more recent focus on governance has started to emerge (Adger et al. 2005; Granberg and Elander 2007; Jordan 2008; Urwin and Jordan 2008; Adger et al. 2009). The complexity of responding to climate change requires engaging different actors, with different agendas to make decisions (Ziervogel and Ericksen 2010; van Asselt and Renn 2011). Unpacking the governance of risk can help to identify these different agendas and institutions in order to explore potential adaptation pathways that can address a multiplicity of goals. Adaptive governance recognises institutional constraints but focuses on the interaction between actors and the environment recognising the need for reflexive processes that incorporate learning (Folke et al. 2005; Kemp and Martens 2007; Tschakert and Dietrich 2010). The focus on learning and iterative risk management provides an opportunity to reassess existing practices and explore opportunities for transformative change given the challenge of addressing and integrating adaptation and disaster risk reduction.

A focus on governance supports a focus on process; a process through which goals are defined and pursued collectively with government being one of the actors in this process (Betsill and Bulkeley 2006; van Zeijl-Rozema et al. 2008). The shift in focus from government to governance encourages the exploration of how government relates to a wider range of actors and institutional contexts (Nelson et al. 1998; Betsill and Bulkeley 2006). Relating this back to reducing disaster risk is important because effective governance systems are necessary to realise adaptive capacity (Moser 2009) which supports the agency of individuals and organisations involved in preparing for more frequent extreme events.

Coordination is one important aspect of climate change governance because of the complexity of the challenge, including the number of actors involved, the temporal and spatial scale, and the need to engage in multi-level governance. However, coordination alone is not enough. Understanding the perception of the need, the regulatory constraints, and the political pressures are important in unpacking the ability to adapt further. Although these issues of policy, process and institutions are critical, it is clear that understanding current projects and action and the knowledge on which they are based is still critical. This has led to a growing focus in the literature on the barriers to and enablers of adaptation (Biesbroek et al. 2009; Inderberg and Eikeland 2009; Moser and Ekstrom 2010). For this reason, an integrated framework is needed to assess barriers to urban adaptation processes that include an understanding of the technical, scientific and institutional challenges that are underpinned by an understanding of the governance context.

Winsvold et al. (2009) suggest two key areas that undermine effective urban adaptation; that of knowledge and that of action. They suggest that the knowledge problem can be addressed by enabling knowledge transfer and collective learning, and that the action problem can be addressed by motivating a wide range of actors and including them in a coordinated process, recognising that innovation can be hampered by the complexity of the policy field. In addition, it is evident that institutional constraints often limit adaptation (Inderberg and Eikeland 2009). Inderberg and Eikeland (2009) suggest that resources and technology are of little use if institutional factors hinder their implementation and undermine proper adaptive measures. They suggest that an organisational ability to implement feasible actions is critical, which is learnt through interacting and doing.

Through the governance lens, Winsvold et al. (2009) focus on knowledge and action, combined with an assessment of institutions including formal regulations, to provide a frame to explore the barriers and opportunities for urban adaptation processes. Included in the knowledge problem is the state of information about climate impacts and risks as well as the perceived need to respond to climate change. The action problem includes inter-governmental cooperation and leadership as well as the capacity to act. This includes action that is project specific as well as action around processes. Lastly, the institutional problem focuses on existing policies, mandates and financial measures that support or undermine desired adaptation responses. The governance lens is then used to pull these three problems together and identify any overarching governance challenges that need to be addressed to enable a holistic urban adaptation response.

This chapter focuses on the two largest coastal cities in South Africa—the City of Cape Town, in the south west of the country and eThekwini municipality, formerly known as Durban, on the east coast. The analysis is based on interviews conducted in 2010 with government actors involved in adaptation. The interviews were undertaken as part of a study to understand current adaptation responses and challenges in five coastal cities in South Africa, with the view to strengthening the coastal cities network and identifying issues to take up with national government in terms of further support for adaptation at the local level.

The aim of the research is to explore how these two cities have institutionalised adaptation within their functions and the challenges they are facing in innovating around adaptation within city policies, plans and practice. First, an overview of the emergence of adaptation is presented before exploring the barriers and opportunities faced. Most of the barriers focus on institutional and governance issues, highlighting the importance of framing this type of research within a governance context that includes space for the challenges of technicalities while recognizing the development context, the relationship between actors and the effects of unspecified mandates that many actors are challenged with.

## 3.3 Case Studies: City of Cape Town and eThekwini

eThekwini and the City of Cape Town are two of the five biggest cities in South Africa, which in total, accounted for 44 % of national employment in 2001 (SACN 2006), highlighting the importance of large cities in South Africa's economy. These cities have diverse economies including private sector services and retail, providing important spaces for black economic empowerment and economic advancement, although many of the residents do not benefit from urban services, despite the intention of the Metropolitan cities to consolidate their fragmented municipal history (SACN 2006). Both cities have high urbanisation rates matched by high levels of poverty, echoing the history of segregation and apartheid. In addition to high poverty levels both cities are also popular tourist destinations, with rich biodiversity (Fig. 3.1).

**Fig. 3.1** The dual nature of the city: Cape Town. Informal settlements with limited infrastructure and poor sanitation (*left*), busy urban centre with port and high levels of tourism (*right*). (Source: Photographs by Sean Wilson)

In South Africa, there is no legal or policy framework dealing specifically with cities. Rather, there is a national framework for local government within which policy and legislation of relevance to municipalities is situated (SACN 2006). Within this context urban agendas are developed, with city government aspiring to urban innovation led by a developmental state, with a focus on environmental sustainability, shared economic growth and social inclusion framed in the context of 'good urban governance' (Boraine et al. 2006).

Both eThekwini and the City of Cape Town are coastal cities, with their ports serving as key hubs for economic activities. eThekwini encompasses the city of Durban, whose port is the busiest on the African continent and the biggest in terms of container capacity. Cape Town's port is smaller but also represents an important source of economic activity.

eThekwini has 35 km of coastline that has experienced severe storm surge in the past and expects more in the future while the City of Cape Town is vulnerable to storm surge and sea-level rise. Exposure to coastal storm surge and sea-level rise are the key disaster risks related to climate change in these two cities. Waves from extreme storms are potentially the greatest threat to the coast and evidence suggests that storminess has increased in the past 50 years (Brundritt and Cartwright 2012) (Fig. 3.2).

**Fig. 3.2** Location of eThekwini and city of Cape Town in South Africa. (Source: Map data AfriGIS (Pty) Ltd, Google)

Although eThekwini and the City of Cape Town share a number of characteristics, there are also a number of differences. eThekwini, in KwaZulu-Natal, sits on the east coast of the country, and experiences an annual summer rainfall of around 1,000 mm, whereas the City of Cape Town, in the Western Cape province in the south west of South Africa, is situated in a Mediterranean climate and receives an average annual rainfall of 550 mm, which falls mainly in winter. A more detailed background on these cities is presented before exploring recent adaptation processes and lessons learned.

## 3.4 Contextualising Adaptation Processes within the City

### *3.4.1 eThekwini*

eThekwini has a population of 3.5 million people, an unemployment rate of 43 %, excluding those employed in the informal sector, and high levels of HIV and AIDS (Roberts 2008). There is a significant proportion of tribal land within its borders which creates challenges for planning, as the formal policies of the city cannot always be followed because of the differences in tribal governance. Climatic change is expected to decrease rainfall runoff by 158 million $m^3$ per annum by 2100 in the Mgeni catchment, where eThekwini lies, because of increases in temperature and changes to rainfall variability, significantly impacting water resources that are currently stressed in terms of both quality and quantity (Roberts 2008). Maximum and minimum temperatures have been increasing and it is expected that the number of hot days will increase in the future (Naidu et al. 2006). Direct impacts on health are expected due to heat waves and extreme weather events. Impacts on biodiversity and agriculture are a concern as well while sea-level rise threatens the infrastructure and ecosystems of coastal communities (outlined in more detail in Roberts 2008).

In eThekwini, the process of developing a municipal response to climate change started in 2004 when the Environmental Planning and Climate Protection Department started an impact analysis to inform a climate change response strategy that ultimately led to their Municipal Climate Protection Programme (MCPP). When they started they did not use the term adaptation but rather began with an analysis of climate impacts that they thought would be the basis for formulating policy and strategy that would work its way through the usual local government cycle. However, council simply wanted to know how they were going to stop all of the climate impacts they had identified and how these impacts would affect the strategic vision for the city. It was clear that an immediate work programme was needed to explore the impacts in more detail and match local government responses to the findings.

A number of extreme events helped to garner support for policies related to managing climate impacts. These included a big storm event along the coast in 2007, rain storms that damaged properties in 2008, and two tornados that wrecked two informal settlements. These extreme events assisted in getting politicians and senior officials to understand the dynamics of the problem that includes both environmental and developmental impacts. The timing also helped to expedite an Integrated Coastal Management Act that gained support from politicians who had just witnessed the extent of the storm damage. More recently there have not been as many extreme events and it has become harder to access funding for pro-active disaster risk reduction.

The Environmental Planning and Climate Protection Department see themselves as an implementing agent but are faced with a resource and skills poor environment. They feel that because of the pressure to ensure development-linked co-benefits adaptation responses have received more support than mitigation (Roberts 2010). There is more political support for protecting the houses they have just built from an increased flood return level, than working with industry to reduce a gas that cannot be seen in the atmosphere. In the developing country context, this driver of adaptation responses should be maximised.

### *3.4.2 The City of Cape Town*

The City of Cape Town has an urban and peri-urban population of near 3 million, with high levels of informality and in-migration. Water security is a concern, partly because of the geographic location of the city on the drier western side of the country and partly because of growing domestic and agricultural demands. Further exacerbating the problem is the slow pace at which water resource management has been to integrate climate change-related issues (Ziervogel et al. 2010). High numbers of informal dwellers have poor access to water and sanitation while experiencing annual flooding, whereas wealthy areas located nearby have access to cheap reliable water supplies (Smith and Hanson 2003). In addition, the exposed coastline makes the city's residents, infrastructure and ecosystems vulnerable to storm surge and sea-level rise (Cartwright et al. 2008). Climate projections suggest that even with similar levels of rainfall, increased temperatures are likely to lead to increased evapotranspiration, which has serious implications for ecosystem services. In addition, an increase in the prevailing summer south-easterly wind is expected to lead to increased upwelling (Tadross et al. 2012). These changes in temperature, rainfall variability and ocean dynamics, on top of attempts to redress inequitable access to water, housing and services amplify the need to focus on adaptation to climate change.

In the City of Cape Town, climate change work started in 2001, driven by the Environmental Resource Management Department, where there was broad thinking around climate change, rather than a focus on adaptation or mitigation specifically. Momentum was gained partly because of the frequently occurring rolling blackout of the city, which provided a key leverage point from an energy perspective. In the early stages there was external finance linked to mitigation projects, including the first Clean Development Mechanism (CDM) project in Kuyasa township, that implemented low-cost urban housing energy upgrades (Sutter and Parreño 2007). In 2005, some local government positions linked to climate change were funded, and through this an adaptation framework was developed by external consultants in 2006. Shortly thereafter, the budget was cut along with the positions.

Although an adaptation framework existed it took a while for proposed adaptation projects to be developed and implemented. Initial adaptation work focused on the coast (Cartwright et al. 2008; Fairhurst et al. 2008), where sea-level rise modeling generated political interest in protecting the coastline. The coastal work has not been driven as an adaptation issue politically but rather it has been pushed as being part of coastal zone protection, citing an enhanced level of resilience to more frequent and extreme storm surge events linked to sea-level rise. This approach has provided a key leverage point to support arguments for climate change adaptation.

Efforts to integrate adaptation more widely across the City saw the Environmental Resource Management Department start to develop a Climate Adaptation Plan of Action. During the process, and drawing on eThekwini's experience, they decided to move away from an overarching plan and rather develop sector specific plans. The plans, developed in consultation with line functions, have been signed off administratively and politically.

Throughout this period there have been links to external academics, consultants and NGOs that have culminated in the recent City of Cape Town Climate Change Think Tank that brings together different groups to explore projects, policies and processes related to climate change in the city. One of the strengths of this approach is that the think tank provides a venue to advance the notion of a collective responsibility around these issues that reaches beyond city officials to include key people in society with the recognition that "we are all in this problem together" and need to find solutions collectively. Although there have been several political champions from the different ruling parties in the province, action has taken a lot of time and effort to get going.

The city currently operates in a crisis manner. Although flooding in informal settlements is a major challenge, many officials say that they expect it to flood every winter and they will respond to the crisis by relocating people and providing shelter and provisions as necessary. Municipal officials note that a proactive response is challenging because if people are moved out of the flood prone areas, others will just move in and there is limited land to move them to anyway. So flood preparedness is seen as an unmanageable objective and it is easier to deal with the crisis on an annual basis. It is clear from this example that socio-economic issues drive action in the City of Cape Town. Adaptation is something that "comes behind trying to shift changes" at present. Those concerned with adaptation maintain that addressing both short and long term climate impacts is key to the city's socio-economic and environmental sustainability.

## 3.5 Urban Adaptation Barriers and Opportunities

### *3.5.1 Knowledge and Understanding*

One of the key issues identified in both eThekwini and the City of Cape Town is the perception within government and civil society that climate change adaptation is an environmental issue. Disaster risk management has been slow to link risk reduction related to extreme events to climate change impacts. Those pushing adaptation however, see it as more of a development issue. As one of the officials stated, she sees it as "an economical issue, social issue, political issue and a sustainability issue, its all of those things". Another official said that "climate change is a big threat and challenge to the social and economic fabric of society", highlighting his desire to not see it positioned as an environmental responsibility. The concern is that if it stays in the environmental realm, which is how it tends to be seen by the public and many politicians, it will suffer, as climate change is a broader, cross-cutting phenomenon. The opportunity here is to illustrate why it is about so much more than the environment and how it impacts on many facets of development, including risk reduction. This is particularly important in the global South, where this argument is likely to hold traction with policy-makers and underpin more holistic development responses.

Another concern is the widespread belief that the impacts of climate can be dealt with as a crisis. When a crisis happens or things change beyond a certain point then an engineering solution will be found. This way of thinking is a barrier to actively engaging with planned adaptation. Again, the opportunity lies in reframing this type of thinking. As an official from the City of Cape Town articulated, adaptation requires a new kind of thinking:

> I think what adaptation really asks is for the first time for us to collectively, as communities, as a city, as administrators, as officials, to try and do something that is not very well done at the local level, which is forward planning—to think ahead and say if we do this now we will benefit later.

But he recognised that this requires a huge shift in mindset, as by human nature we are bad at planning for the future. However, this shift presents an opportunity for more flexible long-term management.

### 3.5.2 Action

One of the barriers identified in the City of Cape Town was an ambiguity as to who is responsible for adaptation, as was the case in earlier research on adaptation in the Cape Town water sector (Ziervogel et al. 2010). This ambiguity spreads across local government departments and spheres, inhibiting action on the ground. Part of the challenge is that there is not enough engagement and interaction between the three spheres of government (national, provincial and local) around the various roles. In order to be effective, roles need to be communicated effectively and well aligned in terms of both responsibility and coordination. This is particularly important in the context of addressing disaster risk that includes reducing exposure to and recovering from adverse climate change impacts that often requires a holistic response rather than responses by one actor or group undermining that of another. Unfortunately there are examples of provincial and local government working at cross-purposes in ways that undermine robust adaptation. Opportunities for collaborative action therefore need to be identified and prioritised in order to reduce inappropriate responses.

Politicians in the City of Cape Town have understood some of the issues around climate change but implementing goals and policies has been slow, although it is picking up speed. Everyone is doing their business as usual and adding another task to that adds additional burdens, so there is the recognition that institutional structure, including resources for personnel and finances for projects, will help to support action in the future if matched with appropriate capacity building measures. New developments related to the CAPA have got different departments thinking about how climate change is likely to impact their line functions and what they might do to prepare. The test will be whether there is the capacity and momentum to implement these ideas.

In eThekwini, a Headline Climate Change Adaptation Strategy (HCCAS) was developed in 2006 using a cross-sectoral approach but did not result in new adaptation actions, partly because the strategy was generic, targeted at a high-level and had to compete with issues perceived as more urgent (Mather et al. 2011). In order to overcome this barrier, adaptation was embedded in and aligned with departmental priorities and capacities. Building on this, a review of current adaptation work has identified three priorities for moving forward, including (1) ecosystem-based adaptation, (2) community-based adaptation and (3) development and monitoring of pilot adaptation plans (Roberts et al. 2012).

### 3.5.3 *Institutional*

According to one City of Cape Town official, "The factor that determines success or progress in the sectors is not about technical skills or money, it's about institutional hurdles".

In eThekwini, the climate function at the local level is not structurally based and not acknowledged. Where it is present is because people have shown leadership on the issue due to their personal convictions rather than a policy mandate. Because of this, if those people disappear then the current climate change work is likely to lose momentum. When this study was conducted, the municipality of eThekwini was running at fifty percent staffing capacity with tight budgets and urgent development pressures. This inhibits the likelihood of proactive risk reduction, as short-term development needs are prioritised rather than planning for longer-term events.

Another institutional constraint on planning for long-term risks is the short political terms held by elected officials that has been recognised internationally, such as in coastal management in the U.K. (Few et al. 2007). This is the case in South Africa, where politicians' long-term strategic thinking is overshadowed by short-term gains that are linked to five-year election cycles and activities that will get them re-elected. This does not support adaptation planning and risk reduction that often requires current development pathways to be reconsidered.

Another key institutional barrier is the limited level of financial support needed to adequately address climate change adaptation. Currently actors are drawing on funding from international donors or from their mandated activities and adding a climate component to the work they do in creative ways, but few are receiving government financing to explicitly work on climate change. A City of Cape Town official suggested that:

> Until the city puts their own money into climate change investments they will remain vulnerable to international trends, to funding models, to donors coming in and out. A sensible city government approach requires the proper investment mechanisms to generate its own funding, or to have its own resources built up to deal with these issues. The ability to do this is very limited in South Africa given the current regulations.

The current financial regulations hinder adaptation. Further exploration is needed into the options for changing financing in the near future. This will require engaging with municipal financial policy as well as focusing on departments having a climate change adaptation mandate so that they can access funds directly.

In South Africa, adaptation is seen as an environmental issue, and is therefore not as salient in the cities or provinces as other issues that compete for attention. In order to develop and institutionalise a holistic approach, adaptation needs to be given greater standing relative to other competing priorities. This is also necessary to institute change across organisations, which is currently undermined because of the silo approach that the current government structure supports. To address this, relationships need to be built and champions found that can work within the current line functions and collaborate across departments and spheres. As adaptation is more systemically considered across municipalities, as is starting to happen in eThekwini and the City of Cape Town, more champions are starting to emerge who are able to tackle sector-specific aspects of reducing risk.

## 3.6 Future Action: Linking Environmental and Developmental Issues

In order to move forward, key knowledge, action and institutions need to be addressed. Seven action points are presented that have emerged from this analysis and should be used to inform future pathways.

**Frame adaptation as a development issue** In terms of the knowledge problem it is critical to broaden the understanding of climate change adaptation so that it is not seen as an environmental issue, but a development issue. This needs to be communicated to politicians, officials and civil society and a discussion needs to happen around how this might be done in a manner that affects policy change. Although politicians and senior executives in the City of Cape Town recognise and understand the role and importance of adaptation, this has been slow to translate into actionable support.

**Enhance the role of local government** Another aspect of the knowledge problem is to improve the understanding of the role that local government can play in adapting to climate impacts. All local government actors were able to identify the importance of their contribution, however they felt there was not recognition of this from the national level. Local government felt that they play a critical role in implementing policy and facilitating action yet they do not feel valued by national government.

**Develop proactive policy** Although crisis mode tends to be the norm for managing disasters, the nature of the adaptation challenge demands support for anticipatory adaptation. The proliferation of new climate change policies presents an opportunity to position cities to take advantage of the post-disaster window of opportunity to link disaster recovery, risk reduction, sustainable development and climate change adaptation.

**Provide space for learning in policy and practice** One of the critical components of the action problem, as highlighted in the literature, is the necessity for adaptation to be addressed as a process where learning is supported (Tshakert and Dietrich 2010). This requires explicit support and leadership that provides a space for reflection. Within the City of Cape Town this has been recognised by those currently engaged in adaptation, as noted by one official:

> I think that the unique thing about adaptation is that everyone is learning as they go. There is going to be a lot of fumbling and a lot of mistakes made. So leadership is going to be critical in taking something forward that allows for errors to be made and for the process to unfold and to learn as we go. For adaptation to be successful it will require strong leadership that will stand in the face of interrogation.

It is hard to plan for impacts that may only happen in thirty years' time or may actually never happen. This means that the success of adaptation is measured over decades as opposed to over a short-term time frame. Although there is recognition of the challenges around learning it is clear that more needs to be done to create a collective learning environment that recognises the long-term time horizon associated with climate change while taking advantage of "teachable moments" following extreme events.

**Provide adequate resources in support of collaborative governance and local capacity building** Another step towards addressing the action problem is to ensure that the correct institutions are in place to support action and emerging leadership around climate change adaptation. Given that there is currently no climate mandate at the local level this makes it hard. Institutional change is required to create mandates that will enable access to staff, budget, operational activities and for the issues to be mainstreamed in a way that does not depend on the initiating department's self-funding all activities. As an eThekwini official said:

> If you get a mandate coming down from national, it gives us the capacity to build the skills, to put the people in place, to answer those questions, and that's the point, we're never going to be able to answer those questions until we have actually got people working full time.

One way of strengthening the mandate of cities is to revise the Constitution to amend the allocation of the powers and functions of government by including climate change and environmental functions at the local government level. Metropolitan municipalities need to be given the same powers and mandates as provincial government. Currently the three-tiered governance system complicates the mandate at a local level, as there is not enough engagement and interaction between the three spheres of government (national, provincial and local) around the various roles and funding streams.

**Local advocacy and disasters as opportunity** In eThekwini, the Environmental Planning and Climate Protection Department recognised that despite there not being a mandate or political pressure, there was an opportunity. They facilitated the integration of adaptation work into the mainstream by working with different departments to assess their current work, resource limitations and business plans and how they might consider climate impacts. eThekwini also had a mayor that took on a stronger leadership role around climate change born out of his experience of the disasters beginning in 2007 and particularly those in 2008/9 which hit local communities and resulted in losses of houses and deaths. Seizing these opportunities has been key to profiling adaptation in the early stages.

**Improve financial support for climate change adaptation** Addressing financial regulations is a significant barrier that has no clear resolution at present. There is no obvious way to change current regulations to build up financial resources beyond the current budget cycle, which requires a shift to longer-term thinking. Based on the interviews conducted it is clear that when municipalities put their own money into action, there is greater ownership but because of the lack of a mandate, money has to be used creatively to support adaptation work. While flexible international funding has enabled activities to get kick-started, it does not currently represent a sustained source of revenue. Although future international funding is seen as a potentially important contribution, there was a strong sense from both eThekwini and the City of Cape Town that municipal budgets need to be ear-marked for adaptation activities to ensure that the city is invested in and supports adaptation over the long term.

## 3.7 Conclusion

The future action points above suggest a number of conceptual areas to develop. Translating theory into practice has numerous challenges, but it is exciting to see how climate change has been used to tackle a number of issues that support the development of cities of the South more broadly. A few of the conceptual action points are expanded on here in order to help provide a set of practical next steps:

- In order to frame adaptation as a development issue, champions within local government need to be more overt in doing this. Preparing for the impacts of climate change now will help to ensure that a range of development priorities are not undermined in the future. This rhetoric needs to be actively developed when engaging with public relations officers, media, and in policy and research documentation.
- The role of local government needs to be given more visibility by provincial and national government. In order for this to happen, local government actors need to proactively pursue opportunities to engage at the national and provincial level and develop channels of communication and cooperation.
- In order for social learning to occur, new skills are needed. This may require facilitators or experts in learning to be brought in to facilitate the process, as currently those skills are limited within government.

The governance of adaptation within cities is complex, hampered by unclear mandates, institutional hurdles accessing finance and staff and political ambiguity about its relevance within cities of the global South. Innovative responses have emerged within eThekwini and the City of Cape Town, primarily because of individual leadership, responding to current climate risks that are threatening development and flexible funding. Although a small number of leaders have started work in these two cities, recognition of the adaptation challenge is growing, albeit in fits and starts. However, the perception that it is an environmental problem is inhibiting the mainstreaming of adaptation across governmental departments. Although there is international funding for some project work it is clear that more fundamental shifts are required to support robust adaptation efforts and that requires changing cities' mandate. This suggests that both internal and external funding needs to be more responsive to supporting process, rather than the current preference for projects. Although the importance of collective learning and adaptive management is recognised in the literature, evidence of how it has been supported in cities of the global South is limited, suggesting this as an important area for future research and action.

## References

Adger, W. N., Arnell, N. W., & Tompkins, E. (2005). Successful adaptation to climate change across scales. *Global Environmental Change, 15*(2), 77–86.

Adger, W. N., Lorenzoni, I., & O'Brien, K. (Eds.). (2009). *Adapting to climate change: Thresholds, values and governance*. Cambridge: Cambridge University Press.

Betsill, M., & Bulkeley, H. (2006). Cities and the multilevel governance of global climate change. *Global Governance, 12*(2), 141–159.

Biesbroek, G. R., Termeer, C. J. A. M., Kabat, P., & Klostermann, J. E. M. (2009). Institutional governance barriers for the development and implementation of climate adaptation strategies. International Human Dimensions Programme (IHDP) conference 'Earth system governance: People, places, and the planet', Amsterdam, The Netherlands, December 2–4, 2009.

Birkmann, J., Garschagen, M. A., Kraas, F., & Quang, N. (2010). Adaptive urban governance—New challenges for the second generation of urban adaptation strategies to climate change. *Sustainability Science, 5*(2), 185–206.

Bloomberg, M. R., Sachs, J., Small, G. (2010). Climate change adaptation in New York City: Building a risk management response. *Annals of the New York Academy of Sciences, 1196*(1), 1–3.

Boraine, A., Cranskhaw, O., Engelbrecht, C., Gotz, G., Mbanga, S., Narsoo, M., & Parnell, S. (2006). The state of South African cities a decade after democracy. *Urban Studies, 43*(2), 259–284.

Brundritt, G., & Cartwright, A. (2012). Understanding the risks to Cape Town of inundation from the sea. In A. Cartwright, S. Parnell, G. Oelofse, & S. Ward (Eds.), *Climate change at the city scale*. (pp. 21–37). Abingdon: Routledge.

Burris, S. C., Drahos, P., & Shearing, C. (2005). Nodal governance. *Australian Journal of Legal Philosophy, 30,* 30–58.

Cartwright, A., Brundrit, G., & Fairhurst, L. (2008). Sea-level rise adaptation and risk mitigation measures for the city of Cape Town. Report prepared for the city of Cape Town. Cape Town.

Corburn, J. (2009). Cities, climate change and urban heat island mitigation: Localising global environmental science. *Urban Studies, 46*(2), 413–427.

Corfee-Morlot, J., Kamal-Chaoui, L., Donovan, M. G., Cochran, I., Robert, A., & Pierre-Jonathan, T. (2009). Cities, climate change and multilevel governance, OECD environmental working papers N. 14, 2009, OECD publishing.

Fairhurst, L. (2008). Sea-level rise risk assessment: Risk and impact identification. Report prepared for the City of Cape Town. Cape Town.

Few, R., Brown, K., & Tompkins, E. L. (2007). Climate change and coastal management decisions: Insights from Christchurch Bay, UK. *Coastal Management, 35*(2–3), 255–270.

Folke, C., Hahn, T., Olsson, P., & Nordberg, J. (2005). Adaptive governance of social-ecological systems. *Annual Review Environment and Resources, 30,* 441–473.

Granberg, M., & Elander, I. (2007). Local governance and climate change: Reflections on the Swedish experience. *Local Environment: The International Journal of Justice and Sustainability, 12*(5), 537–548.

Heinrichs, D., et al. (2009). Adapting cities to climate change: opportunities and constraints (findings from eight cities). Fifth urban research symposium, Marseilles, France. http://go.worldbank.org/P74DFSRCD0. Accessed 10 Dec 2011.

Horton, R., Rosenzweig, C., Gornitz, V., Bader, D., O'Grady, M. (2010). Climate risk information: climate change scenarios and implications for NYC infrastructure. New York City panel on climate change. *Annals of the New York Academy Science, 1196*(1), 147–228.

Inderberg, T. H., & Eikeland, P. O. (2009). Limits to adaptation: Analysing institutional constraints. In W. N. Adger, I. Lorenzoni, & K. O'Brien (Eds.), *Adapting to climate change: Thresholds, values and governance* (pp. 433–447). Cambridge: Cambridge University Press.
Jordan, A. (2008). The governance of sustainable development: Taking stock and looking forwards. *Environment and Planning C: Government and Policy, 26*(1), 17–33.
Kemp, R., & Martens, P. (2007). Sustainable development: How to manage something that is subjective and can never be achieved? *Sustainability: Science, Practice and Policy, 3,* 1–10.
Mather, A., Roberts, D., & Tooley, G. (2011). Adaptation in practise: Durban, South Africa. In K. Otto Zimmermann (Ed.), *Resilient Cities and Adaptation to Climate Change Proceedings of the Global Forum 2010* (Vol. 1, pp. 543–563). Dordrecht: Springer. doi:10.1007/978-94-007-0785-6_53.
Moser, S. C. (2009). Whether our levers are long enough and the fulcrum strong? Exploring the soft underbelly of adaptation decisions and actions. In N. W. Adger, I. Lorenzoni, & K. L. O'Brien (Eds.), *Adapting to climate change: Thresholds, values, governance* (pp. 313–334). Cambridge: Cambridge University Press.
Moser, S. C., & Ekstrom, J. A. (December 21, 2010). A framework to diagnose barriers to climate change adaptation. *PNAS, 107*(51), 22026–22031.
Mukheibir, P., & Ziervogel, G. (2007). Developing a municipal adaptation plan (MAP) for climate change: The city of Cape Town. *Environment & Urbanization, 19*(1), 143–158.
Naidu, S., Hounsome, R., & Iyer, K. (2006). Climatic future for Durban, prepared by CSIR NRE, edited by D. Roberts, A. Mather, & M. Maganlal.
Nelson, L., Robbins, R., & Simonsen, B. (1998). Introduction to the special issue on governance. *The Social Science Journal, 35*(4), 477–491.
Parnell, S., Simon, D., & Vogel, C. (2007). Global environmental change: Conceptualising the growing challenge for cities in poor countries. *Area, 39*(3), 1–13.
Pielke, R. A., Prins, G., Rayner, S., & Sarewitz D. (2007). Climate change 2007: Lifting the taboo on adaptation. *Nature, 445,* 597–598.
Roberts, D. (2008). Thinking globally, acting locally—institutionalizing climate change at the local government level in Durban, South Africa. *Environment and Urbanization, 20*(2), 521–538.
Roberts, D. (2010). Prioritizing climate change adaptation and local level resilience in Durban, South Africa. *Environment and Urbanization, 22*(2), 1–17.
Roberts, D., Boon, R., Diederichs, N., Douwes, E., Govender, N., McInnes, A., McLean, C., et al. (2012). Exploring ecosystem-based adaptation in Durban, South Africa: "learning-by-doing" at the local government coal face. *Environment and Urbanization, 24*(1), 1–29.
Romero-Lankao, P. (2008). Urban areas and climate change: Review of current issues and trends. Issues Paper for the 2011 Global Report on Human Settlements.
SACN. (2006). State of the cities report 2006. Johannesburg: South African cities network (www.sacities.net). Accessed 10 Dec 2011.
Satterthwaite, D., Huq, S., Pelling, M., Reid, H., & Lankao, P., (2007). *Adapting to climate change in urban areas: The possibilities and constraints in low-and middle-income nations*. London: IIED (International Institute for Environment and Development).
Smith, L., & Hanson, S. (2003). Access to water for the urban poor in Cape Town: Where equity meets cost recovery. *Urban Studies, 40*(8), 1517–1548.
Storch, H., Downes, N., Moon, K. (2009). Climate change and the resilience of mega-cities in South-East-Asia: Creating risk-based climate change information for Ho Chi Minh City's Settlements, Proceedings REAL CORP-2009, Tagungsband.
Sutter, C., & Parreño, J. C. (2007). Does the current clean development mechanism (CDM) deliver its sustainable development claim? An analysis of officially registered CDM projects. *Climatic Change, 84*(1), 75–90.
Tadross, M., Taylor, A., & Johnston, P. (2012). Understanding Cape Town's climate. In A. Cartwright, S. Parnell, G. Oelofse, & S. Ward (Eds.), *Climate change at the city scale*. Routledge: Abingdon.
Tschakert, P., & Dietrich, K. (2010). Anticipatory learning for climate change adaptation and resilience. *Ecology and society, 15*(2), http://www.ecologyandsociety.org/articles/3335.html.

Urwin, K., & Jordan, A. (2008). Does public policy support or undermine climate change adaptation? Exploring policy interplay across different scales of governance. *Global Environmental Change, 18*(1), 180–191.

van Asselt, M.B.A., & Renn, O. (2011). Risk governance. *Risk Research, 1*(4), 431–449.

van Zeijl-Rozema, A., Cörvers, R., Kemp, R., & Martens, P. (2008). Governance for sustainable development: A framework. *Sustainable Development, 16*(6), 410–421.

Wilbanks, T., Romero-Lankao, P., Bao, M., Berkhout, F., Cairncross, S., Ceron, J., Kapshe, M., Muir-Wood, R., & Zapata-Marti, R. (2007). Chapter 7: Industry, settlement and society. In M. Parry, O. Canziani, J. Palutikof, P. van der Linden, & C. Hanson (Eds.), *Climate change 2007: Impacts, adaptation and vulnerability, contribution of working group II to the fourth assessment report of the intergovernmental panel on climate change* (S. 357–390). Cambridge: Cambridge University Press.

Winsvold, M., Stokke, K. B., Klausen, J. E., & Saglie, I. (2009). Adapting to climate change: Thresholds, values and governance. In W. N. Adger, I. Lorenzoni, & K. O'Brien (Eds.), *Organizational learning and governance in adaptation in urban development* (S. 476–490). Cambridge: Cambridge University Press.

Ziervogel, G., & Taylor, A. (2008). Feeling stressed: Integrating climate adaptation with other priorities in South Africa. *Environment, 50*(2), 32–41.

Ziervogel, G., & Ericksen, P. (2010). Adapting to climate change to sustain food security. *WIRES Climate Change, 1,* 525–540.

Ziervogel, G., Bharwani, S., & Downing, T. E. (2006). Adapting to climate variability: Pumpkins, people and pumps. *Natural Resource Forum, 30,* 294–305.

Ziervogel, G., Shale, M., & Du, M. (2010). Climate change adaptation in a developing country context: The case of urban water supply in Cape Town. *Climate and Development, 2,* 94–110.

# Part II
# The Nature of Disasters and the Role of Natural Hazards Planning in Building Resilient Communities

# Chapter 4
# Climate Change Adaptation and Disaster Risk Reduction in Highland Peru

**Anthony Oliver-Smith**

**Abstract** The Andean nation of Peru is currently assessed to be particularly vulnerable to climate change. Andean populations are vulnerable to both disasters and climate change effects due to poverty, food insecurity, poor health and marginalisation. Adaptation to climate change and disaster risk reduction in Peru must address systemic vulnerabilities rather than weather related disaster effects only. Policies must integrate measures to address specific hazards with programs to reduce systemic vulnerabilities and societal inequality.

**Keywords** Peru · Adaptation · Climate change · Disaster · Vulnerability

## 4.1 Introduction

The nation of Peru has experienced considerable environmental change in the last quarter century and is currently assessed to be particularly vulnerable to climate change. Indeed, the Andean region of that nation is projected to experience dramatic changes in the relatively near future. The sensitivity of mountain ecosystems and the vulnerability of resident populations to climate change are now only beginning to be appreciated. Mountain regions around the world contain some of the planet's most sensitive and fragile ecosystems that account for roughly half of all biological diversity and almost half of all biodiversity hotspots. Furthermore, mountain regions are home to roughly one sixth of the world's population, who in many cases are among the poorest and most vulnerable to both natural hazards as well as the impacts of climate change.

The sensitivity of mountain ecosystems to climate change will be especially seen in alterations of the cryosphere, hydrology, biodiversity and vulnerability to extreme events and natural hazards (Macchi 2010). Mountain regions, as well, provide important contexts in which to assess climate change impacts because climate, vegetation and hydrology rapidly change with altitude in relatively reduced horizontal distances

A. Oliver-Smith (✉)
University of Florida, Gainesville, FL, USA
e-mail: aros@ufl.edu

B. C. Glavovic, G. P. Smith (eds.), *Adapting to Climate Change,* Environmental Hazards,
DOI 10.1007/978-94-017-8631-7_4, 

(Beniston 2003, pp. 5–6). Mountain systems in South America, however, are not well understood in terms of climate change and vulnerability (Trigoso Rubio 2007).

The Andes, like most mountain chains around the world, constitute a repository of biological and cultural diversity, providing vital environmental services such as water, power, tourism, minerals, medicinal plants and fibers to mountain peoples, as well as nearby heavily populated lowland areas (Macchi 2010, p. 1). However, the Andes are particularly vulnerable because their high relief, steep slopes, shallow soils, geological instability and adverse climatic conditions tend to exacerbate and amplify the effects of environmental changes, with serious implications for both ecosystem health and human well-being. Climate variations are particularly affected by the El Niño Southern Oscillation (ENSO) phenomenon, which periodically produces extreme events throughout the coastal and Andean regions of the Andean Community nations.

Andean populations and their livelihoods are particularly vulnerable to climate change effects due to their disproportionate rates of poverty, prevalence of food insecurity, poor health, high dependency on natural resources, marginalization and limited livelihood diversity. Climate change, thus, both intensifies hazards that Andean peoples have coped with for millennia and may also add to the array of hazards to which Andean peoples are exposed and vulnerable (Vergara 2008). Since climate change contributes to both vulnerability and, in some cases, disaster, climate change adaptation in most instances is a feature of disaster risk reduction and therefore must address systemic vulnerabilities rather than weather related disaster effects only (UNISDR 2011). In that sense, the most effective overall policy for both climate change adaptation and disaster risk reduction is an approach, consistent with sustainable development, that integrates measures designed to addresses specific hazards with programmes to reduce the systemic vulnerabilities and societal inequality. Although the focus of this chapter is the nation of Peru, the highland regions of the other three members of the Andean Community, Bolivia, Colombia and Ecuador, are experiencing similar challenges from climate change.

## 4.2 Adaptation and Mitigation: Toward Risk Reduction

There are two basic and related responses to changing climate processes and the challenges they represent: mitigation and adaptation, both of which in the context of disaster risk reduction entail changes in social, technological and environmental relations. Mitigation is concerned with strategies to minimise impact and loss, and to facilitate recovery. Mitigation, to the degree that it addresses how systemic features contribute to vulnerability, deals with causes, although such action may be undertaken in response to the impact of a hazard event. Mitigation is proactive, aimed at increasing the resilience of a society; that is, increasing the capacity to absorb the impacts of hazards that exist in its surroundings without major disruption of basic functions (Wisner et al. 2004). Such strategies can be of a technological, economic, or social organizational nature.

Adaptation, on the other hand, is a process that offers possible adjustments that may enable people to safeguard livelihoods and welfare. Adaptation, however, because it is deployed in numerous institutional and environmental contexts, is a complex issue. Adaptation used in a formal analytical sense is a concept borrowed from ecology, in which it is defined as the process of developing or enhancing structural, physiological and/or behavioral characteristics that improve chances for survival and reproduction in a given environment. It is a concept that is inherent in natural systems in which those organisms with the characteristics that best equip them for survival in an environment have higher probabilities for both survival and reproduction and hence passing on those characteristics to succeeding generations. Thus, the adoption of a concept used to describe natural systems becomes problematic when used to describe the fundamentally socio-cultural nature of human-environmental relations.

When used in regard to human societies, adaptation becomes the fundamental conceptual nexus in human-environment relations. It is through the process of adaptation that humans and natural systems conjointly construct socio-ecological systems, or environments. Humans interact with and adapt to both a socio-cultural (institutional) environment as well as a natural environment. We adapt to natural features, land or water, for example, but also to human institutions such as labor, economics, markets, schools, governments and churches and the resources and constraints they represent. That is, our institutions are at once part of our overall adaptation, but must be adapted to as well. At the same time, our environments, to which we must adjust, are as much socially as naturally constructed.

From a social scientific perspective, adaptation refers mainly to changes in belief and/or behavior in response to altered circumstances to improve the conditions of life (or survival). In that sense, adaptation in general is reactive, adjusting primarily to environmental conditions within a range of variation. Adaptation in its special application to human beings has a wider number of attendant features for adaptive capacity including complex human cognition, social organization, values and meanings. Human adaptations to environmental change are largely social, organizational and technological. For human beings, decision-making and implementation are central features of adaptation, of every strategy for exploiting the energy potentials of a particular habitat (Holling 1994; Bennett 1996). In a sense, an adaptation is a form of belief, behavior, or technology that is part of the overall "toolkit" of a society that enables it to survive and reproduce in its total environment.

However, some difficulties can be seen here both in the concept as well as in the process of adaptation. In the way the concept has been used to describe such a wide array of changes, it becomes difficult to distinguish it from notions of social and economic development (Lavell 2011). In regard to climate change, Orlove (2009), for example, questions the utility of the term adaptation and particularly, the way it has been accepted as a key dimension of climate change policy. He is concerned with the way it is taken for granted as a key element in climate change policy, but does not capture full impacts of climate change nor does it always represent accurately either the perceptions of the people affected by these impacts or the range of alternatives open to them. Adaptation, furthermore, does not fully engage the issue of systemically imposed vulnerability. That is, vulnerability to disasters or to climate change effects is essentially socially constructed, the outcome of the way wealth and security are distributed in a society. In effect, part of what people will be adapting to is the systemic vulnerability imposed by society. Is adaptation, then, at least in part, adjusting so the status quo can persist? Basically, the question becomes what is being adapted to, climate change or a system of structural disadvantage perhaps made worse by climate change?

With these cautions in mind, if we are to assess the linkages between adaptation to climate change and disaster risk reduction, we must ask whether the impacts of climate change will qualify as disasters. While it is not my intention to digress into the deeply complex definitional debate around the concept of disaster (Oliver-Smith 1999, 2002; Quarantelli 1991), linking the concept of disaster to the process of climate change correctly relocates the focus of analysis from an event to a socio-ecological process of vulnerability construction. However, regardless of how analysts view this process, it is fairly clear that the outcomes of many climate change effects will be seen and felt as disasters by the affected populations.

The concept of coping proves useful in clarifying the relationship between climate change adaptation and disaster risk reduction. Coping essentially refers to decision-making in novel situations for which there is no ready institutionalised response. It involves improvisation and creativity (Bennett 1996). If the novel situation becomes recurrent and coping measures prove effective in dealing with it, they may become absorbed into the "toolkit" of adaptations to a dynamic environment. In that context, then, climate change will in most cases simply exaggerate the effects and frequencies of existing hazards which are largely the outcome of reigning patterns of local vulnerability. Indeed, climate change effects will also increase the vulnerability of people to geological and other hazards not related to climate change. However, even in cases where the climate change driven hazard is novel, its impacts will still be expressed and coped with through local vulnerability patterns. In the final analysis, adaptation to climate change will assist people in adjusting to "new normal conditions" to enhance social and material reproduction. As the pace of climate change increases, coping strategies that prove effective will increasingly become community level adaptations, in effect part of the "toolkit" for sustaining social and material life.

Disaster risk reduction must continue to focus on processes that make people vulnerable to events which fall beyond the capacity of communities to cope which may have been made even more extreme by climate change processes. Thus, both climate change adaptation and disaster risk reduction must be framed and designed to address those social and economic features that render people vulnerable to environmental hazards in general. Climate change adaptation and disaster risk reduction then are related undertakings, and both must address systemic vulnerabilities as well as hazards posed by specific climate change effects as fundamentally development related phenomena (Birkmann and Von Teichen 2009; Kelman and Gaillard 2010; Lavell 2011). In that sense, the most effective overall policy for both climate change adaptation and disaster risk reduction is addressing the systemic vulnerabilities and reducing inequality. This policy in effect is also entirely consistent with a policy of sustainable development.

## 4.3 The Peruvian Andes: The Physical Setting

Three vast interacting natural systems affect all biological life in the nation of Peru: the Amazon rain forest, the Andean cordilleras and the Pacific Ocean. The interaction of these three natural systems create essentially five macro-environmental contexts: coastal, mountain valleys, altiplano (high altitude plateau), humid eastern slopes and tropical forest which, fragmenting into numerous micro-environments have enabled human societies to emerge and proliferate over the past 10,000 years, producing in the last 4,000 years a wide range of complex societies of diverse ethno-linguistic and socio-economic configurations (Moseley 2001). Indeed, Peru contains 28 of the 34 climates of the world (CAN 2008).

Furthermore, the Pacific Ocean, the Andes and the Amazon produce environments characterised by wide variation and frequent ecological extremes. While all three systems are vast in their scope, the Andean cordillera plays the key role in how the other two systems are expressed in the lives of human beings. The Andes are the product of the subduction of the oceanic Nazca plate moving eastward underneath the continental rim of the South American plate. Over many millions of years the Andean cordillera was formed by the tectonic energy that continues to drive the westward moving continental plate to override the eastward moving Nazca (oceanic) plate at rates as high as 15 cm a year. This tectonic uplift eventually produced a chain of mountains high enough to actually split the continental climate into two regimes: an eastern wet environment and a western dry environment (Moseley 2001, pp. 25–26).

The Andes of Peru rise abruptly from the narrow strip of coastal desert and descend somewhat more gradually toward the humid Amazon basin in the east. The position of the Andean cordillera is a key factor in the distribution and seasonality of rainfall. Under normal conditions, the Amazon experiences year-round orographic rainfall and the coastal desert receives almost no rain at all except when the pattern is disrupted by temperature changes occasioned by El Niño (ENSO). Indeed, Peru is blessed with abundant water resources, but 98 % of the Andean runoff, returns east to nourish the rivers of the sparsely populated Amazon basin, while the rest flows through mountain rivers toward the coast where fully two thirds of the population live. Periodically, the ENSO phenomenon subjects Peru to large climatic interannual variations, but their impacts vary a great deal between the coastal regions of northern Peru and the southern highlands and altiplano regions. When it occurs, El Nino tends to reduce the transport of moisture from the Amazon region into the sierra, during which the warming of the lower atmosphere inhibits the air flow (and rain) from the east (Sperling 2008). A reverse process takes place in La Nina years when easterly winds enhance the flow of moist air into the highland regions (Fig. 4.1; Sperling 2008).

The mountains between the jungle and the desert experience wide variations in rainfall and temperature (Winterhalder and Thomas 1978). The surface conformation of Andean environments is the product of geologically more recent processes of glacial and tectonic activity, faulting, volcanoes and erosion and deposition (Winterhalder

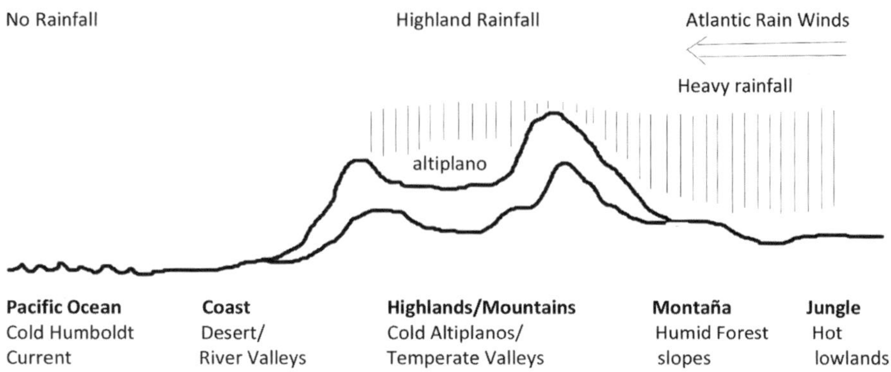


**Fig. 4.1** The coastal-mountain-tropical lowland rainfall regime in Peru.

and Thomas 1978). These processes have produced deep V and U shaped valleys, frequently capped by deep glacial lakes, dammed by terminal morainic loops. These formations are characteristic of strong alluvial and glacial erosion exacerbated by both persistent faulting and uplift of the mountains, processes that continue to be significant forces in Andean cultural time. The majority of the highland populations occupy these intermontane valleys and surrounding slopes where a wide variety of good crops are cultivated, ranging from garden vegetables to corn and higher altitude Andean tuber and grain crops such as oca, olluco and quinoa (Fig. 4.2).

Thus, the Andes, the second highest chain of mountains in the world, are characterised by extreme instability in the form of significant seismic activity, active volcanoes, unstable soils and avalanches of both minor and major dimensions. There have been more than 50 major earthquakes (greater than 7.0 on the Richter scale) and countless smaller ones since historic records began to be kept (Giesecke and Silgado 1981). There are 10 active volcanoes as well. Many of the peaks are sharply angled, ranging from 45 to nearly 90° at their highest reaches which are often over 5,500 m. The combination of their extreme altitude and their location in subequatorial latitudes with sustained high insolation makes for a pattern of rapid alternation of snowfall and melt, which results, on the beneficial side in water availability for valley agriculture in the dry season, but on the negative side in unstable icefields susceptible to even slight tremors, producing ice and rock avalanches of varying proportions (Caviedes 1981). When the path of these avalanches ends in a glacial lake, the morainal loop dam may give way or be washed over resulting in significant flooding of human settlements downstream. Currently, under conditions of climate change there is significant glacial retreat. The loss of these glaciers which supply water for agriculture, energy generation, mining and human consumption will severely impact these sectors as well as local vulnerability patterns both in the highlands and the coast (Figs. 4.3 and 4.4; Leavell 2008).

**Fig. 4.2** The Callejon de Huaylas, an intermontane valley in the Peruvian Andes.

**Fig. 4.3** Mt Huascaran, the highest peak in the Peruvian Andes and the avalanche scar that buried the town of Yungay in 1970.

**Fig. 4.4** The Laguna Paron, an Andean glacial lake.

The Andean highland region is noted for its wide variation of microenvironments with complex gradients of climate and vegetation. Such micro-environments, and even those in close proximity, often experience extreme variation in temperature and rainfall. Very localised frosts are hazards to crops as are hailstorms which can be even more devastating to agriculture. The Andean region has experienced both severe regional and local droughts, with no predictable regularity, which may persist for as long as three years with devastating agricultural losses and famine. The uneven topography of the region also figures importantly as a variable in the distribution of rainfall, producing local high concentrations of rainfall, hail and snow. Heavy rainfall may combine with unstable soils on steep mountain slopes to produce mudslides of varying proportions with frequently tragic results for human settlements, agriculture and infrastructure.

## 4.4 Life in the Andes: The Cultural Setting

As this complex and dynamic region has been the home to human inhabitants for over 10,000 years, and the site of major cultural complexity for the last 4,000 years, the nature of human cultural adaptation to the many environments of the Andes becomes a compelling issue. Long term adaptations to the challenges of life in the Andean highlands, including hazards, revolved around the exploitation of multiple ecological tiers, complex water management systems, dispersed settlement patterns, environmentally appropriate building materials and techniques, communal labor systems, interregional exchange and distribution systems, surplus storage and preparedness and ideological modes of explanation and meaning formulation for frequent environmental disturbances (Oliver-Smith 1994). Although environmental forces, particularly tectonic uplift, seismic activity and El Niño induced droughts and flooding appear to have played significant roles in culture change in the Andes (Moseley 2001), as a whole these adaptations seem to have been relatively effective in enabling Andean peoples not only to survive, but to flourish, if population growth and cultural complexity are any measures of success. However, several post-conquest chroniclers do comment on recurring natural disasters such as floods, volcanic eruptions, droughts and earthquakes with resultant destruction. Nevertheless, there is strong, though not conclusive evidence, that the Inca empire, the last pan-Andean culture before the Spanish conquest, was reaching the limits of its agricultural base (Cook 1981). It is also true however, that another human generated disaster, a devastating smallpox pandemic brought by Europeans, contributed greatly to the vulnerability and downfall of the Inca Empire.

The development of Andean civilization has since the earliest days of habitation involved the control of the multiple ecological tiers available in the abrupt ascent of slopes in the region, usually referred to as the principle of verticality. The availability of multiple ecological niches, often in close proximity, enabled Andean highlanders to spread both risk and resources over wider areas, diminishing the impacts of localised floods, hails, mudslides and fronts, while at the same time producing a varied diet (Murra 1972). Farmers in the Andes have adapted to the spatial heterogeneity afforded by their environment by diversifying domestic economies, using multiple fields, employing numerous seed and crop types, and organizing various strategies of communal labor and sectoral fallowing systems (Young and Lipton 2006; Kraft 1995). While in some cases such strategies may be sub-optimal for production, they are effective in producing varied food products and in averting highly localised risks generated by temperature and rainfall variation (Fig. 4.5).

The arrival of the Spaniards and the establishment of the colonial system that was to rule Peru for almost three centuries disrupted the adaptive strategies developed by Andean populations, installing a system based primarily on the extraction of resources rather than the social reproduction of a society and population, resulting in the demographic collapse of 98 % of the indigenous population in the first 100 years of the colony (Cook 1981; Oliver-Smith 1998). The extractive nature of

**Fig. 4.5** Remains of ancient terracing and peasant agriculture at multiple ecological tiers.

the entire system cast a mould which over the centuries has produced an infrastructure for extraction rather than internal distribution, and ultimately a nation that is food-dependent on outside sources. Since the highland regions, apart from certain mining sectors, both during the colony and well into the republic and contemporary times, provided few resources for foreign exchange, they received little investment in either productive activities or infrastructure. Most development investments focused on the coastal region.

The political system that evolved to support these activities equally directed both power and wealth coastward. The social system which articulated the various segments of Andean society also reflected this concentration of power and wealth on the coast, ideologically buttressed by racial and cultural biases that historically denigrated the highland indigenous population and justified their poverty, rural underdevelopment and vulnerability to hazards (Oliver-Smith 1998). Both dominated and ignored by coastal power and wealth, the highland populations suffered from all the indicators of profound poverty with high rates of malnutrition, infant mortality, and high morbidity. Today, 76 % of Peru's population is urban with 24 % in the rural areas. Coastal rural and urban settlements, including the mega-city of Lima, are home to 54.6 % of the population while 32 % live in the mountains and 13.4 % live in the lowland jungle region (INEI 2009). Currently, more than half of Peru's current population labor to survive below the poverty line. Although most of the poor live in or close to Lima, poverty is deepest in the rural areas of the highlands. Those highland areas most populated by the indigenous Quechua and Aymara speakers, with some 5 million people, are the most destitute. There are also some 40 ethnolinguistic indigenous groups in the Amazonian region many of whom suffer from severe poverty as well.

The legacy of Peru's colonial and republican past continues to manifest itself in conditions of extreme poverty, vulnerability and underdevelopment. Highland people suffer from a lack of property rights to land and water. The state has not provided them with essential services such as health care, education or electrical power. Illiteracy rates, particularly for women, are extremely high. The rates of childhood malnutrition are extremely high and food security is frequently endangered by lack of access to resources, credit, lack of agricultural extension services, or environmental processes (Rural Poverty Portal 2010). In some regions of the Andes, under current economic conditions (costs of inputs, labor, yields and market prices), livelihoods based solely on agriculture are increasingly insufficient to meet family needs (Young and Lipton 2006, p. 71). Socioeconomic changes in land tenure, migration, market competition (often international) and increased costs of inputs are accentuating household and community vulnerability as well. These factors notwithstanding, vulnerability is never distributed equally throughout a community. Within Andean communities, vulnerability will vary according to household resources, such as amount, quality and location of land, livestock owned, family size and ages and educational and health levels. In general terms, however, in the Andes, the higher the community, the fewer resources households will have and the more severe the associated vulnerability will be. Those with the least land and livestock, which may be sold in times of stress, will be the most vulnerable to changing climatic conditions.

However, Andean communities are not entirely without resources with which to confront such issues. Life in the Andes has always dealt with exposures and vulnerabilities both from the natural environment as well as those imposed by the structural characteristics of the larger society. Under "normal" conditions Andean peoples must deal with a wide range of environmental variability and relatively high levels of uncertainty (Gobel 2008). In effect, environmental risks must be coped with and adapted to regardless of the structural disadvantages imposed by the larger society and Andean villages have evolved customs and practices that enable them to survive despite such risks. Indeed, they are matters of daily concern and effort for households and communities and Andean peoples have developed a variety of social and environmental strategies and practices to cope with uncertainty and risk. Principal among these factors is the deep knowledge of their ecologically complex environment managed through the exploitation of multiple ecological tiers, mentioned earlier as verticality, that diversifies production and spreads risk, which is highly localised, over a wide area, thus ensuring that climate extremes may not damage or destroy all current production (Murra 1980). Equally important is the longstanding Andean tradition of communal or cooperative labor, variously known as *faena, minka, ayni,* or *la republica,* that enable households to call on resources from the community in times of need. Indeed community based systems control and regulate the use of multiple natural resources. Andean village social organization reflects an acute awareness that risks and challenges frequently require cooperative

**Fig. 4.6** Peasant agricultural fields in the Callejon de Huaylas.

efforts for community security and welfare. However, a gradual erosion of these social institutions in Andean towns and villages by migration, market integration and increasing social differentiation has been documented and their loss will compound the vulnerability of highland inhabitants (Fig. 4.6).

## 4.5 The Hazards of Climate Change in the Andes

Considerable environmental change has taken place in the Andes since the European conquest, but the rate of change has accelerated in the twentieth century, and especially the last quarter century, some of it attributable to climate change. Troposphere mountain zones, such as the Andes, are warming faster than nearby lowland regions and are projected to experience dramatic changes in the relatively near future.

Over the last several decades Peru has experienced significant changes in precipitation and temperature levels, leading to increased glacial retreat, flooding on the normally arid north coast and droughts in the southern highlands. Increases in mean temperatures have caused Andean glaciers to recede 22% in the last 35 years, some of them disappearing altogether, resulting in significant losses in water reserves. This reduction in glacial mass has caused a concomitant reduction of water available for irrigation in highly populated inter-montane valleys as well as a decrease of 12% in the fresh water supply so essential to the desert coastal region, home to approximately 60% of the nation's population (Espinoza Villar et al. 2009).

Changes in temperature and rainfall regimes are also expected to affect crop growing seasons, crop fungal diseases and human disease vectors as well as increasing the desertification of coastal agricultural lands. Since the effects of temperature change are more pronounced at higher elevations, significant temperature extremes are currently being experienced in the Andes (Sperling 2008, p. 97). Indeed, the extreme cold weather temperatures during the winter of 2010 prompted authorities to declare a state of emergency in several areas and it was reported that at least 409 people died from the extreme weather conditions (The Guardian 2010). However, according to the Andean Community of Nations, average temperature in the Andes has increased 70% more than the global average by 0.34 °C per decade and the number of extreme weather events has doubled (CAN 2008). In addition, average rainfall in the Amazon Basin, so vital to Andean eco-systems, has experienced a 9% loss in annual rainfall over the period of 1975–2003 (Espinoza Villar et al. 2009).

Between 2003 and 2007 extreme temperatures and flooding imposed heavy human and economic costs in Peru. Extreme temperatures have affected roughly five million people (18% of the population) and floods have impacted 0.5 million. The annual cost of disasters between 2000 and 2004 averaged $ 325 million. The number of disasters also rose in the same period with floods increasing over 60% between 1970–1980 and 1990–2000, and landslides increasing by almost 400% for the same period (MINAM 2010).

The interaction and influence between regional macro-climate patterns in coastal (and sea surface temperatures), mountain and jungle environments suggest that adaptations to climate change will be similarly influenced and affected. The major expressions of global climate change that will seriously affect human welfare are loss of ecosystem services, water scarcity, loss of land, changes in rainfall patterns, glacial melt, spread of both human and plant disease vectors, and increased intensity and frequency of climate based natural disasters (Renaud et al. 2007).

## 4.6 Local Perceptions of Climate Change in the Andes

The impacts of these changes on human populations in the Andes present serious challenges to their well-being. Description and analysis of climate change perceptions and impacts in the highland regions of Peru will be drawn principally from information collected on two exploratory research trips in two regions: Ancash and Cusco. This information will be supplemented by material from Puno and from the emerging literature on climate change in the Andes. In Ancash, the valley known as the Callejon de Huaylas in the north central Andes has been the focus of considerable attention (Young and Lipton 2006; Carey 2010; Dunbar 2010; Oxfam 2010). In Cusco, the altiplano regions have been explored for climate change impacts and adaptations (Orlove 2009; Oxfam 2010). The region of Puno has also been researched for the impacts of climate change (Trigoso Rubio 2007; Sperling 2008). All three of these regions are predominately rural with economies based in agriculture and agro-pastoralism with increasing activity in mining and tourism, particularly in Cusco and the Callejon de Huaylas. All three regions have seen considerable rural-urban migration, but continue to contain significant rural populations characterised by extreme poverty.

The perceptions and experiences of people, including both community members and local and regional authorities in these selected regions, bear considerable similarities when it comes to the challenges associated with climate change. While local and regional variations do exist, rural and urban people in the regions studied have been affected by aspects of environmental change, experienced principally in the form of generally warming conditions, but also with extreme variations of temperature, changes in seasons and uncertain water quantity and availability. Interviews and discussions with local people revealed that the major impact experienced in both regions visited was in their livelihoods and health.

People in the selected regions uniformly express serious concerns regarding environmental change and natural resources. Indeed, there is a traditionally high awareness and comprehension of the multiple hazards that exist as well as an appreciation of the seriousness of climate changes (Young and Lipton 2006; Sperling 2008; Orlove 2009; Oxfam 2010). Temperature change has fomented a number of new threats and risks. Temperatures are also seen to be shifting, as experienced in the appearance of environmental features characteristic of lower regions, such as particular plants and insects at higher elevations. Farmers are struggling to respond to new pests and diseases that have appeared, endangering their harvests and threatening social reproduction of their households. For example, a pest locally called kona-kona is threatening quinoa crops in high altitude communities, reducing the high protein grain to dust in the field. Once characteristic of much lower eco-zones, the appearance of these new biological threats challenges the adaptive capacities of higher altitude communities.

People also generally note considerable variability in the weather, particularly in shifting seasonality. In agriculture, the changes in quantity and availability of water were the most frequently mentioned as a source of concern. Normally, the rainy and dry seasons were clearly marked as rains came in October and lasted until April. The dry season was from May to September. Now people lament that it can rain or be dry any time of the year, making it difficult to plan for agricultural activities. Altered rainfall (and hail) regimes have disrupted agricultural schedules and farmers today complain that they can no longer use traditional calendars for sowing, irrigating and harvesting their fields. Rainfall out of season can disrupt other agricultural and economic activities. For example, during specific months after harvest, farmers are accustomed to laying out their corn or other crops (aji peppers, for example) for drying in the sun. However, they run the risk of losing their harvested crops because rainfall, once unheard of in the months of May to August, can now suddenly occur to drench and rot the crop laid out to dry.

In addition, water availability is less dependable now. In the Callejon de Huaylas, rain that normally fell between November and April now seems to be limited to the period between January and March. In the altiplano region of Cusco, rains were seen as adequate in 2010, but have been highly variable in previous years. In both regions as well there were concerns about pollution and contamination of water resources, both from mines as well as from the use of agrochemicals. The rivers are now seen as polluted and the water undrinkable, unfit for animals and dangerous for plants as well. The intense heat experienced during the day is also felt to have led to high rates of evapotranspiration, limiting water resources even more. In the regions studied people generally expressed deep concerns about glacial retreat and the loss of water for agriculture. Water resources and availability in general were seen as becoming extremely precarious and were a source of widespread concern.

Both the extremes of temperature and the lack of water endanger both the reproduction and wellbeing of domestic animals (Orlove 2009; Oxfam 2010). The alpaca herders of Cusco expressed deep concerns about the extremely cold nighttime temperatures. The herders assert that the nightly freezes cause female alpacas to abort and younger alpacas to succumb to both gastro-intestinal and bronco-pulmonary diseases, thus reducing the size and reproduction of their herds. For these high altitude pastoral communities, the herds are their only livelihood resource (Orlove 2010; Oxfam 2010). At the highest altitudes little agriculture is possible and the intense sunlight and daytime temperatures also have dried up pastures. The extreme daytime heat "burns" the pasture, reducing the amount of food available for herds. High altitude pastoralists, although continuing to reside in their communities, are among those groups most threatened by temperature extremes and variability.

In addition to the impacts of environmental change on the health of plants and animals, people in both regions have specific concerns for human health. People in both regions remarked on either the increased incidence of certain diseases or the appearance of ailments that prior to global warming were not present. Of particular concern in the altiplano region of Cusco were the extremely cold nighttime temperatures that are seen to be affecting the elderly and the young with bronco-pulmonary diseases, frequently resulting in death. In the Callejon de Huaylas warmer daytime temperatures were seen as responsible for the increase in such skin diseases as scabies, but people also felt that colder nighttime temperatures were affecting children. The appearance of flies, mosquitos and rats, common in lower elevations, in intermediate and higher zones, was also considered to be a hazard to human health. The contamination and pollution issuing largely from the mines in both regions were also a cause for concern.

Another environmental concern that is present in both regions is the loss of biodiversity. People in both regions noted the disappearance of particular species. Toads, frogs and snakes, once plentiful in both regions, are seen as vanishing from the environment. The decline in water sources is seen as the cause of the disappearance of frogs. There are further concerns about the loss of plant diversity. The general reduction in the number of species of potato cultivated and their excessive dependence on insecticides and herbicides are also worrying to them.

**Fig. 4.7** Peasant farming in the Andes.

In general, there is great concern in both regions about natural resources and the environment. The temperature extremes, the decline in water resources and availability, the changes in rainfall patterns, the changes in seasonality and the loss of biodiversity are all perceived to be serious threats to well being. In particular, local people felt that environmental changes had rendered much of the traditional knowledge of crops, agriculture and climate either inoperable or obsolete. Many people spoke of the difficulty in knowing when to undertake certain agricultural activities that had previously been encoded in traditional understanding of weather and climate. Seasons calculated by a monthly calendar afforded a level of predictability that enabled people to programme their livelihood activities with reasonable margins of error. Today, people complain that anything can happen anytime and they have diminished means of predicting or planning (Fig. 4.7).

## 4.7 Regional and National Perceptions of Climate Change

The perception of environmental change, and in particular climate change, is widespread among formal institutions of local, regional and national governments in Peru. At the level of national government, such as at the Ministry of the Environment and the National Water Authority, there is high recognition of the importance of environmental change, including climate change. At the Ministry of the Environment, concerns were expressed regarding glacial retreat and incipient conflicts between water enterprises and communities, such as has occurred in the dispute between Duke Energy and the communities near the Laguna Paron in the Callejon de Huaylas. There were further concerns about temperature changes and the appearance of new pests and illnesses, such as has been documented in Cusco. There is high appreciation of the significance of climate change relative to water resources and accessibility, in particular, regarding seasonal availability in the context of glacial retreat. However, so far, the National Institute for Civil Defense (INDECI), the principal disaster response agency, has not moved significantly toward integrating climate change risks into its activities. On the other hand, there has been some significant legislative progress toward integrating disaster risk reduction into their portfolio with the passage of Law 29644 which created the National System of Disaster Risk Management (SINAGERD) in May of 2011.

The Ministry of the Environment, in cooperation with the Swiss Agency for Development and Cooperation (COSUDE), has developed the Programme of Adaptation to Climate Change (PACC), designed to assist the poorest highland populations in dealing with climate change. PACC promotes projects of climate change adaptation through municipal governments (PACCPERU 2009). The programme works in four areas: investigation and diagnosis of vulnerability and conditions of adaptation, including risk analysis, risk perception, and traditional adaptive strategies; monitoring and information; support and advice on local climate change adaptation projects; and incorporating climate change adaptation into local development agendas. Currently PACC is operating principally in the regions of Cusco and Apurimac. The programme has also published pamphlets for local use on climate change, water and climate change, food security and climate change and has planned one on climate change and risk management (PACCPERU 2009).

At the regional level, there is also a wide array of governmental organizations, each with their own specialization, that deal with issues of climate change. Their information and expertise tends to be of a technical nature with considerably less attention to social, cultural and economic concerns. They generally work in isolated fashion with few integrated approaches to the issue. For all these institutions the common denominator of climate change is global warming and they are well aware that Peru is the third most vulnerable nation in the world to climate change and water availability.

There is also a wide array of Non-Governmental Organizations (NGOs), each with their own specialization, dealing with issues of climate change. Several of them, such as Oxfam and Practical Action, work specifically on the issue of disaster risk reduction and climate change. Oxfam's project focuses on reducing risk from flash flooding by organizing community warning systems and improving housing (Oxfam 2011). Practical Action is focusing on flood control from avalanches into glacial lakes and improving irrigation systems to improve resilience to future water scarcities (Practical Action 2011).

Concerns among regional authorities in the Callejon de Huaylas focused on climate change effects on tourism, third in importance behind agriculture and mining in the economy of the region. Tourism is seen as diminishing because of increased risks of ascending melting glaciers and glacial retreat. Some tourism businesses have closed down and their personnel have left the region. Of greater concern is the issue of water. Additional concerns focused on the communities of the Cordillera Negra, the drier western range, which are losing people because of diminished supply of water. The majority of these migrants settle on the outskirts of the capital city of Huaraz. Their communities of origin are becoming known as the "towns of the padlocks" because of the number of empty houses. Regional authorities are also concerned about the recent fall of tons of ice from a destabilised melting glacier into a lake above Carhuaz that created an avalanche of water and mud, destroying houses, fields, trails and roads as it descended the hillside. In 1941, 1964 and 1970 similar, although much larger, events killed 4,000 people in the city of Huaraz, 3,000 in Ranrahirca and 4,500 in Yungay (Oliver-Smith 1992; Carey 2010).

## 4.8 Adaptation to Climate Change and Disaster Risk Reduction in the Andes

Adaptation to environmental change has now become a major focus of research in Peru at the national, regional and local levels (Trigoso Rubio 2007; Gallardo et al. 2008), although policy responses and concrete actions are hindered by a lack of funding and institutional inadequacies that inhibit the effective delivery of state services and functions across a wide spectrum of sectors. At the broadest level of analysis, at which the systemic features driving vulnerability to hazards exist, national institutional goals are consistent with international perspectives that frame the necessary tasks to reduce disaster risk and enhance climate change adaptation. These include: (1) tackling the changing nature of disaster risks and uncertainties, (2) enhancing adaptive capacity of local and regional networks and policies and (3) addressing poverty and vulnerability and their structural causes (Mitchell et al. 2010).

Climate change in the Andes promises a relatively wide array of fast and slow onset disasters. In the case of slow onset processes, such as temperature change, seasonal unpredictability, or changes in rainfall regime, adaptation (sometimes in the form of mitigation strategies) will most likely be the mode of response, such as adoption of temperature appropriate crops, improved health care for plants, animals and humans to handle new disease vectors, or water storage technology designed to address water scarcity. Strategies for adaptation for rural peoples should also entail support for enhancing food security through diversified production based on agro-ecological approaches and techniques rather than market centered mono-cropping (Oxfam 2009).

In some cases, changing climate conditions, particularly excessive rainfall and glacial melt may destabilise both glacial and hillside terrain resulting in sudden onset landslide or flood disasters which require mitigation to reduce risk. However, since these sudden onset disasters are not uncommon in the Andes, as the recent catastrophic floods in the Vilcanota-Urubamba watershed in 2010 attest, climate change may make them larger and more frequent. Indeed, as in the previously mentioned incident in Carhuaz, a chunk of ice from a climate change destabilised glacier fell into the lake below it, breached 23 m high levees and sent a muddy wave down the canyon above the city. One person was killed and 50 houses and a water processing plant serving 60,000 people were destroyed.

In the case of sudden onset disaster processes, climate change adaptation will more closely resemble traditional disaster risk reduction strategies involving soil stabilization measures, community relocation, or improved damming or lake level reduction. In addition, development oriented measures aimed at mitigating the underlying structurally based vulnerabilities of the highland populations, such as increasing food security, improving land use and planning and improving governance, will also lead to reducing exposure to both slow and rapid onset climate hazards. Overall, Andean climate change adaptations must improve management capabilities of those organizations and institutions addressing both current slow and fast onset as well as future climate risks. Strategies, policies and investments undertaken at the national level should also focus on providing assistance to local communities in developing responses according to local capacities and priorities (Sperling 2008, p. 21).

At the national and some regional levels authorities are clearly focusing on the issue of drought. Various strategies ranging from alternative crops to large scale dam and irrigation projects are being developed to cope with the challenges represented by climate change. To some extent, regional authorities in the areas studied reflect concerns that have been communicated by local people and authorities, although local authorities express a more urgent need for risk mitigation. Local authorities stress the need to build dams and reservoirs of various sizes to store water in the face of uneven and inconsistent, and most especially, unpredictable rainfall. Local authorities and populations, as might be expected, also tend to focus on developing responses to the problems of inconsistent seasonal variation, the increase in pests and diseases and the abrupt diurnal shifts between extreme heat and extreme cold. Some local authorities on the altiplano stressed the need for improved approaches to irrigation systems to contend with water scarcity. They also felt that the health centers need to develop strategies to protect the vulnerable elderly and children from the intense nighttime cold that, according to them, has produced much higher incidences of bronco-pulmonary diseases. Improved veterinary care for the high altitude pastoral economy was also needed.

Government officials at the local level are also promoting locally specific adaptations to assist people in their efforts to adapt to climate change. Small dams and reservoirs to help people cope with water scarcity are seen as essential by local authorities. Municipal leaders spoke of the urgent need to collect and store water from the rainy season to cope with extended dry periods. The development of alternative or improved strains of crops that can withstand the temperature and rainfall changes is also given high priority. Municipal leaders also spoke of the importance of health programmes to assist people to adapt to the extreme night time cold.

In many respects, the concerns of local authorities are a reflection of needs expressed by their constituencies. In local communities, people articulate very clearly the need to develop adaptations to the changing conditions. Dams, reservoirs and irrigation to collect, store and distribute water are a very high priority for local communities as are alternative or improved crop strains. People spoke of the need to protect the young and the elderly from the nighttime cold. Pastoralists at high altitudes, fearing for the health and survival of their herds, expressed the need for better veterinary services, specifically antibiotics to help combat the illnesses that are appearing in their herds. Particularly vulnerable are the young animals who succumb easily to cold temperatures. In some communities, herders are wrapping the young animals in blankets to help them survive the cold. Others spoke of the need to improve homes and to develop stables to protect humans and animals from the cold. Generally, the Peruvian highlanders interviewed feared that they might be forced to change livelihoods to adapt to the changing conditions.

In that context, migration is seen by some to be a possible option. Migration may be a coping response or an adaptation to events and processes aggravated by climate change. In Peru the latter half of the twentieth century witnessed a massive rural-urban migration, principally to Lima. Peru is now a predominantly urban nation with over a third of the population residing in Lima. However, over the last decade there has been significant growth of secondary cities. Twenty-one cities now have a population of 100,000 or more. Rural migration increased the urban population from 35.4 % of the total population in 1940 to an estimated 74.6 % as of 2005 (State Department 2011). While much of this growth of secondary cities has been on the coast, highland regional capitals have also grown, although often migrants treat these cities as temporary way stations on the journey to larger coastal cities.

Nonetheless, the growing population of highland cities, often in informal and unplanned ways, has placed more people in harm's way from hazards exacerbated by climate change effects. Deglaciation has created short term rapid onset risks for urban people residing on river banks and unstable hillsides and longer term risks of water scarcity threaten not just highland populations but the nation as a whole. The increasing demand for water from growing highland cities may threaten that resource for farmers (Young and Lipton 2006). The nearly nine million people of Lima depend on the highland lakes for almost half their water. It is in this context that climate change caused water shortages in the highlands and may result in adaptations aimed at retaining and storing water which in turn will seriously impact the coast that depends on the highlands for its water supply.

Although there seem to be relatively few cases of environmentally driven migration to date, the potential in the future for such displacement is significant. For the high altitude pastoral communities, further losses in herd size could leave people with few resources on which to subsist. Agriculture at those elevations is marginal at best, and usually dedicated to producing fodder for animals. The endangerment of herd reproduction caused by excessive night time freezes and the loss of pasture by searing daytime temperatures constitutes a double-sided threat to the existence of these communities. Barring some form of livelihood substitution for pastoralism, mariculture (trout farming) has been suggested to people in these high altitude communities, or they may have little recourse but to migrate in the face of continuing environmental stressors.

In addition, small-holder farmers in the intermediate zones of the region face continuing water scarcity due to glacial retreat and unstable rainfall regimes that may stimulate migration in the future as well. For example, communities in the Cordillera Blanca normally enjoy a year round supply of water provided by glacial melt and the rainy season and is therefore more highly populated than the drier Cordillera Negra. If glacial retreat continues at current rates, water supplied by glacial melt during the dry season will eventually diminish, essentially reducing the agricultural cycle by half, unless measures for the collection and storing of water are installed. Such a reduction in water will impair the capacity of local agriculture to support the denser populations of the Cordillera Blanca and may trigger significant migration.

## 4.9 Conclusion

Three basic interconnected conditions make disaster risk reduction and climate change adaptation absolutely essential and urgent in highland Peru: the sensitivity of mountain ecosystems to climate change; Peru is assessed as the world's third most vulnerable country to climate change; and social, economic and political structural disadvantages have rendered the highland populations of Peru extremely vulnerable to both the systemic hazards of their environment and to climate change. Climate change is a dynamic process whose effects will make people more vulnerable to existing hazards as well as present them with new challenges as it advances.

As an aspect of disaster risk reduction, climate change adaptation must address the systemic vulnerabilities of the society as well as the specific hazards. Strategies, policies and investments must be implemented at the national level with the goal of supporting local level efforts to respond according to local capacities and goals (Sperling 2008, p. 21). It is clear that local institutions will play a central role in most adaptation to climate change within the context of overall disaster risk reduction strategies (Agrawal 2008). Indeed, although technological and infrastructural measures may help to reduce vulnerabilities and enhance adaptive capacity, these innovations will be energised, directed and implemented through local social and cultural institutions and values toward strategies most appropriate to address the needs of communities (Smith and Lenhart 1996). Given that climate change is estimated to increase poverty in Peru, and therefore, in most cases accentuate the vulnerability of the poor (Anderson et al. 2009, p. 22), the most effective approach for both climate change adaptation and disaster risk reduction must address the systemic vulnerabilities embodied in structural inequality of the society. Locally specific disasters and global climate change take place in the context of local vulnerability. Any attempt to address the risks of disasters and climate change that does not put that condition at the forefront, will reduce its chances of any durable success.

## References

Anderson, L. E. (2009). *Social impacts of climate change in Peru*. Policy Research Working Paper 5091, Sustainable Development Department, World Bank.

Agrawal, A. (2008). *The role of local institutions in adaptation to climate change*. Paper prepared for the Social Dimensions of Climate Change, Social Development Department, The World Bank, Washington, DC (March 5–6).

Beniston, M. (2003). Climatic change in mountain regions: A review of possible impacts. *Climatic Change, 59*, 5–31.

Bennett, J. (1996). *Human ecology as human behavior*. New Brunswick: Transaction Publishers.

Birkmann, J., & von Teichman, K. (2009). *Addressing the challenge: Recommendations and quality criteria for linking disaster risk reduction and adaptation to climate change*. Bonn: DKKV Publication Series 38. (In J. Birkmann, G. Tetzlaff, & K.-O. Zentel (Eds.)).

CAN (Comunidad Andina). (2008). El cambio climático no tiene fronteras. Impacto del cambio climático en la Comunidad Andina. http://www.comunidadandina.org/public/libro_84.htm. Accessed 25 June 2012.

Carey, M. (2005). Living and dying with glaciers: Peoples' historical vulnerability to avalanches and outburst floods in Peru. *Global and Planetary Change, 47*, 122–134.

Carey, M. (2010). *In the shadow of melting glaciers: Climate change and Andean society*. Oxford: Oxford University Press.

Caviedes, C. (1981). Natural hazards in Latin America: A survey and discussion. In T. L. Martinson & G. S. Elbow (Eds.), *Geographic research in Latin America: Benchmark 1980, Conference of Latin Americanist Geographers* (pp. 280–294). Indiana: Ball State University, Muncie.

Cook, N. D. (1981). *Demographic collapse: Indian Peru*. Cambridge University Press: Cambridge.

Dunbar, K. W. (2010). *Perceptions and realities of changing water availability: Responding to climate change in the Peruvian Andes*. Paper presented at the Annual Meeting of the American Anthropological Association. New Orleans, LA.

Espinoza Villar, J. C., Ronchail, J., Guyot, J. L., Cochonneau, G., Naziano, F., Lavado, W., de Oliveira, E., Pombosa, R., & Vauchel, P. (2009). *Spatio-temporal rainfall variability in the Amazon basin countries (Brazil, Peru, Bolivia, Colombia, and Ecuador). International Journal of Climatology, 29*(11), 1574–1594.

Gallardo, M., Gomez, A., Torres, J., Walter, A. (2008). *Directorio Nacional Cambio Climático en el Peru: Instituciones, investigadores, políticas, programas y recopilación bibliográfica. Primera aproximación*. Lima: Soluciones Prácticas-ITDG.

Giesecke, A., & Silgado, E. (1981). *Terremotos en el Peru*. Lima: Ediciones Richkay.

Gobel, B. (2008). Dangers, experience and luck: Living with uncertainty in the Andes. In M. J. Casimir (Ed.) *Culture and the changing environment: Uncertainty, cognition and risk management in cross-cultural perspective*. New York: Berghahn Books.

Holling, C. S. (1994). An ecologist view of the Malthusian conflict. In K. Lindahl-Kiessling & H. Landberg (Eds.), *Population, economic development, and the environment*. New York: Oxford University Press.

Instituto Nacional de Estadistica e Informacion (INEI). (2009). *Estado de la Poblacion Peruana*. Fondo de Poblacion de las Naciones Unidas: Lima.

Kelly, A. (2010). Peru's mountain people face fight for survival in a bitter winter, The Guardian (3 Jan.). http://www.guardian.co.uk/world/2010/jan/03/peru-mountain-farmers-winter-cold/print. Accessed 25 June 2012.

Kelman, I., & Gaillard J. C. (2010). Embedding climate change adaptation within disaster risk reduction. In R. Shaw, J. M. Pulhin, & J. J. Pereira (Eds.), *Climate change adaptation and disaster risk reduction: Issues and challenges* (Chap. 2, pp. 23–46). Community, Environment and Disaster Risk Management, Emerald Group Publishing Limited: Bingley.

Kraft, K. (1995). *Andean fields and fallow: communal land management under pressure for land use intensification*. Unpublished PhD. Dissertation, Department of Anthropology, University of Florida.

Lavell, A. (2011). Unpacking climate change adaptation and disaster risk management: Searching for the links and the differences: A conceptual and epistemological critique and proposal. IUCN-FLACSO Project on Climate Change Adaptation and Disaster Risk Reduction.
Leavell, D. N. (2008). The consequences of climate change for the water resources of Peru. Proceedings of the XIII World Water Congress, International Water Resources Association, Montpelier, France.
Macchi, M. (2010). *Mountains of the world-ecosystem services in a time of global and climate change: Seizing opportunities-meeting challenges*. International Centre for Integrated Mountain Development (ICIMOD), Kathmandu.
Magrin, G., García, G., Cruz Choque, D., Giménez, J. C., Moreno, A. R., Nagy, G. J., Nobre, C., & Villamizar, A. (2007). Contribution of working group II to the fourth assessment report of the intergovernmental panel on climate change. In M. L. Parry, O. F. Canziani, J. P. Palutikof, P. J. van der Linden, & C. E. Hanson (Eds.), *Latin America. Climate change 2007: Impacts, adaptation and vulnerability* (pp. 581–615). Cambridge: Cambridge University Press.
MINAM. (2010). Segunda Comunicación Nacional del Perú a la Convencion Marco de las Naciones Unidas sobre Cambio Climático.
Mitchell, T., Ibrahim, M., Harris, K., Hedger, M., Polack, E., Ahmed, A., Hall, N., Hawrylyshyn, K, Nightingale, K., Onyango, M., Adow, M., Phukan, I., & Sajjid Mohammed. S. (2010). *Climate smart disaster risk management, strengthening climate resilience*. Institute for Development Studies: Brighton.
Moseley, M. (2001). *The Incas and their ancestors: The archaeology of Peru*. New York: Thames & Hudson.
Murra, J. V. (1972). El control vertical de un maximo de pisos ecologicos en al economica de las sociedades andinas. In I. Ortiz de Zuñiga (Ed.), *Visita de la provincia de Leon de Huanuco: Tomo II* (pp. 429–476). Huanuco: Universidad Hermilio Valdizan.
Murra, J. V. (1980). The economic organization of the Inca State. Greenwich: JAI Press.
Oliver-Smith, A. (1992). *The Martyred city: Death and rebirth in the Andes*. Prospect Heights: Waveland Press.
Oliver-Smith, A. (1994). Peru's five hundred year earthquake: Vulnerability to hazard in historical context. In A. Varley (Ed.), *Disasters, development and environment* (pp. 31–48). London: Wiley.
Oliver-Smith, A. (1998). Global challenges and the definition of disaster. In E. L. Quarantelli (Ed.), *What is a disaster? A dozen perspectives on the question* (pp. 177–195). New York: Routledge.
Oliver-Smith, A. (1999). Peru's five hundred year earthquake: Vulnerability in historical perspective. In A. Oliver-Smith & S. M. Hoffman (Eds.), *The angry earth: Disaster in anthropological perspective* (pp. 74–88). New York: Routledge.
Oliver-Smith, A. (2002). Theorizing disasters: Nature, culture, power. In S. M. Hoffman & A. Oliver-Smith (Eds.). *Culture and catastrophe: The anthropology of disaster*. Santa Fe: The School of American Research Press.
Orlove, B. (2009). The past, the present and some possible futures of adaptation. In W. Neil Adger, I. Lorenzoni, & K. O'Brien (Eds.), *Adapting to climate change: thresholds, values, governance*. Cambridge University: Cambridge University Press.
Orlove, B. (2010). Glacier retreat: Reviewing the limits of human adaptation to climate change. *Environment, 51*(3), 22–34.
Oxfam. (2009). *People centered resilience: Working with vulnerable farmers towards climate change adaptation and food security*. Oxfam Briefing Paper 135. Oxfam International.
Oxfam. (2010). Environmental change, climate change and migration in Latin America, Final Report UNU-EHS, IOM and Oxfam.
PACCPERU. (2009). Adapting to climate change in the Andes. http://www.adaptationlearning.net/sites/default/files/boletinPACC-COP15-ingles.pdf. Accessed 28 June 2012.
Practical Action. (2011). Disaster preparedness and climate change. http://practicalaction.org/ancash-peru.
Quarantelli, E. L. (1991). *More and worse disasters in the future*. Paper presented at the UCLA International Conference on the impact of natural disasters: Agenda for future Action, Los Angeles, California, on 10–12 July 1991.

Renaud, F., Bogardi, J, Dun, O., & Warner, K. (2007). Control, adapt or flee: How to face environmental migration? *Intersections No. 5*. Bonn: United Nations University Institute for Environment and Human Security.
Roche, M. A., Bourges, J., Cortes, J., & Mattos, R. (1992). Climatology and hydrology of the lake Titicaca Basin. In C. Dejoux & A. Iltis (Eds.), *Lake Titicaca: A synthesis of limnological knowledge* (pp. 63–88). Boston: Kluwer Academic. (H. J. Dumont & Werger, M. J. A., *Monographiae Biologicae*, *68*).
Rural Poverty Portal. (2010). http://www.ruralpovertyportal.org/web/guest/country/home/tags/peru. Accessed 1 Dec 2011.
Smit, B., & Wandel, J. (2006) Adaptation, adaptive capacity and vulnerability. *Global Environmental Change*, *16*(3), 282–292.
Smith, J. B., & Lenhart, S. S. (1996) Climate change adaptation policy options. *Climate Research*, *6*(2), 193–201.
Sperling, F., Valdivia, C., Quiroz, R., Valdivia, R., Angulo, L., Seimon, A., & Noble, I. (2008). *Transitioning to climate resilient development: Perspectives from communities in Peru*. Washington, DC: Environment Department Papers #115 (Climate Change Series) World Bank.
State Department 2011-Peru. http://www.state.gov/r/pa/ei/bgn/35762.htm. Accessed 30 June 2012.
Trigoso Rubio, E. (2007) Climate change impacts and adaptation in Peru: The case of Puno and Piura. Human Development Report 2007/2008 Human Development Report Office Occasional Paper. UNDP. http://hdr.undp.org/reports/global/hdr2007–8/papers/Trigoso_Erika.pdf.
UNISDR. (2011). Global assessment report on disaster risk reduction. Geneva: United Nations Strategy for Disaster Risk Reduction.
Vergara, W. (2008). *World Bank climate change program in Peru*. Washington, DC: The World Bank.
Winterhalder, B., & Thomas, R. B. (1978). Geoecology of Southern Highland Peru. Occasional Paper No. 27, Insitute of Arctic and Alpine Research. University of Colorado, Boulder.
Wisner, B., Blaikie, P., Cannon, T., & Davis, I. (2004) *At risk: Natural hazards, people's vulnerability and disasters* (2nd ed). New York: Routledge.
Young, K., & Lipton, J. (2006) Adaptive governance and climate change in the Tropical Highlands of Western South America. *Climate Change*, *78*, 63–102.

# Chapter 5
# Firm Foundations or Castles on Sand? The Shifting Sources of Flood Risk and the Implications for Flood Governance: An English Case Study

**Iain White**

**Abstract** This chapter will provide an overview and analysis of the experiences of flooding in England since the end of the twentieth century with a view to informing debate in other countries. A critical reflection on the events of the last decade is particularly illuminating; encompassing a complete readjustment of our understanding of the sources of risk and subsequently exposing deficiencies in the ability of related governance frameworks to respond. The response to a series of damaging events from the scientific and policy making community was relatively swift, and included a significant change to the dominant paradigm from flood defence to flood risk management. This fundamental transition did, however, lead to a cascading series of interrelated governance implications and the development of new socio-technical assemblages, some of which were easier to anticipate than others. The effects encompassed alterations to the related methodology and a more neoliberal approach to risk management with new responsibilities distributed amongst a wider array of people and professions. The shift of the main source of flood risk from the rivers and sea towards surface water and drains was sudden and largely driven by forcing trends in climate and urbanization, creating the potential for lessons to be passed on to other countries who may experience similar pressures in the future. A key finding was inadequate flood governance which struggled to adapt to the changing sources of flood risk.

**Keywords** Flooding · Surface water flooding · Risk management · Governance · Spatial planning

## 5.1 Introduction

If one thinks about flooding, what perceptions and images does it conjure? For most, if not all, it will be related to the sea or rivers being driven by powerful natural forces creating the potential to detrimentally impact people and places. Similarly,

---

I. White (✉)
University of Waikato, Hamilton, New Zealand
e-mail: iainw@waikato.ac.nz

B. C. Glavovic, G. P. Smith (eds.), *Adapting to Climate Change,* Environmental Hazards,
DOI 10.1007/978-94-017-8631-7_5, 

if one considers how to manage flood risk, attention will logically focus on the provision of hard defences with a view to 'holding back' water. This chapter aims to confront these deeply ingrained perceptions by providing an insight into how a fuller understanding of the shifting sources of flooding in England has presented a fundamental challenge to understanding the risk, the most appropriate governance frameworks and the efficacy of the selected managerial responses; an experience that should provide valuable lessons for other countries.

An accurate knowledge of the extent of the differing *sources* of risk is critical—not only may they demand completely different mitigation and adaptation measures, but, as this chapter will explain, they may operate outside of traditional scientific, financial and managerial frameworks, which have been mainly designed for structural responses. For example, defending the line against storm surges presents a very different proposition to addressing emerging threats generated from an excess of surface water in urban areas. Perhaps surprisingly, it is the latter that is now by far the biggest source of flood risk in England; a change that has only recently been recognised and its implications are still being considered within academia and practice. The effects of this shift in flood sources should not be underestimated—as it challenges hundreds of years of perception underpinned by supportive and interconnected governance paradigms, methodologies, professions and financial systems.

This chapter will firstly outline the logical reasons for developing the conventional approach to flood defence before analysing why it became subject to such rapid and sustained critique. It will then discuss the implications for governance that this shift in the sources of flood risk presented. Significantly, the potential forcing effects of climate change and urbanization on surface runoff means that the debate and inherent managerial difficulties experienced within the highly developed built environment in England since the turn of the twenty-first century may provide valuable lessons for other countries. As will be explained in the following section, there has been a tendency to first experience detriment before action occurs, but this reactive approach does not necessarily have to be replicated elsewhere. This does, however, entail critical reflection upon the limits to knowledge, engaging with uncertainty and re-examining systems of governance—all of which are currently in process within England.

## 5.2 Firm Foundations or Castles on Sand?

The United Kingdom (UK) is subject to both capricious seas and volatile weather patterns, which has contributed to the view that we have more 'weather' than most countries; with its sheer changeability leading to the well worn truism regarding it being a peculiarly British obsession. Like many aphorisms, there is more than a kernel of truth in this assertion and one unexpected benefit of this national fixation is a rich source of flooding records stretching back over 1,000 years.

Within his eighth century *Ecclesiastical History of the English People* the Venerable Bede, a monastic scholar, emphasised the geographical vulnerability of the country to the extremes of weather, detailing amongst other factors the effect of the seasons and the behaviour of rivers and the sea (White 2010). A further, more vivid example is provided by the *Anglo Saxon Chronicle*, which provides an important account of British history from the Middle Ages up to 1154 (Britannia 2007). Flooding events are recorded frequently throughout the document, for example one of the notable events of 1014 was described as:

> This year, on the eve of St. Michael's day, came the great sea-flood, which spread wide over this land, and ran so far up as it never did before, overwhelming many towns, and an innumerable multitude of people.

In addition to coastal flooding, there is also a long record of flood events with regard to individual watercourses. The river Severn, one of the more notoriously flood prone watercourses in England, has been recorded as flooded 697 times since the first incident was noted in 1236 (Law et al. 2010). The view that inundation is driven from watercourses and the sea is part of the national psyche in many countries beyond England, and, as we can see, there is usually good historical evidence why this should be the case. The reasons for the perception of flooding as being directly related to these sources are therefore understandable, having their roots in cultural history and harsh experience.

Considering this viewpoint, the approaches to managing flood water also developed logically, focused on increasing the defence against the undesirable intrusion of water onto land. A discrete and conveniently spatial risk, such as from these clearly definable sources, could therefore be effectively addressed via a strong wall alongside the sea or river. Similarly, an excess of precipitation in an urban area could be managed by an engineered drainage network designed to efficiently transport water to safe outfalls. In order to operationalise this approach, professions were gradually developed to specialise in better understanding the natural world, ascertaining probability to risks and implementing these engineering-led solutions. Over time, institutions and governance frameworks were also established to support flood defence, with responsibilities allocated to dedicated expert agencies. When considering the growth in knowledge, scientific expertise, finances and technology for much of the last century there is little doubt that great progress has been made in protecting people and property from the risk of flooding, and the success of this approach has helped it permeate across the globe.

Yet, given these advances one could be forgiven for assuming that flooding should be lessening, its risk increasingly understood and the threats better managed. This view is, however, far removed from reality. Although the science and technologies used to control water in England gained in sophistication, the actual experience of flooding *increased*—and not just in isolated areas, but in a consistent manner across the globe. While questions may be legitimately raised concerning population growth and the volume of people exposed, managerial strategies, particularly those in more technologically and scientifically advanced nations, should still be able to cope with long term, predictable social and land use trends. This is an important point to make, and provides the initial challenge to our perception of both the causes of flooding and the subsequent efficacy of the managerial response. If societies are so adept at managing these events why do they keep experiencing severe and highly damaging floods? And why do they appear to be increasing in frequency?

A common problem associated with policies to hold back water are that they may also increase risk via the 'escalator effect' (Parker 1995) or 'safe development paradox' (Burby 2006), whereby the provision of flood defences may make the land behind them appear more attractive for development. Allied to this is the driving role played by climate change and urbanization trends. However, given the progress in many fields over the twentieth century this chapter argues that it is also due to the highly imperfect understanding of the *shifting* nature of flood risk. In short, one of the reasons why flooding has continued to occur is that despite many perceptions it is not *one* homogeneous risk; a simple inundation of land may be the common result, but in practice there are an array of distinct sources, each subject to differing drivers which are not analogous. In practice, flood risk comprises a number of spatially variable sources; from the well-understood threats emerging from rivers and the sea to emerging risks from urban runoff, infrastructure failure and rising groundwater levels—each of which may demand differing adaptation strategies. The modern pluralistic nature of the threat challenged the long-held narrow hegemony of perceptions, governance processes and appropriate intervention measures in England, and deserves equal consideration elsewhere.

## 5.3 Shifting Sources: Global, National and City Perspectives

Perhaps counter intuitively, given the rise in knowledge concerning flood defence, countries all over the world are becoming increasingly subject to flooding, making it one of the most frequent and widespread natural hazards. The second half of the twentieth century witnessed a rising number of flood events worldwide; a general trend being experienced on all continents, regardless of how advanced they may be (White 2010). If technology and science was so effective one would expect that these advantages would be reflected spatially, yet from a global perspective floods are well distributed. The uniform nature of the rise in flood events provides a challenge to how this hazard has been addressed, questioning the long-term efficacy of the dominant flood defence approach that has been applied on an international basis.

Although this information provides an argument for more effective intervention, there are a number of caveats. The first concerns the theory of time-space compression (Harvey 1989). Here, it is a feature of modernity that time appears to accelerate and and space appears compressed—or put more simply, technology makes it easier to quickly document events, regardless of where in the world they may occur. From this perspective, contemporary societies may seemingly experience a rise in the recording of floods in comparison with less information-rich times. Secondly, there may be disparities in defining what exactly constitutes a flood particularly when comparing records collated from differing spatial and temporal sources. Is it a simple inundation of water onto land? Do people and properties have to be affected? If so, how many would be required to turn a natural event into a disaster?

While we should problematise the argument for a long-term and global rising trend in flood events, there is greater confidence at the national scale. From an English perspective, recent data clearly shows an increasing number of damaging floods, particularly since 1998, thereby helping to counter the possible obfuscatory effects of space and time. These flood events resulted in three long established perceptions being confronted: the first concerning the actual sources of flood risk; the second relating to how floods have been governed and the closely related third, regarding the most effective intervention approaches to be adopted. During the transition from the late twentieth century to the early twenty-first these long held views were gradually undermined, and then with a startling swiftness, almost completely reconstructed. And crucially, it was driven by a repeated series of extremely damaging national-scale flood events—most notably in 1998, 2000, 2004 and 2007, supported by more localised, but very high profile incidents occurring in areas such as Boscastle, Carlisle and Cockermouth.

In the aftermath of each serious flood incident, reports were commissioned which were not only helpful in determining the issues connected with a specific event, but from a recent historical perspective taken together provide an insight into the gradually changing awareness of the nature of risks and its rapid and significant impact on the policy narrative. While each new flood event can prove devastating for the communities affected, it also presents an opportunity to cross reference existing information on risk and help provide additional data on ascertaining a truer picture of the sources of, and exposure to, inundation. The lessons of the last decade demonstrate how the agenda has quickly developed. For example, the report investigating the Easter 1998 flood exposed deficiencies in areas such as forecasting, warning, emergency response, standards of defense, severity assessment and management structures and skills (Bye and Horner 1998). Yet, it was the cumulative effect of the more widespread and severe Autumn 2000 flood suffered just two years later that Deputy Prime Minister John Prescott described as a 'wake-up call'. Experiencing two huge events in quick succession that in some geographical areas the science estimated should only occur at least once in every 150 or 200 years inevitably casts doubt on the entire flood management process—particularly the methodologies associated with forecasting, prediction and exposure.

In the aftermath of the 2000 flood, measures were taken to improve these issues with, for example, a review determining that 1,724,225 properties were at risk of fluvial, tidal and coastal flooding (National Audit Office 2001). During this period there was no real acknowledgement of *other* sources of flood risk, which provides an interesting snapshot of just how far, and fast, flood risk management has progressed in a decade. A further step forward occurred a few years later with the publication of the Foresight Future Flooding report (Evans et al. 2004). This wide ranging document provided the scientific evidence base to drive the significant policy shift from flood defence towards flood risk management discussed in more depth later. Although at this time around two million homes were again identified as being at risk from flooding from rivers and the sea, climate change and urbanization were highlighted as powerful forcing drivers with exposure from surface water (incorporating urban runoff and local drainage failure) predicted to rise sharply from the highly uncertain level detailed in the report of around 80,000 properties.

The need not just for better information but more sharing of data on this source was also now becoming recognised amongst stakeholders. For example, figures on drainage were not held by the same agency as conventional sources of flooding and were not even within the public sector; the privatisation of the water industry in 1989 created ten separate companies each holding what may be commercially sensitive information. Although there was a suggestion in the Foresight report that the number of properties exposed to surface water flooding was almost certainly underestimated, it is doubtful that even the most pessimistic commentator would have predicted the extent to which this would prove the case.

In a related development, longer term and more sustainable water management methods began to influence the policy arena, underpinned by a move towards risk based approaches supported by scientific evidence. The ability to provide a consistently successful *defence* against flooding was now openly questioned, with a new pragmatism extolled and, despite the recent advances in knowledge and management, the message was that society should expect to experience periodic flooding, be prepared to 'live with water' and develop new working partnerships (White 2013). Notwithstanding this newfound realism, the rapidly developing scientific evidence base, supported by national, regional and local policy initiatives, led to a growing sense that although modern flood risk was more unpredictable and complex than in the past, society was beginning to master some of the intricacies. For example, flood risk maps were now available online and as part of the planning process new developments may be subject to Flood Risk Assessments, incorporating consideration of modeling and downstream impacts (Department of Communities and Local Government 2006).

However, the summer 2007 flood inundated whole areas that were previously deemed to be 'safe'. This is a key point to pause upon. While the transition from defence to risk management appears a logical response, how successful can a risk based approach be if it is so difficult to be accurate? And to what extent is the uncertainty inherent in this strategy understood and communicated to decision-makers? This wasn't just a matter of a few degrees of probability, or a difference between a medium or high risk but a failure to recognise whole swathes of the country as actually being exposed to flooding. As a result there was an increasing recognition that to date scientists and policy-makers had concentrated on compiling information on the risk from coastal, estuarine or fluvial sources and there was a gap in knowledge concerning the extent of flooding from elsewhere, in particular from surface water and inadequate drainage (Pitt 2008).

The most current data has been revised upward to suggest that 5.2 million properties in England, or one in six of the total housing stock, are now at risk from flooding, with 3.8 million of those newly recognised as being exposed to inundation from surface water (Environment Agency 2009). The data continues to evolve with

**Table 5.1** The changing knowledge of the sources of flood exposure in England between 2001 and 2011. (Source: Data from National Audit Office 2001; Evans et al. 2004; Jacobs Engineering UK 2004; Environment Agency 2009, 2011)

| Year | Estimated properties at risk by source | | | | Total |
|---|---|---|---|---|---|
| | Rivers and sea | Surface water | Groundwater | Reservoir failure | |
| 2001 | 1,724,225 | 0 | 0 | 0 | 1,724,225 |
| 2004 | 1,740,000 | 80,000 | 1,700,000 | 0 | 3,420,000[a] |
| 2009 | 2,400,000 | 3,800,000 | 1,700,000 | 0 | 6,800,000[b] |
| 2011 | 2,400,000 | 3,800,000 | 1,700,000 | 1,100,000 | 7,900,000[b] |

[a]Note that 112,855 properties are at risk from both rivers and the sea and groundwater flooding. (Jacobs Engineering 2004)

[b]Note that 112,855 properties are at risk from both rivers and the sea and groundwater flooding, and 1 million properties are at risk from both rivers and the sea and surface water flooding. (Jacobs Engineering 2004; Environment Agency 2009)

one recent report adding an estimated 1,100,000 properties threatened by reservoir failure (Environment Agency 2011), a new risk which nearly occurred in 2007 at the Ulley reservoir in South Yorkshire after heavy rainfall. Further, this exposure will continue to rise even if all unsafe construction stopped tomorrow, due to a gradual, incremental rise in urbanization elsewhere in the catchment and possible increases in rainfall intensity due to climate change. Indeed, it has been estimated as 'very likely' global anthropogenic greenhouse gas emissions substantially increased the risk of the Autumn 2000 event occurring (Pall et al. 2011).

Flooding from surface water is now identified as *the* main source of flood risk in the country. While changes in the definitions of 'risk' can vary slightly over time and between documents, Table 5.1 displays an accessible comparative overview of how quickly knowledge has changed regarding both the estimated number of people at risk and from which source, encapsulating data from advances in ascertaining exposure and the reviews of recent flood events. The table is designed to serve as a didactic device; not necessarily with regard to attaching significance to the number of properties, but rather as a demonstration of the way that seemingly firm foundations can quickly be undermined and how academics and policy-makers in countries outside England should learn from this experience and question what may ostensibly appear to be sound scientific datasets.

Within the space of 10 years the overall estimated exposure to flood risk in England has more than quadrupled, with surface water and drainage emerging from being completely unacknowledged to accounting for around half of the total risk. While reflecting on the rapidly changing data compiled during this time, the speed of development does inevitably expose flaws in both the evidence base and governance of flooding. However, the relatively quick response of the scientific and policy making community to identify, and adjust to, new and emerging risks should also be recognised. The shifting sands of what were thought to be relatively firm foundations in such a short space of time exposed, perhaps understandable, deficiencies in fundamental aspects of flood risk management; knowledge gaps that had inhibited the ability to respond to *all* sources of flood risk. The financial and governance focus on structural defences merely addressed one part of the picture,

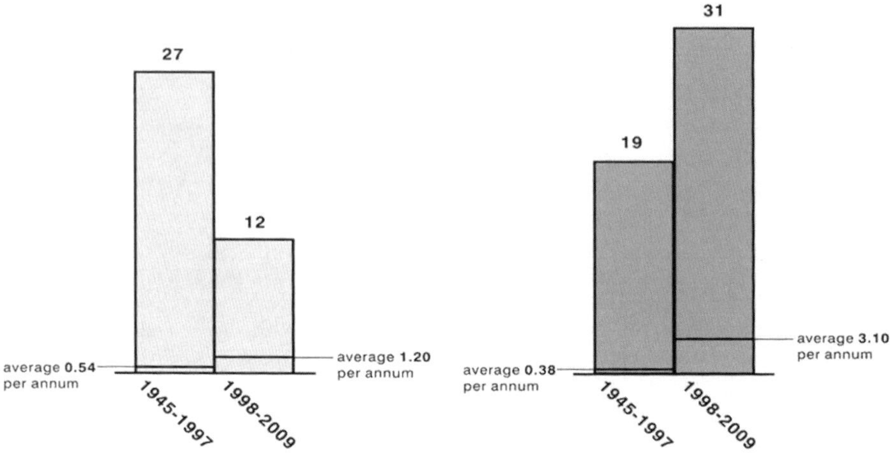


**Fig. 5.1** Riverine (*left*) and surface water (*right*) flood events in Greater Manchester. (Source: Lawson and Carter 2009)

and it was only after the inadequacies in this approach were revealed that changes were initiated. The events therefore provided the evidence *and* the momentum for what will be argued was a fundamental shift to how flood water was both perceived and managed. It may be that the experiences of policy-makers in England, whereby damaging events demanded reviews, which in turn led to wide-ranging recommendations and responses, can provide salutary lessons to differing countries subject to similar climatological and societal drivers, particularly where there is a heavy focus on pursuing 'defend the line' approaches.

Reducing further in scale to the city level, data from the ten local authorities that make up the Greater Manchester city region provides a similar reinforcing story of changing flood sources; albeit a more gradual and incremental shift in contrast with the succession of sharp shocks from the national perspective. The inland area is accustomed to managing a relatively high rainfall and has experienced floods from both riverine and surface water sources, yet analysis of the events and their causes provides supporting evidence of a localised shift in risk that reflects the national trend, with similar managerial and adaptation implications.

Figure 5.1 outlines the differing frequency of these two flood types between two periods: from 1945 to 1997; and from the first of the recent major national floods in 1998 until 2009. Overall, flooding from watercourses has steadily decreased, which has been more than offset by a rapid increase in surface water events, with the average up from 0.38 to 3.10 events per annum. While the localised nature of these floods may not be as devastating as those on a national scale, their similar rising frequency and changing composition provides evidence of both the shift towards surface water flooding and the difficulties within the governance systems to both recognise, and adjust to, inundation from this source. On reflection it could be argued that there is some truth in the perception introduced at the start of the chapter that societies *are* getting better at managing floods—but as these figures suggest, just from what could be termed 'traditional' sources. The investment in river and sea defence has brought benefits, but while the focus has been centered here not only has the risk shifted elsewhere but it has grown in strength. The overall number of floods discussed at the start of the chapter does suggest that there have been governance inadequacies, but this view is reinforced when analysing how governance frameworks have addressed the possible *sources* of inundation, where there has been a clear emphasis on floodplain protection via structural defences.

At its simplest level, the lack of success in preventing flood events, regardless of their scale, could be considered a series of governance failures—with an array of culpable stakeholders identified by the press and public from the government to planners to the Environment Agency. Yet, in the absence of experiencing serious and widespread flood events, and with no real international lessons to draw upon, one could be forgiven for assuming that the threat of inundation was both driven by rivers and the sea and under relative control. In reality, the evidence argues for a more pluralist approach; one that extends the view of flooding beyond the geographical floodplain and therefore challenges the traditional hegemony of existing governance frameworks that are overwhelmingly designed to deliver structural defences. As the chapter will also show, this shift is a work in progress and not only may there be an implementation gap between what a policy advocates and its translation into practice, but also a 'governance deficit' with surface water in particular being subject to a complex framework of responsibilities (Douglas et al. 2010).

## 5.4 Cascading Governance Implications: Paradigms, Methodologies and Responsibilities

The changing sources of flooding and the resulting paradigm shift from defence to risk management has had significant effects on fundamental governance issues such as the desired managerial approach and the actors and agencies charged with determining intervention. The operationalisation initially required a change in methodological emphasis from the probability based technique developed throughout the twentieth century to a risk management approach focused on a broader scientific understanding of causes and solutions. This in turn required a challenge to the professions, moving flooding beyond the realm of engineering towards other experts, most notably spatial planning. The changes in governance also had an effect on who is responsible for managing flood risk, with power and accountability devolved away from the State and towards a wider array of stakeholders, including communities and individuals.

In just over a decade, managing flooding in England has moved to being within the remit of almost everyone, from government ministers to local authority planners to homeowners. Considering the inertia and conservatism inherent in many of the areas of governance affected the response has been brisk, but the sheer scope of areas that required attention was undeniably challenging. Figure 5.2 summarizes the key governance implications which have been gradually revealed and implemented

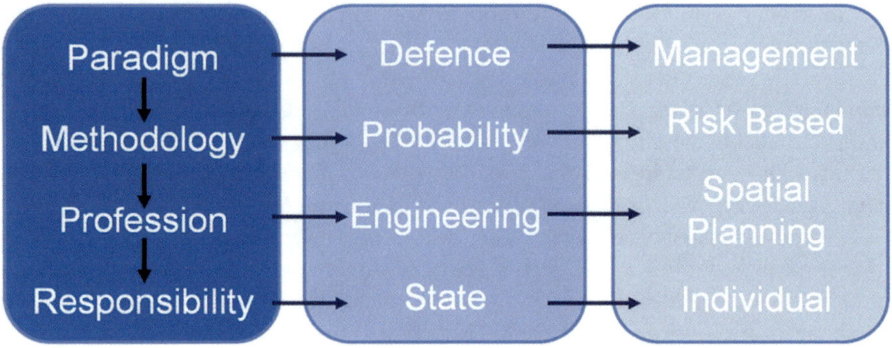


**Fig. 5.2** The interconnected cascade of recent and emerging shifts in the governance of flooding

by the recent failure to protect people and places against flooding—a cascading series of managerial shifts that occurred once the move from flood defence to flood risk management was operationalised at the start of the twenty-first century. The information contained in the figure is not designed to delineate an absolute transfer of power or a dialectic between the technical and social, it is more about revealing new hybrids of knowledge, or even more simply, a direction of travel. This shift has provided the foundation of a more effective management of flooding and will be of relevance beyond the English case study.

Each key aspect is discussed in more depth in the following sections which detail the interrelated cascading response to the realisation that the governance foundations designed to address flood risk were not as secure as previously thought. This involves analysis of the changing flood management paradigm, the methodology employed, a widening of the professions connected and a devolution in stakeholder responsibilities.

## 5.5 Paradigm Changes

The succession of flood events combined with an awareness of the difficulties in managing water led to an acknowledgement within government that it was unrealistic to assume that all floods could be prevented. The long established 'flood defence' paradigm could not be effectively applied to the highly uncertain events of the twenty-first century, and there was a distinct shift towards 'flood risk management'—an alternative, more comprehensive approach that aims to avoid new risk where possible and includes wider human and socio-economic factors (White 2008). Moreover, recognizing that the problem was not just one of controlling water—as risk could also be beneficially influenced by managing people—effectively widened the stakeholders involved and the scope of policy influences with links to flooding. While hard engineering and restrictions on floodplain development would continue to be important, non-structural measures began to be seriously discussed—ranging from green infrastructure to planning policies to insurance provision. The move towards pluralism also meant that the blame for any future floods would be both harder to assign and more widely distributed; within a defence paradigm arguably *any* flooding is a failure and the people responsible for implementing this approach culpable; within risk management some flooding could be expected and multiple stakeholders bear a share of the responsibility if any is assigned.

A significant development driving this change was the gradual acknowledgement that flood risk, such as expressed in Table 5.1, occurred from sources *beyond* rivers and the sea. It should be recognised that this threat still exists, and that structural defences are very effective mechanisms to address this source, but that the spatial and managerial certainty inherent in such an approach could not be easily transferred to the problem of surface water flooding. Essentially, the hard surfaces of a city serve as a multiplicity of artificial water pathways, operating in a complex and seemingly chaotic manner. The urban streets can therefore be considered as a part of the drainage infrastructure, and like any system can fail under high pressure.

The difficulties assigned to the flood defence approach connect to the ability to accurately predict the interaction between precipitation and the urban environment: where will the water pond or flow? What size of event will trigger dangerous levels of runoff? Which spatially discrete location will be exposed and how will climate change and urbanisation affect this? Regardless, all these questions should be considered in the context of the dominant managerial paradigm of the time—if both the volume of runoff and the areas that generate and receive it are unpredictable, irregular and multiplex how do you respond via structural measures? Or in short, how can a society adopt a 'defend the line' approach when a line can't be identified?

The shift towards flood risk management could better address the twin issues of urban runoff and the rise in uncertainty, including those floods that cannot be reasonably prevented, as their effects can be prepared for and the impacts lessened. To operationalise the new paradigm therefore required a further reassessment of the methodology determining where and how we should intervene: flood defence is tightly connected to an engineered response supported by probabilistic calculations; flood risk management is more diffuse and combines wider non-structural approaches. To be effective required a significant re-examination of the methodology employed by the scientific and policy making communities to inform intervention.

## 5.6 Methodological Challenges Associated with Flood Risk Assessment

The way we interpret risk shapes responses. From a flooding perspective, it is the recent paradigm challenging awareness that the majority of flood risk in some cities may not come from the readily identifiable and geographically fixed floodplains and coasts that presents the strongest argument for a new approach. For example, as outlined in Table 5.1, in 2004 in England there were an estimated 80,000 homes at risk from surface water flooding, a figure that rose to 3.8 m in the space of 5 years. It is helpful to reflect on the implications that this fundamental shift may have had on those concerned with flood defence. How can flooding be managed when traditional hard defence engineered approaches may not only be completely irrelevant to this risk, but there may potentially be hundreds of separate sources of flood risk which defy quantification? Moreover, how can probability realistically be considered as a basis for decision-making in this situation?

The first lesson to take from the English experience is regarding the concept of certainty. It is dangerous to assume that knowledge is established and that the threat of flooding operates in a definable and predictable milieu. Ascertaining the extent of flood risk necessarily demands an engagement with uncertainty; in reality we use unreliable and imperfect knowledge to provide a basis for how we use and manage land. Moreover, although the calculations may appear precise this is a highly transient position; in practice they are temporally static yet may be expected to be applied within a highly dynamic environment and be accurate many years into the future. We know climate change and urbanization will exert a significant, yet elusive, forcing effect on risk, but how effective will decisions be in 5, 10 or 20 years time? This quandary may be understood as challenging the principle of *stationarity*, a central tenet around which the analysis of hydrological time series is founded. While it may be expected that precipitation varies daily, seasonally and annually, over a longer time series it has been assumed to be stationary—that is one record should be comparable to another (Zevenbergen et al. 2010). This data has informed flood defence and drainage strategies, has underpinned decision-making and links well with a probabilistic engineered-led methodology. Yet surface water flooding transferred the threat of inundation to the urban environment, a place that alters over time and space, sometimes significantly. Further, when you also consider the unpredictability of climate change, it is clear that applying a static, conservative methodology to what may be a highly dynamic environment can underestimate the threat due to a failure to incorporate forcing drivers altering the properties of a system.

While information on recurrence intervals can be of use in providing a retrospective indication of the relative strength of an event in comparison to past floods, its veneer of scientific certainty regarding *future* risk should be viewed as illusory. In reality, the urban system is subject to such significant variability that its value in aiding strategic decision-making is actually of limited value. Indeed, recognition of this statistical uncertainty has helped drive the shift from probability based approaches towards risk and resilience, as has happened in many countries in incremental stages over the last decade (O'Hare and White 2013).

The second issue to note is related to the concept of certainty and the subsequent desire for quantification. The development of flood management as being within the remit of experts, and in particular within the auspices of professions such as engineering and scientific modeling, created an understandable desire for confidence and surety closely associated with these fields. Quantification is not a natural bed fellow of complexity however; and as the uncertainty associated with flood risk management rose, so errors with this approach were exposed. In practice, the precise volume and location of urban land at risk from flooding may be highly tentative, raising questions as to the extent to which uncertainty regarding the existence or strength of a hazard may be reduced to a calculation of probability (Hanssen 2009). Therefore a methodology centered on ascribing a figure to communicate a notional risk and inform decision-making may have flaws both externally, by creating spatial inequalities based on the availability of appropriate data, or internally, by only considering certain aspects of the risk, such as those which are most easily quantifiable.

The simplistic language of risk calculation may falsely reduce uncertainty to a comforting illusion of deterministic, probabilistic processes within which the inherent gravitas of scientific calculations can attach a misleading confidence to what may be very cautiously derived outcomes. This was a point recognised by Wynne (2009, p. 308) who argued that this methodology for managing risk is erroneous and: "the dominant risk science approach is more than a method; it is a misbegotten culture which inadvertently but actively conceals that ignorance". While understanding hazard management in this manner can be useful as a simple communication tool, it is helpful to move away from a dominant view that risk can be definitively measured; it can't. In reality, attaching a probabilistic value to a highly uncertain event may actually be an extremely tentative judgement masquerading as hard scientific truth; as Adams (1995, p. 29) puts it: "risk is constantly in motion".

The acknowledgment that the management of natural hazards is connected to our perceptions and constrained by information forms the basis of the more contemporary standpoint that responses are socially and culturally constructed. Beck (1992, p. 99) argues that the perception of risk has become shaped by the desire for quantification and the expansion of the insurance industry and referred to:

> … systematically caused, statistically describable and, in this sense, 'predictable' types of events, which can therefore also be subjected to supra-individual and political rules of recognition, compensation and avoidance.

However, it was these 'rules of recognition, compensation and avoidance' that helped drive the shift away from a narrow, reductionist probabilistic methodology, as the impacts of the floods damaged the very political, scientific and financial sectors that underpinned the approach. The legitimate questions, as to the extent to which uncertainty regarding the existence or strength of a hazard may be distilled to a calculation of probability, drove the move towards more holistic risk management, where probability was offset by a more precautionary approach and a wider array of social, economic and environmental factors were considered. Achieving this shift required a much broader engagement with flooding than ever before, within which the management of water now includes professions beyond the traditional spheres, such as engineering and modeling, to incorporate all those with influence over support systems and particularly those connected with where people live and how they act.

## 5.7 Widening Professions

As the previous sections argue, one of the reasons why the effective management of floods has been placed under pressure is that their causes have been influenced by the dynamic and uncertain processes of urbanization and changing weather patterns. Yet, it is too simple to suggest that these influences should equal more damaging events as there has been a huge increase in investment in flood risk management over recent years. According to the Department for Environment, Food and Rural Affairs (Defra 2010) the flood defence budget has risen sharply since the first of the major national floods in Easter 1998, with an estimated £ 765 million forecast in the 2010/2011 financial year, in comparison with £ 310 million in 1997/1998 and £ 590 million in 2007/2008. This also partly explains why flooding from the rivers and the sea has arguably become better managed, as this spending was mainly focused upon new capital projects, and to a lesser, although annually rising extent, on the maintenance of existing structural defences.

This also helps clarify why technologically, financially and scientifically well resourced nations, such as those within Europe and North America in particular, are experiencing a similar rising frequency of floods as countries elsewhere, as human activity—from the burning of fossils fuels to urban development—can create new risks or shift existing ones. The move in the sources of risk towards threats from surface water, which are closely related to a highly developed built environment, did provide a specific funding problem, however. The agencies with power over flood defence spending were not configured to positively influence the flow of water in urban or rural areas, which mainly rests with the local water and sewerage provider, the agricultural sector and local authorities. The need to address the issue of stationarity also ensured that those professions with influence over future risks, from controlling new development to increasing the ability to store water within the catchment, also became important actors in flood risk management; with spatial planners now firmly at the forefront.

Although it was the Easter 1998 floods that proved the catalyst for the first planning policy guidance on flooding released a few years later (Department of the Environment, Transport and the Regions 2001) the role of planning in managing this risk was still in its infancy and in comparison to the established professions it was very much the junior partner. Over the decade the role of planning steadily grew, however, as its contribution became more widely accepted and planners became more adept at exerting influence in this unfamiliar sphere (White and Howe 2002; White and Richards 2007). The new emphasis on the planning profession was also reinforced in related scientific research (Evans et al. 2004), new working relationships (Potter et al. 2011) and policy documents, with, for example, Defra (2004, S. 7) stating:

> The Government is committed to ensuring that its development and planning policy seeks where possible to reduce, and certainly not to add to, the overall level of flood risk.

The growing threat of surface water flooding continues to bring a real critical mass to this shift however, and is essentially helping to widen responsibility from the engineering-led professions towards those concerned with people and places, and in particular those with influence over the interaction between the built and natural environments. Spatial planning is now accepted as one of the key mechanisms to avoid, reduce and manage flood risks (Communities and Local Government 2010) and operationalise the paradigm shift initiated in the early twenty-first century. Structural and non-structural options are now routinely considered as viable measures in both reducing risks and adapting in the future, although the differing actors and agencies involved increase the potential for more effective strategies. They also instill a greater complexity of policy choices and the need to construct negotiated policy agreements across this growing network of stakeholders.

The role of the professions is not just an external relationship, where the involvement of more skills and knowledge should lead to better decisions, but also an internal dialogue—wherein discourse within and between people and organisations in key sectors takes place. Enacting such a shift in governance from the technical towards more complex socio-technical assemblages inevitably takes time to become effective in practice, particularly where the risk is subject to uncertain spatial and temporal influences. To operationalise the new paradigm therefore requires changes to how professionals understand, and act upon, flood risk across a broad range of areas, and potentially within sectors without the long history of flood management as detailed in the first part of the chapter. As the next section will emphasise however, not only do multiple agencies communicate more effectively, but this approach also means that responsibility now goes way beyond the formal professions towards the citizen and their relationship with both the State and the market.

## 5.8 Devolved Stakeholder Responsibilities

One significant implication of the move from flood defence to flood risk management was that the simple chain of responsibility between the State and its key managerial organization in this area, the Environment Agency, was unraveled. The 'living with risk' agenda, within which prevention is advocated alongside an ability to cope with any possible impacts, could be pursued across multiple scales and professions; including the public sector, the private sector, communities, households or even individuals.

As problems could be exacerbated by forces beyond the control of any one agency there was also an onus on shared responsibilities. Although the newer strategy adopts a similarly spatial approach as previously identified, the re-examination of scale better reflects the realities of the risk and, significantly, artificial administrative boundaries became challenged as inappropriate with new plans advocated at the city, catchment or regional scales. While recognising the wider variety of professions with the ability to influence flood risk and providing a broader array of mechanisms to influence policy options and outcomes was a clear step forward, there is still an ongoing need to work together better, as actions in one part of the

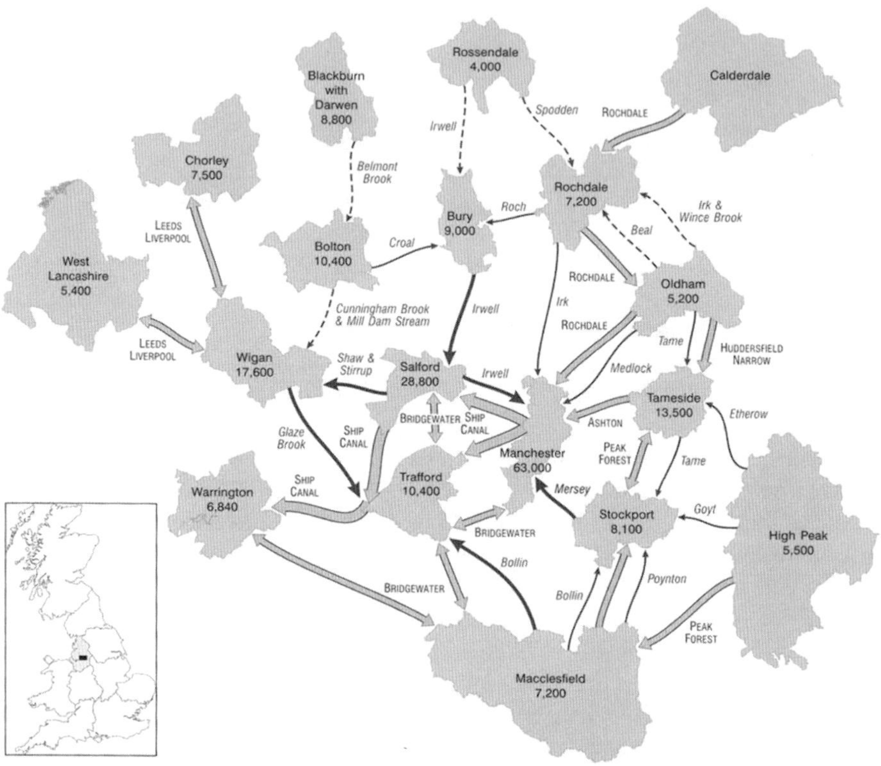


**Fig. 5.3** Housing allocations and hydrological links schematic of Greater Manchester in the North West UK. (Source: Adapted from Scott Wilson and AGMA 2008)

catchment can have a significant affect elsewhere. A good example of this is detailed in Fig. 5.3 where the Greater Manchester region in North West England has been broken into its ten local authorities to better demonstrate the interconnected hydrology. Each area details both the future housing targets to 2021 *and* the direction of water flow between the most significant rivers and canals.

Figure 5.3 provides an understanding of the shape of the catchment, with water from the higher surrounding areas flowing towards the western side of the conurbation. The diagram outlines how the outer authorities essentially export their runoff into Salford, Trafford and the outlying area of Warrington, emphasising the cumulative nature of risk and the need for flood risk management to be a collective responsibility. Indeed, if all the housing targets are met, a total of 145,100 new homes, and their associated infrastructure, will be constructed upstream of Salford by 2021, an area already at relatively high risk of flooding. The externally driven hazard argues for inclusive partnerships and spatial policies that are coherent with the priorities of neighbouring authorities.

The new roles also extend beyond the public sector and related professions. The rise in information enabled the State to devolve responsibility away from the centre, with both the private sector encouraged to play a part, such as via insurance and

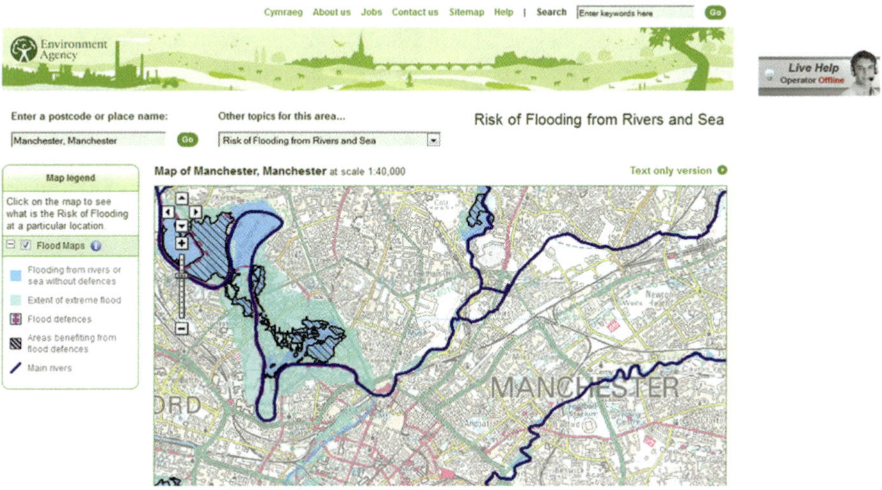


**Fig. 5.4** A screenshot of the Environment Agency's flood map service, which encourages individuals and communities to become more aware of flood risk. (Source: Contains Environment Agency information © Environment Agency and database. 2012)

flood resilience products, and communities and people expected to consider how to reduce their own exposure and vulnerability. Facilitating the shift required information on flood risk to be collected and made publicly available, with, for example, people being encouraged to consider their own acceptable level of risk by checking online maps provided on the Environment Agency's website when purchasing a house (Fig. 5.4). However, as with other risk management techniques these tools may not effectively communicate the certainty of calculations, utilise unfamiliar nomenclature and focus on spatially representing the more easily quantifiable risks from rivers and the sea (Merz et al. 2007).

The wider implications of the trend to devolve power and responsibility away from the State towards communities and individuals are still emerging. Risk management (like 'sustainability' or 'progress') is one of those agreeable, pliable and nebulous concepts that is difficult to argue against. Although a contrary position may appear to be counter-intuitive, the lack of contention does not mean that the concept should be uncritically unpacked, automatically promoted or unthinkingly applied. Far from being a universal good, when this approach is translated into practice there may be significant spatial inequalities within society, many of which may be unwitting externalities. Once individuals and communities are encouraged to take responsibility for the level of risk they wish to be exposed to, and protection from flooding commoditised via the public sector, there will inevitably be winners and losers. There are also wider questions about the long term social justice implications, of both the initial paradigm shift and its resultant effects. The move to flood risk management was a pragmatic policy decision reflecting the unpredictability of a flood defence approach, but would communities actually want to live with water or do they just want the state to keep them safe? The implications of this final shift in the governance chain are only just being considered within academia and policy and, reflecting on recent history, there is a possibility it may require more serious flood events to shed light on the social impacts of devolving responsibility from the State towards the private sector, communities and individuals.

## 5.9 Discussion and Conclusion

This critical analysis of both the shifting sources of flood risk and the governance implications of a change in approach reveal a cascading series of impacts. Some of these were anticipated and have been already experienced, others were difficult to recognise and are still being unpacked. Although flood defence decisions were made using the best available information within a long established scientific approach, this techno-rational process of considering hazards in this narrow probabilistic manner and using it as a basis for judgement not only fails to adequately address the differing sources of flooding, but effectively helps to immunize decision-making from failure (Reith 2009). A reductionist probabilistic approach can provide a comforting veneer of efficacy and certainty but it may be a deceptive state where flaws can continue until the strength of detriment demands the need for a re-evaluation of the entire methodology. That is not to say that modeling is not useful, particularly at a large catchment scale, but it should be reinforced by broader knowledges and new understandings of local level risk management. In hindsight, the focus for intervention has been quantitative and based squarely on what is known; when in reality more attention could have been profitably spent on a precautionary approach cognisant of what is unknown. Indeed, the search for certainty may be impossible in this sphere (White 2013).

The shift in risk from rivers and the sea towards surface water will be one of the emerging challenges of the early twenty-first century in many developed areas, particularly when considered in conjunction with the rise of more extreme climatic events and burgeoning urbanization. Flooding is subject to differing geographies, where, for example, the behavior of a coastal floodplain is eminently more predictable and manageable than intra-urban flooding; a landscape subjected to a wide array of drivers and affected by the powerful binary influences of the natural and built environments. Both risks are relevant to a modern flood risk management approach however, and this chapter doesn't aim to provide a discussion of the dialectics of defence versus risk management, more a tempering of the former and recognition of its inherent governance constraints as a tool to address surface water.

The hydraulics beyond the boundaries of the sea or watercourses are subject to the multiplex forces of urbanity, where water flows in unpredictable ways and can be constrained or released according to what may be indeterminate pinch-points. One logical way of reacting to this rise in uncertainty inherent in threats from surface water is to make our cities more aware of, and resilient to, flood risks—and in this respect proactively and strategically influencing the current and future use of land is key. Therefore, it may be that the most significant shift in the management of water in England and beyond will not be the one from flood defence to flood risk management, or from probability to risk based approaches, but rather the emerging transition in power and responsibility from engineering to spatial planning and from the State to the individual. While there will always be a role for providing hard structures to defend against flooding, there will clearly be a stronger responsibility to link water concerns with where and how people live. The implementation of new, more appropriate policy initiatives, such as the European Flood Directive (European Commission 2007), may provide the basis of firmer governance foundations but their implementation still needs to influence a complex managerial arena.

Each disaster drives a spectrum of possible action from reflection to revolution; improvements which may not be achievable, without first experiencing these potentially catastrophic drivers. Yet the extent and consequences of flood risk argue against such a reactive approach; should we only change in direct response to disasters, or should we use scientific knowledge to help predict and adapt? While it is acknowledged that damaging events do have the potential to set the agenda (Kingdon 1984) or create momentum for policy change (Johnson et al. 2005), this is an unsustainable, reactive process within which changes occur after experiencing detriment. Within risk and disaster management, all too often suffering damaging events is a key phase as they stimulate reflection and reaction, but this doesn't have to be experienced in each country individually and it is here that this chapter may provide key lessons. From a period of relative stability England became quickly subject to frequent damaging floods to the extent that the collective awareness is moving from a feeling of safety toward the more realistic viewpoint of 'before the next catastrophe' (Perrow 2007), galvanising action to protect people and places and challenging managerial norms.

It may well be that just as the flood defence approach spread across the globe during the last two centuries, the direction of travel from defence to management, from probability to risk, from engineering to spatial planning, and from the State towards the individual, will resonate with many different cities and countries. Therefore the strategies adopted in England, and the resulting series of governance changes gradually developed to operationalise the shift in approach, may help others break the closed cycle between detriment and policy change. The new hybrids of knowledge and understanding being developed represent a logical response, but it should be noted that these changes were not all implemented as part of a comprehensive strategic redirection, but have rather grown organically as the implications of the initial paradigm shift adopted to manage this hazard became apparent in practice. Other localities could use this international comparison to plan intervention more strategically; however—the perception of flooding as one homogenized risk emerging from rivers and the sea has created governance frameworks which may not be well suited for responses beyond this conveniently quantifiable and spatial area.

While the move towards flood risk management greatly enhances the ability to both protect people and lessen future impacts, far from being a universal good, in reality when such a managerial shift occurs there are often significant governance challenges or spatial and social inequalities, many of which have only been revealed through experience. In reality, what may appear to be a simple, benign and progressive paradigm change has actually transpired to be a cascading series of powerful governance changes, the implications of which are still being analysed. Moreover, these transformations may require new socio-technical assemblages developed to manage the interaction between flood governance and human behaviour. Adaptation to climate change is not a predictable, neutral, value free process; natural hazards can challenge long held conventions and adjusting intervention measures can bring new and often difficult to anticipate problems. In hindsight, the governance of flooding in England has been revealed to be on less firm foundations than envisaged, yet the heuristic and sometimes harsh experience in addressing natural hazards has the potential to provide valuable lessons for elsewhere.

**Acknowledgements** Thanks to Professor Graham Haughton, Dr. Yasminah Beebeejaun and Dr. Angela Connelly of the University of Manchester for their comments. I also would like to extend my gratitude to the Environment Agency, Scott Wilson and AGMA for allowing the reproduction of their images.

## References


Adams, J. (1995). *Risk*. London: University College London Press.

Beck, U. (1992). *Risk society—towards a new modernity*. London: Sage.

Britannia. (2007). Anglo Saxon chronicles. http://www.britannia.com/history/docs/1014-17.html Accessed 12 Oct 2009.

Burby, R. J. (2006). Hurricane Katrina and the paradoxes of government disaster policy: Bringing about wise governmental decisions for hazardous areas. *The Annals of the American Academy of Political and Social Science, 604*(1), 171–191.

Bye, P., & Horner, M. (1998). *Easter 1998 floods: Volume 1. Report by the independent review team to the board of the environment agency*. Bristol: Environment Agency.

Communities and Local Government. (March, 2010). *Planning policy statement 25: Development and flood risk*. London: The Stationery Office.

Defra. (2004). *Making space for water: Developing a new government strategy for flood and coastal erosion risk management in England*. London: Defra.

Defra. (2010). Flooding—investment allocation. http://www.defra.gov.uk/environment/flooding/funding/allocation.htm Accessed 29 Nov 2010.

Department of Communities and Local Government. (December, 2006). *Planning policy statement 25: Development and flood risk*. London: Department of Communities and Local Government.

Department of the Environment, Transport and the Regions. (2001). *Planning policy guidance 25: Development and flood risk*. London: HMSO.

Douglas, I., Garvin, S., Lawson, N., Richards, J., Tippett, J., & White, I. (2010). Urban pluvial flooding: A qualitative case study of cause, effect and non structural mitigation. *Journal of Flood Risk Management, 3,* 112–125.

Environment Agency. (2009). *Flooding in England: A national assessment of flood risk*. Bristol: Environment Agency.

Environment Agency. (2011). Understanding the risks, empowering communities, building resilience: The national flood and coastal erosion risk management strategy for England. http://www.official-documents.gov.uk/document/other/9780108510366/9780108510366.pdf. Accessed 11 Dec 2012.

European Commission. (2007). Directive 2007/60/EC of the European parliament and of the council of 23 October 2007 on the assessment and management of flood risks. http://ec.europa.eu/environment/water/flood_risk/index.htm Accessed 8 Feb 2011.

Evans, E., Ashley, R., Hall, J., Penning-Rowsell, E., Saul, A., Sayers, P., Thorne, C., & Watkinson, A. (2004). *Foresight: Future flooding. Scientific summary: Volume 1, Future risks and their drivers*. London: Office of Science and Technology.

Hansson, S. O. (2009). A philosophical perspective on risk. In R. E. Lofstedt & A. Boholm (Eds.), *The earthscan reader on risk*. London: Earthscan.

Harvey, D. (1989). *The condition of postmodernity: An enquiry into the origins of cultural change*. Cambridge: Blackwell.

Johnson, C. L., Tunstall, S. M., & Penning-Rowsell, E. C. (2005). Floods as catalysts for policy change: Historical lessons from England and Wales. *International Journal of Water Resources Development, 21*(4), 561–575.

Kingdon, J. (1984). *Agendas, alternatives and public policies*. New York: Harper Collins.

Law, F. M., Black, A. R., Scarrott, R. M. J., Miller, J. B., & Bayliss, A. C. (2010). Chronology of British hydrological event. http://www.dundee.ac.uk/geography/cbhe/#searching Accessed 29 Nov 2010.

Lawson, N., & Carter, J. (2009). Greater Manchester Local Climate Impacts Profile (GMLCIP) and assessing Manchester City council's vulnerability to current and future weather and climate. Manchester: Ecocities Project, University of Manchester. http://www.sed.manchester.ac.uk/architecture/research/ecocities/library/documents/GMLCLIP.pdf Accessed 29 Nov 2010.

Merz, B., Thieken, A. H., & Gocht, M. (2007). Flood risk mapping at the local scale: Concepts and challenges. In S. Begum, M. J. F. Stive, & J. W. Hall (Eds.), *Flood risk management in Europe*. Dordrecht: Springer.

National Audit Office. (2001). *Inland flood defence*. London: The Stationery Office.

O'Hare, P. & White, I. (2013). Deconstructing resilience: Lessons from planning practice, planning. *Practice & Research, 28*(3), 275–279.

Parker, D. J. (1995). Floodplain development policy in England and Wales. *Applied Geography, 15*(4), 341–363.

Pall, P., Aina, T., Stone, D. A., Stott, P. A., Nozawa, T., Hilberts, A. G. J., Lohmann, D., & Allen, M. R. (2011). Anthropogenic greenhouse gas contribution to flood risk in England and Wales in autumn 2000. *Nature, 470,* 382–385.

Perrow, C. (2007). *The next catastrophe: Reducing our vulnerabilities to natural, industrial and terrorist disasters*. Princeton: Princeton University Press.

Pitt, M. (2008). *Learning lessons from the 2007 floods, an independent review by Sir Michael Pitt*. London: Cabinet Office.

Potter, K., Ward, S., Shaw, D., Macdonald, N., White, I., Butler, D., & Kellagher, R. (2011). Engineers and planners—the role of partnerships in pursuing sustainable water management objectives. *Engineering Sustainability, 164*, 239–247.

Reith, G. (2009). Uncertain times: The notion of 'risk' and the development of modernity. In R. E. Lofstedt & A. Boholm (Eds.), *The earthscan reader on risk*. London: Earthscan.

Scott, W., & AGMA. (2008). *Strategic flood risk assessment for Greater Manchester*. Manchester: Scott Wilson.

White, I. (2008). The absorbent city: Urban form and flood risk management. Proceedings of the Institution of Civil Engineers: Urban Design and Planning, December (DP4), 151–161.

White, I. (2010). *Water and the city: Risk, resilience and planning for a sustainable future*. London: Routledge.

White, I. (2013). The more we know, the more we don't know: Reflections on a decade of planning, flood risk management and false precision. *Planning Theory and Practice, 14*(1), 106–114.

White, I., & Howe, J. (2002). Flooding and the role of planning in England and Wales: A critical review. *Journal of Environmental Planning and Management, 45*(5), 735–745.

White, I., & Richards, J. (2007). Planning policy and flood risk: The translation of national guidance into local policy. *Planning, Practice and Research, 22*(4), 513–514.

Wynne, B. (2009). Uncertainty and environmental learning: Reconceiving science and policy in the preventative paradigm. In R. E. Lofstedt & A. Boholm (Eds.), *The earthscan reader on risk*. London: Earthscan.

Zevenbergen, C., Cashman, A., Evelpidou, N., Pasche, E., Garvin, S., & Ashley, R. (2010). *Urban flood management*. Oxford: Taylor and Francis.

# Chapter 6
# Planning for Resilient Coastal Communities: Emerging Practice and Future Directions

**Timothy Beatley**

**Abstract** In this chapter the author argues that resilience must be the primary goal for coastal communities in the future, as they face a series of daunting challenges and future shocks. The chapter examines what resilience means in practice, and identifies a variety of measures and strategies by which coastal communities can become more resilient.

**Keywords** Coastal resilience · Sea-level rise · Climate change · Principles · Passive survivability

## 6.1 Introduction

Coastal communities around the world face immense pressures and shocks across a complex mix of biophysical, social and economic dimensions. Coastal communities are cities and human settlements literally "on the edge" and face a host of natural hazards including hurricanes and coastal storms, earthquakes and tsunami, riverine flooding and tornadoes, among others. The impacts of global warming, especially sea-level rise, will present perhaps the most severe challenge to the livability, livelihoods and long term viability of coastal cities and communities.

Much of the world's urban population sits along coastlines, for instance, where future sea-level rise will be a significant problem. A recent study by researchers at the International Institute for the Environment and Development (IIED) estimates that about 10% of the world's population, much of it urban, now resides in highly vulnerable Low Elevation Coastal Zones (LECZ), or locations less than 10 m in elevation (a mere 2% of the world's land area; McGranahan et al. 2007). A specific look at the world's largest port cities find many of them highly vulnerable. Nich-

---

This chapter draws heavily from and expands upon Beatley, *Planning for Coastal Resilience*, Island Press, 2009.

---

T. Beatley (✉)
University of Virginia, Charlottesville, VA, USA
e-mail: beatley@virginia.edu

B. C. Glavovic, G. P. Smith (eds.), *Adapting to Climate Change,* Environmental Hazards,
DOI 10.1007/978-94-017-8631-7_6, © Springer Science+Business Media Dordrecht 2014

olls et al. (2007) conclude that 40 million urban residents are already subject to a 100-year storm surge and that by the 2070s this could triple in number to around 150 million (and assets at risk approximating US $ 35 billion). As urbanisation and climate change progress together, in combination with subsidence, the exposure in many cities will grow dramatically. In Mumbai, for instance, an exposed population today of about 2.8 million, is projected to rise to 11.4 million in 2070, while in Dhaka numbers are expected to increase from 844,000 today to a projected 11.1 million. For cities in the developed world, property damage and exposure are greater concerns as evidenced by Miami (which ranked number 1 in total exposure to expected future losses), whereas New York and Virginia Beach (USA) ranked within the top 20 most asset exposed cities in the world.

Cities like Miami and New York, in the USA, or London in the UK, will see significant expansion of the areas prone to coastal inundation in the future. How to plan for and adapt to these sea-level rise predictions will be one of the most serious questions to be faced: will cities be able to (afford to) build new coastal defenses, or alternatively should cities engage in long term shoreline "retreat," steering urban growth away from flood-prone locations and relocating homes, business and infrastructure when opportunities arise? Should cities look for new and creative ways to design buildings and infrastructure that are better able to withstand future flooding, for instance, following the Dutch who are experimenting with floating forms of homes in some areas?

Global warming will also likely exacerbate many other existing problems faced by coastal cities and regions. A recent study by Bell et al. (2007) of 50 eastern U.S. cities predicts that urban concentrations of ozone will increase there by 2050, and the greatest increases will occur in cities that already have the highest levels of ozone pollution. The number of days that exceed federal ozone standards will increase and the respiratory and other associated health implications will be significant.

The health impacts and implications of natural disasters extend well beyond immediate mortality and injury-related statistics. Surviving families and individuals are often displaced, living in challenging and often unhealthy circumstances. Following Katrina such conditions included inadequate housing (including formaldehyde emitting FEMA trailers), inadequate access to food and nutrition and unemployment, among others (Beatley 2009). As in Katrina, the emotional and psychological impacts are substantial as well.

The long term health implications of climate change will create new challenges for communities and local governments. Rising urban temperatures and heat waves create real dangers for especially vulnerable populations, such as the elderly and socially-isolated. The devastating 2003 heat wave that gripped much of Europe resulted in nearly 15,000 deaths, most of whom were older residents. While a few cities have developed policies for sheltering residents in air-conditioned public buildings in such emergency heat conditions (e.g. Toronto), cities and communities have relatively limited plans or capabilities for preventing such impacts.

It is also clear that the degree of social isolation in a neighbourhood or community will influence vulnerability, and in this regard the overall trends in the USA are not encouraging. Recent evidence suggests that Americans exhibit an even greater degree of social isolation today than just two decades ago (McPherson et al. 2006).

A significant dimension of coastal vulnerability is economic as well. Is the community's or the region's economic base robust and sufficiently diverse that it will be possible to weather an economic downturn or global economic shock? Is the economic base of a community or region dependent on one or a small number of businesses and industries, and susceptible to the economic devastation that occurs when a manufacturing plant closes or relocates? As an example, the impacts associated with Hurricane Katrina and the difficult recovery that followed, are strongly foreshadowed by a weak and struggling economy, including high unemployment and low economic productivity. Weak coastal economies do not bode well for resilient responses and quick recoveries from a major hurricane. In turn, stronger social cohesion, networks and capital can work together to enhance resilience in the face of these economic trends, highlighting the important interconnectivity of these and other variables discussed throughout this chapter.

Communities also face a host of related natural resource constraints and limitations, often leaving a community and region more susceptible to the negative effects of disasters when they strike. Many communities around the world face serious and long-term shortages of potable freshwater. A combination of profligate use and waste, water shortages are exacerbated by drought, though the underlying causes may be more complex (a combination of high usage, population growth and unsustainable sourcing). Similarly, a variety of other community resources, from forests and farmland, and the timber, food and other goods and products generated from them, to fisheries, are now often in short supply and in decline. This decline may be gradual, but often results in episodic crises. One of the most serious long term resources in decline will likely be oil, as arguments that we are or have reached "peak oil" are increasingly compelling (see Bill McKibben's 2011 summary of evidence in his book *Eaarth*, for instance). Decline in global oil will challenge communities to become more resilient in many ways and may in the long run help communities to develop local capacities and capabilities (e.g. producing more energy from local, distributed sources) that are less vulnerable to disasters.

It is indeed a perfect storm in many ways for coastal communities. It is argued here that the lens and framework of "resilience" must become the primary organising concept in coastal planning in the years and decades ahead. While its precise meaning is still debatable, it represents a powerful framework for organising coastal policy and planning. What follows explores some of the potential ways a coastal city or community might become profoundly more resilient.

## 6.2 The Vision of Resilient Coastal Communities

There is a growing consensus among planners and community leaders in the U.S. and around the world that resilience ought to be a primary goal and objective, and a larger organising concept for all future planning and development. The nature of looming disasters and future shocks suggests that resilience is an appropriate vision for cities, towns and communities in the future. Much of what follows in this chapter is a discussion of what resilience means in practice and the measures and strategies that can be undertaken to bring it about.

*Resilience* as a concept and term has emerged as an important new way of thinking about the design and planning of communities and regions, and a number of competing definitions have been put forth. C.S. Holling's work on ecological resilience is often identified as the starting point. Holling speaks of the resilience of ecosystems as "the capacity of a system to absorb and utilise or even benefit from perturbations and changes that attain it, and so persist without a qualitative change in the system's structure" (1973, p. 9).

The word resilient actually derives from the Latin *resiliere,* meaning to "jump back" and has a useful common meaning that describes an ability to easily or quickly "bounce back" from a disturbance or crisis (Paton 2006). A number of common language themes emerge when reviewing the resilience literature. Words like flexible, adaptable, durable and bendable emerge as important descriptors of a resilient coastal community or region. While much of the contemporary planning meaning is captured in this common language usage, resilience has come to mean more than this. Here is a sampling of definitions found in the literature:

> Measure of how quickly a system recovers from failures. (Emergency Management, Australia 1998)

> Capacity to draw upon personal and social resources to manage the consequences of disasters. (Paton et al. 2006, p. 106)

> The ability of a community to not only deal with adversity but in doing so reach a higher level of functioning. (Kulig, as cited in Pooley et al. 2006)

> The potential of a system to remain in a particular configuration and to maintain its feedbacks and functions, and involves the ability of the system to reorganise following disturbance-driven change. (Walker et al. 2002)

> Achieving resiliency in a disaster context means the ability to survive future natural disasters with minimum loss of life and property, as well as the ability to create a greater sense of place among residents; a stronger, more diverse economy; and a more economically integrated and diverse population. (Berke and Campanella 2006, p. 193)

As Godschalk et al. note, "A resilient community is one that lives in harmony with nature's varying cycles and process." (2003, p. 137). Godschalk argues compellingly for a vision of resilient *cities*, and that resilience should be the "overriding goal" of urban hazard mitigation:

> Such cities would be **capable of withstanding severe shock** without either immediate chaos or permanent harm. Designed in advance to **anticipate, weather, and recover** from the impacts of natural or terrorist hazards, resilient cities would be built on principles derived from past experience with disasters in urban areas. While they might **bend from hazard forces, they would not break**. Composed of **networked social communities and lifeline systems, resilient cities would become stronger by adapting and learning from disasters**. (Godschalk 2003, p. 136–137)

As Godschalk (2003, p. 137) continues:

> Resilient cities are constructed to be **strong and flexible**, rather than brittle and fragile. Their lifeline systems of roads, utilities, and other support facilities are designed **to continue functioning in the face of rising water, high winds, shaking ground**, and terrorist attacks. Their new development is **guided away from known high hazard areas**, and their vulnerable existing development is relocated to safe areas. Their buildings are constructed or retrofitted to meet code standards based on hazard threats. Their natural **environmental protective systems** are conserved to maintain valuable hazard mitigation functions. Finally, their **governmental, non-governmental, and private sector organizations are prepared with up-to-date information about hazard vulnerability** and disaster resources, as linked with **effective communication networks**, and are experienced in working together.

*Resilience* is often viewed as an important antidote or response to vulnerability. Resilient communities work to reduce or even eliminate vulnerability. Vulnerability can be defined "as the conditions determined by physical, social, economic, and environmental factors or processes, which increase the susceptibility of a community to loss from hazard impacts" (Buckle 2006, p. 90).

The notion of *adaptive capacity* is often a key feature in definitions of resilience, the idea that it is not simply possible or even desirable to return to a former condition or state, and that entities (people, organisations and communities) should strive to learn from and creatively respond to disasters and disruptive events and trends and that they should evolve and move from a crisis or disaster to a new and perhaps improved set of circumstances (but undoubtedly different). Resilience, then, according to Paton (2006), is "a measure of how well people and societies can adapt to a changed reality and capitalise on the new possibilities offered" (p. 8).

Implicit in the notion of resilience is an emphasis on taking actions and steps to build adaptive capacity, to be ready ahead of a crisis or disaster. It is *anticipatory*, *conscious* and *intentional* in its outlook—while much cannot be known about future events, much can, and planning ahead becomes a key aspect of resilience.

Hazard *mitigation* has for several decades been the more common term within the natural hazards community for describing long-term anticipatory planning. More specifically, it refers to all the actions, steps, programmes and policies that can be adopted today that will reduce loss of life and property damage later when a natural event occurs (Godschalk et al. 1999). Mitigation is often contrasted with preparedness and response activities, in that the focus is on long-term, proactive steps (such as adoption and implementation of a building code or construction standards, or prohibiting building in a high-risk coastal hazard zone). Preparedness refers to those short-term activities undertaken immediately in advance of a natural hazard event (e.g. evacuation in the face of an approaching hurricane), and response, those actions taken immediately following an event (e.g. search and rescue, debris removal). By contrast these are short-term in duration and usually aimed at addressing fairly immediate health and safety concerns.

To a considerable extent, then, *resilience* has become the new way of talking about mitigation, essentially new language for talking about and advocating long-term mitigation. Resilience is different, however, in at least *two* ways: its focus on creative *adaptation* and learning, and on developing an underlying *capacity* mark a difference in emphasis. While mitigation has historically meant physical changes (a stronger building) resilience is broader, and connotes stronger social and community systems, larger processes and mechanisms for facilitating effective response and recovery. There are certainly many physical design and building responses, for instance, elevating structures in the floodplain, or setting-back from areas subject to sea-level rise, but community resilience must also be about developing supportive community institutions and networks that help families and individuals prepare for and respond to disaster events.

## 6.3 What Are the Qualities of a Resilient Coastal Community?

Walker and Salt in their important book *Resilience Thinking* identify nine qualities or values that characterise a "resilient world" (Walker and Salt 2008). Resilience is characterised by *diversity* (biological, landscape, social and economic), and by *ecological variability* (i.e., allowing ecosystems to change and move and "probe their boundaries"). A resilient world reflects a degree of *modularity* so that shocks and perturbations are controlled or contained. *Slow, controlling variables* receive emphasis in a resilient world; these are the ecological conditions or processes that help to control or stabilise change (and may be such things as the density of a key predator, or the nitrogen level in soil, or the frequency of hurricanes). *Tight feedbacks* (how quickly and strongly the impacts of a change are felt) are an important quality in resilience, because they allow us to take actions and response steps before ecological and other thresholds are crossed (e.g., learning that loss of coastal wetlands results in increasing coastal flooding, might permit timely actions to prevent future losses).

A high degree of *social capital* is also viewed by Walker and Salt as an important factor in promoting resilience. "Resilience in social-ecological systems is very strongly connected to the capacity of the people in that system to respond, together and effectively, to change any disturbance. Trust, strong networks, and leadership are all important factors in making sure this can happen" (2008, p. 47). *Innovation* (placing "an emphasis on learning, experimentation, locally developed rules, and embracing change," (2008, p. 147), *overlap in governance* (redundancy in governance structures), and finally *ecosystem services* (including the otherwise un-priced services provided by nature in our policy and planning deliberation), are also identified as key qualities.

Resilience in coastal environments, indeed in all environments, can be understood as occurring at multiple geographical scales. Resilience can apply at an individual or family level, but also at larger social or societal levels. This can be viewed as a nested model of resilience that understands individual/family resilience is both constrained by and influenced by large societal and environmental settings, but that the latter are in turn affected by resilience, or lack thereof, at smaller scales.

Resilience can be seen to exist, and can be nurtured, at both individual and collective levels, and at a number of geographical scales (from neighbourhood to region and beyond). Buckle (2006) identifies certain elements that support resilience at an individual level, including: information and advice, resources (including financial), management capacity, personal and community support and involvement (p. 96). At the community level, Buckle identifies the following as elements supporting resilience: Knowledge of hazards, shared community values, established social infrastructure (including information channels, social networks and community organisations such as churches and supporting clubs), positive social and economic trends (e.g. viable economy, stable or growing population), partnerships and resources and skills (pp. 97–98).

The resilience of a coastal community can be viewed in terms of the resilience of its physical and built environments—the ability of a coastal community's homes and buildings and built infrastructure to withstand and adapt to natural forces and changing circumstances, as well as similar performance by ecosystems and the natural environment. Will homes sustain wind forces, are buildings and urban form located outside floodplains and high risk locations and so will respond well to future flood events? Are the region's ecosystems and natural systems sufficiently intact and healthy that they will be equally resilient?

A community's resilience can also be understood as a function of its social systems and networks, and levels of social and community support. Personal and community support include, according to Buckle, "post event personal support, such as outreach services, advocates and gatekeepers and community support, for example community development officers" (2006, p. 96). Buckle uses the term "involvement" to indicate the broad importance of social networks and relationships. "Involvement" includes, according to Buckle, "linkages with other people, with a wide network of family, friends and acquaintances shown to be critical in supporting and sustaining resilience" (2006, p. 96).

Advancing true resilience in coastal communities will require more than just a single or a handful of mitigative projects or actions. Rather, resilience requires thinking holistically, and taking many steps to *grow* community culture and society that are resilient.

Resilience and sustainability are highly related concepts, and indeed the former is often viewed and stated as a foundation or "cornerstone" of the latter. Sustainability of an ecosystem, or landscape, or city, *requires* resilience. Resilience in and of itself is not an intrinsically desirable goal—maintaining or enhancing the resilience of a depleted resource (a fishery or forest), is not a worthy goal. Enhancing the resilience of a desired ecological or built form—a diverse ecosystem providing important benefits and services, a city or community providing a high quality of life, buildings and urban form providing shelter, jobs and income—are indeed very desirable and appropriate things to *sustain*, or to apply resilience tools and thinking to. How valued or valuable the regime or system is that we're attempting to make more resilient is important.

Economic resilience is a major and important aspect of disaster resilience in the sense that a resilient coastal community will require businesses to be able to quickly adapt, rebuild and reopen. To this end, how many local businesses, how much of a community's economic sector is situated in high-risk locations? How many businesses have prepared business continuity plans, and how many have generally taken steps to think ahead of time about the likely impacts of a future natural event? As well, what mechanisms—economic, social, organisational—have been put in place by local governments to help businesses following a major disaster?

## 6.4 Some Principles of Coastal Resilience

What follows is a set of broad principles to help coastal planners, decision-makers and citizens begin to think about how to design and plan for greater resilience. These principles are not mutually exclusive, but rather reinforcing and complementary. The principles are synthesised and drawn from several key sources including: the coastal planning and resilience literature, insights gleaned from interviews of key coastal planners and leaders, and the author's extensive professional experience working in coastal planning.

### *6.4.1 Take a Long Term, Multi-scaled Approach*

It perhaps goes without saying that coastal resilience requires a long term temporal frame of reference. Attention should be given to short term problems and solutions (buying that backup generator), but longer term trends and problems must be confronted. How will sea-level rise, or drought and climate conditions change the nature of the coast and impact coastal communities in 50 years, 100 years or even longer?

And many of the most effective strategies will require long term action, such as purchasing and setting aside coastal retreat zones, for instance, and coastal ecosystem repair and restoration efforts. While adoption of longer planning time-frames is difficult, there are local examples to cite of communities, especially in developing climate change adaption plans, that have expanded their temporal frame. Sarasota County, Florida's new comprehensive plan, "Sarasota 2050", takes a 50-year time horizon as do several other communities (See Laushe 2009).

The Dutch, often viewed as the gold standard in sea-level adaptation and planning, have chosen to develop a national strategy with a 200-year timeframe. For a country where half the land lies below sea-level, a long term strategy is essential and will require concerted efforts at land planning, coordination of investments and raising funds (e.g. Wolman 2008). The 200-year strategy calls for a mix of measures: building new land at the ocean's edge, raising and reinforcing dikes and levees and constructing new flood barriers. While the mix of tools will vary from place to place, a longer term perspective is essential to achieving coastal resilience.

Coastal resilience requires action at several different geographical scales: from the individual or household level, all the way to regional and bioregional scale. Coastal resilience is best thought of as a *nested* planning and policy system: actions can and must be taken at the individual level (e.g. strong home construction standards), but many things must occur at broader geographical scales (e.g. regional land use and growth patterns that steer population and development away from high-risk locations, regional systems of green infrastructure). Planning interventions can (and must) occur at multiple scales and ideally lead to an integrated, interlocking set of resilience measures, from rooftop to region.

### 6.4.2 Coastal Resilience Requires a Compelling Vision of the Future

Any coastal resilience strategy is doomed to ultimate failure unless citizens and businesses and public officials embrace it as a positive and compelling vision of the future. Partly this is about the community dialogue and frank discussion about the community's future exposure and vulnerability and partly about putting the pieces together in a way that conveys this positive future in a visual and compelling way. Is this a community you and your family want to live in, a place that values the unique and special place attributes? It is necessary to have a vision that conveys the possibility of dramatically improving quality of life and at the same time reduces the vulnerability to natural disasters.

This long term coastal vision can be expressed and conveyed in different ways. Maps can significantly shape perceptions of the future, as can renderings and visuals that convey a sense of what is possible. A community's comprehensive plan is a key document for presenting this future vision, in words, maps and images, but certainly not the only tool where this can occur. And every effort should be made to capture local knowledge of hazards and to incorporate that into the planning process.

### 6.4.3 Guide Growth and Development Away from High-Risk Locations

*Avoidance* is ultimately the most effective and sensible approach to resilience in the face of physical forces. Land use planning, and a variety of implementation tools from zoning to transfer of development rights to conservation easements and land acquisition, can be used to steer development and people *away from* and *out of harm's way*. A variety of coastal hazards are already mapped and delimited—high-erosion zones, floodplains, areas prone to subsidence, earthquake fault zones and high-slope areas subject to slides and mass movements—and these areas should be left undeveloped ideally, or developed at low densities. These are areas, moreover, where opportunities will exist for more resilient sustainable relocation following a hurricane, earthquake or other disaster event.

A community's long range spatial vision and planning can do much to minimise future risk and enhance resilience and there are many positive examples of coastal communities moving in this direction. Worchester County, Maryland, that state's only oceanfront county, has adopted a comprehensive plan that seeks to steer future growth into its historic towns, away from high-risk oceanfront locations (Beatley 2009). In Virginia Beach, designated strategic growth centers, including a new town center where higher density growth is envisioned are largely located outside of areas that will be affected by sea-level rise and inundation (UVA 2011). Coastal communities such as Collier County, Florida, have utilised transfer of development rights (TDR) to shift growth away from coastal wetlands to desired growth areas landward, and King County, Washington, has actively purchased the lands most vulnerable to flooding. There are many different tools available to influence and steer growth, but a community's comprehensive plan should lay the foundation.

### *6.4.4 Re-imagine the Edge*

Sea-level rise, both current and projected, has engendered a host of new ideas about what cities and communities can plan for, including the appropriate occupation of the ocean and shoreline edges. This requires a new view of these "edges" as more fluid and dynamic, necessitating more creative adaptation measures.

New ideas have been proposed for New York City, for instance, that accept the reality that periodic flooding will likely be a common occurrence in many of the city's low-lying areas and edges. A recent exhibit at the Metropolitan Museum of Art (MoMA) called Rising Tides, explored these future possibilities for a New York City 70 years into the future. The vision is, as New York Times architectural critic Nicolai Ouroussoff said, one of a more "watery" New York, where hard edges and infrastructure give way to those that are more "soft" (Ouroussoff 2011). Architectural and landscape design teams fleshed out future options and visions for five different parts of the New York waterfront, proposing a variety of soft edge alternatives, including new artificial islands with housing floating on top, while much of the Manhattan waterfront would become wetlands and watery edged. The city would become much more harbour-oriented, and there would be many new opportunities, for instance, to install new oyster nurseries (see Bergdoll 2011).

Part of the task of re-imagining the edges will involve developing new patterns of usage and new notions of the temporary occupancy of edge spaces. Here there is inspiration in the longstanding adaptation practices and behavioral resilience in cities like Venice, where the impacts of high water conditions (*acqua alta*) are a common condition, and where residents can often expect to have to cross parts of the city on elevated scaffolds and carry boats to traverse inundated streets.

In less urban environments, other opportunities for re-imagining the edges will exist. Already, many coastal states and communities impose setback requirements, restrictions on building non-movable structures, and various versions of rolling easements (e.g., the Texas Open Beaches Act and the common law doctrine it codifies; see Beatley et al. 2002). These existing coastal adaptations will need to be expanded and strengthening in the decades ahead.

### 6.4.5 *Locate Critical Facilities Out of or Away from High-Risk Locations*

The ability of a city to weather a major natural event will depend greatly on the design and siting of critical facilities—these include basic infrastructure such as municipal sewage collection and treatment, water supply systems, roads and highways, shelters and critical medical facilities among others. The first priority ought to be to ensure that these are sited to avoid or minimise exposure. Local case studies described here from Florida discuss, for instance, that the state's hazards planning system prohibits the siting of new critical facilities in Coastal High Hazard Zones.

There are a number of positive examples of coastal communities that have taken steps to ensure that their lifelines and critical facilities are resilient. Cannon Beach, Oregon, for instance, has been gradually taking steps to move its critical facilities outside of its high-risk "local" tsunami inundation area (Beatley 2009). Worcester County, Maryland, has undertaken an inventory of critical facilities, and most, including municipal sewage treatment plants, are well away from floodplains and are located on upland, in-town sites. Ocean City, Maryland, has been gradually placing power and telephone lines underground (Beatley 2009).

### 6.4.6 *Plan Ahead for a Resilient Recovery and Growth*

Resilience requires advance planning and being prepared ahead of an event. This means the need to think systemically before the storm or earthquake about how the community might rebuild and redevelop in ways that will reduce exposure and enhance long term resilience, and in ways that will allow for adaptation and learning while taking advantage of post-disaster opportunities. Are there especially dangerous areas where rebuilding should not occur? How will the community ensure that local businesses are able to recover and cope and where will they function and operate during the period of reconstruction?

New planning instruments and approaches will be helpful in taking this longer-term view, including the preparation of disaster recovery and rebuilding plans, and long-term growth and land use modeling that will help to show what the implications are of maintaining a business-as-usual approach to community and regional growth. Some communities, such as Palm Beach County, Florida, have developed ahead of time detailed business recovery strategies, including identifying business relocation and recovery sites in the community and other steps to prevent the loss of core businesses following a major disaster (Beatley 2009).

### 6.4.7 Preserve and Restore Ecosystems and Ecological Infrastructure

A city or region's natural ecosystems and green infrastructure represent one of the clearest and most important lines of defense against many natural hazards (Walker and Salt 2008). Protecting natural coastal marshes and wetlands that soak up and absorb flood waters, dune and beach systems that act as natural seawalls, trees and healthy tree canopy that shields homes against wind are all positive steps that increase long term resilience.

The ecosystems and natural environments of coastal regions are at once *subject to* perturbations and impacts of natural events such as hurricanes (and longer term changes such as global climate change and sea-level rise) and important *moderators of* the impacts of these forces on people and built form. We can meaningfully speak, then, about both of these important dimensions: the resilience of these ecosystems and natural systems in the face of perturbations, and the important role they play in enhancing the resilience of built environments and human communities.

Examples of planning actions that would be taken to ensure the ecological resilience in the former category might include: actions that ensure sufficient wetlands buffers, to permit coastal wetlands to migrate landward in response to long term sea-level rise, or to protect ecological systems and land areas (landscape) that are sufficiently large, complex and diverse that any particular perturbation (storm, wild fire) will not cause irreversible harm (e.g., extinction of a species, complete loss of a biological community).

Examples of planning for ecological resilience in the latter category might include: ensuring the existence and health of beach and dune systems recognising that they are effective flood barriers, and preserving extensive coastal marsh systems because they act as natural sponges, retaining large amounts of flood waters. Indeed many of the actions that could be taken to enhance ecological resilience of one type will help to advance the others.

### 6.4.8 Promote a Diverse Local Economy

The ability of a city or community to spring back quickly from a disaster will depend heavily on the business sector and the extent to which business are able to re-open expeditiously. There is a growing sense that a local economic base characterised by a diverse number of locally owned stores and businesses, committed to the community and region, will promote local sustainability overall but will also likely make the community more resilient following a hurricane or other significant hazard event. According to Campanella: "A city with a robust, diversified economy, for example, will rebound much more quickly than a city with a narrowly specialised or weak economy" (2006, p. 143). The U.S. Green Building Council's New Orleans Principles strongly advocate support of locally owned sustainable businesses including reconstruction-centered businesses, waste-based industries, solar roofing and other sustainable businesses, agriculture and eco-tourism, among others (U.S. Green Building Council 2005).

A community will also respond more quickly and resiliently if it is prepared in advance to assist local businesses in recovering (and staying) and dealing with the host of issues that they will confront. To the extent the businesses have planned for disaster ahead of time, this will increase the likelihood that residents are in turn able to recover more quickly.

Kathryn Foster has constructed an aggregate measure of the adaptive capacity of regions, called the Resilience Capacity Index. The index, which is part of the Building Resilient Regions initiative, reinforces the importance of economic (and social) variables. "Regional economic capacity" is, in Foster's framework, a key determinant of resilient regions, including economic diversification (Building Resilient Regions, undated).

### 6.4.9 Work Towards a Landscape of Resilience

In line with the above principle of multi-scale strategy there are many ways in which urban and suburban landscapes can be designed to be more resilient in the long run. The use of Low Impact Development (LID) and other innovative urban greening and stormwater management strategies are perhaps the best examples. LID can be achieved by designing buildings with green rooftops, reduced and permeable paving, rain gardens, Xeriscaping (planting yards, gardens and public spaces with drought-resistant native plants) and other natural and green features. In many parts of the U.S., drought and summer heat are significant hazard events and these techniques help make urban landscapes more resilient. The concept of Xeriscaping can be expanded to include the use of wind, flood, fire and ice-storm resistant native plants.

Coastal localities such as Seattle have pioneered LID ideas, investing in green streets and the concept of green grids, where conventional stormwater engineering and hard surfaces are replaced with bioswales, permeable paving and other decentralised stormwater management techniques. In part this represents an important change in philosophy about stormwater—that it is not something to be whisked away, but something visible and should be celebrated at an urban neighbourhood level.

The American Society of Landscape Architects (ASLA), in collaboration with the Lady Bird Johnson Wildflower Center, and the U.S. Botanic Garden, has been developing a comprehensive set of guidelines and standards for sustainable landscapes that will likely prove highly useful in coastal areas. An emphasis in the standards is placed on preserving soil, avoiding use of chemicals, using native plant species, designing landscapes to minimise consumption of energy and water and utilising sustainable planting materials (local materials with low embodied energy; see Venhaus 2008). Called the Sustainable Sites Initiative, a landscape rating system is envisioned (similar to LEED for buildings, and will likely be integrated into this programme of the U.S. Green Building Council). A preliminary report identifying some 200 sustainable landscape design strategies has been released, with a final set of guidelines issued in 2009 (ASLA 2008).

### *6.4.10 Design for Passive Survivability and Sustainability*

There are many new and interesting ideas for designing homes and buildings in ways that better respond to changing circumstances, including climate change and rising sea levels. The Dutch have been building floating homes and entire neighbourhoods that can withstand severe fluctuations in water levels. Rotterdam leads the way in this thinking, and has recently designed a “floating pavilion,” powered by the sun and made from a material lighter than glass, the structure is in part a signal of the city’s intention to develop floating neighbourhoods in areas outside protective levees (Rotterdam Climate Initiative, undated).

Homes and buildings should be designed to ensure “passive survivability”. At the level of an individual home or building the idea is in many ways to return to the old design ideas as homes used to be constructed with high ceilings, utilising natural ventilation and natural light that permitted their functioning as living spaces in the absence of electricity.

Following Hurricane Katrina, there has been much discussion of how homes and buildings could be designed to not just fall down, or fly apart during a disaster event, but to ensure conditions of livability for its occupants following the event. Can buildings be designed to be “survivable” or inhabitable for some decent period of time, and under conditions where usual public services and facilities (power, water) have been disrupted? Hence the notion of *passive survivability*, which Alex Wilson, of the Environmental Building News and one of the most vocal advocates for the concept, defines as the “ability of a building to maintain critical life-support conditions for its occupants if services such as power, heating fuel, or water are lost for an extended period” (Wilson 2005).

Passive survivability is a major theme in “The New Orleans Principles,” derived from a 3-day charrette convened by the U.S. Green Building Council in 2005. As recommended in the principles, such buildings:

> “should be designed to maintain survivable thermal conditions without air conditioning or supplemental heat through the use of cooling-load avoidance strategies, natural ventilation, highly efficient building envelopes, and passive solar design. Schools and other public buildings should be designed and built with natural daylighting so that they can be used without power during the daytime. Co-locate healthcare facilities with schools as part of the community anchor and to strengthen survivability.”

Other specific recommendations include encouraging installation of emergency water systems and rooftop rainwater harvesting systems in homes and public buildings, backing up electricity generation for municipal sewage systems and installation of solar electric and solar hot water heating systems.

One result of designing for “passive survivability” might be a rediscovery and return to some of the building vernacular and wisdom found around the country. As Alex Wilson says:

> "There was a reason why homes in the Southeast had wide porches and large roof overhangs 200 years ago, why the New England saltbox had most of its windows on the south, and why homes in the Midwest's tornado belt were so often bermed into the ground. A design criteria of passive survivability would bring back these vernacular styles" (Wilson 2005).

Passive survivability has been a key design feature in several major projects in New Orleans in the aftermath of Katrina. The Holy Cross project, planned and funded by Global Green, in the lower Ninth Ward, explicitly includes these design features. Among other features of the homes they are located on higher ground, living spaces are elevated substantially above what is required by code, and materials used are intended to limit unhealthy living conditions in the homes—for instance using rigid foam insulation that dries more quickly, and paperless drywall that limits the formation of mold, among other features.

In addition to designing passive survivability at the building level, *neighbourhood* passive survivability should also be considered. Extending this idea might suggest some additional elements to take into account: fruit trees and edible landscaping, though subject to damage in flood events, might serve as a part of a resilient neighbourhood food system, when as in New Orleans it may take months for conventional grocery stores to reopen or restock their wares sufficiently. There are undoubtedly other food dimensions to passive survivability, and it is important to understand this concept as broadly as possible.

### *6.4.11 Design and Build Decentralised Resilient Infrastructures*

Consistent with the notion of "passive survivability" for individual homes and buildings, emphasis can and should be given to replacing highly vulnerable centralised and "rigid" infrastructure systems, such as centralised power production and distribution, with distributed and decentralised systems that reduce exposure and vulnerability during a disaster and that also provide robustness and redundancy. And consistent with Walker and Salt's criteria, such decentralised forms of infrastructure would exhibit greater modularity (e.g. one unit's loss or failure need not affect others).

Greater reliance on localised energy production from photovoltaics and other building and neighbourhood-based renewables, for instance, will help reduce vulnerability following a storm or earthquake, and reduce the likely time of service interruption. On-site stormwater collection and treatment, through low-impact development techniques (rain gardens, green rooftops), reduce reliance on centralised stormwater collection systems vulnerable to damage and failure. Many other examples of more sustainable and decentralised forms of coastal infrastructure can be cited.

A major trend involves the construction of more decentralised or "distributed" forms of infrastructure, and cities and communities that have experienced the damage and loss of service from disasters are at the forefront in investing in this infrastructural approach. Houston is a case in point, having experienced significant damage and (relatively) long electrical service disruptions to its citizenry, following Huricane Ike, causing thousands of residents to lose power for weeks. Still reeling from the impacts of the coastal storm, which struck in September 2008, the Mayor appointed a task force to review the resilience of the city and region's electricity system. The task force's report and recommendations, issued in April of 2009, further buttress the shift towards sustainability (City of Houston 2009). Among the key recommendations is the need to move towards more resilient distributed energy systems such as solar and combined heat and power production, as well as investments in a more intelligent grid. Developing a master list of vulnerable populations and critical facilities in the city and region, encouraging personal readiness (including personal investments in solar and including the intriguing possibility of plug-in hybrids along with two-way inverters helping to power homes in the aftermath) as well as smart vegetation management will likely pay dividends in future hurricanes, but will also make the city safer, more sustainable and better able to adapt to a range of future circumstances.

Resilient forms of urban infrastructure increasingly recognise that multiple objectives and functions must be accomplished, and that single-purpose infrastructure will be too costly in the future. Innovative examples exist in many cities. Sydney, for example, is pioneering the concept of "green transformers," in which CHP power stations serve as sites of stormwater retention and water reclamation and reuse, as well as new public parks (see Beatley and Newman 2008). The City of Rotterdam, as part of a larger climate adaptation programme, is designing new parks and play areas that will also serve as temporary floodwater detention facilities. Called "water plazas" they address the need to find places in dense urban settings for floodwater retention. A water plaza is described as "a below-grade plaza designed to accommodate and temporarily retain excess water from the surroundings, to relieve pressure on the urban water system and mitigate flooding in the streets" (Rotterdam Climate Proof 2010).

How communities respond to rapidly rising sea levels represent a special infrastructural challenge. Hard engineering solutions, hard adaptations as they are called by some, are still viewed favorably. This is despite the fact that dikes, levees, seawalls and floodgates are extremely expensive, take years to complete and are highly damaging ecologically. Nevertheless, they remain a favored option especially in the engineering community. Softer alternative approaches, such as beach renourishment, are also expensive, and temporary in nature. Some form of long-term retreat is likely inevitable (or strategic retreat, that is retreat from particularly vulnerable locations such as Galveston Island), and this could happen through active relocation (e.g., moving structures) or long term land use planning and regulation (e.g., steering development and infrastructure away from high-inundation zones). Relatively recent hurricane events, such as Ike, that devastated the Bolivar Peninsula, resulted in some promising efforts to relocate a large number of coastal properties away from high-risk beachfront locations.

### 6.4.12 Plan for Long Term Community Sustainability

A desire for a more sustainable future is one increasingly expressed at individual and collective levels and community sustainability has been gaining much traction in recent years. Sustainability and resilience are usually mutually-reinforcing and should be viewed as such. Protecting landscapes and natural systems, a goal of sustainability for many different reasons, usually helps to preserve the long-term resilience of these systems and the human and built environments around them.

Designing energy-efficient homes and buildings will often tend to make them more resilient in the face of hazard events, as well as help to make the broader coastal region more resilient (i.e., reduced energy loads, chances of brown-outs and allowing for the transition to cleaner, healthier, and more renewable forms of energy production). Investing in greater community food security (e.g., programmes for supporting local farms and farmers, community markets, etc.) will advance a number of sustainability goals, and also enhance the resilience of the community in the face of a future major event or perturbation such as a hurricane or a rapid decline in global oil supplies.

### 6.4.13 Think Holistically

The ability of a community to respond to and adapt successfully to a major disaster requires a level of holistic thinking not yet common, even though natural hazards touch every aspect of daily life as well as a broad range of environmental, social, and economic issues that define a community and region. Resilience is also about a number of non-traditional planning subjects such as the availability of food and community food systems, and provision of energy and water.

For instance, resilience thinking involves addressing underlying community trends that exacerbate social and economic vulnerability in the face of natural hazards. These include existing inadequacies in housing quality and availability, underlying poverty and food insecurity. Enhancing and building long term resilience in the face of hazard events will also require a concerted effort to address the underlying patterns of vulnerability that are predictive of post-event conditions.

### 6.4.14 Achieve Equity Through Resilience

Steps taken to make coastal communities more resilient can also make them more equitable in the long run. Efforts to achieve coastal resilience must acknowledge the profoundly differing vulnerabilities of individuals and groups, as well differing distributions of benefits and costs associated with resilience outcomes. Resilience policies and projects should aim to reduce burdens on the most vulnerable in the community, and seek to distribute the positive community and neighbourhood benefits and amenities associated with resilience (e.g., greener, more resilient landscapes) fairly and in ways that fully benefit all socio-economic groups. In short, coastal resilience must acknowledge an ethical obligation to address the profound distributive implications of these efforts and strive to treat all individuals and groups fairly. While what specific ethical or equity standard to apply remains an open question, coastal resilience should be understood as a unique opportunity to raise the life prospects and living conditions of the most disadvantaged in the community and remove or at least reduce vulnerabilities to the next storm or hazard event.

Efforts to include passive survivability features in new or renovated housing for lower income residents offers the chance, for instance, to combine equity and resilience. Finding new ways to integrate food production into neighbourhoods, as a further example, provides the chance to increase resilience and food security together. *Equitable Resilience*, then, should be a key goal.

### 6.4.15 Promote Social Resilience by Nurturing Critical Social Networks and Institutions

A coastal community has little hope of achieving or substantially advancing resilience without placing adequate attention on the social realm. Indeed, a resilient coastal community is one that has a strong social system and network of social relationships. We know that in times of stress and crises these social networks and relationships can provide important support systems. Research increasingly shows the value of deep and extensive friendship patterns in recovering from disease (e.g. lower mortality and higher recovery rates for cancer patients with deeper more extensive friendship patterns). Friendships, knowing ones' neighbours, well-developed patterns of community and neighbourhood socialising and sharing represent significant and important ways that a community can be prepared for a future hazard event or crisis.

Resilience in recovery and reconstruction will depend heavily on this social realm. As Campanella notes,

> "Broken highways can be mended, buildings repaired and made taller than before, communications systems patched back together. But cities are more than the sum of their buildings. They are also concatenations of social and cultural matter, and it is often this that endows a place with its defining essence and identity. It is one thing for a city's buildings to be reduced to rubble; it is much worse for a city's communal institutions and social fabric to be torn apart as well. To enable total recovery, familial, social and religious networks of survivors and evacuees must be reconnected" (2006, p. 142).

Putnam and others refer to this as the "social capital" of a community, and building up and strengthening this capital may be an equally important strategy for enhancing resilience as strengthening and building up the homes and buildings (see Putnam 2000; Putnam et al. 2005). Social capital "includes all the formal and informal networks among people: family, friends and neighbours, as well as social institutions at all levels, like churches, social clubs, local, state, and national governments, NGOs, and international organisations" (Costanza et al. 2006, p. 318).

A robust and extensive set of social networks and institutions will both help in achieving a more effective and equitable recovery and will also allow a community to resiliently weather the event and to perhaps inoculate itself against the most severe impacts. Extensive and healthy "social capital" can assist in each disaster phase—if neighbours know each other, and if communication and social networks are strong between and among neighbours, locally owned neighbourhood stores and religious institutions.

### *6.4.16 Work to Build Resilience at the Grassroots and Neighbourhood Levels*

While all geographical scales are important, effective building of more resilient coastal communities will often require bringing these issues down to the level of a neighbourhood. Engaging at the grassroots level, finding meaningful ways in which neighbourhoods can work together to become more resilient in the face of climate change, natural hazard events and other shocks represent an important ingredient of success. The concept of passive survivability at the neighbourhood level is important and already mentioned. A number of examples exist of coastal communities creating other ways in which neighbourhoods can participate in and organise around resilience. Opportunities to participate in Community Emergency Response Teams (as is evident in communities such as Charleston County, South Carolina, USA) are helpful and plans to develop a neighbourhood based network of "tsunami buddies" in Cannon Beach, Oregon, USA, provides another example. Programmes that provide opportunities for direct hands-on participation in local environmental restoration and cleanup efforts offer similar benefits of place-strengthening and relationship and friendship-building that will help in times of disaster or stress (e.g. programmes such as CoastCare and Sustainability Street in Australia).

A recent study by Oxfam of Cuba's resilience in the face of hurricanes provides further evidence of the essential value of strengthening the community capacity to plan for and respond at the grassroots to hazards and disasters. While the country has experienced six major hurricanes between 1996 and 2002, only 16 deaths occurred, a much different story than Katrina, for instance. The Oxfam study attributes Cuba's success to its multi-dimensional approach and to benefits of social development (high levels of literacy, access to health care, rural infrastructural development) and effective preparedness. Especially interesting is Cuba's demonstration of the benefits associated with Community Based Disaster Management (CBDM), which "focuses on strengthening capacity and building skills for risk reduction at the community level" (Oxfam 2004, p. 4). Oxfam describes this as resulting in the development of a "culture of safety."

> "Many Cubans with whom we spoke saw themselves as actors with important roles to play in preparedness and response. Education and training, a culture of mobilisation and social organisation, and a basic trust in the government to prioritise human life in an emergency situation promote this vision" (Oxfam 2004, p. 27).

Neighbourhood risk mapping is at the core, in which there is a specific knowledge of which individuals and families are vulnerable, and what will need to be done to evacuate or take care of them.

### 6.4.17 *Encourage an Active, Healthy Community and Citizenry*

Coastal populations and communities that suffer from high rates of obesity, sedentary lifestyles and unhealthy diets, will not be very resilient in the face of future stressful and challenging natural hazard events and will not likely cope or respond well. One of the best and most effective strategies for promoting long term community resilience is to encourage and facilitate healthier lifestyles, including community and neighbourhood land use patterns that allow for opportunities for outside recreational activities. Resilience, then, will be enhanced through investments in such things as bike and walking trails, water and beach access points, sidewalks and pedestrian and biking facilities of various kinds.

In this way, programmes intended to promote and enhance community resilience, must begin to explicitly connect with the literature and research on healthy communities, which has been growing in significance in recent years (e.g. see Dannenberg et al. 2011).

### 6.4.18 *Engage the Community by Nurturing Forward-Looking Leadership*

Coastal resilience will depend on the active involvement and participation of the community; plans for community resilience (including the community's land use plan to guide future development in more resilient ways) will require community "buy-in" to be effective and for any hope of long term implementation. Mechanisms for citizen participation and community engagement, moreover, are important in deepening a sense of rootedness and caring for place, a sense of bonding and trust between and among residents and technical, political and collaborative leadership (Smith 2011).

There is little doubt that coastal resilience can also be advanced through forward-looking and responsible leadership, especially from elected officials. Many examples exist of bold mitigation and resilience measures that succeeded because they were championed by a mayor or county commissioner. Strong leaders have the potential to form coalitions, build bridges, and work to overcome the usual objections and political impediments that exist to thinking and acting in ways that take the long view and that advance a larger and broader notion of the public interest.

## 6.5 Conclusions

In this chapter an effort was made to identify some key Principles of Coastal Resiliency. These should not be viewed as exhaustive or definitive but as a first attempt at articulating some principles to guide development and planning in coastal environments. Each principle is relatively general and its precise application or meaning "on the ground" will vary. The main intent is to stimulate thinking and discussion about what resilience requires of us: think of them as starting points from which to modify, adapt and develop your own specific community resiliency principles.

## References

American Society of Landscape Architects (ASLA). (2008). *Landscape structures commits $ 200,000 to sustainable sites initiative*. Washington, DC: ASLA. (Press release, June 17).

Beatley, T., Brower, D., & Schwab, A. (2002). *An introduction to coastal zone management* (2nd ed.). Washington, DC: Island Press.

Beatley, T., & Newman, P. (2008). *Green urbanism down under: Learning from Australian sustainable communities*. Washington, DC: Island Press.

Beatley, T. (2009). *Planning for coastal resilience: Best practices for calamitous times*. Washington, DC: Island Press.

Bell, M. L., et al. (2007). Climate change, ambient ozone, and health in 50 U.S. cities. *Climate Change, 82,* 61–76.

Bergdoll, B. (2011). *Rising currents: Projects for New York's waterfront*. New York: MoMA.

Berke, P. R., & Campanella, T. J. (2006). Planning for postdisaster resiliency. *The Annals of the American Academy of Political and Social Science, 604,* 192–207.

Buckle, P. (2006). Assessing social resilience. In D. Paton & D. Johnston (Eds.), *Disaster resilience: An integrated approach* (pp. 88–103). Springfield: Charles C. Thomas.

Building Resilient Regions. (Undated). Resilience capacity index. University of California at Berkeley. http://brr.berkeley.edu/rci/.

Campanella, T. J. (2006). Urban resilience and the recovery of New Orleans. *Journal of the American Planning Association, 72*(2), 141–146.

City of Houston. (2009). *Mayor's task force report: Electric service reliability in the Houston Region*. http://www.houstontx.gov/mayor/taskforce-electricity.pdf. Accessed 7 Jan 2014.

Costanza, R., Mitsch, W. J., & Day, J. W., Jr. (2006). Creating a sustainable and desirable New *Orleans. Ecological Engineering, 26*, 317–320.

Dannenberg, A., Frumkin, H., & Jackson, R. (2011). *Making healthy places: Designing and building for health, well-being, and sustainability*. Washington, DC: Island Press.

Godschalk, D. R. (2003). Urban hazard mitigation: Creating resilient cities. *Natural Hazards Review, 4*(3), 136–143.

Gornitz, V., Horton, R., Siebert A., & Rosenzweig, C. (2006). Vulnerability of New York city to storms and sea level rise. *Geological Society of American Abstract Programmes, 38*(7), 335.

Holling, C. S. (1973). Resilience and stability of ecological systems. *Annual Review of Ecology and Systematics, 4,* 1–23.

Laushe, B. (2009). *Policy tools for local adaptation to sea level rise, Technical report.* Sarasota: Mote Marine Laboratory.

McPherson, M., Lynn S.-L., & Bashears, M. (2006). Social isolation in America: Changes in core discussion networks over two decades. *American Sociological Review, 71,* 353–375.

McGranahan, G., Balk, D., & Anderson, B. (2007). The rising tide: Assessing the risks of climate change and human settlements in low elevation coastal zones. *Environment and Urbanization,* 19(1):17–37.

McKibben, B. (2011). *Eaarth: Making a life on a tough new planet*. New York: St. Martins.

Nicholls, R. J., Hanson, S., Herweijer, C., Patmore, N., Hallegatte, S., Corfee-Morlot, J., Chateau, J., & Muir-Wood, R. (2007). Ranking of the world's cities most exposed to coastal flooding today and in the future. Paris: OECD. https://support.rms.com/publications/OECD_Cities_Coastal_Flooding.pdf. Accessed 7 Jan 2014.

Ouroussoff, N. (2011, March 25). Imagining a more watery New York. *New York Times*.

Oxfam (2004) "Weathering the Storm: Lessons in Risk Reduction from Cuba," found at: http://www.eird.org/isdr-biblio/PDF/Cuba%20Weathering.pdf. Accessed 7 Jan 2014.

Paton, D. (2006). Disaster resilience: Building capacity to co-exist with natural hazards and their consequences. In D. Paton & D. Johnston (Eds.), Disaster resilience: An integrated approach (pp. 3–10). Springfield: Charles C. Thomas.

Paton, D., McClure, J., & Burgelt, P. T. (2006). Natural hazard resilience: The role of individual and household preparedness. In D. Paton & D. Johnstone (Eds.), *Disaster resilience: An integrated approach*. Springfield: Charles C. Thomas.

Putnam, R. D. (2000). *Bowling alone: The collapse and revival of American community*. Simon and Shuster.

Putnam, R., Feldstein, L., & Cohen, D. (2004). *Better together: Restoring the American community*. New York: Simon and Shuster.

Rotterdam Climate Initiative. (undated). Floating pavilion in the centre of Rotterdam. http://www.rotterdamclimateinitiative.nl/documents/Persberichten/RCP-08102009-English-persbericht-pavilion.pdf. Accessed 6 Jan 2012.

Smith, G. (2011). *Planning for post-disaster recovery: A review of the United States disaster assistance framework*. Fairfax: Public Entity Risk Institute.

Tobin, G. A., & Montz, B. E. (1997). *Natural hazards: Explanation and integration*. New York: Guildford Press.

University of Virginia (UVA). (2011). Changing tides: *A sea level rise planning analysis for Virginia Beach, VA*. Charlottesville.

U.S. Green Building Council. (2005). The New Orleans principles. New Orleans planning Charrette, Nov. pp. 9–11.

Venhaus, H. (2008). Sustainable sites initiative, home energy, July/August. www.homeenergy.org.

Walker, B. S., et al. (2002). Resilience management in social-ecological systems: A working hypothesis for a participatory approach. *Conservation Ecology, 6*(1), 14.

Walker, B., & Salt, D. (2008). *Resilience thinking*. Washington, DC: Island Press.

Walker, B., Holling, C. S., Carpenter, S. R., & King, A. (2004). Resilience, adaptability and transformability in social-ecological systems. *Ecology and Society, 9*(2), 5.

Wilson, A. (2005). Passive survivability. *Environmental Building News, 14*(12). http://www.buildinggreen.com/auth/article.cfm/2005/12/1/Passive-Survivability/. Accessed 7 Jan 2014.

Wolman, D. (2008). Before the levees break: A plan to save the Netherlands. *Wired Magazine*. http://www.wired.com/science/planetearth/magazine/17-01/ff_dutch_delta?currentPage=all. Accessed 7 Jan 2014

# Chapter 7
# Adaptation to Seismic Risk and Climate Change: San Francisco and Berkeley, California, USA

**William Siembieda**

**Abstract** This chapter examines the multi-sector resiliency and climate adaptation approaches undertaken by the San Francisco Bay Area cities of Berkeley and San Francisco. The Paton and Johnston disaster resiliency model is used to assess the adaptive capacity of these two cities. Findings show that "partnering" with civic, business and community based organisations (CBO) is a central element for producing sustainable resiliency and lowering disaster risk. San Francisco adopted a Climate Action Plan in 2004, followed by Berkeley in 2009. Both cities use state government-generated data to assess risk and their own funding to lower risk and adapt to climate change. The key lesson uncovered in this analysis is that it matters less which level of government (local, state, federal) provides guidance; rather it matters more that guidance is based on accepted science that can be easily accessed and used by anyone interested in resiliency and climate change adaptation.

**Keywords** Resiliency · Mitigation planning · Social capital · Risk reduction · Climate change adaptation · San Francisco · Berkeley

## 7.1 Introduction

The San Francisco Bay Area has many natural hazard risks and climate change adaptation challenges. Among the most vulnerable United States metropolitan areas for seismic events with active faults in all nine counties, it is also subject to climate change impacts from sea-level rise. The Bay Area has many forms and layers of governance and urban services providers (transport, electric, gas and water) that operate semi-independently. By necessity, actions addressing hazard risk and climate adaptation require the use of multi-stakeholder solutions.

Using organisational and behavioral constructs this chapter examines how two San Francisco Bay area cities (San Francisco and Berkeley) have developed their seismic risk reduction approaches and how these, in part, have paved the way for

W. Siembieda (✉)
California Polytechnic State University, San Luis Obispo, CA, USA
e-mail: wsiembie@calpoly.edu

B. C. Glavovic, G. P. Smith (eds.), *Adapting to Climate Change,* Environmental Hazards,
DOI 10.1007/978-94-017-8631-7_7, 

them to address climate change adaptation. The organisational construct allows linkages between resources and understanding of local needs to be examined in terms of risk reduction actions and recovery (Smith 2011; Wisner et al. 2012). The behavioral construct also provides a way to link the personal, community and institutional environments together in terms of self and community efficacy (Paton and Johnson 2006; Norris 2008). These approaches overlap in their premise that well developed horizontal and vertical integration of collective efforts on the part of participating stakeholders yields lower risks, be they natural hazard or climate change-induced.

### *7.1.1 Climate Adaptation Regulatory Context*

California has a progressive state government that pays attention to global warming, carbon emissions and greenhouse gases. State legislation in 2006 (AB32 California Global Warming Solutions Act) set carbon reduction targets, including a rollback to 1990 emissions levels no later than 2020. This has resulted in various actions by state agencies, as well as by cities and counties. Cities and counties have responded by developing Climate Action Plans that include an inventory of emissions sources and actions to meet identified targets. Since 2006 all nine Bay Area counties and 34 cities (representing one third of all municipalities in the Bay Area) have completed and implemented Climate Action Plans. In 2004, two years before the state legislation was passed, the city of San Francisco set its own target to be 20 % below 1990 emission levels by 2012. This early adopter stance on the part of San Francisco made it a leader among US cities addressing climate change.

California state government does not tell municipalities or counties how to reach their greenhouse gas emissions targets, nor does it provide direct funding to cities to reach the targets. Rather, the state provides guidance and information in various forms; among them the Energy Commission that funds science studies related to greenhouse gas emissions and modeling of climate change. California's Natural Resources Agency funds applied research that has resulted in the California Adaptation Planning Guide series of documents used by professionals at the municipal and local levels (see http://resources.ca.gov/climate_adaptation/local_government/adaptation_policy_guide.html). California State agencies also provide cities with examples of positive action. For example, the State Department of Water Resources, which operates the 700 mile long state water project (bringing water from the north to the south through what is known as the Central Valley), will reduce annual greenhouse gas emissions by more than 1 million metric tons by 2020, and by more than 2.5 million metric tons in 2050 by introducing new technology in water transport, the use of renewable energy and the purchase of carbon offsets. These actions are targeted to reduce greenhouse gas emissions to 50 % below 1990 levels. It is believed that municipalities will learn from the state water agency experience in terms of programme design and implementation efforts (see http://www.water.ca.gov/climatechange/CAP.cfm).

**Fig. 7.1** San Francisco bay area counties. (Source: Association of Bay Area Governments 2013)

In 2008 California strengthened its commitment to managing the impacts from sea-level rise, increased temperatures, shifting precipitation and extreme weather events when Governor Arnold Schwarzenegger signed Executive Order (EO) S-13-08. The order called on state agencies to develop California's first strategy to identify and prepare for these expected climate impacts. The California Natural Resources Agency (CNRA) produced the California Climate Adaptation Strategy (CAS) through the work of seven sector-specific working groups: public health; ocean and coastal resources; water supply and flood protection; agriculture; forestry; biodiversity and habitat; and transportation and energy infrastructure (CNRA 2009). The strategy contains three broad directions regarding sea-level rise, and each represents a means to address this threat (Johnson and Tam 2012). First, the strategy directs state agencies to avoid permitting or siting new development in areas that cannot be adequately protected and that are at high risk of flooding, wildfire or erosion due to climate change. For instance, the California Coastal Commission uses their oversight of development in coastal areas of the state for this purpose. The second part of the strategy is to amend the California Environmental Quality Act (CEQA) so project review considers the potential impacts of locating new projects, including infrastructure, in areas susceptible to climate change impacts. The third part requires local governments to consider climate change impacts and identify vulnerable areas when updating general plans and coastal plans. The state however, does not tell any city or county how to address climate change or natural hazard risk.

### *7.1.2 San Francisco Bay Area Characteristics*

The 4.39 million acre San Francisco Bay Area is comprised of 101 municipalities, nine counties (government units that provide public health, courts and social support services above the municipal level) and semi-independent districts that manage rapid transit, air quality and wastewater services (see Fig. 7.1). The region contains

7,150,000 people, with growth rates between 2000 and 2010 of 0.5%. Its economy ranges from world-class technology firms in the South Bay (Google, Apple, Cisco Systems, Facebook, etc.) to world-class wine growing regions in the North Bay (Napa and Sonoma Valleys). It has 10 universities, including the University of California, Berkeley and Stanford. Of the 7 million people, 30% are foreign born, while 23% are Asian and 23% are Hispanic, making the region quite diverse. Nearly 3.5 million people are employed, with per capita income at $ 38,294, and management, professional and related work make up 44% of the workforce (American Household Survey 2010).

### *7.1.3 Sea-level Rise*

Historic records show that sea-level in the San Francisco Bay has risen by as much as 7 in. in the past century. Based on research conducted by scientists at the U.S. Geological Survey (USGS), a sea-level rise of 16–55 in. over the next century will affect the shoreline of the Bay and Delta, including an increase in the risk of levee failure. The San Francisco Bay Conservation Development Commission (BCDC) has developed maps depicting the areas that are most vulnerable to sea-level rise.[1] The maps depict 16 in. of sea-level rise at midcentury and 55 in. at the end of the century, respectively. Associated impacts are most extensive in the low lying north and south bay areas, as shown in Fig. 7.2.

The BCDC maintains a special permitting plan within the sea-level rise zone. One high impact scenario estimates 270,000 people and $62 billion in economic resources are at risk of flooding by the end of the century due to a projected 55-in. sea-level rise. Some critical facilities at risk from sea-level rise include the Oakland and San Francisco Airports, and the Port of Oakland, as well as marinas, piers, walking and biking trails and natural habitat. Of the 840 critical health care facilities in the Bay Area, 16 are in the 55-in. inundation zone, while 10 are in the 16-in. inundation zone. Of the 6,153 critical facilities owned by cities, counties and other special districts in the Bay Area, 228 are in the 16-in. inundation zone, while 291 are in the 55-in. inundation zone. In general, sea-level rise adaptation will require a focused set of targeted actions over the next 90 years, however, there are no near term large-scale threats.

### *7.1.4 Earthquake Risk*

Earthquake is the natural hazard with the largest potential to negatively impact the San Francisco Bay Area. Most seismic damage is due to ground shaking, with relatively little due to liquefaction and land sliding (see Fig. 7.3). For example, in the

[1] The 27-member San Francisco Bay Conservation and Development Commission (BCDC) was created by the California Legislature in 1965 in response to broad public concern over the future of San Francisco Bay. Its jurisdiction includes the open water, marshes and mudflats of greater San Francisco Bay and the first 100 feet inland from the shoreline.

**Fig. 7.2** Sea level rise estimate, 2050 and 2100. (Source: Bay conservation and development commission, http://www.bcdc.ca.gov/planning/climate_change/index_map.shtml)

1989 Loma Prieta earthquake, only 1.6 % of the $6 billion in losses could be attributed to liquefaction, and an even smaller percentage to landslides. Surface fault rupture can do significant damage to infrastructure systems, depending on the earthquake (Multi-Jurisdictional Local Hazard Mitigation Plan, ABAG 2010 Update). Of the 4.39 million acres of land in the Bay Area, 2.2 % is located in areas mapped as having very high liquefaction susceptibility, while 23.8 % of the land area is mapped as falling within the moderately high-very high liquefaction susceptibility category. Existing urban land uses with the highest percentages in those areas mapped as having very high liquefaction susceptibility are mixed commercial-industrial complexes (17.8 %), military use (15.1 %) and industrial (11.3 %).

### *7.1.5 Applying the Conceptual Framework*

A premise evaluated in this chapter is that resiliency capacity is the sum of efforts between government and its citizens expressed through shared responsibilities and acceptance of the risk conditions they understand at any point in time. It reflects a

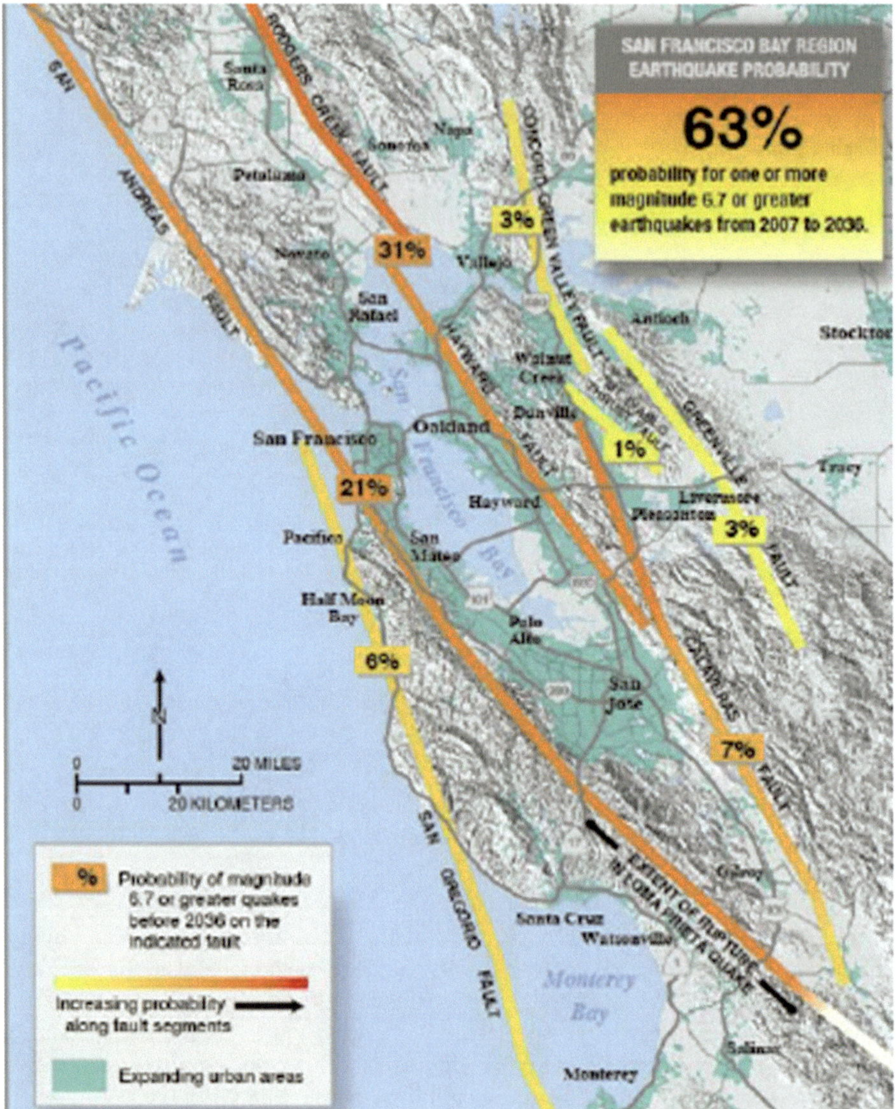


**Fig. 7.3** Earthquake risk in the San Francisco bay area. (Source: USGS 2007)

paradigm shift from government "managing the city" to government sharing decisions and responsibilities with its citizens. The actions in these cities are taken with a view of influencing the post-disaster condition by addressing the pre-disaster condition. The acceptance of risk becomes the most important step to reducing risk and adapting to climate change. Risk acceptance, as an implementation strategy, means that science and engineering information is continuously collected and disseminated to multiple stakeholders as part of an ongoing policy dialogue and is used

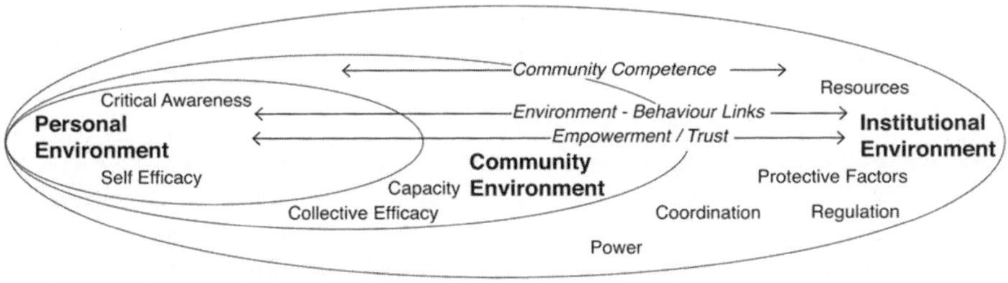


**Fig. 7.4** Disaster resiliency model. (Source: Adapted from Paton and Johnston 2006)

to inform adaptation actions. The cities of San Francisco and Berkeley represent examples of the movement to recognise social as well as the physical infrastructure resiliency needs and to take action based on this recognition (Siembieda 2010; Alesch and Siembieda 2011; Alrich 2012; Wisner et al. 2012).

To promote cohesive action, Paton and Johnston put forth an adaptive capacity model that includes individual, community and societal resources, coupled with the mechanism that facilitates interaction within and between those resources (2006, p. 309). This is shown in Fig. 7.4 along with three cross-domain activities that need to occur to support the capacity to address an adverse event. This model allows resilience to be understood from the individual, community (collective) and societal (institutional) domains. [2] Building adaptive capacity according to Paton and Johnston requires a corresponding level of reciprocity between and among the domains. This builds trust. Each must, in various ways, listen and support each other.

This chapter examines, in part, the model through an institutional and pragmatic framework lens. Paton and Johnston provide no metrics (measures) to calibrate their model, and in fact state that resilience is about "nurturing and sustaining the capacity of people, communities and societies to adapt…and this realisation cannot be prescribed" (2006, p. 315). While this statement makes systematic analysis difficult, it does recognise that socio-economic systems are in a state of constant change and adaptation. Using an inductive approach we explore resiliency through the examination of the actions taken by stakeholders in case study cities.

Harvard University's Acting in Time (AIT) Disaster Recovery Project could be considered a Paton and Johnson variant. AIT provides a more prescriptive approach to adaptive capacity.[3] The AIT project promotes specific project actions, mostly at

[2] Another way to approach community resilience is provided by Norris (2008). Under the Norris et al. model, community resilience emerges from four primary sets of adaptive capacities—Economic development, social capital, information and communication and community competence—that together provide a strategy for disaster readiness. They present a theory of resilience that encompasses contemporary understandings of stress, adaptation, wellness and resource dynamics. Community resilience in their view is a process linking a network of adaptive capacities (resources with dynamic attributes) to adaptation after a disturbance or adversity. The San Francisco Office of the Administrator uses the work of Norris as a guiding construct.

[3] John F. Kennedy School of Government. (2009) Acting in time disaster recovery project. Cambridge, MA: Harvard University.

the institutional level, to assure response capacity is present after a disaster event. These actions reach down to the community level by focusing on social capital found in neighbourhoods.

The cases presented in this chapter fall into two main streams of actions, or strategies, taken by local governments to adapt to risk and to build resilience prior to a disaster event. The first strategy involves the use of laws, ordinances and directives that are derived externally from higher levels of government. The second strategy involves actions by individual units of local government (e.g., fire department, emergency management, public works department) to address risk and adaptation using their own missions and external requirements for guidance. The institutional environment portrayed by Paton and Johnston is not unitary in and of itself and therefore is subject to variation in practices. For example there is no assurance that different government departments will act in coordinated ways, or share the same values (Smith 2011).

## 7.2 San Francisco

Located on the western edge of the San Francisco Bay area region, this municipality of 120.95 km$^2$ (96.7 square miles), 807,000 people and 368,346 dwelling units, is the most densely populated city in California.[4] It also has high exposure to natural and anthropocentric risk. Having experienced a major earthquake in 1906 and a serious earthquake in 1989 (Loma Prieta), the city is aware of its risks. In the 1990s, it began to develop seismic programmes to strengthen its buildings, but it was not until after Hurricane Katrina (2005) struck that it began to adopt an expanded approach focused on the dual capacity of local government and its citizens to address future seismic events. The goals of this expanded approach were to lower its risk, to lessen the impact of a seismic event it knew would happen at some point in the future and to adopt a more systematic process than ever taken before. To do this, the City created a multi-part strategy:

- Address risk in housing and small commercial buildings through seismic retrofits,
- Strengthen links with neighbourhoods as partners in resilience,
- Involve all major City departments in resiliency and post-event recovery planning,
- Engage and respond to community-based organisations, and
- Build new alliances with private utility companies who are the operators of many critical lifelines for the city.

The Paton and Johnston disaster resiliency model calls for a boundary-spanning empowerment dimension between the institutional, community and personal environments. San Francisco's Resilient SF approach depicts government, community

[4] In 2010, the San Francisco housing characteristics were as follows: 30% single-family homes, 33% buildings with two to nine units and 37% buildings comprised of 10 or more units (San Francisco Planning Department 2010).

and the private sector as being at the same level in a non-hierarchic status (Resilient SF 2011; SF Resilient Initiative 2011). This "flat government approach" includes neighbourhoods and civil society (containing Community Based Organisations, Faith Based organisations and Non-profit Organisations). The model assumes that resilience is something local government does not do by itself. This provides evidence that a component of resilience is the existence of a broad set of stakeholders that promote strong horizontal integration (Smith 2011, p. 23). Horizontal integration means people know what other people, or units, are doing and make appropriate connections that create trust at some level.

In 2009, the San Francisco Mayor's office announced a commitment to cultivate a citywide community disaster preparedness ethos based on a *Culture of Preparedness* (San Francisco Department of Emergency Management 2009). This announcement does not represent a new direction for the City, but rather a clearer means to expand and build on years of previous work dating back to the late 1970s. In 2004, San Francisco Department of the Environment and the Public Utilities Commission issued the first Climate Action Plan in California. The 2012 City General Plan Safety Element update focuses on seismic hazards, but only considers sea-level rise in terms of flood management issues near the bay. One of the 25 mitigation policies in the Safety Element is climate change focused, while three emphasise strengthening the capacity of vulnerable communities.

### *7.2.1 Citywide Post-disaster Recovery Initiative.*

When Hurricane Katrina occurred along the U.S. Gulf Coast, the inability of the City of New Orleans' local government to operate and carry out its normal functions in the aftermath of this extreme event signaled to San Francisco's elected leaders and senior administration that there was a clear need to engage in pre-event planning beyond seismic strengthening efforts. More specifically, the Mayor, the Chief Executive and local business officials witnessed the vast fiscal and operational problems and wanted to ensure a different outcome for their city should a disaster occur in the Bay Area. Thus an external event gave rise to the contemporary San Francisco Resiliency movement. The objectives sought were to lower the impact of an event, to prevent cascading damages and to return to acceptable operating levels as quickly as possible. The General Services Agency (GSA), the Department of Emergency Management (DEM) and the Office of the Controller coordinated this project. The inclusion of the financial and administrative arms of local government established its importance as a citywide effort.[5] The Controller is a fiscal agency, not an operational service agency such as police, fire, parks and

[5] San Francisco is both a city and a county, the only such consolidated unit of government in California. This requires San Francisco to operate the courts, the health services, the tax collection services and other government functions that cities do not provide.

recreation or the superior court.[6] Its broad reach to other agencies is important and the Controllers inclusion in the process is a signal that pre-disaster activities intended to speed post-disaster recovery are taken seriously by the administrative and legislative leadership. The General Services Agency (GSA) comprises a broad array of departments, divisions, programmes and offices reporting to the Office of the City Administrator. The majority of the services provided by the GSA support the effective operations of other city departments.

The San Francisco resiliency approach is comprised of five components:

- Work with independent community-based organisations that are interested in resilience,
- Expand disaster preparedness beyond the traditional emergency services agency to an all-agency approach,
- Involve the private utility sector in the dialogue,
- Support neighbourhood involvement, and
- Seek to instill behavioral change in the populace.

These components have their own sub-parts and task lists such as making sure that private donations can be quickly accepted by the City after an event (a fiscal activity) to upgrading buildings that house social service agencies so that they can quickly return to service.

San Francisco has a wealth of social capital. In addition to having many community-based organisations (CBOs), the city is fortunate to have an independent community-based, member-supported organisation solely devoted to urban policy issues. Begun in 1910, San Francisco Planning and Urban Research Association (SPUR) is an open membership organisation that promotes good planning and good government in the San Francisco Bay Area through research, education and advocacy (SPUR 2008). More recently, it has dedicated professional staff to climate change monitoring and resiliency. In 2007, with the 20th anniversary of the 1989 Loma Prieta earthquake soon to occur, the SPUR Board of Directors began to discuss the notion of a safer and more resilient city. A series of forums were held and in 2008, the SPUR Board adopted the following statement advancing resilience: "What we do before the disaster determines what happens after the disaster. We cannot control or precisely predict what nature will do, but we have it within our power to take steps to make this city resilient in the face of a major earthquake" (SPUR 2008, p. 28). SPUR also set its own targets for recovery by setting a reference point for municipal officials and other community groups (see http://www.spur.org/documents/article020109/target.gif). Since 2008, SPUR has issued a series of topical disaster reports on issues such as the transportation sector and mitigation requirements while sponsoring open public talks.

[6] The San Francisco Controller is responsible for all financial management systems, procedures, internal control processes and reports that disclose the fiscal condition of the city to managers, policy-makers and citizens. The controller is also the auditor for the city and county performing financial and performance audits of departments, agencies, concessions and contracts.

**Table 7.1** SF Resilient programme areas. (Source: ResilientSF (2011))

| |
|---|
| Community Engagement, Capacity Building and Communications |
| Community Infrastructure and Lifelines |
| Finance, Budget and Risk Management |
| Housing and Shelter |
| Economic and Community Development |
| Citywide Planning |
| Emergency Planning and Response |
| Governance, Legislation and Intergovernmental Coordination |
| Environmental Impact and Restoration |
| Resiliency Indicators Project |

Other efforts to strengthen CBOs focus on creating a disaster resilience "standard" for community and faith based service providers (Eisner 2010). This effort, supported by the Fritz Institute, included training in resilience and understanding of the post disaster challenges these organisations would face. Topical areas discussed included how to staff their offices after a seismic event and assessing if the buildings they occupied would remain functional given different earthquake damage scenarios. This type of planning expanded awareness and established improved communications between CBOs and city government. Such efforts to build community effectiveness by supporting new partnerships with city agencies furthers horizontal integration at the institutional level as was evidenced by increased community competence and strengthened ties between the community and other institutional actors, thereby supporting the "flat government model".

The SF Resilient Programme action areas are shown in Table 7.1. Each of these areas, which includes a series of projects associated with it, strengthens the networks inside the government, improves citizen connections, promotes private utility sector participation, helps identify areas of risk related to these organisations and links government programmes to personal interests. Staff resources have been allocated to these areas, and aside from the resiliency indicators, contributed to improved horizontal and vertical integration.

Establishing a useful resiliency indicator is a difficult task to achieve, as no consensus exists on what the term means and how the baseline data is to be derived. This is not surprising given the challenges associated with how measurements are made among various sectors. While indicators would help in testing policy alternatives, their construction and their data requirements for local government have yet to be clearly developed (Cutter et al. 2010).

### 7.2.2 *Community Engagement, Capacity Building and Communications*

The city's Empowered Communities Programme relies on an asset-based organising model, comprised of several components. This effort uses existing networks of CBOs, faith-based coalitions, special interest coalitions and local universities (mainly service learning units) to communicate the resilience message and to deepen participation in a programme to promote technical resilience improvements as part of each group's activities. This process has also helped to build a thread of common action between what were at one time a disparate set of stakeholders. In many ways this ongoing exercise has fostered greater critical awareness, while building trust within and among the community and the local government.

San Francisco has a long history of working with people on behavioral changes. One of its greatest challenges involved the public health sector working with AIDS victims. Through the AIDS project, San Francisco agencies learned how to form behavioral messages including the differences between telling people what to do and having people decide to act on their own behalf. This understanding is used in the SF Neighbourhood Empowerment Programme Culture of Preparedness strategy (DEM 2009). The key elements are directed at one's overall sense of self-efficacy, a key component of the Paton and Johnston resiliency model.

People need to believe their actions can make a difference. For San Francisco, preparedness includes drawing on broad themes tied to knowledge, connectedness, awareness and strength.[7] The focus here is to engage the whole community over the long-term through a social movement strategy that includes: (1) consistent visual cues in neighbourhoods and communities, (2) environmental messaging reframed to align with positive images, and (3) green behavior techniques (e.g., recycling of materials, lower energy and water use).

The San Francisco engagement strategy has three parts. First, align the efforts of the Neighbourhood Empowerment Network (NEN) to build capacity for recovery. Second, use social media tools to script messages and instill prevention and preparedness-related actions as part of community behavior. Third, develop a public relations campaign with a firm specialising in disaster messaging. For instance, the social media group produced a new disaster game (SF Heroes, http://sfheroes.com/) used on smart phones and directed at 19–25 year olds. When sufficient points are earned the player becomes an SF Hero. This behavioral tool is designed to build awareness, and in this case, support self-efficacy. As cell phone use is a dominant communications medium for young people, it also takes the message to the user directly rather than having the user find it in another communications medium.

In 2006, the Mayor's office and other government officials began a dialogue with the community on how the government could better meet local needs by working in tandem with others. The platform emerging from the dialogue is called the Neighbourhood Empowerment Network (NEN). The NEN (www.empowersf.org) offers a paradigm of collaboration that seeks to establish a "flatter government" with less hierarchy and increased consensus-derived actions.

Operating out of the Mayor's office means that the NEN represents political and functional interests. NEN strategies for community efficacy building include social and spatial dimensions. For example, it supports San Francisco's Lesbian Gay Bi Transgender Queer (LGBTQ) community to address existing and emerging

[7] Department of Emergency Management. (2009). San Francisco preparedness strategy. City and County of San Francisco.

challenges, such as cuts to social and health services and thinking about how they would be involved in a recovery effort. The NEN effort builds alliances with local universities that are a source of technical assistance and student labour to assist communities and people. Getting universities involved means more attention to neighbourhoods, the introduction of more systematic techniques of information collection and distribution and the involvement of a younger labour pool in disaster risk reduction activities.

Diamond Heights, located south of the San Francisco Central Business District represents an example of a neighbourhood involved in NEN activities. The St Aidan's Church in Diamond Heights is the institutional home for a neighbourhood resilience pilot plan. St. Aidan's will leverage its social capital to convene stakeholders across all sectors in the community to increase disaster resilience. A cohort of NEN Members, including the City Administrator's Office, SF State University, SF Community Agencies Responding to Disasters (CARD) and the SF Interfaith Council, offer technical support and other resources. Advanced administrative science tools, including network analysis, are being used to identify the linkages between individual and community groups and assist stakeholders determine what is working in the areas of individual and community efficacy.

### *7.2.3 San Francisco and the Built Environment*

San Francisco has a long history of addressing seismic safety. In the late 1970s the Building Department began a programme of reinforcing building parapets that were weakly attached and represented a clear risk of failure in a large seismic event. The 1989 Loma Prieta earthquake spawned new studies on risk and loss to existing housing stock. With an estimated residential loss of 92,000 units (about 24 % of the total stock) projected in a hypothetical 7.2 magnitude event, seismic safety became a people's issue. In 1992, the citizens approved a Seismic Safety Loan programme with funds provided by general obligation bonds. There have been two bond issues of US$ 35 million each with funds going to support loans to 13 commercial projects at market rates and 16 projects at below-market rate, with a deferred loan payable at the end of the structure's mortgage. The largest loan was used to address a US$ 15 million below-market rate project. In 2010, US$ 25 million of these funds still remained.

There is also an effort underway to address "soft story" structures. There are 4,400 wood-frame buildings in San Francisco that were built before May 21, 1974 with five or more residential units and three or more stories that might include a "soft story" (Applied Technology Council ATC-3 2010a). In an initial effort to address this need, San Francisco tried to fund retrofits through a public bond issue. This direct subsidy, which did not receive the needed votes to pass, was replaced by a regulatory requirement in 2013, when legislation was passed mandating an evaluation of 3,000 buildings by the owner and a subsequent retrofit of vulnerable structures as needed. All work is to be completed by 2020. This programme is a key physical component of the city's larger residential resiliency strategy.

### *7.2.4 Community Infrastructure and Lifelines*

In 2008, the SPUR resiliency dialogue identified the need to bring private utility companies into the discussion along with public utility providers. This need arose from the realisation that having safe buildings without electric power, natural gas or phone service was not an acceptable post-disaster condition. The Mayor responded by supporting the Resilient SF Lifelines Council to focus on the interdependency of lifeline service providers and the continuity of government services. The Council discovered that the public and private utilities were much more dependent on each other than previously thought. The council chose to use "interdependent system" modeling techniques to help define system performance goals for municipal water, electric power, natural gas, telecommunication, highways and roads, ports, transit and airports. The modeling and the discussion of the model results served to establish gaps to be filled and to build a stronger network between providers. The Lifeline Council began to address the seismic as well as the climate adaptation issues, as each related to their core business in different ways. The creation of the Lifelines Council also highlights the importance of fostering a broader, regional perspective that is an important precondition of building a functional network at the metropolitan scale. This work supports an integration of effort approach that is a part of San Francisco's regional resiliency strategy.

### *7.2.5 Community Action Plan for Seismic Safety*

One San Francisco Bay Area earthquake scenario involves a potential 7.2 Mw event, resulting in over one-quarter of the residential housing stock (~92,000 units) suffering damage, thereby requiring major repair or replacement (Applied Technology Council 2010). Given the possibility of large scale damage, the Community Action Programme for Seismic Safety (CAPSS), a volunteer group of seismic experts formed in 2001, proposed a three step strategy: (1) engage market forces to encourage structural retrofits, (2) enact measures to reduce fire damage, and (3) promote non-structural risk reduction measures. The strategy is directed at expanding the stock of resilient houses and lessening the need for temporary post-disaster shelter.[8] The strategy involves the following steps: facilitate a market in which earthquake performance is valued; nudge the market by requiring an evaluation upon sale, or by a deadline; and require the retrofitting of a structure based on the findings of the assessment. Programmes related to each part of the strategy are now being implemented. The CAPSS project is overseen by the San Francisco Department of Building Inspection, which contracts with the non-profit Applied Technology Council (ATC) to manage this effort.

[8] The purpose of the CAPSS project is to recommend specific, comprehensive mitigation efforts for the privately owned buildings in San Francisco in order to reduce the impacts from future earthquakes.

In 2010, the ATC issued ART-52-2, “Here Today-Here Tomorrow: The Road to Earthquake Resilience in San Francisco, A Community Action Plan for Seismic Safety”. Here Today-Here Tomorrow proposes some important goals including the following: (1) residents will be able to stay in their own homes, (2) no building will collapse catastrophically, (3) business and the economy will quickly return to functionality, and (4) the City’s sense of place will be preserved (ATC-52 2010). The aspirational statements are based on taking action before a disaster occurs. In 2011, the city adopted the 30-year CAPSS work plan with six of the tasks in phase one slated to begin in 2012. An Earthquake Safety Officer was placed in the office of the City Administrator, signaling the importance of this effort.

The CAPSS volunteer group consists of leading seismic engineers from the academy and the consulting industry. This committee, like SPUR, is an enlightened and activist civic group willing to engage the city and the community over the long term. A series of experiments in assessing soft-story buildings began in 2013 to determine the lowest assessment cost method, and less intrusive ways to determine a buildings structural status.

### *7.2.6 San Francisco Community Agency Capacity*

Community efficacy takes several forms. The San Francisco Community Agencies Responding to Disasters (SFCARD) programme works with human service agencies serving vulnerable populations in San Francisco to ensure operational continuity after a disaster. It provides disaster preparedness training to support the capacity of local agencies and the vulnerable populations that they serve. In partnership with HELPLINK and the Volunteer Center, SFCARD is creating a Disaster Database to assist Health and Human Service agencies before, during and after a disaster. SFCARD can be viewed as a “non-structural” risk reduction initiative (people and process focused) whereas CAPSS can be viewed as a “structural” (built environment focused) effort to reduce risk.

### *7.2.7 Summary*

The City established operating programmes supporting resilience in its institutional, community and personal environments, beginning in the early 1990’s. It promotes pre-event planning in administrative units outside of the safety and emergency management departments. Institutionally, the city’s approach represents an attempt to work outside traditional single function silos and is exemplified by placing the director of Earthquake Safety in the office of the City Administrator and not under the Department of Emergency Management. San Francisco has supported community level resiliency planning that includes faith based organisations, social support non-profits, universities, as well as community organisations. In addition, the city has developed direct social media tools to reach out to a wide range of people and promote self-efficacy. Attempts at tying these environments together through empowerment and trust are supported by the NEN, the Lifelines Council and civic organisations such as SPUR and SFCARD. The adoption of a 30 year plan for CAPSS demonstrates a long term commitment to build community and institutional competency in seismic residential safety.

## 7.3 Berkeley

Situated between the San Andreas and Hayward fault lines, the City of Berkeley contains 113,000 people and 49,500 housing units. The city's 27.7 $km^2$ (10.5 square miles) jurisdiction, is at risk from hazards such as urban and wild land fires, flooding, earthquakes and landslides. The devastation caused by the 1989 Loma Prieta earthquake and the 1991 firestorm in the East Bay Hills of Oakland and Berkeley signaled a need to address a number of disaster resiliency issues. Moreover, the experience of these two disasters demonstrated that local government could not do the job acting in isolation. As a result, Berkeley has built resiliency into the fabric of local government operations, its neighbourhoods and its citizens. According to Topping (2010), Berkeley has distinguished itself by integrating mitigation and preparedness into city policy. Berkeley's resiliency approach has three major components: (1) recognise risk reduction as part of local governments' work, (2) strengthen public buildings and assist private owners to do the same, and (3) involve the community in preparedness efforts, working in tandem with the city. Berkeley's 2004 Hazard Mitigation Plan identifies the following hazards: earthquakes, wildfires, landslides, floods, hazardous materials accidents and acts of terror.[9] Earthquakes and wildfires are the hazards most likely to cause significant damage in the city. The United States Geological Survey (USGS) calculates that there is a 62 % chance that a 6.7 magnitude earthquake will strike the Bay Area in the next 30 years and a 27 % chance that it will occur on the Hayward/Rogers Creek fault system that runs directly through Berkeley (Hazard Mitigation Plan 2004).

In 2009, Berkeley adopted a separate Climate Action Plan. The Berkeley Climate Action Plan is the result of the community-based climate action campaign that the Berkeley voters set in motion. The Plan is rooted in a vision for a sustainable Berkeley that will emerge from the climate action planning process. The community's target for the year 2020 is to reduce community-wide GHG emissions 33 % below

[9] In order to comply with federal regulations pertaining to disaster relief funding, the city adopted a Hazard Mitigation Plan in 2004. The plan includes objectives linked to the city's General Plan statements. These include: (1) reduce the potential for loss of life, injury and economic damage from earthquake, wildfires, landslides and floods; (2) increase the ability of the city government to serve the community during and after hazard events by mitigating risk to key city functions; (3) protect the community's unique character and values from being compromised by hazard events; and (4) encourage mitigation activities to increase the disaster resilience of institutions, private companies and lifelines systems that are essential to Berkeley.

**Table 7.2** Berkeley's main climate action strategies. (Source: Berkeley Climate Action Plan progress 2011 update)

| 2009 Climate Plan Adoption | Progress as of 2011 |
|---|---|
| Sustainable Transportation & Land Use | Created 694 bike spaces, improved pedestrian safety at intersections, trail plans completed |
| Building Energy Use | A 20% reduction in 2020 electricity targets met, 6% of 2020 natural gas targets met |
| Waste Reduction and Recycling | 50% of 2020 waste reduction and recycling target met |
| Community Outreach and Empowerment | 777 people participated in climate change action groups. Green Business participation lower than expected |
| Adapting to a Changing Climate | Increased tree planting annually |

2000 levels. The plan rests on five strategies shown in Table 7.2 along with a 2012 issued progress report. (Berkeley 2009 Climate Action Plan).

A notable aspect of the Plan is a separate strategy for community outreach and empowerment. This indicates a recognition and desire to create community efficacy as part of the overall effort. Sea-level rise is not discussed as a separate area of concern in this plan, but will be addressed in its 2013 Plan update. Berkeley will be subject to Bay Area Conservation and Development Commission (BCDC) oversight for development adjacent to its bay shoreline.

### 7.3.1 Recognising Risk Reduction as Part of Local Governments' Work

California law does not promote a statewide growth-management system as found in the states of Florida or Hawaii. Instead, it emphasises local accountability for coordinated planning and implementation actions, which must meet broad state standards. In California, a city General Plan is a legal document required by state law. In 2001, Berkeley's General Plan underwent a comprehensive revision with the update identifying seven major goals related to city form, governance and safety. The commitment to inclusion is evidenced by its General Plan goal to maximise and improve citizen participation in municipal decision-making. This reflects a commitment to a high level of citizen involvement in civic matters.

Several hundred citizens serve on boards and commissions to help formulate policy and advise the City Council on all matters including social as well as physical form-related issues. There are many active neighbourhood associations, merchant groups and advocacy groups. The separation between civic and government functions is not large in Berkeley. The effort at connecting, in Paton and Johnson's terms, the personal to the community to the institutional environments is continuous, thereby contributing to efficacy and trust as demonstrated by the expansion of neighbourhoods choosing to participate in hazard-response training.

A distinct commitment to resiliency appears in the General Plan goal to make Berkeley a disaster-resistant community that can survive, recover from and thrive after a disaster. Operationally this is a climate change type goal. It is also a bold statement establishing a framework for action by city government and the community. The Plan highlights the need and the opportunity to ensure that new construction reduces rather than increases risk. This stance is an indicator of overall risk-reduction behavior and demonstrates internal consistency within the plan. Policies also call for improving the identification of the locations of hazards through the designation of flood, landslide, or earthquake zones; improving awareness of their presence and consequences; and adopting and enforcing regulations to minimise the exposure to such risks.

In California, all municipalities are required to adopt a General Plan with seven required elements.[10] Berkeley chose to add a separate *Disaster Preparedness and Safety Element* that has six objectives, 28 policies and 73 specific actions. This Plan element specifically includes *Neighbourhood Preparation and Education* (to continue to provide education, emergency preparedness training and supplies to the community at the neighbourhood level to support neighbourhood- and community-based disaster response planning). There are also policies addressing *Special Needs Communities* (continue to work with the social service community to ensure the safety of special needs' populations) and *The City's Role in Leadership and Coordination* (ensure that the City provides leadership and coordination to the private sector, public institutions and other public bodies in emergency preparedness).

Berkeley's Disaster and Fire Safety Commission (DFSC) participates in the review of emergency, disaster and mutual aid plans and agreements and makes recommendations to the City Council regarding legislation and regulations needed to implement such plans and agreements. All major projects in the City are subject to review by this commission that is appointed by the mayor and approved by the City Council. The Disaster and Fire Safety Commission is an additional part of the check and balance system built into local government operations that improve transparency. The Hazard Mitigation Plan, which originates in the Fire Department, receives its initial review from this commission.

### *7.3.2 Strengthening Public Buildings and Assisting Private Owners*

Urban development prior to the advent of modern building codes resulted in an inventory of structures highly vulnerable to earthquakes. These include unreinforced masonry (URM), concrete frame, tilt-up buildings built before the mid-1970s and buildings with "soft" stories (such as multi-story apartment buildings

[10] The required elements of a California General Plan are: land use, housing, circulation, safety, open-space, conservation and noise. Others can be added, such as urban design, economic development, historic preservation, climate action planning and sustainability.

with ground-level parking and units above). The Association of Bay Area Governments (ABAG) estimated in 2002 that more than 28 % of Berkeley's housing stock (13,300 units) would be rendered uninhabitable by a major earthquake. Berkeley's approach has been to commit to seismically retrofit schools and public buildings, create programmes and provide monetary and nonmonetary incentives for housing improvements.

Between 1991 and 2000, Berkeley's citizens approved municipal bond issues to fund retrofitting of schools and public building totaling approximately US$ 336 million. Citizens were asked six separate times to tax themselves to protect children first and the institutional offices second. All six-tax measures were approved. At the beginning of the 2008 US and California economic crisis an increase in property tax on residential and commercial buildings was enacted that provides US$ 3.6 million per year to keep the fire stations open and to preserve existing levels of risk reduction services and training. Such support represents a bond of trust and empowerment between the institution and individual citizens, especially those who are property owners (Chakos et al. 2002).

The building of community efficacy starting in 1991 carried through for the next 20 years as evidenced by Berkeley's ability to forge continued gains in support of its Hazard Mitigation Plan and its Climate Action Plan. Berkeley's broad and deep community involvement through boards, commissions, special interest groups and the sharing of fiscal resources with people (e.g., retrofit grants) show how the Paton and Johnson *community competence* dimension becomes real in the Berkeley case.

### *7.3.3 Retrofitting of Houses*

In 1992, City Council established the Residential Seismic Retrofitting Incentive Programme providing two types of incentives for seismic upgrading. This programme waived one third of the transfer tax on home sales if the funds are used for seismic upgrading on that specific unit.[11] Between fiscal years 1992 through 1998, approximately 7,600 properties had taxes waived under this programme totaling US$ 3.5 million (Berkeley General Plan, Disaster Preparedness and Safety Element 2011). This policy resulted in upgrades to 15 % of the city's housing stock. In addition, between 1992 and 1999, the city waived building permit fees for retrofitting existing homes and unreinforced masonry structures in the amount of approximately US$ 1 million which was applied to 4,100 homes, or 8 % of the housing stock.

By the mid 2000's, the City of Berkeley experimented with a behavioral approach to lowering risk in soft story buildings. This action was due in part to declining

[11] In California a municipality is allowed to receive a 1.5 % property transfer fee on the sale of real property within its boundaries. The city can use these funds any way it wishes. In Berkeley, the council decided to allocate 1/3 of the revenues to address seismic safety issues identified in the house being sold. This approach serves to upgrade older units built under previous codes and is socially equitable in that the policy returns some of the fee revenue to the owner of the unit.

**Table 7.3** Berkeley soft-story retrofit programme. (Source: Rabinovici 2012)

| |
|---|
| 1. Public inventory is conducted and an official list is made |
| 2. Mandatory seismic evaluation must be made, approved and kept on file within two years of the inventory notice |
| 3. After seismic evaluation and acceptance of condition is made, the building is removed from the inventory for 15 years |
| 4. Mandatory signage posted that the building may not be safe during or after an earthquake |
| 5. Notify tenants of the buildings status throughout the process |
| 6. Notice is recorded on the property deed and remains there until an action is completed |

municipal revenues to fund a monetary subsidy programme, but also due to a desire to continue the seismic strengthening of existing housing stock. A 2005 amendment to the Municipal Code (Chap. 19.39) established an ordinance directed at assessing building condition without cost to the city. For those properties that fit the soft story profile, the new ordinance placed notice on the property title requiring owners to inform tenants, post warning signs on-site and hire a structural engineer to evaluate their property. Of the 320 buildings originally in the soft story-building inventory, 20% conducted a voluntary seismic upgrade, and 79% of the buildings were in compliance with the engineering report requirements by 2010 (Rabinovici 2012). Seventy-nine building owners voluntarily applied for building retrofit permits, in lieu of, or in addition to, producing the required engineering report. The greatest implementation challenge involved the development, communication and consistent application of technical standards for the evaluating engineers to use.

By having buildings tagged with a warning sign and notice placed in the property title documents, the programme raised awareness on the part of property owners as well as alerting building tenants of possible danger. The possible risk issue became one shared by the occupant, owner and city. A cloud of uncertainty was created that motivated a percentage of owners to have an inspection conducted, and if needed, to undertake seismic upgrades. This is not a voluntary programme in that penalties could be assigned to an owner if the inspections were not made. The success of the programme rests on motivation by the owner to determine actual building condition and take some type of action (Table 7.3).

### *7.3.4 The University Connection*

The City of Berkeley is fortunate to have a University of California campus (UC Berkeley) within its boundaries. In 1995, a group of three UC Berkeley professors were appointed to a municipal seismic safety panel that helped the City structure a framework for safety and establish standards for companies implementing seismic safety improvements funded through bonds. The work of this panel sparked a University programme to lower seismic risk of campus buildings. The Berkeley SAFER programme, funded by a US$ 1.2 billion reconstruction plan for buildings and facilities on campus, is designed to protect research projects, labs and data. The principal impetus for this large infrastructure upgrade was the realization among faculty that the loss of data, experiments and other materials would be extremely detrimental to the campus and the careers of many scientists. Faculty simply could not accept the risk of losing years of work.

### 7.3.5 Berkeley Community Engagement Efforts

Implementing the goals in the General Plan, the Climate Action Plan and the Hazard Mitigation Plan required more than money. What the city needed was an enhanced level of community support and engagement. Berkeley knows that there will be cascading events related to any major disaster such as breaks in water mains and potential fires after a seismic event. To address these risks, a decentralised approach, called the Community Resiliency Programme, has been put in place.

Begun in 2002, the Fire Department operates the programme which emphasises a two-part approach that is comprised of training and equipment-related elements. First, neighbourhood groups are invited to a general meeting where training options are discussed. The groups then voluntarily seek training in a number of possible areas, including hazards identification, use of two-way radios and the implementation of Community Emergency Response Team (CERT) activities. For example, CERT training involves educating people about disaster preparedness initiatives that are germane to hazards that may impact their area and teaches them basic response skills, such as fire safety, light search and rescue, team organisation and disaster medical operations. When the neighbourhood group completes the training, it receives a cache of equipment (generators, hoses, first aid, power saws, etc.) to keep within the neighbourhood at a location of their own choosing. Refresher courses are offered to keep groups involved and to build networks between neighbourhoods. In 2007, there were 32 groups and caches in place; in 2008, 40 groups and caches; and in 2010, 46 groups and caches. Over 1,400 citizens have been trained and networked, thereby contributing to community competence. These groups are a source of social capital and are likely to be the first responders on scene following a disaster.

The first neighbourhoods to participate in the programme were middle-class and active in local affairs. Over time, the programme expanded to include more working class and low-income neighbourhoods located in the City's western flatlands areas which are more subject to flooding and minor sea-level rise as indicated in the 2012 BCDC Climate Change Plan. The 2012 BCDC Climate Change Plan indicates that any building within the 100 foot bay line edge would be subject to mitigation of sea-level rise impacts as well as compliance with Berkley's own risk reduction initiative tied to new buildings.

In 2008, the CERT training of neighbourhoods was extended to 30,000 students located on or near the UC Berkeley campus. To address this transient population, the Community Resiliency Programme entered into arrangements with sorority houses as local training sponsors. Sorority houses were chosen because each has a resident adult advisor on the premises which provides a consistent contact and a programme advocate. The city continues to provide training to those houses that wish to participate.

## 7.4 Discussion

This chapter is concluded by positing two questions. One, are we seeing sustained actions in San Francisco and Berkeley that lead to resiliency and adaptation that is informed by natural hazard planning practices? Two, are we seeing changing relations between citizens, civil society and their government as described in Paton and Johnson's disaster resiliency model? Both cities operate within a broader governmental context, with the State of California providing general guidance and scientific information, while offering no statutory requirements. Regional agencies, such as ABAG, BCDC and the Metropolitan Transit Commission provide avenues for government-to-government support that is a form of vertical integration and can provide a unified voice on matters such as climate change adaptation.

Given this context, it appears that both cities have chosen their own paths to advance the broader aims of adaptation and resiliency. Their efforts, which emphasise a bottom up approach supported by higher levels of government, have shaped the nature of the Association of Bay Area Governments regional resiliency plan (ABAG 2013). The boundary-spanning plan calls for cooperation, integration and risk reduction initiatives targeting critical facilities that serve the region.

The San Francisco resiliency effort is significant due to its initial location in the Office of General Services (GSA) and the provision of Controller oversight of fiscal issues in the City as well as support for many operating agencies such as the courts, purchasing and public works. The inclusion of neighbourhood, faith-based and the community-based organisations puts in place a framework for building more social capital that can support technical, social and political issues that are certain to arise in the post-disaster time period. Elements of the 2004 San Francisco Climate Action Plan appear in the Berkeley 2009 Plan but include a stronger articulation of strategies that are bolstered by a more extensive community outreach effort. Berkeley learned from the San Francisco experience and sought to improve its institutional capacity to address larger resilience-based goals. For example, in each revision of its local hazard mitigation plan (a Federal requirement for post disaster financial assistance) and the General Plan safety elements (a California requirement), a more in-depth analysis of natural hazards risk has been undertaken and programmatic efforts are more clearly articulated. These actions have led to more internal consistency across plans, which are an important aspect of organisational capacity (Smith 2011, p. 289).

The use of social media strategies directed at younger age groups may be the single most important long-term action identified in the case studies. This effort is sophisticated in scope and action and relies on a different form of relations and information sharing. It opens up new avenues for organising, sharing and feedback. The San Francisco social media actions are supplicated, multi-level, adaptive and flexible. They help build constituencies that can take action now and after a disaster occurs. Statewide, natural hazards planning practice is behind the curve in terms of social media use, and thus San Francisco is more representative of a model city rather than an adaptor of practices from other locales.

The Berkeley approach demonstrates that it is possible to integrate resilience in many ways including legislation, risk assessments and actions and neighbourhood-shared responsibility. Innovation in the Berkeley case includes legislative leadership, citizen advocacy, strong partnering with the local university and a willingness to engage the community and its residents. The Berkeley effort, which spans more than 20 years, includes a series of advocates who have served as leaders over time. The city continues to use its social capital to produce new champions as conditions warrant.

Broad-based involvement is a characteristic of both cities at the community and the city department levels. In Berkeley, for instance, there is "buy-in" from groups as diverse as property owners to the rent control board. Berkeley demonstrates that a variety of approaches are required to obtain "buy-in," so the development of a flexible implementation strategy is needed. Berkeley exhibits a capacity for experimentation, innovation and using feedback to improve its practices. They are open to criticism and improvement and can act in the "flat government" mode as well as a hierarchical mode. The use of property tax transfer funds to help buyers of older homes invest in seismic improvements has been copied, on a voluntary basis, by the Bay Area cities of Palo Alto and Oakland.

The propinquity factor has also played a role in advances made by both cities. Berkeley and San Francisco pay attention to one another's resiliency programmes, and learn technical and political lessons,[12] making adjustments based on the experience of the other. Other cities in the region (e.g., Oakland, San Jose, Palo Alto) also replicate programmes based on those formulated by San Francisco and Berkeley.

The test of resiliency, however, is best measured after a hazard event occurs. Then we will have a better idea if the efforts described in this chapter yield less loss of life, less damage and allows each city to recover more quickly and return to its next period of normalcy.[13] Berkeley appears to have built resiliency capacity in the way Paton and Johnson depict it, using a broad range of actors aware of the risks, while empowering their neighbours in the social and business sectors to take actions that are linked to identified climate change targets. The large sums of bond funds and special assessments that citizens of these cities have made available to support resiliency in buildings alone demonstrates that they are adaptive and take action in relation to risk reduction.

The evidence demonstrates that resiliency and adaptation are being practiced and sustained by San Francisco and Berkeley. The hazards model encompassing assessment, dialogue and action (Wisner et al. 2012) is in place and being practiced, although the process can take decades to achieve and is subject to adjustments over time based on external conditions such as the status of the economy. Climate change

[12] It is not uncommon to find Berkeley municipal staff that previously worked in similar positions in San Francisco, which allows those individuals to draw on the experience and broader perspectives gained in programme design and implementation.

[13] Testing the efficacy of resiliency programmes will require the development of appropriate study parameters, metrics and analytical procedures that can measure the cause and effect relationships between elements of resiliency programmes and tangible outcomes.

actions, starting with the development of the 2004 San Francisco Climate Action Plan are taking place. By the end of 2012, 34 of the 101 Bay Area cities completed a Climate Action Plan and have begun to implement actions.

Systematic information is a powerful tool, and the State of California is a leader in providing such information and guidance to its constituents, cities and counties. The primary policy lesson identified in these case studies is that it matters less which level of government is providing guidance; it matters more that such guidance is based on sound science and that it can be easily accessed and used by anyone interested in resiliency and climate change adaptation.

## References

Alesch, D. J., & Siembieda W. (2011). The role of the built environment in the recovery of cities and communities from extreme events. *International Journal of Mass Emergencies and Disasters, 30*(2), 197–211.

Alrich, D. P. (2012). *Building resilience: Social capital in post-disaster recovery*. Chicago: The University of Chicago Press.

Applied Technology Council. (2010a). Here today-here tomorrow: The road to earthquake resilience in San Francisco: A community action plan for seismic safety. ATC 52-2 Redwood City.

Applied Technology Council. (2010b). Here today-here tomorrow: The road to earthquake resilience in San Francisco- earthquake safety for soft story buildings. Seismic safety. ATC 52-3. Redwood City.

Association of Bay Area Governments. (2013). *Regional resilience initiative: Policy agenda for recovery*. Oakland. March.

CNRA (California Natural Resources Agency). (2009). 2009 California climate adaptation strategy. A report to the Governor of the State of California in response to executive order S-13-2008. http://www.climatechange.ca.gov/adaptation/.

Chakos, A., Schultz, P., & Tobin, L. T. (2002). Making it work in Berkeley: Investing in community sustainability. *Natural Hazards Review, 3*(2), 55–67.

City of Berkeley, CA. (2009). *Climate action plan*.

Cutter, S., Burton, C. B, & Emrich, C. T. (2010). Disaster resilience indicators for benchmarking baseline conditions. *Journal of Homeland Security and Emergency Management, 7*(5), Article 51.

Department of Emergency Management (DEM). (2009). *The preparedness movement communications strategy*. City and County of San Francisco, September.

Eisner, R. (2010). Building disaster resilient organisations in the non-government (NGO) sector." *Journal of Disaster Research, 5*(5), 503–508.

Johnson, L. A., & Tam, L. (2012). San Francisco. In Edward J. Blakely & Armando Carbonell. *Resilient coastal city regions: Planning for climate change in the United States and Australia* (Chap. 5, pp. 117–144). Cambridge: Lincoln Institute of Land Policy.

Leonard, H. B., & Howitt, A. M. (2010). Advance recovery and the development of local hazard mitigation plan. City of Berkeley.

Meyer, S., Henry, E., Wright, R. E., & Palmer, C. A. (2010). Post-disaster redevelopment planning: Local capacity building through pre-event planning. *Journal of Disaster Research, 5*(5), 552–564.

Mitchell, J. K. (2010). *Changing knowledge about disaster recovery: Bridging gaps between research and action in light of hurricane Katrina and other policy-forcing events*. University of Leipzig Presentation, 600th Anniversary.

Norris, F. H., Stevens, S. P., Pfefferbaum, B., Wyche, K. F., & Pfefferbaum R. L. (Mar 2008). Community resilience as a metaphor, theory, set of capacities, and strategy for disaster readi-

ness. *American Journal of Community Psychologist, 41*(1–2), 127–150. PMID: 18157631 [PubMed—indexed for MEDLINE].
Paton, D., & Johnston, D. (2006). *Disaster resilience: An integrated approach.* Springfield: Charles C. Thomas Publishers.
Rabinovici, S. J. M. (2012). *Motivating Private Precaution with Public Programs: Insights from a Local Earthquake Mitigation Ordinance*. PhD. Dissertation, School of Public Policy, University of California, Berkeley. Fall.
ResilientSF. (2011). http://resilientsf.org/partners. Accessed 20 June 2011.
The ResilientSF Initiative. (2011). Office of the Mayor, City and County of San Francisco.
Siembieda, W. (2010). Towards a concept of disaster resilience: A commentary on behalf of the editorial committee. *Journal of Disaster Research*, Special issue on "Building Local Capacity for Long-Term Disaster Resilience", *5*(5), 487–493.
Smith, G. (2011). *Planning for post disaster recovery: A review of the United States disaster assistance framework.* Arlington: Public Entity Research Institute.
SPUR. (2008). file:///Users/wsiembie/Documents/SPUR%20Resilinet%20SF%20part-i-condensed-version.html. Accessed 20 Apr 2011.
Topping, K., & Schwab, J. C. (Eds.). (2010). *Hazard mitigation: Integrating best practices into planning*. Planning advisory service report no. 560. Chicago: American Planning Association Press. Chap. 7, Case studies in intermediate jurisdictions.
U.S. Geological Survey, Field, E. F., and Milner R.F., and the 2007 Working Group on California Earthquake Probabilities. (2007). Forecasting California's Earthquakes—What Can We Expect in the Next 30 Years? Fact Sheet 2007–3027, version 1.0.
Walter, R., Young, R., & Husain, K. (2012). *Technical review of the 2010 community wide GHG inventory for the city and county of San Francisco*. San Francisco: ICF International. April 12.
Wisner, B., Gaillard, J. C., & Kelman, I. (2012). *The Routledge handbook of hazards and disaster risk reduction*. London: Routledge.

# Chapter 8
# Rising to the Challenge: Planning for Adaptation in the Age of Climate Change

**Philip R. Berke**

**Abstract** The traditional planning paradigm is insufficient to address the challenges of climate change. Planners must confront heightened uncertainty and longer time horizons. A new model of planning is offered to make the transition to plans that account for multiple futures, monitoring and more flexible strategies needed for adaptive responses.

**Keywords** Urban planning · Climate adaptation · Public risks · Environmental planning · Natural hazards

## 8.1 Introduction

Consensus statements among leading scientists are increasingly unambiguous that the climate is warming and that warming during the prior 50-years is primarily caused by human-induced emissions of heat-trapping gases (c.f., IPCC 2007; GCRP 2009). Warming during the twenty-first century is projected to be greater than over the last century. Climate induced changes have already been observed in all regions of the world, including sea-level rise, more intense hurricanes and precipitation events and extended droughts and heat waves. The impacts will adversely affect human health, availability of water, food production, risk of life and property from hazard events and many other social and natural environmental systems. Further hindering our ability to deal with the consequences is the reality that various climate changes have been more rapid than predicted by prior assessments.

Considerable scientific and technical effort has been directed toward climate change mitigation to reduce heat-trapping $CO_2$ emissions (Schipper 2006; Wheeler 2008). These include improving energy efficiency, using non-carbon based fuels and capturing and storing carbon dioxide from fossil fuel use. Over the long-term, lower emissions will lessen the magnitude of climate-change impacts and the rate at which they appear. However, much less attention has been given to climate adaptation by planners and policy-makers. Adaptation refers to changes made to better respond to present and future climatic and other environmental conditions.

P. R. Berke (✉)
University of North Carolina at Chapel Hill, Chapel Hill, NC, USA
e-mail: pberke@unc.edu

B. C. Glavovic, G. P. Smith (eds.), *Adapting to Climate Change,* Environmental Hazards,
DOI 10.1007/978-94-017-8631-7_8, 

No matter how aggressively heat-trapping emissions are reduced, some amount of climate change and resulting impacts will continue due to the effects of gases that have already been released. Many gases are long-lived and levels will remain elevated for centuries. Further, the earth's oceans have absorbed much of the heat added to the climate system due to increases in heat trapping gases, and will retain that heat for many decades. Consequently, there is also a need for adaptation.

It is becoming evident that the concepts and methods of contemporary planning will not be adequate to address the highly complex and uncertain issue of climate change. Uncertainty and the extended planning horizon require developing new conceptual foundations about how plans should be prepared, implemented, monitored and evaluated.

In this chapter, I initially review the conceptual foundations of new approaches to planning that are emerging in research and practice in the age of climate change. I then discuss the critiques of contemporary planning in light of the emergence of climate change. I contend that the field of climate adaptation has much to gain from over 30-years of research and practice in natural hazard mitigation (which means 'adaptation' in climate change terminology). Next, I draw on the literature in natural hazard mitigation planning and distill key insights to derive a set of principles of plan quality that are uniquely designed to assess the quality of climate adaptation plans. To illustrate how the principles are operationalised, I then apply them to case studies of climate adaptation plans developed for London, United Kingdom and The City of Punta Gorda, Florida, U.S.A.

## 8.2 Conceptual Foundations for Adaptation in the Age of Climate Change

Adaptation is rooted in theories of resiliency, adaptability and the sustainability of a human-ecological system. *Resiliency* is the ability of a community or society, along with the bio-physical systems upon which they depend, to resist or absorb the impacts (deaths, damage, losses, etc.) of hazards, to rapidly recover from those impacts and to reduce future vulnerabilities through adaptive strategies (Godschalk 2003; Peacock et al. 2008; Chapin III et al. 2009). *Adaptability* is the capacity of the human community to anticipate change and successfully manage coupled human-ecological systems to change, and is major feature of resiliency (Chapin III et al. 2009). *Sustainability* addresses the interconnectedness of social, economic and environmental values, as adaptive actions must strike a balance in deriving solutions that build resiliency of economic, built environment and ecological systems upon which human communities depend, and ensure the socially equitable distribution of benefits generated by an improved level of resiliency (APA Sustaining Places Task Force 2011).

## 8.3 What is Known About Hazard Mitigation Planning? How does such Knowledge Apply to Climate Adaptation Planning?

As noted, most planning that deals with climate change has focused on the reduction of heat trapping gases, notably $CO_2$ (Wheeler 2008). Notwithstanding the success of reduction strategies, there is a growing concern that some amount of global climate change is going to occur. Therefore, attention to climate change adaptation is as important as reducing $CO_2$ emissions.

Although the field of city and regional planning during the twentieth century has evolved from elite and rigid visions of physical development to more participatory, broader based strategies for managing change (Kaiser and Godschalk 1995), planning for climate change requires a fundamental transformation which poses a particularly difficult and far reaching challenge. Cities and regions cannot rely on traditional approaches to planning to tackle ever increasing threats and uncertainties posed by climate change.

Of all the specialisations in the field of city and regional planning (e.g., hazard mitigation, environment, land use, transportation, health), research and practice on hazard mitigation planning is perhaps the most instructive in understanding community behavior in adapting to risk posed by climate change. Both hazard mitigation and climate adaptation deal with events that are rapid-onset (tornadoes, hurricanes) and slow-onset (sea-level rise, drought). Both are oriented toward the future. Both are focused on anticipating future needs and impacts, rather than responding to yesterday's events. Thus, the relatively well-developed knowledge on the human dimensions of natural hazard events can be useful in formulating effective climate adaptation plans and implementation actions.

Persistent obstacles revealed by over four decades of extensive investigations in hazard mitigation offer insight to the immense challenges for effective adaptive action to climate change. Consensus-based documents by leading experts in natural hazard research agree that among the major obstacles to local mitigation planning, and likely similar hindrances to climate adaptation, four are particularly important: (1) low priority that local officials give to reducing future threats given the lack of a public constituency; (2) costs of reducing risk are immediate, benefits are long-term and uncertain and may not occur during the tenure of elected officials; (3) the physical manifestations of improved public safety are not visible; and (4) failure of planning to effectively engage marginalised population groups who are most vulnerable to future threats (Mileti 1999; NRC 2006).

These obstacles create what urban planning scholar Raymond Burby refers to as the local government paradox wherein vulnerable communities fail to enact effective planning programmes to prevent disaster losses even though they have much to lose from an extreme event (Burby 2006). As a result, the effectiveness of mitigation planning is limited. Such plans have consistently been found to be too expert-driven and too narrowly conceived as indicated by goals that concentrate on economic efficiency but not social and ecological issues, fact bases that are based on vulnerability assessments of existing conditions but not future possibilities, and policies that support individual projects but not integrated mitigation strategies (Burby 1998; NRC 2006). These findings are troublesome since goals, vulnerability data and policies serve as the critical direction-setting framework of hazard mitigation plans.

A weak direction-setting framework means that a community is less likely to exert control over its planning agenda and ensure that long-range public interests supersede short-range interests and private concerns. Hazard mitigation is more likely to be reduced to a series of disconnected "projects" intended to address past "mistakes," and not premised on an integrated and flexible forward-looking approach needed for anticipating and adapting to an uncertain future. A weak direction setting framework also means that plans will not provide a clear, relevant basis for implementation, monitoring and learning by doing and evaluating.

Consistent with hazard mitigation research findings, Quay (2010) observes that "predict and plan" describes the current practice of urban and regional planning. He maintains that most planning for physical development forecasts future trends to determine a single scenario of a desired future and then identifies the infrastructure needed to create this future. This approach leaves communities vulnerable to the limitations of forecasting. Further, in planning efforts where scenario planning is considered a success it is has been criticised as too dominated by an expert-driven agenda that inadequately facilitates public participation which limits the exploration of the full range of possible futures, and opportunities for raising awareness and deriving tangible solutions that address locally defined issues (Bartholomew 2007).

Limitations in climate science also raise concerns about the efficacy of traditional planning approaches. While there is general agreement that the global-scale climate is warming and that global-scale warming over the past 50-years is due primarily to human-induced emissions, there is uncertainty about the specific causes and consequences of climate change over time at regional and local spatial scales. Currently there exists a web of science-based models that represents a wide range of disciplines for predicting the effects of climate change at various timeframes and spatial scales (IPCC 2007). Each set of models introduces more variance and uncertainty when integrated with other models. The high level of uncertainty, combined with a 100-year time frame needed to estimate climate change impacts, serve as major constraints to adaptation and resiliency.

Given the limitations of hazard mitigation and urban planning in general, and the scientific uncertainties about climate change predictions, particularly at the local level, traditional planning is not able to deal with problems related to climate change. Instead, traditional planning is better equipped to deal with human-ecological systems that are relatively stable and predictable over relatively short periods of time. In its 2009 report, *Global Climate Change Impacts in the United States,* a group of leading scientists agreed that,

> Humans have adapted to changing climatic conditions in the past, but in the future, adaptations will be particularly challenging because society won't be adapting to a new steady state but rather to a rapidly moving target. Climate will be continually changing, moving at a relatively rapid rate, outside the range to which society has adapted in the past. The precise amounts and timing of these changes will not be known with certainty. (U.S. Global Climate Research Program 2009, p. 11)

The high levels of uncertainty, combined with a 100-year or more time-frame for making predictions, further exacerbate the challenges that confront contemporary planning practices to climate adaptation. Prior traditional planning approaches are inadequate to overcome the obstacles.

## 8.4 The Quality of Climate Adaptation Plans

Adaptation to climate change calls for a new paradigm of planning. Plans should take into account a range of future climate scenarios and associated policy alternatives to guide adaptation instead of reliance on historical trends, present conditions and a single predicted future. Plans should rely on a collaborative process that engages both experts and stakeholders. Lay knowledge is important as it reveals public preferences about how to respond to uncertain futures, but also for weighing in on analyzing a situation and formulating policies given the limitations of climate science. Viewed as emerging from a process of collaborative rationality (Norton 2008; Innes and Booher 2010), climate plans should be grounded on a collaborative process between experts and stakeholders that fosters exploration about future uncertainties, policies to adapt to a range of possible futures and recognition that over time local conditions will change and plans will need to be revised.

Research on the strengths and weaknesses of municipal climate adaptation plans is almost nonexistent. The Pew Foundation (2012) and Climate Adaptation and Knowledge Exchange (CAKE 2011) catalogue and describe the policies and programmes of a range of state and local climate adaptive planning efforts. Quay's (2010) study of three case studies of exemplary climate adaptation planning efforts (Denver, New York City and Phoenix) is the only systematic comparative effort in evaluating climate adaption plans in the United States. It concluded that climate adaptation plans should be derived from three steps: (1) anticipation and futures analysis, (2) creation of flexible adaptation policies, and (3) and monitoring and action (Quay 2010).

The past two decades of research in natural hazard mitigation planning has yielded a well-developed literature of evaluating the content and quality of mitigation plans (e.g., Berke and French 1994; Deyle and Smith 1998; Brody 2003). The mitigation work along with other investigations focused on plan quality culminated in a comprehensive review of the literature on plan quality evaluation by Berke and Godschalk (2009). The review identified the major principles of plan quality.

These principles and metrics have been used to test the extent to which plans are successful in supporting the public interest across multiple domains (such as biodiversity, housing and transportation). As discussed in Berke and Godschalk (2009), many investigators have examined the relevance of these principles for helping to explain the performance of plans in local adoption and implementation of hazard mitigation practices throughout the world.

The principles are achieved consecutively in the plan making process. The sequence starts with (1) issue identification and visioning, followed by (2) direction-setting elements that include goals, (3) fact base for policy selection, and (4) policies for guiding future settlement patterns. Principles 1 through 4 provide the foundation for (5) plan implementation actions, and (6) monitoring and evaluation that tracks and assesses the effectiveness of the plan in resolving issues and achieving goals.

For purposes of this study, four principles of plan quality are particularly relevant for climate adaptation: (1) a fact base premised on anticipation and alternative futures analysis that provides the empirical foundation for adaptive selection and prioritisation of policies; (2) flexible policies that guide decisions to ensure that desired outcomes are achieved; (3) flexible plan implementation actions that assign organisational responsibilities, timelines, and funds to implement a plan; and (4) monitoring of both climate changes that impact environmental, social and economic conditions, and the performance of plan policies in achieving desired outcomes.

### 8.4.1 Fact Base: Scenarios to Inform Alternative Futures

The fact base of a climate adaptation plan requires that some aspects of the future are of high uncertainty and that any forecast of impacts should account for multiple possible futures. In an early consideration of scenario planning, the classic planning text, *Urban Land Use Planning*, Chapin (1965) recognised the shortcomings of forecasting and that scenarios are needed to explore future uncertainty by suggesting that plans should be "the culmination of the repeated tests and retests of alternative arrangements in the pattern of uses" (p. 466). In the context of climate change, advanced scenario planning methods for considering uncertain futures has reached a new level of prominence (Quay 2010). Chakraborty et al. (2011) proposes that analyses should involve a variety of constituent organisations, and these investigators have proposed and used integrated models to assess impacts of alternative scenarios in the fields of transportation, land use and environmental planning. They suggest development of a range of possible scenarios for long-range time horizons, sensitivity analysis of factors or decisions driving the scenarios, identification of common and different impacts among the scenarios and the identification of unacceptable scenarios. They emphasise integration of extensive quantitative evaluation and public participation. Thus, the fact base for a climate adaptation plan should be derived by experts and stakeholders working in concert, rely on the development and evaluation of impacts from multiple possible scenarios and avoid use of one scenario based on a single forecast.

### 8.4.2 Flexible Policies: Contingent and Robust

Policies and strategies (an integrated set of policies) represents the heart of a plan because they guide public and private decisions to achieve a desired state of resiliency, but climate change policies must be designed to be adaptive. Using a fact base of anticipated alternative futures and associated impacts, *contingent* policies to adapt to one or more of these possible futures should be developed. By stipulating that policies are to anticipate and respond to a variety of future possibilities, then policies must be tailored contingent to specific futures. If a particular policy is preferred under a set of changes but not the other, then the policy is contingent. However, if a future outlined by a particular scenario does not materialise, then the policy aligned with that scenario will remain unused; yet, without such a policy, a community risks being unprepared. *Robust* policies are those that have a positive impact across many possible futures and can preserve future options. These policies yield preferable results under multiple scenarios.

Chakraborty et al. (2011) suggests that a combination of *contingent* and *robust* policies offer a flexible approach that can be implemented as needed over time. This approach allows for distribution of costs over time as opposed to one-time lump sum investments to carry out a particular policy that might be abandoned (Quay 2010). Contingency policies are to respond to specific scenarios like a worst case policy that addresses the worst scenario. Robust policies are to guide low-regrets and no-regrets actions in the short-term that can be adapted over time to address several possible scenarios.

### 8.4.3 Accountable Implementation

The accountable implementation principle states that a plan should ensure that organisational responsibilities for carrying out each policy are clearly stated, along with a timeline of when policies are to be implemented (APA Sustaining Places Task Force 2011). An implementation element of a plan should articulate mechanisms and procedures to implement the plan once it is adopted. Implementation depends not only on the ability of a community to implement its plan in a timely fashion, but also to designate responsibility for actions, enforce adopted standards and sanction those who fail to comply. Given that climate change will have variable impacts over the next 100-years, flexibility is an important attribute of the implementation principle. Changing conditions will require a parallel commitment to changing organisational responsibilities, timelines for action and allocation of resources committed to action.

### 8.4.4 Monitoring and Evaluation

Planners must be able to react to constantly changing conditions, including acceleration of climate change and associated impacts, sudden shifts in public interests and objectives and a continuous barrage of new and often uncertain information. The monitoring process is relatively straightforward; new information is identified, evaluated and used to adjust policies. Measurable indicators of change should be created to monitor on a regular basis, and decisions to implement anticipated adaption policies considered in light of actual trends. Monitoring and evaluation represents a break in the traditional “predict and plan” approach that has resulted in the widespread failure among local communities to monitor change and thus embrace an adaptive approach in which a community learns by doing and adjusts to change (e.g., Berke et al. 2012; Seasons 2003). Given that climate change will have variable impacts on communities over the next 100-years, decisions on and implementation of policies will need to be adaptive.

## 8.5 The Quality of Adaptive Plans for Climate Change: Case Studies

While attention to climate change adaptation planning has been increasing world-wide, most municipalities and metro areas are focused on reviewing vulnerabilities and adaptation strategies and have not engaged in implementation and monitoring. Further, among those places that have created adaptation plans, only a few address uncertainties in projections of future hazards and vulnerabilities through scenario planning. I present two innovative case studies of a large "world city" (London, UK) and small city (Punta Gorda, Florida, USA) undertaking climate adaptation planning. I apply the principles of plan quality for climate adaption to assess the progress of these ongoing planning efforts.

Data for the cases was derived from reviews of the plans and documents produced by each planning process focused on the efforts and results. Unstructured interviews with knowledgeable planners and technical staff active in the planning efforts were also conducted. While the plans are still in the formative stages, my hope is that other places will benefit from these efforts and will save them time in making effective plans in light of the fact that human vulnerability to climate change is rising and in many locations the rates are faster than has been anticipated. Both case studies have undertaken vulnerability assessments and used alternative future scenarios to address uncertainties and are in the process of developing flexible adaptation policies. Both have begun implementation of some policies, but both have not progressed to monitoring of outcomes and evaluation of plan performance.

### *8.5.1 London, UK*

The most recent round of London's climate change adaptation efforts began in 2008 in response to the UK 2008 Climate Change Act which requires local governments assess the risks that climate change pose to their operations and develop a climate adaptation plan. The city, which comprised a 2009 population of 7,753,600, initiated the Greater London Corporation partnership in order to produce a management strategy. The partnership included representatives from 30 local, metro and national government agencies as well as members of the private and non-profit sectors, and was charged to study and produce a management strategy to deal with changing climate risks, and to ensure that the city's assets, services and infrastructure continue to function appropriately, and that the city as a whole continues to thrive (Davoudi et al. 2009).

In 2010, the partnership published *Rising to the Challenge: The City of London's Climate Change Adaptation Strategy* (City of London 2010). The City of London's plan focused on five sets of management initiatives: flood risks, water resources risks, heat risks and air pollution, landscaping (or ground conditions) to reduce subsidence in highly vulnerable areas and cross-cutting issues involving identification of how any of the risks can be addressed through the review of development and design standards and codes. After completion of the plan, it will be evaluated against 188 national performance indicators which are used to measure progress in addressing the risks and opportunities posed by climate change.

London's process for developing its climate adaptation plan included assessments of climate vulnerabilities, opportunities to take action through current plans and regulations and adaptation responses to address the risks and opportunities. It also covered development of an adaptation action and implementation programme, and a monitoring and evaluation programme to ensure progress. As of 2010, the city has completed all phases up to monitoring and evaluation.

#### 8.5.1.1 Futures Analysis

London initially explored four major climate risk factors: temperature, precipitation, sea-level rise and extreme events. To define these factors, climate specialists examined temperature and precipitation results from multiple climate change models for two emission scenarios defined by the IPCC *Working Group III* report (Nakićenović and Swart 2000) in each of two future 30-year time periods. The IPCC's (2007) global sea-levels projection of 0.2–0.6 m by 2095 was considered to be too low for the London region. To account for this, the UK Environment Agency projection was used which estimates that sea-level rise in the River Thames to be in the range 0.2–0.9 m by 2100 with a worst-case scenario of 2.7 m.

The committee estimated the most likely range for each factor in each future time period, including the 33–66% likely events range, and the 10–90% extreme events range. The probabilistic projections were used to illustrate ranges of future changes in climate variables over a selected location. The information they provide on low probability (extreme) events will be of particular relevance to those involved in contingency planning.

To illustrate how the average change and probabilistic ranges for factors are depicted, Fig. 8.1 shows the average projected future increase and possible ranges in the wettest winter day for London under a high greenhouse gas emissions scenario. Under this scenario, the likely range of change in average winter rainfall is +1 to +10% by the 2020s and +1 to +24% by the 2050s. Experts from across city, metro and national agencies, as well as other stakeholder groups were then asked to identify which social, economic and infrastructure systems would be vulnerable to impact associated with the climate risk factors using these probabilities.

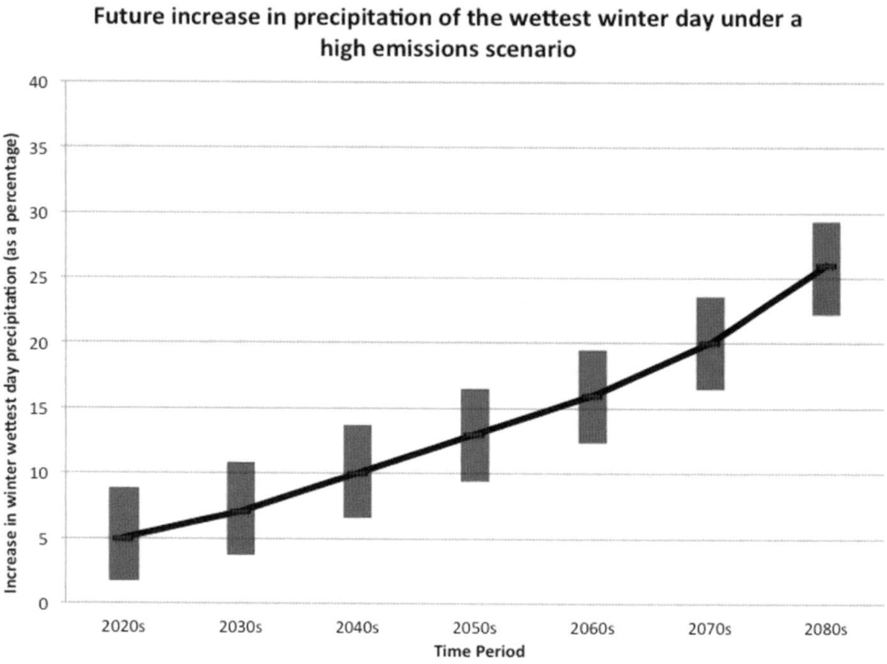


**Fig. 8.1** Rise in winter precipitation in City of London due to climate change. The black line shows the central estimate (50th percentile) of the increase in precipitation on the wettest winter day for the high emissions scenario. The wide grey bars show the likely range of change (33rd to 66th percentiles). The error bars show the 10th and 90th percentile events (future increase in precipitation on the wettest winter day is very unlikely to be outside this range). (Source: City of London 2010)

#### 8.5.1.2 Flexible Policies

London planners developed flexible adaptive policies which emphasise that actions are to be taken in small incremental steps, or in some cases eliminated over time, with the aim of minimising costs and lost assets. The adaptation options are grouped into research and monitoring, policy and practical actions, and are categorised under the following decision criteria:

- *No-regrets* measures that deliver benefits that exceed their costs, whatever the extent of climate change;
- *Low-regrets* measures that are low cost, and have potentially large benefits under climate change;
- *Win-win* measures that contribute to climate adaptation and also deliver other benefits; and
- *Flexible* measures that are useful for dealing with uncertainties associated with longer-term climate change.

Table 8.1 illustrates how the criteria are applied to adaptation policies.

**Table 8.1** Adaptation actions to manage flood risks in the city of London's climate adaptation plan. (Source: City of London 2010, p. 17)

| | |
|---|---|
| *Research and Monitoring* | *No Regrets*. The City of London should work to identify and map flash flood 'hotspots' and assign responsibility for coordination and liaison on flood risk management in order to ensure its practical implementation |
| | *Low Regrets*. The City of London should improve the monitoring and recording of gully overflows linked to heavy rainfall events and assess the capacity of sewers managed by the City of London to cope with increasing rainfall due to climate change, as well as coordinating with the Thames Tideway Tunnel project |
| *Policy* | *No Regrets*. The draft LDF (Local Development Framework) includes policies on Flood Risk and Sustainable Design and Climate Change, which promote the use of sustainable drainage systems, such as green roofs, in developments and street enhancements.Sustainable drainage systems (SuDs) such as green roofs should be encouraged as part of new developments, redevelopments and major refurbishments through the LDF planning agreements should be used to secure long-term commitment to the management and maintenance of SuDs |
| | *Low Regrets*. The City of London LDF should require that drainage systems in all developments have the capacity to cope with heavier rainfall events expected over their lifetimes, taking account of climate change |
| *Practical Actions* | *Low Regrets*. The City of London should encourage businesses to consider relocating flood-sensitive IT equipment and archives to areas with low risk of flooding. The Contingency Planning Department should encourage businesses with assets and equipment that need to be on-site, to move them away from locations at higher risk of flooding, such as basements |
| | *No Regrets*. Developers should be encouraged to install sustainable drainage systems and green roofs in targeted flash flood 'hotspots' for new developments, redevelopments or major refurbishments |
| | *Win-Win and No Regrets*. The City of London Corporation should consider installing sustainable drainage systems, green roofs or green walls on City of London-owned car parks and buildings, when they are refurbished or replaced |
| | *Low Regrets*. The City of London should examine a range of incentives to encourage sustainable drainage systems and green roofs |

London is in the midst of developing a broader regional approach to adaptation policies. Given its broad regional foundation of social, economic and infrastructure systems, the adaptation policies will be coordinated with other localities beyond London's jurisdiction, and will suggest a regional governance approach for other governmental and private sector organisations. In general, London is suggesting that any activity that reduces vulnerability to climate change-associated impacts makes the city more resilient and sustainable.

Climate adaptation policies will be further advanced via integration with the city's comprehensive plan, called the Local Development Framework (LDF) plan, which is to be completed in 2012. LDF sets out the planning vision for the city and the key policies to deliver this vision. LDF integrates and coordinates the spatial planning aspect of all the City's other policies, including those associated with adaptation. The achievement of sustainable development is a central consideration in the LDF by setting out policies with the intent of increasing prosperity of the city's economy while protecting the environment and ensuring the well-being of all the city's communities.

#### 8.5.1.3 Implementation and Monitoring

The adaptation plan gives considerable attention to coordinating city organisations and agencies with plan implementation responsibilities. It assigns responsibilities and sets a timeline when each action is to be taken. The actions are prioritised based on the likelihood of occurrence and the expected level of impact associated with each risk. Progress on implementation will be monitored at six-month intervals by a cross departmental officers working group.

The indicators and metrics to measure the degree to which the adaptation policies achieve risk reduction outcomes are not identified in the plan, but are under development. The city will also gauge progress in the context of The Local Government Performance Framework, introduced by the national government in 2008. The framework includes a set of 188 national indicators that UK local governments must use to measure the accomplishment of national priorities. The City of London will be reporting against national indicators on an annual basis.

### *8.5.2 Punta Gorda, USA*

The development of *City of Punta Gorda Adaptation Plan* (Beever et al. 2009) was completed in 2009 as initiated in 2007 by the USA Environmental Protection Agency and the National Estuary Programmes. The City of Punta Gorda had a 2007 population of 16,262 and is located in Charlotte County with a 2007 population of 151,184. The county and city, located within one of six major estuaries were selected to develop plans to help protect sensitive coastal ecosystems, infrastructure and economies from the effects of climate change.

On December 17, 2008, the Punta Gorda City Council voted unanimously to participate in the CHNEP-CRE (Charlotte Harbor National Estuary Program-Climate Ready Estuaries) pilot programme. One of the reasons that Punta Gorda was amenable to working with Charlotte Harbor to develop a climate adaptation plan was the fact that Southwest Florida is one of the most vulnerable areas in the world to the consequences of climate change, especially sea-level rise and increased hurricane activity and severity. Another reason was the community was devastated by Hurricane Charley in August 2004. At its peak intensity, Charley attained 150 mile per hour winds, making it a strong category 4 on the Saffir-Simpson hurricane intensity scale. The storm made landfall in southwestern Florida at maximum strength, and at that time making it the strongest hurricane to hit the USA since Hurricane Andrew struck the state in 1992. Charley caused US$ 14.6 billion in property damage in Florida alone.

After extensive public, city staff and council review the plan was unanimously accepted by the city council in November 2009. The council directed local agencies to incorporate components of the plan into their normal work process, including Punta Gorda's comprehensive planning update.

#### 8.5.2.1 Futures Analysis

In developing the 2009 plan, a series of public workshops, presentations and resident and business surveys were held between April and July 2009 by the City of Punta Gorda. Through this process 54 vulnerabilities were identified and combined into eight major priority areas of climate change vulnerability: (1) fish and wildlife habitat degradation; (2) inadequate water supply; (3) flooding; (4) unchecked or unmanaged growth; (5) water quality degradation; (6) education, economy and lack of funds; (7) fire; and (8) availability of insurance. Planners and experts from federal, state, regional and city agencies then defined ranges of future possible conditions under each of these categories of vulnerability. For example, predictions of habitat loss were developed using IPCC (2001) findings and more fine grained estimates derived from a model called SLAMM (Sea Level Affecting Marshes Model). Working with the state and university experts, the City of Punta Gorda estimated losses for 14 types of habitats under minimum, mean and maximum sea-level rise conditions. In another example, local planners and hazard mitigation experts drew from the joint county-city 2005 hazard mitigation plan to consider three categories of hurricane strength (Category 1 = lowest, 3 = middle and 5 = highest). These factors were used to generate scenarios of an estimated number and value of structures at risk from flooding for five classes, including residential, historic, structures owned by top 10 employers, repetitive loss structures and critical facilities.

As part of a pilot project in climate change adaptation planning, involving urban planning and transportation experts from Florida State University (Chapin et al. 2010), the City of Punta Gorda and Charlotte County extended the scenario development by exploring the range of sea-level rise and storm surge impacts based on variations of three future urban growth patterns for the year 2050, as illustrated in Fig. 8.2:

- Policy Scenario: current land use and development policies that tend to emphasise low density and sprawl;
- Smart Growth Scenario: land use and development policies that shifts higher densities to existing core urban areas; and
- Resiliency Scenario: land use and development policies that establish new core urban areas to accommodate future growth that avoids flood hazard areas.

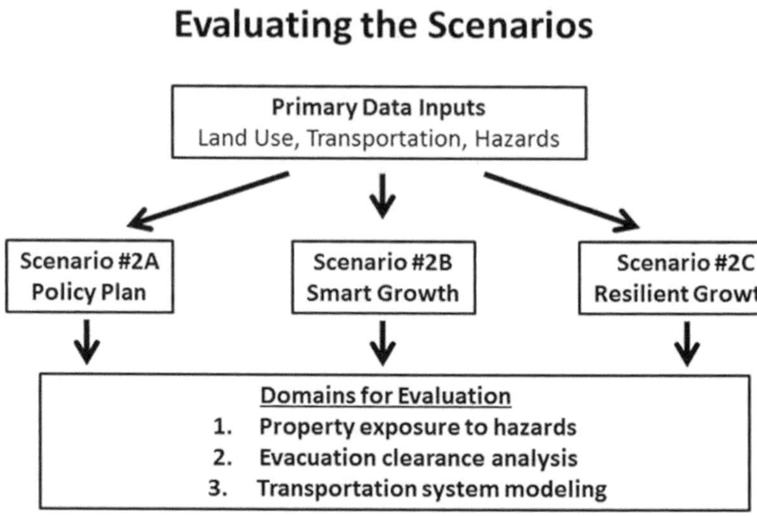


**Fig. 8.2** Evaluating the impacts of future scenarios, Punta Gorda, Florida.1. (Source: Chapin et al. 2010)

Primary data inputs included projections of population and economic growth that were used to drive projections of land use and transportation infrastructure investments. Mapped flood data derived from projections of sea-level rise and storm surge were then overlaid on the three growth scenarios. Alternative growth patterns were derived based on assumptions about shifts to higher densities in core urban areas, and the degree of avoidance of flood hazard areas. Integrated land use allocation, transportation and loss modeling was then employed to determine the number of people and structures exposed to hazards, and evacuation clearance times of vulnerable populations.

#### 8.5.2.2 Flexible Adaptation Actions

The City of Punta Gorda is still discussing its adaptation strategies and decision framework. Given the wide range of potential vulnerabilities and effectiveness of potential adaptive actions, the city adopted a flexible two-step approach to acting on various types of climate change impacts based on a series of public meetings. Initially, Punta Gorda identified and classified three categories of impacts that require action (or no action) as indicated in the 2009 plan:

- Tolerate impacts. No reasonable options will be found (e.g., accept the loss of its coral reefs).
- Mitigate impacts. Actions are needed to compensate for some of the adverse effects (e.g., set aside coastal lands so that tidal wetlands can migrate inland as sea level rises).

- Adapt to impacts. Actions that will change the way of life, design of infrastructure and/or economy (e.g., buildings may need to be designed to new standards or located farther from vulnerable shorelines).

Given uncertainties and the long time frame of climate change impacts, the next step taken by the city was to identify a flexible "no- and low-regrets" approach to taking adaptive action. The "no-regrets" option is justified by current climate conditions, and further justified when climate change is considered. For example, reducing water pollution could improve potable water supplies. The pollution reductions may be even more valuable should climate change reduce water supplies or degrade water quality. The "low-regrets" actions are those taken because of climate change, but at a minimal cost. Thus, there is "low regret" if the investment proves not to be needed under future climate conditions. For example, incorporating risks of climate change in design of infrastructure may offer improved protection against current extreme climate events, as well as potential future events under climate change, while increasing costs only marginally.

In the pilot project with Florida State University researchers, Punta Gorda generated findings that identified two types of adaptation strategies: robust strategies that would work well across all scenarios; and a worst-case future land use pattern for sea-level rise and hurricane surge penetration. One example of a robust strategy arose from the finding that all scenarios that allocate future land development showed evacuation times to be very high and roughly equivalent (25–37 h), and evacuation routes in need of expansion to ensure safe evacuation under a *trigger point* clearance time of populations living in Punta Gorda and Charlotte County. In all cases, it was new development to accommodate growth in population that would trigger the need for expansion of the capacity to transport people (e.g., more bus service, more lanes).

The pilot project included a worst-case future portraying the severity of property loss from the Smart Growth scenario to be greater than the other two development scenarios (Policy Scenario and Resiliency Scenario). Specifically, the Smart Growth Scenario consists of three conditions: (1) development is constrained to a smaller urban service area; (2) a range of housing types are developed in or near identified urban centers; and (3) commercial development is targeted in identified urban centers. This scenario reveals that conventionally accepted model development regulations and investment schemes that support compact urban form can dramatically increase vulnerability compared to other models of urban form if hazard mitigation is not incorporated into the model development design standards (Berke et al. 2009; Stevens et al. 2010).

#### 8.5.2.3 Implementation and Monitoring

As noted, the 2009 plan was unanimously adopted by the Punta Gorda city council as official local government policy. The city council indicated that a major step in implementation is to incorporate components of the plan into day-to-day decision-making among local agencies. Another major step will be to incorporate the proposed actions of the 2009 adaptation plan into the envisioned 2025 city comprehensive plan to be developed in 2010–2011. However, the 2009 climate plan is not specific in assigning organisational responsibilities tied to specific actions and setting a timeline that outlines when each action is to be taken.

Punta Gorda is in the midst of developing indicators to monitor health of wildlife habitats, sea-level rise, drought, exposure of development to flooding, certainty of water supplies and other factors. Both the selection of indicators for monitoring and the frequency of monitoring can evolve over time as the adaptation process matures; this evolution may continue as the adaptation process is incorporated into a city's overall policy mix.

## 8.6 Discussion of Case Studies

While there are obvious differences in London's world city status and Punta Gorda's small city position as well as their geographic and governance arrangements, the case studies have many commonalities in support of the principles of plan quality for climate adaptation planning:

- Both use a fact base premised on scenarios to anticipate a range of possible future climate change impacts and avoid single "best guess" scenarios;
- Both have begun to formulate flexible policy frameworks that emphasise low-cost/low-regrets policies and win-win/no-regrets policies that are to be integrated into broader day-to-day decision-making and comprehensive spatial plans;
- Both are in the initial stages of developing implementation and monitoring programmes; and
- Both rely on stakeholder participation for scenario and policy formulation, and plan to do so for implementation and monitoring.

There are several differences:

- Punta Gorda's scenarios vary based on population growth and the impacts of alternative land use policies, while London assigned probabilities of climate change factors (e.g., temperature, precipitation, sea level) to its scenarios;
- Punta Gorda's policy framework gives more attention to policy options for a worst-case scenario than London, which led to a modification of smart growth policies to account for hazard mitigation on the Florida coast;
- London has a better developed implementation programme with more accountability through tracking progress over short-time intervals (6 months), and more flexibility with adjustable actions based on likelihood of occurrence of changes, but Punta Gorda has not yet specified organisational responsibilities and a timeline for action; and
- London is developing a monitoring programme under a national mandate that requires the city to annually track progress against a set of national indicators, but Punta Gorda does not operate under such requirements which creates uncertainty about if, when and how often the city will actually conduct monitoring.

## 8.7 Conclusions and Recommendations

Traditional planning is not well-equipped to build community resiliency to climate change. Experience with natural hazard mitigation planning clearly demonstrates that local governments do not act when costs of reducing vulnerability are immediate and the benefits are long-term and uncertain. As a result, such mitigation plans have consistently been weak and inconsequential. Integration of knowledge about collaborative rationality, hazard mitigation and scenario planning offers a plan making framework that can overcome the obstacles to formulating plans that matter in reducing vulnerability to climate change. The case studies from London and Punta Gorda represent cutting edge planning practice in incorporating plan quality principles for climate adaptation planning and implementation. Although the outcomes of the case studies could not be assessed since these efforts are not complete, their experiences provide useful lessons for policy and future research.

### *8.7.1 Policy*

First, uncertainty in climate science should not pose a barrier to seeking planned solutions to reducing vulnerability. A range of scenarios collaboratively defined by experts and stakeholders and related impacts tied to each scenario can help communities to anticipate and adaptively respond to climate changes. Collaborative engagement with stakeholders can also build a sense of ownership and support for the climate adaptation plan which is critical for plan implementation in the long-run.

Second, flexible policies adapted to changing conditions make climate adaption plans more economically and politicallly feasible. *Robust* policies that support low- and no-regrets actions are important because they have a positive impact across multiple scenarios and can be implemented incrementally in modules to limit costs and demonstrate benefits in the short-run. For example, large infrastructure projects can be built one module at a time and at the same time remain functional as modules are added. *Contingent* policies are specifically tied to a particular scenario and can be used to acknowledge and prepare for worst-case scenarios. Whenever possible, both types of adaptation policies should also deliver other social, economic or environmental benefits.

Third, an implementation programme is a critical element of adaptation plans. A plan should include clear identification of organisations responsible for carrying out specific actions, and timelines for action. However, unlike less flexible traditional plans, adaptation plans should account for change in responsibilities, timelines and commitment of resources that parallels variability in climate conditions. Whenever possible, sanctions and/or information about compliance should be communicated to the public and elected officials to coerce and/or encourage accountability.

Fourth, development of a strong monitoring programme for climate adaptation planning is critical. Monitoring detects change in impacts and increases the likelihood that a community will have sufficient time to act in identifying which scenarios and associated actions are relevant and which are not in the short-term.

Monitoring is important to gauge the performance of adaptation policies. Coupled with a clear public communication effort, monitoring contributes to social learning as plan policies are tested and revised over time.

### *8.7.2 Future Research*

First, comparative research on the causes and consequences of plan quality for climate adaptation is needed. Such research would improve the identification of the key elements or principles and related best practices that define plan quality for climate adaptation. It would also support development of metrics and methods to detect how well plans support the principles, and to assess the degree of success in plan implementation efforts. Policy-makers at the national and provincial (including state) levels could then have an even better knowledge-base to design interventions that create higher quality local adaptation plans and effective plan implementation.

Second, the methods of scenario formulation and testing are only in the early stages of development. Future research should focus on better technical integration of multiple models needed to assess impacts and reduce uncertainty. This will also require better data to improve modeling. Development of indicators and collection of data for monitoring at multiple spatial scales and collaborative arrangements among multiple levels of government is critical.

Third, improving techniques for engaging stakeholders in using the results of modeling is important in formulating realistic scenarios that are relevant to the realities of urban development place making. Excessive reliance on expert-driven scenario development is more likely to produce superficial "straw dog" scenarios that lack relevancy. As noted, public support for plans and implementation is more likely to be sustained in the long-run when stakeholders are actively involved in participating in exploring causes and consequences of alternative policies.

Fourth, the planning field needs an improved understanding of how to design and finance large infrastructure projects incrementally. Major transportation, wastewater treatment, water supply and flood control installations need to be designed in ways that ensure they remain functional while expanding incrementally.

In sum, as the natural hazard mitigation field has demonstrated after 30-years of research and practice, all too often communities remain inactive in the face of pending threats. However, to be resilient in the face of ever increasing threats in the age of climate change, action is needed more than ever. Society must be better prepared to anticipate alternative futures to adapt now and long into the future.

**Acknowledgments** This material is based upon work supported by the U.S. Department of Homeland Security, Coastal Hazards Center of Excellence under Award Number: 00313690. The views and conclusions contained in this document are those of the authors and should not be interpreted as necessarily representing the official policies, either expressed or implied, of the U.S. Department of Homeland Security.

## References


APA Sustaining Places Task Force. (2011). *Sustaining places: The role of the comprehensive plan*. Chicago: American Planning Association.

Bartholomew, K. (2007). Land use-transportation scenario planning: Promise and reality. *Transportation, 34*(4), 397–412.

Berke, P., & French, S. (1994). The influences of state planning mandates on local plan quality. *Journal of Planning Education and Research, 13*(4), 237–250.

Berke, P., & Godschalk, D. (2009). Searching for the good plan: A meta-analysis of plan quality studies. *Journal of Planning Literature, 2*(3), 227–240.

Berke, P., Song, Y., & Stevens, M. (2009). Integrating hazard mitigation into new urban and conventional developments. *Journal of Planning Education and Research, 28*(4), 441–455.

Berke, P., Smith, G., & Lyles, W. (2012). Planning for resiliency: An evaluation of state hazard mitigation plans under the disaster mitigation act. *Natural Hazards Review, 13*(2), 139–149.

Beever, J. W., Whitney G., Daniel T., Dan C., Jason U., David H., John G., Tim W., Moji A., Lisa B. B., & Judy O. (2009). *City of Punta Gorda adaptation plan*. Charlotte Harbor National Estuary Program, Technical Report 09-4. Punta Gorda, FL: City of Punta Gorda.

Brody, S. (2003). Are we learning to make better plans? A longitudinal analysis of plan quality associated with natural hazards. *Journal of Planning Education and Research, 23*(2), 191–201.

Burby, R. (Ed.). (1998). *Cooperating with nature: Confronting natural hazards with land use planning for sustainable communities*. Washington, D.C.: Joseph Henry Press.

Burby, R. (2006). Hurricane Katrina and the paradoxes of government disaster policy: Bringing about wise governmental decisions for hazardous areas. *The Annals of the American Academy of Political and Social Science, 604*(March), 171–191.

Chakraborty, A., Kaza, N., Knaap, G.-J., & Deal, B. (2011). Robust plans and contingent plans scenario planning for an uncertain world. *Journal of the American Planning Association, 77*(3), 251–266.

Chapin, F. S. (Ed.). (1965). *Urban land use planning*. Urbana, IL: University of Illinois Press.

Chapin, S. F. III, Carpenter, S. R., Kofinas, G. P., Folke, C., Abel, N., Clark, W. C., Olsson, P., Stafford Smith, D. M., Walker, B., Young, O. R., Berkes, F., Biggs, R., Grove, J. M., Naylor, R. L., Pinkerton, E., Steffen, W., & Swanson, F. J. (2009). Ecosystem stewardship: Sustainability strategies for a rapidly changing planet. *Trends in Ecology and Evolution, 25*(4), 241–248.

Chapin, T., Deyle, R., & Higgins, H. (2010). Integrating hazard mitigation and sea level rise adaptation into long-range transportation planning. Paper presentation at the ACSP Annual Conference Minneapolis, Minnesota, October, 2010.

City of London. (2010). *Rising to the challenge: The city of London's climate change adaptation strategy*. London: Sustainability Unit, City of London.

Climate Adaptation and Knowledge Exchange (CAKE). (2011). http://www.cakex.org/case-studies. Accessed 4 Sept 2011.

Davoudi, S., Mehmood, A., & Brooks, L. (2009). *The London climate change adaptation strategy: Gap analysis*. London: Global Research Unit of the ARCADIA Project, Electronic Working Paper No. 44.

Deyle, R. E., & Smith, R. A. (1998). Local government compliance with state plan mandates: The effects of state implementation in Florida. *Journal of the American Planning Association, 64*(4), 457–469.

Global Climate Research Program (GCRP). (2009). *Global climate impacts in the United States*. New York: Cambridge University Press.

Godschalk, D. (2003). Urban hazard mitigation: Creating resilient cities. *Natural Hazards Review, 4*(3), 146–143.

Innes, J., & Booher, D. (2010). *Planning with complexity: An introduction to collaborative rationality for public policy*. Oxford: Routledge/Taylor and Francis.

Intergovernmental Panel on Climate Change (IPCC). (2001). *Climate change 2001: Impacts, adaptation, and vulnerability*. Cambridge: Cambridge University Press.

Intergovernmental Panel on Climate Change (IPCC). (2007). Summary for policymakers. In S. Solomon, D. Qin, M. Manning, Z. Chen, M. Marquis, K. B. Avery, & H. L. Miller (Eds.), *Climate change 2007: The physical science basis. Contribution of working group I to the fourth assessment report of the intergovernmental panel on climate change* (pp. 1–18). Cambridge: Cambridge University Press.

Kaiser, E., & Godschalk, D. (1995). Twentieth century land use planning: A stalwart family tree. *Journal of the American Planning Association, 61*(3), 365–385.

Mileti, D. S. (1999). *Disasters by design: A reassessment of natural hazards in the United States*. Washington: Joseph Henry Press.

Nakićenović, N., & Swart, R. (Eds.). (2000) *Special report on emissions scenarios. A special report of working group III of the intergovernmental panel on climate change* (p. 599). Cambridge: Cambridge University Press.

National Research Council (NRC). (2006). *Facing hazards and disasters: Understanding human dimensions*. Washington: National Academy Press.

Norton, R. (2008). Using content analysis to evaluate local master plans and zoning codes. *Land Use Policy, 25*(3), 432–454.

Peacock, W. G., Kunreuther, H., Hooke, W. H., Cutter, S. L., Chang, S. E., & Berke, P. R. (2008). *Toward a resiliency and vulnerability observatory network: RAVON*. Report # 08-02R. College Station: Hazard Reduction and Recovery Center.

Pew, F. (2012). State adaptation plan. http://www.c2es.org/what_s_being_done/in_the_states/adaptation_map.cfm. Accessed 8 Apr 2012.

Quay, R. (2010). Anticipatory governance a tool for climate change adaptation. *Journal of the American Planning Association, 76*(4), 496–511.

Schipper, E. L. F. (2006). Conceptual history of adaptation in the UNFCCC process. *Review of European Community & International Environmental Law, 15*(1), 82–92.

Seasons, M. (2003). Monitoring and evaluation in municipal planning. *Journal of the American Planning Association*, *69*(4), 430–440.

Stevens, M., Berke, P., Song, Y. (2010). Disaster resilient subdivision design: Evaluating the potential and performance of new urbanism for reducing flood risks. In F. Columbus (Ed.), *Floodplains: Physical geography, ecology and societal interactions* (p. 17). New York: Nova Science Publishers.

U.S. Global Change Research Program. (2009). *Global climate change impacts in the United States*. Cambridge: Cambridge University Press.

Wheeler, S. M. (2008). State and municipal climate change plans: The first generation. *Journal of the American Planning Association, 74*(4), 481–496.

# Part III
# Lessons from Disaster Experience

# Chapter 9
# Applying Hurricane Recovery Lessons in the United States to Climate Change Adaptation: Hurricanes Fran and Floyd in North Carolina, USA

**Gavin P. Smith**

**Abstract** Hurricanes Fran and Floyd, the two most costly disasters in the history of the state of North Carolina (USA), struck within a three year time span. This provides a unique opportunity to assess the degree to which lessons were transferred from one extreme event to another in terms of the state-level approaches taken to address recovery needs at the local level, including the proactive adoption of supportive state policies advancing hazard mitigation and sustainable development measures in anticipation of the next disaster. The lessons uncovered in a review of these two events include three important themes that are relevant to climate change adaptation: (1) large-scale disasters in the U.S. trigger considerable amounts of post-disaster assistance and attention; (2) the manner in which the "window of opportunity" to garner and effectively coordinate the distribution and use of these resources through changes in policies, plans, public perceptions and programmes varies significantly over time; and (3) the transfer of lessons from one event to another, including the institutionalisation of policies and the hiring of personnel needed to sustain the implementation of these policies requires a significant commitment of political, financial, technical and administrative resources and is difficult to achieve in practice.

**Keywords** Disaster recovery · Hazard mitigation · Housing relocation · Disaster assistance · State recovery programmes

## 9.1 Introduction/Setting

This chapter describes the actions taken by the State of North Carolina (USA) to recover from two major hurricanes and apply the lessons identified to a number of key issues that are directly pertinent to climate change adaptation. Hurricanes Fran (1996) and Floyd (1999) each represented at the time they struck, the worst disaster in the State's recorded history. The short time period between events, the flooding of many of the same communities and the magnitude of these two storms highlight a number of factors that would seem to facilitate lesson drawing across a network

G. P. Smith (✉)
University of North Carolina at Chapel Hill, Chapel Hill, North Carolina, USA
e-mail: gavin_smith@unc.edu

B. C. Glavovic, G. P. Smith (eds.), *Adapting to Climate Change,* Environmental Hazards, DOI 10.1007/978-94-017-8631-7_9,  

of participants, including communities, the state and members of the larger national policy framework. Although, as Tom Birkland describes in his book *Lessons of Disaster: Policy Change After Catastrophic Events* (2006), disasters, which are defined as low probability high consequence events, can serve to focus attention on a number of problems, this condition does not necessarily result in the adoption of well-informed policy change guided by participatory and deliberative processes (pp. 7–8).

In practice, the policies adopted and institutions created by local, state and the federal government to deal with the aftermath of the two storms discussed in this chapter are similar in some ways and very different in others. North Carolina was led by the same charismatic, politically astute governor during both events and he was closely aligned with those in positions of influence and power in Washington, D.C., including the Director of FEMA and the President of the United States. The Governor also proved to be a highly effective politician in the North Carolina statehouse. Working with the Governor, state officials took advantage of these relationships, resulting in the acquisition of significant post-disaster federal assistance. In addition, the state sought to develop an enhanced emergency management capability, placing a strong emphasis on hazard mitigation and disaster recovery, including the development of new programmes addressing sustainability.

Next, a discussion of the destructive impacts associated with hurricanes Fran and Floyd is intended to provide a contextual background for the description of the hazard mitigation and disaster recovery polices adopted by the state and to help frame recommended policy lessons that are applicable to climate change adaptation. Embedded in these lessons are three important themes: (1) large-scale disasters in the U.S. trigger considerable amounts of post-disaster assistance and attention; (2) the manner in which the "window of opportunity" to garner and effectively coordinate the distribution and use of these resources through changes in policies, plans, public perceptions and programmes varies significantly over time; and (3) the transfer of lessons from one event to another, including the institutionalisation of policies and the hiring of personnel needed to sustain the implementation of these policies requires a significant commitment of political, financial, technical and administrative resources and is difficult to achieve in practice.

### 9.1.1 The Historical Context of Hurricanes in North Carolina

In North Carolina, Hurricane Floyd, which struck the state in 1999, caused unprecedented flooding, primarily in small to mid-sized rural communities, many of which were already facing a loss of jobs and an outmigration of residents associated with a declining, largely agricultural-based economy (Delia 2001). The ensuing flood approximated a 200-year event in many places and devastated a number of cities and towns located adjacent to rivers that flow toward the State's coastal sounds (Barnes 2001, pp. 230–231).[1] Flooding was due to two principal factors, including heavy

[1] The media and government officials widely reported that Hurricane Floyd's rainfall produced a "500-year" flood. The United States Geological Survey later revised these estimates based on a

rainfall amounts over a two month period and the human modification of North Carolina's coastal drainage systems (Riggs 2001, pp. 30–44). Hurricane Floyd, which dropped up to 15 in. of rain in the eastern third of the state was preceded by heavy rains associated with Hurricane Dennis which made landfall 11 days earlier. In total, five events; including Hurricanes Dennis, Floyd and Irene; a tropical depression; and a frontal system resulted in rainfall totals approximating 40 in. of rain in the coastal plain during this time period (Riggs, pp. 30–31).[2] Compounding the effects of the large amounts of rainfall over a relatively short time period was the widespread channelisation and drainage of wetlands, the construction of elevated roadbeds in floodplains that included inadequately sized culverts and bridge openings to convey floodwaters and the development of riverine communities in areas prone to flooding (Riggs, pp. 37–44) (Fig. 9.1).[3]

Hurricane Floyd, which caused an estimated US$ 6 billion in damages in North Carolina, resulted in 52 deaths, 35 of which were associated with freshwater drowning and of those, 24 occurred as people were trying to cross flooded roadways in their vehicles. Over 45,000 homes were flooded, 7,300 of which were destroyed (i.e., more than 50 % damaged). Two hundred and thirty-five Red Cross shelters were opened, temporarily housing approximately 50,000 individuals. More than 30,000 hogs, 2.4 million chickens and approximately 700,000 turkeys died in the flood. One thousand four hundred roads were closed (including highways and interstates), 24 municipal waste water treatment plants were flooded and over 1.5 million customers were left without electricity (Barnes 2001, p. 260). Particularly hard hit were communities like Kinston and Goldsboro located on the Neuse River, and Rocky Mount, Tarboro and Princeville located in the Tar River-Pamlico Sound watershed. In many cases, floodwaters exceeded 10–15 ft. in depth, requiring the extraction of some residents from their attics as they sought to escape the rapidly rising water (Fig. 9.2).[4]

Many of eastern North Carolina's communities were also flooded by Hurricane Fran, the state's worst disaster to date, until Floyd struck.[5] Prior to Hurricane Fran,

more careful review of what were outdated and inaccurate Flood Insurance Rate Maps (Barnes, 2001, pp. 230–231).

[2] The all-time record for rainfall from a single tropical system in the U.S. is Tropical Storm Claudette, which dropped 42 in. of rain on Alvin, Texas in 1979 (Hearn 2004, p. 72).

[3] Stanley Riggs refers to more recent development patterns in eastern North Carolina as "contoured urban sprawl." While many of the initial settlements were built on higher ground, more recent growth has occurred in lower-lying areas as development pressure increased. The growth pattern in eastern North Carolina is similar in some ways to that found in New Orleans, Louisiana (USA) as the city core was initially platted on ridges of higher elevation and future development occurred in increasingly vulnerable, low-lying areas (Colten 2005). This development, coupled with the widespread destruction of protective wetlands and Cypress forests and the overreliance on poorly constructed levee systems presaged the Hurricane Katrina disaster in this major U.S. city, which is discussed in Chap. 15.

[4] During and in the immediate aftermath of Hurricane Floyd, over 1,500 people were rescued (Barnes 2001, p. 260).

[5] The 1990's was an active period for hurricanes in North Carolina as Emily (1993); Bertha and Fran (1996); Bonnie (1998); and Dennis, Floyd and Irene (1999) struck the state (Barnes 2001).

**Fig. 9.1** New development in the floodplains of Greenville, North Carolina that was flooded following Hurricane Floyd. (Source: Photograph by Dave Saville, FEMA)

**Fig. 9.2** The small town of Pactolus, North Carolina was severely flooded following Hurricane Floyd, requiring this family to revisit their home by boat. (Source: Photograph by Dave Gatley, FEMA)

the state had not experienced a major disaster since Hurricane Hazel in 1954, which hit the North and South Carolina border with 150 mile per hr winds.[6] Hurricane

[6] In the 1950's, the role of the federal government in emergency management was ill-defined, as the Federal Emergency Management Agency (FEMA) did not exist (it was established in 1979) and the State of North Carolina (nor the rest of the country for that matter) did not have in place hazard mitigation programs like those that existed when Fran hit.

Fran, unlike Hurricane Floyd, brought with it higher winds (maximum sustained winds of up to 115 miles per h) and more damaging storm surge, which was similar in many ways to that experienced during Hurricane Hazel.[7] This caused significant coastal damage to low lying barrier island communities, like North Topsail Beach and sound side towns such as Belhaven, North Carolina.

Heavy rains caused major flooding in the cities of Kinston, Goldsboro, Rocky Mount and others. More than seven inches of rain fell in parts of eastern North Carolina, four inches in the central part of the state and up to 12.5 in. in several mountain communities in western North Carolina prior to the storm. Fran was also recognised for the duration of its hurricane force winds and heavy rains, which were felt in the more urbanised areas of the central Piedmont cities of Raleigh and Durham that are located hundreds of miles inland and did not believe they were vulnerable to the impacts of coastal storms. Hurricane Fran resulted in over US$ 5 billion in damages in North Carolina and approximately US$ 1 billion in damages in states along the eastern seaboard including South Carolina, Virginia, Maryland, Pennsylvania, West Virginia and Ohio (Barnes 2001, pp. 202–203).

According to the National Oceanic and Atmospheric Agency (NOAA), North Carolina is among the most vulnerable coastal states in the U.S. to the damaging effects of land-falling hurricanes, due in large part to the state's coastal geology, which is defined by a low lying chain of barrier islands and a number of shallow sounds that jut into the Atlantic Ocean and the frequent path of hurricanes that tend to track up the southeastern coast of the U.S. When hurricanes make landfall they often bring torrential rains that flood coastal and inland communities. Coastal storms also play an important part in maintaining the health of coastal ecosystems through high rainfall which recharges rivers and periodically flushes the wetlands and coastal sounds, while high winds and storm-related waves and surge affects the shoreline's geophysical dynamics, as evidenced by the migration and accretion of sand and the formation and maintenance of dune fields (Pilkey et al. 1998, pp. 34–35). This assumes that coastal systems have not been excessively modified; otherwise these interdependent processes cease to function naturally, ultimately leading to increased vulnerability for coastal communities. Thus, coastal storms, best described as natural hazards, are part of the natural environment and become the construct we refer to as disasters when hazard events impact human settlements and the resources on which people depend and exceed the coping capacity of those impacted (White 1945; Mileti 1999; Reice 2012).

Several factors make North Carolina an important case to assess the effects of climate change, including the manner in which this coastal state may or may not choose to adapt to these changes. For instance, the increasing modification of the low lying barrier islands and riverine systems makes existing and future human settlements on barrier islands, along coastal sounds and in the coastal plain highly

[7] The 18 ft storm surge was worsened by the fact that it struck on a full moon in October during the highest lunar tide of the year, raising the flood elevation by several feet (Barnes, 2001, pp. 83–84). This reality provides some parallels with climate change-induced impacts as hurricanes of greater intensity are likely to strike areas that are also prone to rising sea levels.

**Fig. 9.3** Housing damage in Long Beach, North Carolina. The coastal home in the foreground was destroyed by coastal storm surge whereas the home in the distance survived due to the fact that it was elevated. With the advent of rising sea levels and increased coastal storminess the appropriateness of elevation as a hazard mitigation/adaptive strategy, including the height to which it should be elevated is uncertain. (Source: Photograph by Dave Gatley, FEMA)

vulnerable to rising sea levels and the predicted increase in more severe hurricanes (Riggs et al. 2011). As a result, the state and communities face major challenges including increasingly intense conflicts tied to the preservation, conservation and allowed migration of two key natural systems that have historically protected coastal communities. The effective management of wetlands, for instance, will necessitate developing a land use strategy that will allow for marshlands to move landward with rising sea levels. In the case of barrier islands, a long-term adaptation strategy to address the ongoing migration of barrier islands towards the mainland and the predicted collapse of these islands as rising seas threaten their very existence will require a sober assessment of resettlement options and water-based transportation systems that connect remaining islands to the mainland (Riggs et al. 2011).

North Carolina is facing rapid development along the coast, resulting in increased densities, larger supporting physical infrastructure (e.g., roads, bridges and public facilities) and the construction of bigger rental properties (that often replace smaller units) and hotels, both of which are tied to the growing tourism industry. The escalating investments in known high hazard areas not only speaks to the growing vulnerability of a state that is already among the most susceptible in the nation to the effects of coastal storms, it also points out how current policy decisions and investment choices can limit the types of adaptive strategies that may be available to the state and coastal communities as noted by Timothy Beatley in Chap. 6 (Fig. 9.3).

For instance, a uni-dimensional commitment to physical infrastructure and engineering-dominant solutions that attempt to alter or minimise the natural dynamism of the coastal zone hinders adaptability and once constructed, requires increasingly large investments in these protective measures. A heavy dependence on protective measures, often referred to as "armoring the coast" has the effect of encouraging further development in known hazard areas and sets the stage for larger disasters when the design parameters of the protective infrastructure is exceeded (Pilkey and Dixon 1996). Historically, North Carolina has been considered a national leader in the development of coastal management policies that significantly limit the use of armoring measures on the coast. As development pressures increase and a new political regime attempts to roll back environmental regulations that they view as hindering economic development, past policies and new initiatives are in jeopardy as will be discussed later in this chapter.

Inland riverine communities in eastern North Carolina face a different set of challenges. In most cases, this region is not experiencing rapid growth, but rather an outmigration of people in an area subject to high unemployment and limited job opportunities. In many of these communities, a primary concern is the identification of additional sources of revenue and economic development. The ability to retain population and attract businesses and their associated tax base while decreasing flood hazard vulnerability represents a significant challenge and one confronted with varying degrees of success following Hurricanes Fran and Floyd.

## 9.2 Drawing Lessons from the North Carolina Experience

The North Carolina experience offers insights into the underlying political, financial and administrative conditions present before and after disasters, including how these conditions can influence the policy and institutional choices made by the state, and how such choices affect the actions of other stakeholders. The role of states in disaster recovery, which remains less studied when compared to federal and local efforts in the US is perplexing as states formulate policy; coordinate resource delivery between federal and local actors; and provide training, outreach and educational efforts before and after disasters (Smith 2011, p. 45). These same characteristics are highly relevant to the study of climate change adaptation. The U.S. National Research Council, for instance, has noted the importance of state action regardless of whether the federal government assumes a leadership role in climate change adaptation-related activities, including those actions that address natural hazard threats (2010, p. 229).

Later in this chapter, the choices made by the state are framed in terms of how they present potential barriers and opportunities to climate change adaptation. In this case study, emphasis is placed on the post-disaster environment as this is where many of the most significant policy changes occurred that are directly relevant to issues surrounding disaster recovery, hazard mitigation and climate change adaptation. Recommendations focus on developing a new state organisational framework to manage post-disaster recovery and hazard mitigation efforts and developing a robust and comprehensive assessment of current and future risk (including climate change-induced threats). Ideally this information is used to guide future land use, shape building codes and design parameters, limit public expenditures in areas subject to the effects of natural hazards and inform anticipatory plans that guide the actions of an expanded network of relevant stakeholders to include those advocating for more robust climate change adaptation strategies. In practice, the degree to which the substantial federal and state investment in post-disaster, project-specific hazard mitigation measures and the state-of-the art assessment of flood hazard risk has led to significant changes in land use policy remains highly variable at the community level.

The North Carolina case study also describes the implications of adopting a mix of reactive and proactive policies, including the uneven degree to which lessons were learned from Hurricane Fran and applied following Hurricane Floyd. The ability to develop what amounts to an unprecedented state commitment of resources, including the creation of 22 new state disaster recovery programmes in the aftermath of Hurricane Floyd and the later institutionalisation of many of these programmes for use in future disasters represent a unique effort and is instructive for those grappling with climate change adaptation as such efforts will necessitate changes in existing policy and the creation of new policies based on future-oriented scenarios. The difficulties associated with the transfer and translation of lessons from one event to another in the throes of a disaster highlights the need to adopt a more proactive, systemic and sustained set of actions.

Pre-event planning for post-disaster recovery has the *potential* to empower states to make informed, albeit sometimes difficult decisions surrounding land use and codes and standards, advance a greater collective understanding of hazard risk and ways to reduce it before the next disaster strikes, capitalise on the window of opportunity to affect policy change and more effectively and equitably coordinate what amounts to substantial resources after a federally-declared disaster (Smith and Wenger 2006). Pre-event decision-making as well as what Bacharach and Baratz (1963) term "non-decision-making" are equally important realities in that addressing important, often complex issues (like the identification of hazard risk, effectively communicating that risk to broad audiences or "publics" and taking action through modified policies and plans) can be scuttled through inaction. In other cases involving natural hazard risk, those in positions of power, who often benefit from unregulated development in known hazard areas, successfully limit the adoption of policies that may hinder short term profits but significantly increase the vulnerability of the larger community (Freudenburg et al. 2009).

During the intervening time period between Hurricane's Fran and Floyd, the state learned a number of valuable lessons. Perhaps the most important was the need to expand the state's pre-event capacity to more effectively recover from and mitigate against future events. The effect of two disasters striking a state in close temporal and physical proximity to one another was magnified as the storms flooded many of the same communities. This resulted in the state, affected communities and those residing in these locales to gain a greater appreciation of flood hazard risk and the need to take actions to reduce future damages to those structures directly impacted through the large-scale use of post-disaster hazard mitigation grants. The degree to which these disasters caused a major shift in local land use planning, however, remained mixed. In many cases, the post-disaster hazard mitigation grants, which were triggered by a federal disaster declaration and used to purchase flood damaged properties, were principally seen by officials and residents as a way to assist individual homeowners recover from the event, and did not necessarily serve as part of a larger strategy to reduce overall flood hazard vulnerability at the community level (Fig. 9.4).

The low level of state and local preparedness for an event of this magnitude and the identification of significant gaps in federal assistance resulted in enhancements

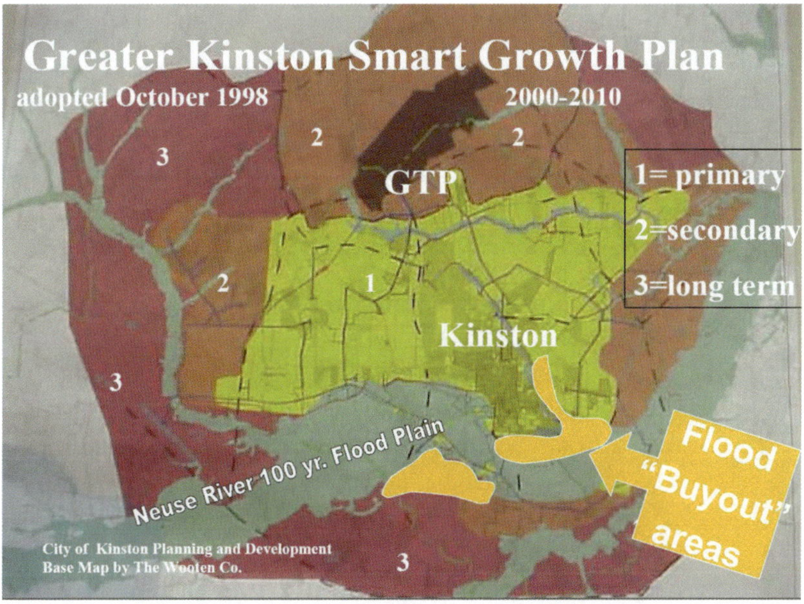


**Fig. 9.4** The City of Kinston Smart Growth Plan. Unlike many communities flooded after Hurricanes Fran and Floyd, the City of Kinston began an impressive effort to clear the floodplain of development while encouraging those involved in the post-Fran and Floyd acquisition programmes to relocate within the city limits on in-fill lots thereby reducing flood risk and managing growth in accordance with their smart growth plan (FEMA 2012). Note the image shows that their efforts began prior to Hurricane Floyd, which struck 1 year after this graphic was developed. Eventually 1,747 homes in the city and surrounding county were acquired and removed from the floodplain and turned into open space (personal communication with Chris Crew, North Carolina State Hazard Mitigation Officer). (Source: Image from Benchmark Consulting)

to existing state programmes, policies and personnel actions. New responsibilities, including those typically assumed by FEMA, were undertaken by the state (e.g., floodplain mapping, the management of emergency housing) and new long-term recovery programmes were created and codified. Related actions included the building of an enhanced state emergency management capacity, an expanded commitment to hazard mitigation and the creation and funding of state-level recovery programmes advancing the concept of sustainable disaster recovery. Specific actions included one of the largest single-state acquisition of flood-prone properties in the country, the development of a state-wide hazard mitigation planning programme and the remapping of the state's floodplains. The ability to sustain this high level of capability over time has been problematic as many of the positions created were temporarily funded using post-disaster federal assistance. The episodic rise and fall in state capabilities to administer hazard mitigation and disaster recovery programmes is a major problem nationwide (Smith 2011; Smith et al. 2013).

Hurricane Fran exposed the state's limited capacity to effectively manage a large disaster recovery operation. As a result, the state hired a new Emergency

Management Director who built a larger and more comprehensive emergency management programme.[8] Much of this growth was tied to the hiring of staff capable of administering large post-disaster recovery and hazard mitigation grant programmes, a competency that has become increasingly important in the emergency management profession (Smith 2002). Once hired and trained, the staff, which was in the midst of recovering from Hurricane Fran, faced an even greater challenge in Hurricane Floyd.

### *9.2.1 Hazard Mitigation in North Carolina: The Relocation and Elevation of Flood-Prone Housing and the Creation of the Hazard Mitigation Planning Initiative*

After presidentially declared disasters in the United States, several FEMA-funded programmes are triggered, including the Hazard Mitigation Grant Programme (HMGP). HMGP funding is based on 15 % of total federal disaster costs and requires a 25 % non-federal match. The post-disaster grant programme can be used to fund a range of hazard mitigation projects, including, for example, the acquisition[9] or elevation of flood-prone housing, the retrofitting/strengthening of public facilities to better withstand the forces associated with natural hazards (e.g., high winds, flooding, ground motion and fire), education and outreach initiatives, and planning. It is the state's responsibility to develop priorities as specified in their HMGP Administrative Plan.

Hurricane Fran HMGP funds were used to acquire or elevate flood-prone housing; floodproof or relocate critical public facilities; floodproof, windproof, or relocate businesses; conduct preliminary scoping studies and implement stormwater management projects; and fund what are referred to as "category 5" projects (North Carolina Division of Emergency Management 1999, p. 17). The state chose to use category 5 funds to assist communities prepare local hazard mitigation plans, establish local warning systems, conduct research, implement dune revegetation and sand fencing programmes, administer education and training programmes for local officials, acquire undeveloped property and develop zoning and building ordinances tied to risk reduction (p. 17).

Hurricane Fran HMGP funds led to the acquisition or elevation of 1,146 and 401 flood-prone homes respectively (Personal communication with Chris Crew, North Carolina State Hazard Mitigation Officer). Prior to the event, the state, FEMA, and

[8] Eric Tolbert, originally an emergency manager from North Carolina, was hired by the State of Florida following Hurricane Andrew (a category 5 storm which struck the state in 1992) to assist them deal with the recovery process and establish an enhanced response capability. After Hurricane Fran, the State of North Carolina hired him to serve as the Director of the Division of Emergency Management.

[9] Homes acquired using HMGP funds are either purchased and demolished or physically relocated to an area located outside the floodplain. In both cases, the purchased land on which the structure was located is converted to open space and maintained as such in perpetuity.

local governments did not possess the ability to quickly write grant applications, get them approved and implement them in a timely manner. The approval of the first round of acquisition and elevation projects, for instance, was announced on the one year anniversary of Hurricane Fran. Many local governments were in the process of administering large-scale acquisition and elevation projects when Hurricane Floyd struck three years later.

Following Hurricane Fran, some residents in flood-prone areas were skeptical of the programme and chose not to participate. However, as communities began to make offers on homes and they were demolished and turned into open space, nonparticipants began to express greater interest. Attention was further stoked as many residents living in or adjacent to areas that were part of the "buyout," as it was commonly referred to in eastern North Carolina, realised that their neighbourhood was disappearing and they might be one of the few remaining homes in the area. Eventually, federal and state funds used to implement the programme were committed to existing proposals and funding demands exceeded the supply of available resources.

This led communities to develop additional grant applications in anticipation of future funding. Three years later, Hurricane Floyd's heavy rains flooded many of the same communities that were inundated after Hurricane Fran. During the intervening period between the two storms, federal, state and local officials, as well as private sector contractors tasked with grants management at the local level, had gained greater experience administering what is a complex, highly bureaucratic and time consuming programme. As a result, several large grants were approved one week after Hurricane Floyd struck, thereby facilitating the release of funds to acquire more than 600 homes. Additional factors that increased the speed of project development and implementation included: (1) pre-event hazard mitigation planning and the identification of potential participants, (2) a strong interest among homeowners and owners of rental properties located in flood-prone areas to participate in the programme, and (3) a negotiated agreement with FEMA that streamlined the determination of eligible applicants.

The flooding associated with Hurricane Floyd proved much worse that Fran. Residential communities located along the Neuse, Tar-Pamlico, Cape Fear, Roanoke and Chowan Rivers were particularly hard hit. As a result, the state chose to dedicate all of its HMGP funds toward the acquisition of flooded housing and link this effort to a larger set of new state recovery programmes. In many ways, the HMGP was viewed by the state and impacted communities as a recovery programme as it enabled many grant recipients who were disproportionately low and middle income residents a chance to use the money to purchase a home outside the floodplain.[10]

Congress often appropriates additional funds to pay for the implementation of hazard mitigation projects beyond that available under the HMGP. Typically referred to as "supplemental funding," the monetary assistance tends to be admin-

[10] Of the 45,000 homes flooded during Hurricane Floyd, approximately 80 % did not maintain flood insurance (North Carolina Division of Emergency Management 2000, p. 21).

istered by the Department of Housing and Urban Development, not FEMA. As a result, the rules associated with HMGP and supplemental funding often differs across agencies. The state recognised that this could be a problem as the two funding sources were to be dedicated to housing relocation efforts and as such the state developed eligibility requirements and administrative rules that were identical for the two programmes. This streamlined the process and reduced confusion at the local level.

The state also negotiated an agreement with FEMA that applied a heretofore new method for determining cost effectiveness of individual structures that greatly sped up the approval process. FEMA typically requires the use of benefit-cost analysis to determine whether it is cost effective to invest hazard mitigation funds on a given project as part of a larger eligibility determination process. Conducting benefit cost analysis normally involves comparing the cost to mitigate (in this case purchase) a structure and the land on which it sits relative to the future expected losses avoided due to flooding (or other natural hazard-related impact). The traditional method is time consuming due to a number of factors, including the collection of information associated with the type and amount of damages sustained to the structure (which typically requires a field assessment). When spread across large projects totaling several hundred homes, it can take months to determine the benefit to cost ratio. At the same time, the state was attempting to build their case to members of the U.S. Congress for additional mitigation funds through the supplemental appropriation process.

Given the experience gained through conducting thousands of benefit cost analyses, the state recognised that homes inundated by two foot or more of floodwater and located within the floodplain were proving to be cost effective using traditional benefit cost analysis methods. The state, working with FEMA, developed a proxy cost effectiveness determination that included homes located in the 100-year floodplain that received over two foot of water above the first finished flood, or were deemed "uninhabitable" by a local building inspector. This approach was also used to demonstrate the need for additional funding from Congress beyond that available under the HMGP. Eventually Congress appropriated an additional US$ 400 million to acquire flood-prone housing.

At the same time, the state sought to more closely associate access to post-disaster hazard mitigation funds to better land use planning. Thus in order for communities to be eligible for Floyd HMGP funds, they were required to develop a local hazard mitigation plan as part of the state's Hazard Mitigation Planning Initiative (HMPI). The HMPI, which pre-dated the federal Disaster Mitigation Act of 2000 (DMA) by four years, represents a state-led effort to encourage communities to develop pre-disaster hazard mitigation plans.[11]

[11] The Disaster Mitigation Act of 2000, which was passed by Congress, in part due to long-standing issues (e.g., the need to initiate more proactive hazard mitigation measures through planning and pre-event grants) that were further highlighted by Hurricane Floyd, represents an important shift in federal hazard mitigation policy. Specific changes included the provision of pre-event hazard mitigation funds to states and communities to implement a range of risk reduction projects and the development of hazard mitigation plans. Elements of the HMPI and the State of Florida's Local Mitigation Strategy influenced the makeup of the Disaster Mitigation Act of 2000's enabling

The HMPI has been a moderately successful programme when assessed relative to its intended purpose, which is to develop a systematic risk reduction strategy at the state and local level that will lead to measureable risk reduction. At the local level, mitigation plans have tended to focus on existing at-risk properties and much less emphasis has been placed on modifying or adopting land use measures that proactively limit development in known hazard areas, even though the state initially made this an explicit focus of the programme. In many cases, local plans have not effectively linked the findings of their risk assessment to the selection of hazard mitigation policies or projects. This is troubling given the high quality of state-created flood insurance rate maps, the provision of state technical assistance and the number of major disasters that have struck the state during this time. Similar findings have been discussed as a part of a national phenomenon (Smith 2009b, pp. 253–264) and further documented as part of an empirical evaluation of the quality of state (Berke et al. 2012; Smith et al. 2013) and local hazard mitigation plans (Lyles et al. 2013). In this same study, the authors found that few coastal states or local governments have begun to explicitly address climate change in hazard mitigation plans. More recent anecdotal evidence observed through the identification of plans and presentations at conferences and workshops suggests that this is slowly changing across the United States.

The failure to link elements of a plan's fact base, including, in particular, the risk assessment to the selection of policies intended to reduce future hazard exposure has significant implications for states and communities that are now beginning to address climate change adaptation. In the U.S., the current reluctance of state and local officials to tie the findings of the natural hazard risk assessment to tangible risk reduction strategies in hazard mitigation plans may portend even more reticence to make these connections to climate-change induced hazards (e.g., sea-level rise, drought, heat waves, wildfire). This became evident in North Carolina when an initial investment of five million dollars in federal funding, obtained through a Congressional appropriation to develop a state sea-level rise adaptation study, was openly challenged by members of the state legislature, a newly elected Governor and a non-profit group representing coastal property owners.

Some U.S. politicians, developers and members of the public refuse to accept the notion of climate change, or are unwilling to confront it during their term of office, as part of their company's investment strategy or individual choices regarding the purchase of a home or taking the initiative to mitigate future expected losses to their personal property. At the national level a federal planning effort that offers the means to address climate change-induced natural hazards remains underemphasised. On the other hand, a growing coalition of environmental, academic, social justice and select government officials in the U.S. have begun to advocate for an improved level of cooperation between hazard mitigation, disaster recovery and climate change adaptation. The influence of these coalitions, including the degree to which the natural hazards risk management and climate change adaptation communities join forces in North Carolina remains to be seen.

rules, including the stipulation that local hazard mitigation plans are required to be developed and adopted by their governing body in order to remain eligible for pre- and post-disaster hazard mitigation assistance.

## 9.3 State Recovery Programmes

After Hurricane Floyd, the most significant recovery policy change adopted by the State of North Carolina included an expanded grants management strategy. More specifically, the state used US$ 836 million to create 22 state recovery programmes guided by the principles of sustainable disaster recovery and identified shortfalls in federal recovery programmes.[12] The focus of state programmes included the relocation of low income residents (including renters) out of the floodplain, providing more affordable housing alternatives for those affected by the flood, stimulating economic development, addressing environmental concerns, remapping the state's floodplains and hiring housing counselors tasked with explaining the myriad federal and state recovery programmes to individuals. For instance, the North Carolina State Acquisition and Relocation Fund (SARF) provided up to US$ 75,000 to homeowners participating in the acquisition programme in order to augment the federal dollars available under the HMGP as the grant programme only covers the pre-disaster fair market value of homes slated for purchase and demolition. The genesis of this state programme emerged following discussions with local officials and flood-affected homeowners. Specific concerns included: (1) the HMGP's rules limited participation among poor residents whose homes had been repeatedly flooded and their associated pre-disaster fair market value was very low and (2) if low income homeowners accepted an HMGP offer this left them with limited alternatives other than purchasing another substandard home or buying a mobile home (which is vulnerable to high winds). The SARF allowed participants in the programme to purchase a "safe and sanitary" home of comparable size to their previous dwelling in an area that was located outside of the floodplain. For some of the low income residents whose homes were damaged, but were not eligible for the buyout (often because they were located outside of the 100-year floodplain), the state's Repair and Replacement Programme provided grants to assist them with the repair of their home if the costs did not exceed the value of the structure.

Prior to Hurricane Floyd there was already a lack of affordable housing in eastern North Carolina, many of which subsequently flooded and became part of the buyout. The state also purchased low income rental units in the floodplain.[13] While the buyout programme reduced flood risk, it had the unintended consequence of further reducing the amount of affordable housing in the region. In order to address the shortfall, the state developed a neighbourhood redevelopment programme that

[12] In 1988, the U.S. Congress passed the Robert T. Stafford Disaster Relief and Emergency Assistance Act to improve the delivery of federal disaster assistance through the consolidation of existing programmes and the creation of several new recovery and hazard mitigation programmes. Key programmes administered under the Stafford Act include Public Assistance, Individual Assistance and the Hazard Mitigation Grant Program. The Public Assistance programme funds activities such as the personnel costs associated with state and local response and recovery efforts, the clean-up of disaster-generated debris and the repair and reconstruction of damaged public infrastructure. The Individual Assistance programme pays for the temporary repair of minor damages to housing.

[13] While the proceeds from the purchase of rental housing stock was paid to landlords, tenants were provided relocation assistance equivalent to no more than nine month's rent.

provided local governments with the funds needed to buy land and build new subdivisions and rental units outside the floodplain. This programme proved difficult to administer and few local governments applied for the funds.

In addition to housing-related issues, the state identified other inappropriate land uses in the floodplain that had deleterious environmental effects, including hog farms as well as a number of automotive junkyards.[14] State recovery funds were used to acquire these properties and remove the hog farms and junkyards from the floodplain, returning the land to open space. These projects reduced point source pollutants, returned the floodplain to a more natural state and increased the floodplains capacity to absorb future floodwaters and therefore reduced the effects of flooding on downstream communities.

North Carolina officials recognised that some of the land use choices made by local governments were tied to inaccurate and outdated Flood Insurance Rate Maps (FIRMs), which depict the extent of the estimated 100-year floodplain. In order to rectify this situation, the governor persuaded the legislature to use state recovery funds to remap North Carolina's floodplains (typically a federal responsibility under the National Flood Insurance Program) starting in the eastern third of the state, which was the area most heavily impacted by Hurricane Floyd. The North Carolina Floodplain Mapping programme has since become the most advanced and comprehensive effort of its kind in the country, creating digital flood insurance rate maps based on new hydrological studies and mapped two foot elevation contours of the entire state. The information can be deployed in concert with other digital data layers thereby enhancing the utility of the maps for risk-based planning activities, including climate change adaptation and the state's Hazard Mitigation Planning Initiative (HMPI).

### 9.3.1 *The Future of Climate Change Adaptation in North Carolina: Building on the Strength of Existing and Emerging Programmes, Polices and Stakeholders*

The State of North Carolina made significant investments in hazard mitigation and disaster recovery programmes and policies following Hurricanes Fran and Floyd, including the codification of these programmes so they could be used following future disasters. More recently, the state has begun two pre-disaster initiatives-the North Carolina Sea Level Rise Study (NCSLRS) and the Integrated Hazard Risk Management (IHRM) programme. Both state-led efforts build on the post-Floyd creation of the North Carolina Floodplain Mapping programme and serve to link the acquisition of hazards data and the assessment of current and future risk that

[14] In North Carolina, industrial hog farms pump animal wastes into large pits, commonly referred to as "lagoons." During heavy rains the lagoons can overflow their banks or be overtopped by adjacent rivers and streams. As a result, large amounts of untreated animal waste can flow into nearby water bodies, as was the case following Hurricane Floyd.

is directly relevant to many of the most pressing climate change-related hazards, including sea-level rise, flooding and the increased intensity of coastal storms. The degree to which this information is used to develop new and modified land use measures and climate change adaptation policies and plans at the local level, merits continued study.

The state codified several recovery programmes, including some of those developed after Hurricane Floyd as well as others intended to mimic federal infrastructure and housing repair initiatives that are triggered by smaller events that occur with greater frequency and do not merit federal assistance. This raises an interesting issue as the nation's current set of disaster policies do not yet address climate change-induced hazards. North Carolina's more flexible policies that recognise differing hazard scenarios and state-specific disaster declaration procedures (and the ensuing assistance it triggers) provide important lessons for other sub-national units of government to emulate when they consider developing state-led climate change adaptation strategies.

For instance, the lessons derived from the creation of state programmes intended to address gaps in federal aid across differing disaster types is significant for several reasons. Perhaps most significantly, it demonstrates an effort by a sub-national institution to begin moving beyond the federally-dominated, post-disaster-focused policy milieu as practiced in the U.S. Federal programmes tend to drive hazard mitigation and disaster recovery actions at the state and local level as stakeholders often adhere to the notion that it is principally the federal government's responsibility to assist communities recover following disasters or provide hazard mitigation assistance to address poor land use decisions made in the past that place people and property in harm's way. This has created a nationwide sense of dependency rather than the proactive development of plans and policies that emphasize greater self-reliance and an expanded notion of collective action (Smith and Wenger 2006). North Carolina provides an example of a state that has created unique policies based on identified problems not addressed by others while at the same time aggressively pursuing external post-disaster funds to achieve identified goals.

Disasters, including those that are slow onset and do not meet clearly identifiable federal thresholds, may provide a venue for state-level solutions as they are not constrained by the Stafford Act and other federal enabling legislation. Less prescriptive programmes, or those intended to address gaps in the delivery of federal assistance following Presidential declarations, may be modified to include climate change adaptation measures that are not currently recognised by FEMA or other post-disaster relief agencies. In addition, programmes and initiatives may be undertaken by non-profits, members of the private sector, quasi-governmental organisations and individuals that have begun to recognise the need to adapt to changing conditions. Lessons from hazard mitigation practice point to the influence of the nongovernmental community in several key areas including issue advocacy and the mobilisation of public opinion; collaborative problem solving; partnership building; trusted, innovative and flexible service delivery; leveraging resources; and research, education and information dissemination (Patterson 1998, pp. 206–207).

Linking natural hazards risk management and climate change will require a new way of thinking and one that seeks areas of mutual interest. For instance, suppose a community applies for and receives HMGP funds to elevate flood-prone coastal properties, under the HMGP rules, the costs of elevation must prove "cost-effective" (i.e., the costs of elevation cannot exceed the expected benefits derived from future losses avoided for a flood event of a given magnitude or return period). This has the effect of limiting the ability to pay for the incorporation of an additional measure of protection beyond that which can be paid for under the federal programme. Elevating a structure to what is currently designated the 100-year flood event may prove cost effective today but does not account for what constitutes a sound future prediction of flood return periods in light of a changing climate.

Lessons like the negotiated agreement with FEMA that developed a proxy for cost effectiveness should be revisited in order to establish new and innovative ways to assess long-term future risks associated with climate change and incorporate them into the techniques used to make cost effectiveness determinations. Another option includes the creation of state, local or nongovernmental programmes as well as the financial commitment of individual property owners to pay the difference between the costs to elevate a home to the 100 year event and the costs associated with an additional height (referred to as "freeboard") that accounts for projected changes in future flood hazard return periods tied to greater expected rainfall amounts and rising sea levels. Two state programmes currently under development in North Carolina (North Carolina Sea Level Rise Study and Integrated Hazard Risk Management) provide examples of how states can play a greater role in the blending of hazard risk assessment, mitigation and climate change adaptation. They also highlight the politicisation of climate change and its effect on emerging policy choices.

#### 9.3.1.1 North Carolina Sea Level Rise Study

In 2009, the North Carolina Division of Emergency Management received a US$ 5 million Congressional appropriation to conduct an assessment of the state's vulnerability to rising sea levels and increased coastal storminess and to develop a series of possible adaptation strategies. More specifically, the study involves the construction of three sea-level rise scenarios; an assessment of the environmental, economic and social impacts associated with these scenarios; and the development of a series of potential state adaptation policy recommendations based on current and future land use projections (North Carolina Division of Emergency Management 2011). The degree to which the recommended state-level adaptation policies are integrated into ongoing hazards management strategies like state and local hazard mitigation plans under HMPI, the Integrated Hazards Risk Management initiative, state and local floodplain management activities, and the adoption of pre- and post-disaster recovery policies and plans, will provide a good measure of how well climate change adaptation policies build on what amounts to an already substantial commitment to natural hazards risk management activities in North Carolina.

In 2013, as the study was nearing completion, the state legislature sought to prohibit the use of the findings by limiting the use of the sea-level rise scenarios that projected an increased rate of rise in water levels. In addition, the Governor and members of his cabinet have sought to limit the further study or dissemination of state reports developed prior to their administration that seek to identify ways the state and local governments can adapt to a changing climate.

#### 9.3.1.2 Integrated Hazard Risk Management

The Integrated Risk Management (IHRM) study is a state effort to develop a multi-hazard assessment of natural hazard risk in North Carolina. IHRM, which builds on the work of the North Carolina Floodplain Mapping initiative, strives to identify ways to develop comparable methods to assess other natural hazards that are prevalent in North Carolina including high winds, wildfire, earthquakes, winter storms, drought and landslides. As part of this study, the state has digitised and geo-referenced the building footprints for all structures larger than 1,000 sq.ft. in the state, and collected data about each structure, including, the age of construction and its first floor ground elevation which is critically important when assessing flood hazard risk.

Recognising the variability of information, the capacity of local governments, and the range of risk assessment tools that are available, members of the IHRM team are developing a three tiered (gold, silver and bronze) approach to risk assessment. The gold level assessment requires the use of the latest available technology and risk assessment techniques. It also necessitates access to high quality data to perform the analysis. The silver level uses more readily available data and widely recognised risk assessment tools. The bronze level assumes that information is provided as part of pre-packaged datasets and risk assessment tools require limited technical and administrative sophistication to utilise and operate.

The data collected and the methods under development through the IHRM are designed to complement the North Carolina Sea Level Rise study and the Hazard Mitigation Planning Initiative. For example, the information used to improve the assessment of coastal flooding and storm surge and the mapping and geo-referencing of structures in North Carolina will be used to evaluate the vulnerability of buildings in coastal counties to the effects of sea-level rise and other hazards prevalent in the state. The state will also make the data available to local governments and others on an open source website. This will enable communities to update their Local Flood Damage Prevention Ordinances, hazard mitigation plans, comprehensive plans and other public policies using this information. The manner in which the IHRM and sea-level rise study informs climate change adaptation and hazards management policies represent a critical stage in the historical evolution of the state's programme from one that embraced taking advantage of the post-disaster window of opportunity following two major disasters to affect significant state-level policy change, to one in which a more proactive approach is taken to manage natural hazards, including those induced or exacerbated by climate change. The degree to which this occurs and the way it follows or differs from the policy lessons derived from Hurricanes Fran and Floyd remains to be seen. The recent opposition to the development of a proactive climate change policy in the state appears to have short circuited the ability to link these complementary aims by drawing on what amounts to a unique combination of post-disaster experience, innovative state programmes, world class mapping and analytical tools and access to flood hazard data that is unmatched anywhere else in the United States.

## 9.4 Key Lessons Learned from the North Carolina Experience

A number of important lessons have emerged from the North Carolina case study that is directly relevant to climate change adaptation. Key lessons include: (1) developing and sustaining flexible state-level programmes aimed at addressing gaps in federal post-disaster assistance, including those that deal with socially vulnerable populations; (2) investing in a state-wide reassessment of flood hazard risk that is designed to be accessible, updateable and serves as a platform for broader risk assessment programmes, including sea-level rise; (3) implementing an aggressive suite of hazard mitigation activities, including the large-scale relocation of flood-prone housing and the conversion of high-hazard areas to open space; and (4) the translating and integrating of disaster recovery and hazard mitigation experience and programmes with climate change initiatives can be hindered by the lack of political support.

### *9.4.1 Develop and Sustain Flexible State-Level Recovery Programmes that Address Gaps in Federal Assistance and Inform the Nature of Future Climate Change Adaptation Policies and Programmes*

In the United States, the federal government often manages post-disaster programmes that are guided by highly prescriptive rules. This has had the effect of hindering recovery and the achievement of state and community-level goals like sustainability and disaster resilience (Smith 2011). The creation of more flexible state programmes intended to address gaps in federal assistance is significant as climate change adaptation scholars point to the importance of developing programmes with the flexibility necessary to adapt to an outcome that still carries with it a high level of uncertainty, and as a result, can lead to inaction (Adger et al. 2009). In many ways this condition points to the earlier reference to Bacharach and Baratz's (1963) term "non-decision-making" writ large as greenhouse gas emission reduction and climate change adaptation will require cooperation on global, national, sub-national, regional and local scales. In the U.S., as well as other countries, the ability to recover from the disaster at hand while adopting hazard mitigation strategies that account for worsening, albeit future hazardous conditions provides an important, yet currently under-emphasised aspect of climate change adaptation practice.

Recognising the unique opportunity to inject hazard mitigation/climate change adaptation measures into recovery operations through robust pre-event planning and post-disaster implementation requires a new way of thinking at the national, sub-national and local levels of government. Specific tasks involve altering programmes and policies in order to facilitate this dual objective. It also requires developing an improved understanding of the hazards we face today, developing methods to better predict how these hazards will change over time in the face of climate change and more effectively conveying these realities to those who remain unaware of the nexus between natural hazards risk management and climate change adaptation as well as those who are skeptical of climate change in general. When a disaster strikes, policies should be in place that include reconstruction practices (e.g., where and how communities will be rebuilt) that reflects current and projected changes in natural hazard risk including those hazards caused or exacerbated by climate change.

### *9.4.2 The Post-Disaster Remapping of Floodplains and Other High Hazard Areas Requires a Strong Champion, a Clear Risk Communication Strategy and Should be Used to Inform Hazard Mitigation, Disaster Recovery and Climate Change Adaptation*

In the U.S. there is a significant number of people who remain unconvinced that climate change is occurring and as a result do not recognise the need to develop adaptation strategies to address a risk that they do not believe exists. The analogous example of remapping areas prone to flooding and conveying an increased flood hazard risk to communities in North Carolina provide a number of relevant lessons. Issues surrounding the pre-event accuracy and post-disaster update of Flood Insurance Rate Maps played a significant role in North Carolina's recovery following Hurricane Floyd and several more recent flood events. The maps also have helped guide future building codes and standards in flood-prone communities, although it has not led to the widespread discontinuation of development in floodplains.

The extensive data collection and analysis undertaken following Hurricane Floyd, and the re-mapping of the state's floodplains that followed, have resulted in the development of a state capability that is unsurpassed in the U.S. The state has continued to build on the state-wide datasets and use of advanced mapping technologies, more recently embarking on the Integrated Hazard Risk Management programme and the North Carolina Sea Level Rise Study. Yet the use of the emerging risk assessment findings, including the greatest climate change-induced threat to North Carolina (sea-level rise) remains elusive and the results are being discounted and in some cases withheld by state legislators and state agency administrators.

The lessons associated with the remapping of flood hazard areas in North Carolina include: (1) garnering the political and financial support needed to embark on a complex and time consuming effort; (2) utilising a publicly available digital platform that can be modified as new data becomes available (including sea-level rise), thereby enabling its use across a range of local planning and policy making efforts;

and (3) emphasising a concerted and ongoing emphasis on education and outreach initiatives to inform communities about the importance of adopting new standards tied to the latest understanding of risk.

The dynamism of hazard risk is perhaps best exemplified in coastal regions of the world. The climate change induced factors associated with rising sea levels and increased storminess, heighten this growing vulnerability. The ability to convey this reality, map future hazard risks based on an accepted degree of uncertainty and incorporate this information into actionable policies is critically important as investments and development decisions made today and in the near future influence growth in areas that are likely to become increasingly vulnerable over time.

The development of improved hazard mitigation and disaster recovery plans and policies should recognise the changing nature of coastal hazard vulnerability, including our emerging understanding of sea-level rise. Ultimately, the mix of policies adopted should include an assessment of the feasibility of relocating flood-prone communities and abandoning supporting infrastructure while simultaneously adopting a disinvestment strategy in areas prone to coastal erosion, storm surge and rising sea levels (Neuman and Price 2009). An important part of this larger strategy should include a strategic readiness to implement these ideas after a major disaster.

### *9.4.3 Emphasise the Implementation of Post-Disaster Hazard Mitigation Programmes*

An important issue facing governments is the identification of where funds to pay for climate change adaptation will come from (Smith et al. 2009, p. 58). This chapter has shown that the delivery of post-disaster resources, defined as funding, technical assistance and policy change should be an important part of a national, sub-national and local climate change adaptation strategy. The State of North Carolina, working in partnership with FEMA, local governments and individual residents has made a major commitment to relocate and elevate flood-prone homes. This has led to the removal of repetitively flooded structures out of the floodplain and the conversion of these areas to open space-dependent uses like public parks, stream buffers and greenways. However, all of the nearly 5,000 properties purchased in North Carolina were located in riverine communities not the state's barrier islands.[15] This speaks to the unwillingness of property owners in the most vulnerable coastal high hazard areas to sell their property (Fig. 9.5).[16] While many barrier island property owners were interested in procuring federal and state funds to elevate their homes in these

[15] By 2013, more than 8,000 flood-prone homes located in riverine floodplains have been acquired through a series of pre- and post-disaster hazard mitigation grant programs in North Carolina (State Hazard Mitigation Advisory Group meeting, 2013).

[16] Many of the homes located on barrier islands and along the coastal sounds are vacation rental properties which serve as a money making venture for investors. In other cases, property owners desire to live on the coast due to the quality of life or they work in water-dependent jobs like tourism and the seafood or recreational fishing industry. Hurricane Irene, which struck North Carolina in 2011, significantly impacted a number of small sound-side communities, many of

**Fig. 9.5** Home elevation project on the Outer Banks of North Carolina. The image demonstrates the extreme measures individual property owners will take in an effort to protect their investment including elevating their home on a highly dynamic barrier island in front of what remains of a protective coastal dune. (Source: Photograph by Gavin Smith)

areas, the state's policy precluded the use of these funds to elevate homes in coastal v-zones (e.g., those areas subject to destructive hurricane-induced storm surge). After Hurricanes Fran and Floyd, only six vacant lots located in a coastal v-zone were purchased under the HMGP and converted into a public beach access point.[17]

The buyout and relocation of flood-prone structures offers lessons tied to the technical aspects of the process (i.e., identifying applicants, writing an eligible grant, financial management, making offers on the property, debris removal and site remediation) as well as political, social, environmental and economic considerations. The complexity of technical matters and the challenges associated with larger decision-making processes, including their societal implications, point to the importance of good pre-event planning for post-disaster recovery and hazard mitigation. It is increasingly evident that these issues will be confronted at some point in the future on a much larger scale by coastal communities, states and nations that are pondering adaptation measures including retreat and resettlement options.[18] Capturing the lessons and avoiding the mistakes associated with large scale voluntary buyout programmes should be an important part of a national sea-level rise adaptation strategy.[19]

which included high levels of low income residents living in low lying areas that are extremely vulnerable to rising sea levels.

[17] Relocation has become an increasingly appropriate strategy when compared to elevation given the inherently dynamic coastal environment and associated factors such as coastal storm surge, erosion, subsidence and threats tied to rising sea levels and more intense storms. On the other hand, questions can be raised about the potential feasibility of this approach given a reluctance to sell ocean-front property and the merits of using federal and state funds to pay for the relocation of development that has been placed in such precarious locations.

[18] Resettlement is already occurring in some island nations due to sea-level rise. The lessons derived from these examples may be instructive as communities in the U.S. begin to examine similar scenarios.

[19] One of the most significant challenges associated with the large-scale, voluntary relocation of flood-prone housing involves the ability to garner the participation of entire neighborhoods or clusters of contiguous properties thereby avoiding what is often referred to as "checkerboarding"

In North Carolina following Hurricane Floyd, the state's State Acquisition and Relocation Fund provided an incentive for low-income residents to move to higher ground. During the initial development of the programme, local officials expressed concerns about losing their tax base as a result of the large-scale acquisition and demolition of existing housing stock. As a result, the state developed a policy that required homeowners who took the additional SARF money to relocate within the county (the largest local administrative division of most states) in which their home was located and outside of the mapped flood hazard area. This policy merits further study and potential use in other states and nations as the resettlement of communities, including large numbers of poor, socially vulnerable populations face rising seas. Future challenges will include addressing social justice issues among the poor (many of whom depend on the coast for their livelihood) and counterarguments against relocation among elected officials who may cite the loss of an often lucrative tax base associated with wealthier coastal property owners.

While the resettlement of large number of people occurred during Hurricane Floyd in North Carolina, most communities did not embrace a comprehensive set of local land use policies and future-oriented hazard mitigation plans that limited future development in floodplains. Rather, they tended to rely upon more stringent flood ordinances and building codes that stressed how structures were built in known flood-prone areas. Much less emphasis was placed on where housing, infrastructure and public facilities should be built in the future.

This raises an important question that has implications for other nations that may adopt similar hazards risk management programmes. Following federally-declared disasters, and the release of hazard mitigation funding that targets existing at-risk structures, it remains uncertain how much the actual flood hazard risk has been reduced in the U.S. For instance, over US$ 500 million has been invested in hazard mitigation projects following Hurricanes Fran and Floyd that was used to relocate 5,196 homes, elevate 443 others, and undertake a range of additional risk reduction projects and initiatives (personal communication with Chris Crew, North Carolina State Hazard Mitigation Officer). Yet development continues to occur in flood-prone areas.[20] Thus when flood risk is evaluated in an aggregated manner (i.e., measuring losses avoided tied to hazard mitigation measures such as housing relocation plus the effects of ongoing development in and around the floodplain and its effect on flood exposure) we don't know this value, nor can we effectively track it over time.

---

where some properties are acquired while other property owners choose to remain in the area. The purchase of larger land areas allow for the removal of supporting infrastructure like roads, water and sewer and eliminates the need to maintain other public services in the area like street maintenance, police protection and the rescue of people trapped in their homes following future floods.

[20] This is particularly true on barrier islands in North Carolina where older, generally smaller beach cottages are being replaced by much larger vacation rental properties. Additional development trends include the increased construction of mid-rise condominiums and hotels, which once sited, limits hazard mitigation or adaptation alternatives to abandonment and demolition (which is unlikely given the up-front investment), armoring the shoreline or the regular renourishment of beaches. The latter strategies are expensive, tend to encourage greater investments in these areas, and offer a temporary fix to a problem that is getting worse over time.

The effective use of hazard mitigation plans to more systematically reduce future hazard vulnerability and assist communities and nations adapt to climate change, including the degree to which this can be attained through modified policies that address future land use decisions, remains a significant challenge in North Carolina and the U.S. The large-scale relocation of vulnerable coastal communities may become more common in the hurricane-prone Gulf Coast and Southeastern U.S. as these areas are also vulnerable to rising sea levels. At the same time, market forces and new national policies, including, in particular, flood insurance rate increases, and more stringent pre- and post-disaster hazard mitigation and reconstruction standards have the potential to reshape the U.S. coastline. The passage of the Biggert-Waters Act, for instance, has set in motion a process that will include charging actuarially sound flood insurance rates. This will dramatically increase flood insurance premiums among policyholders as the National Flood Insurance Programme has historically relied upon lower rates in an effort to encourage participation. Thus the occurrence of low probability, high consequence events and slow onset disasters like sea-level rise present an increasingly complex set of challenges for sub-national governments who strive to assist local governments develop an increased capacity to plan for the future.

### *9.4.4 Create and Sustain a State Disaster Recovery Organisation Committed to Hazards Management and Related Elements of Climate Change Adaptation*

It is common practice in the United States following a major disaster to create a new, often temporary, state-led organisation tasked with addressing recovery issues. Indeed, this has been done following the Iowa (USA) floods of 2010, the 1989 Loma Prieta earthquake in California (Olson et al. 1999) and Hurricane Katrina in both Mississippi and Louisiana (Smith 2011). In North Carolina, this happened after both Hurricanes Fran and Floyd. In the case of the Fran-based organisation, the North Carolina Disaster Recovery Center was created and tasked with: (1) organising a collection of state agencies referred to as the Disaster Recovery Task Force to develop policy recommendations and identify unmet state and local needs and (2) soliciting federal support to address these needs (State of North Carolina 1997a). The Disaster Recovery Center was disbanded after the state recommendations were implemented and unmet needs funding distributed to communities (State of North Carolina 1997b). Following Hurricane Floyd, the Redevelopment Center was created to identify unmet needs and seek federal assistance. Given the size of the disaster, the state realised that available federal assistance, including supplemental appropriations from the U.S. Congress, would not address all identified needs in the state. Thus, the Redevelopment Center led the initial development of the state programmes designed to address gaps in federal assistance. The Redevelopment Center has been maintained over time in the department that oversees emergency management, while several of the state programmes have been institutionalised through the creation of a state tiered disaster declaration process.

The programme instituted by the State of North Carolina and maintained over time, is unique. The principal reason for the approach taken by most states, and one that has important implications for climate change adaptation, is that state emergency management agencies are often unprepared to deal with the complexities of disaster recovery. Rather, they tend to be more focused on preparedness and response-related activities. Closely related to this problem is the failure of most state emergency management agencies to effectively plan for the breadth of issues tied to recovery. Instead, the actions of state emergency management officials during the recovery process are principally driven by the administration of post-disaster FEMA programmes rather than the resources provided by the much larger array of relevant stakeholders in the public sector, private sector, quasi-governmental organisations, non-profits, international aid organisations and nations, emergent groups and individuals (Smith 2011). In the United States, this unfortunate reality is often compounded by the fact that it is often unclear who will foster the creation of collaborative relationships that are such an important part of an effective disaster recovery process (Smith 2011).

The state emergency management community is most familiar with federal programmes triggered by a Presidential disaster declaration. It is the collection of aid providers and resources that are present among other groups that state emergency management agencies are less equipped to deal with, much less able to coordinate the actions of these organisations to achieve a common aim. In many cases, environmental and social justice non-profits, science-based foundations, quasi-governmental planning organisations and university research organisations are more aggressively confronting climate change adaptation-related issues. Given these trends, it is incumbent on state emergency management agencies to embrace the active involvement of new partners in what amounts to an expanding natural hazards risk management planning milieu. This task requires not only the development of agreements, new policies and implementation procedures, it also necessitates building relationships with non-traditional partners that take time to mature.

Disaster recovery is achieved during a time in which there is intense pressure placed on elected officials by members of communities to return to "normal" as quickly as possible. The importance of engaging in a deliberative process among the multitude of stakeholders after a disaster strikes is often discounted. The failure to plan for recovery, before or after disasters, can lead to missed opportunities to implement policies and land use measures that address socially vulnerable communities, preserve natural systems that provide important protective features and adopt new construction standards and land use policies that reduce future losses, including those that may be exacerbated by the effects of climate change.

While the Redevelopment Center coordinated much of what remains one of the most significant state-level commitments to post-disaster assistance in U.S. history, it did not develop an overarching state recovery plan to guide these efforts, which ultimately limited the effectiveness of the overall effort (Smith 2011, pp. 57–58). Specific problems included a lack of inter-agency coordination, the creation of programmes with competing objectives and the failure to account for the capacity of local governments' ability to implement both federal and state recovery programmes simultaneously (Smith 2011, pp. 57–58). At the time this chapter was being written, the State of North Carolina was in the process of revising the North Carolina Recovery Guide, which was originally published in 2006 and later revised in 2007 and 2009 (North Carolina Office of the Governor 2009). The more recent document is intended to adhere to the emerging federal guidance associated with the National Disaster Recovery Framework (NDRF). The degree to which this document and other state recovery plans that are being developed across the U.S. in response to the NDRF, adhere to emerging planning guidance developed by FEMA and the principles described in Chap. 5, are worthy of continued observation and evaluation. And like in the case of state and local hazard mitigation plans, emphasis should be placed on the connectivity between planning for post-disaster recovery and climate change adaptation planning.

Following Hurricane Floyd, the state codified many of the state recovery programmes through a tiered disaster declaration process that defines three disaster types and their associated levels of assistance. A Type I disaster is classified as not meeting federal declaration criteria, but is deemed by the state to merit assistance. North Carolina has developed state programmes including a state fund used to pay for the repair of damaged infrastructure and a programme used to assist homeowners make repairs to their homes. Federally declared events trigger a Type II designation and involve the use of traditional FEMA programmes as defined by the Stafford Act as well as supporting state programmes like the provision of non-federal match for certain grants. Had the typing system been in place during Hurricane Floyd, the storm would have triggered a Type III designation and the deployment of state recovery programmes like the State Acquisition and Relocation Fund.

The Hazard Mitigation Section within the North Carolina Division of Emergency Management represents another type of organisational model created after a major disaster. Further, it demonstrates the difficulties associated with building and sustaining capacity over time when one depends primarily on post-disaster resources. Prior to Hurricane Fran, the state had in place a State Hazard Mitigation Officer (SHMO) (a position responsible for the administration of state hazard mitigation efforts), a state National Flood Insurance Programme Coordinator, an NFIP Planner and an engineer on staff to assist with floodplain management initiatives. After Hurricane Fran and the influx of HMGP funds (approximately US$ 115 million), the state hired 15 temporary staff to assist with the administration of the HMGP, supplemental appropriations and the newly formed Hazard Mitigation Planning Initiative. Following Hurricane Floyd, the size of the state's hazard mitigation staff rose to 50 personnel working across four branches, including Grants Management, the Real Estate and Legal Team, Risk Assessment and Training and Hazard Mitigation Planning.

The increase in staff enabled the state to develop a comprehensive, albeit disaster-based, hazard mitigation programme. Among the most difficult activities included assisting communities develop and administer hazard mitigation grants, many of which were comparable to or exceeded the annual operating budgets of local governments. In some cases, the state took on some of the roles of local governments in those cases where they lacked the capacity to manage the grants. Efforts of the Real Estate and Legal Team, for instance, involved conducting title searches, identifying heirs to properties slated for purchase and assisting with property closings. The Hazard Mitigation Section also invested substantial resources in training, education and outreach efforts tied to assessing hazard risk, writing and implementing grants, informing individuals and businesses about the actions they could take to better protect their property and assisting communities develop hazard mitigation plans.

The funds used to hire the majority of those working in the Mitigation Section were derived from administrative dollars tied to federal hazard mitigation grants. The temporary positions were relatively low paying and did not include state benefits. As a result, there was significant job turnover as staff developed highly marketable skills in hazard mitigation grants management and most eventually sought permanent jobs in the private sector, FEMA, other federal agencies and state government. This not only hindered the continuity of grants administration and relationships with local government officials with whom state hazard mitigation staff worked, it also affected the ability to institutionalise knowledge gained over time. According to other SHMO's, creating a sustained level of state capacity necessary to adequately assist local governments develop a robust hazard mitigation strategy is among their greatest challenges (Smith et al. 2013).

Like disaster recovery and hazard mitigation, the complexities of climate change adaptation require the formation of multi-organisational, collaborative networks that are ideally created before an event occurs and sustained over time (National Research Council 2010). Given the complementary objectives found in what remain two broad, sometimes ill-defined camps, it behooves both the natural hazards risk management and climate change adaptation communities to join forces with the expressed intent to identify common concerns and objectives and develop a well-constructed national climate change adaptation strategy (National Research Council 2010, pp. 221–226). This strategy should hold state and local governments more accountable for policy choices that affect their level of hazards risk, while emphasising the importance of building the pre-event capacity of this larger network to proactively plan for and act on agreed upon policies.

### 9.4.5 Integrate State Hazard Mitigation and Disaster Recovery Planning into State Climate Change Adaptation Policy

Each of the lessons discussed in this chapter stand to benefit from the development, monitoring, implementation and update of high quality hazard mitigation and disaster recovery plans that identify common problems, utilise existing bases of knowledge and pinpoint complementary policies that span natural hazards risk management and climate change adaptation. Hazard mitigation and disaster recovery plans are uniquely positioned to address the rise in the number and severity of disasters through the injection of climate change adaptation into recovery and reconstruction processes and reduce the likelihood of disasters occurring in the first place through the more proactive use of risk reduction measures.

Those tasked with recovery in North Carolina developed post-disaster strategies and organisations responsible for the implementation of state programmes. Yet the state did not develop what could be construed as a disaster recovery plan that adheres to widely accepted plan quality principles (Smith and Flatt 2011; Smith 2011, p. 57). Following Hurricanes Fran and Floyd, North Carolina developed a series of policies and programmes intended to address needs that were not met by federal programmes. State officials also developed quasi-governmental agencies to implement these policies. While following Hurricane Floyd, programmes intended to achieve a more sustainable disaster recovery were not explicitly linked in a manner that guided the actions of those responsible for their implementation under a commonly recognised vision and organising set of goals.

At the federal level, recovery planning has remained a largely reactive enterprise. Following Presidentially-declared disasters, FEMA dispatches a long-term recovery team comprised of federal agency officials and a national cadre of private sector contractors to assist communities develop post-disaster recovery plans. Following Hurricane Katrina, the U.S. Congress passed the Post-Katrina Emergency Management Reform Act (PKEMRA) which required FEMA and HUD to develop a national disaster recovery strategy, which prior to the event remained unclear (Government Accountability Office 2008, 2010). The resulting National Disaster Recovery Framework was finalised six years later. The creation of enabling rules was beginning to be discussed during the writing of this chapter. The final NDRF guidance, including the establishment of planning requirements, capacity building initiatives and the implementation process should be followed closely. Without a clear national recovery strategy, how can we expect states and local governments to plan for recovery with limited guidance or training required to help build the capacity that is so desperately needed in order to address what still amounts to an uncoordinated, largely ad-hoc process? When one adds the additional stakeholders involved in climate change adaptation, including those organisations tasked with developing plans for slow onset events like sea-level rise, loss of permafrost/subsidence, drought and other climatic changes in areas where human settlements were not designed to confront these challenges, the importance of developing enduring partnerships becomes even more prescient.

The Disaster Mitigation Act of 2000, which has been in place for more than 10 years represents a federal programme that is more mature than the nascent NDRF. While state and local plans have improved over time, they remain generally weak and provide modest guidance shaping future decisions affecting hazard risk, particularly those exacerbated by poor land use decisions (Berke et al. 2012; Smith 2009a). These realities expose a major weakness in the current U.S. system, exemplifying both barriers and opportunities to the type of changes needed in natural hazards risk management policy to more effectively address climate change adaptation (National Research Council 2006, 2010). They also provide important policy lessons to learn from as climate change plans emerge and attempts are made to implement them over time.

## 9.5 Barriers and Opportunities for Mainstreaming Hazard Mitigation and Disaster Recovery Policies into Climate Change Adaptation

Based on the findings derived from the North Carolina case study, it is clear that there are a number of issues that represent barriers, opportunities and a combination thereof to meld natural hazards risk management and climate change adaptation. Understood in the context of hazard mitigation and disaster recovery policy-making, two extreme events served as a powerful agent of change in the state and fostered the development of impressive project-focused programmes designed to address gaps in federal assistance as well as perhaps the best floodplain mapping programme in the country. However, the failure to develop a state-level post-Floyd recovery plan to better coordinate the distribution of assistance both across state programmes as well as within the larger governance network and facilitate or mandate the local adoption of more stringent land use plans that limit or guide future development away from known flood hazard areas remains a missed opportunity.[21] In the post-Floyd era, the state has remained committed to acquiring and relocating flood-prone housing and strived to build on the Floodplain Mapping Programme to include the pursuit of a more sophisticated all hazards approach to risk assessment (IHRM) and the initiation of a state sea-level rise study. More recently, a shift in political parties has resulted in a steadfast denial of climate change in the North Carolina statehouse and the limited use of sea-level rise scenario outputs to inform future policy.

### *9.5.1 Sustaining the Capacity to Address Episodic and Long-Term Natural Hazard-Based Threats*

The National Research Council of the National Academies Report *Adapting to the Impacts of Climate Change* emphasises the importance of developing the adaptive capacity of multiple stakeholders to take action even without a clear national strategy (2010, pp. 160–179). Related challenges face the more mature natural hazards risk management community as federal guidance and their associated policies do not place sufficient emphasis on pre-event capacity building initiatives in either the hazard mitigation or disaster recovery arena (Burby 1998; Mileti 1999; Mileti and Gailus 2005; Smith 2009b, pp. 261–262; Smith and Wenger 2006; Smith 2011). While pre-disaster state and local hazard mitigation plans are required in order to remain eligible for pre- and post-disaster hazard mitigation funding, plans remain generally weak. In the case of disaster recovery in the U.S., an unequal emphasis is placed on the management of federal grant programmes when compared to pre-event planning for post-disaster recovery, including the deliberative process

[21] Linking the receipt of state-level disaster assistance to the development of more stringent local land use policies represents one way to achieve this objective.

required to contemplate multiple opinions and possible recovery scenarios. As the effects of climate change lead to more frequent disasters, coordinative planning not only becomes more complex, it becomes more imperative to join forces to address complementary aims before a disaster occurs.

Unfortunately, natural hazards risk management officials and researchers, as well as climate change adaptation experts and practitioners, have yet to develop strong and enduring partnerships, due in part, to the lack of good planning. Many of the stakeholders involved in natural hazards risk management, such as environmental groups, scientists, regional planning organisations and design professionals, among others, are also actively involved in climate change-related activities. The ability to draw on this wealth of expertise requires the repeated, ongoing exchange of information. It also requires dual learning about complementary goals and the bases of existing knowledge and programmes that are relevant to both parties and form the basis of good collaboratively-derived plans. One of the first steps needed to address this divide involves a discussion surrounding natural hazards risk, including that risk tied to a changing climate.

### 9.5.2 Understanding the Changing Face of Risk: Disaster as an Opportunity to Alter Future Settlement Patterns

Post-disaster conditions as well as an established, albeit, underperforming national hazard mitigation programme (due in large part to its limited emphasis on guiding the type, location and density of human settlements relative to known hazard areas) provide several points of intervention in the climate change adaptation conversation, none of which may be more important than the role of land use. In the U.S., land use planning practices are used to address a myriad of issues including sprawl, growth management, the public and private investment of capital and natural resource protection, all of which are relevant to addressing the impending threats associated with climate change. The natural hazard mitigation community has been less successful in using land use to confront rising disaster costs, even though land use planning is among the most effective measures to reduce long-term risk (Burby 1998; Mileti 1999; Burby 2006). Hazard mitigation and disaster recovery plans developed in advance of disasters and during the age of climate change should draw on existing policy frameworks, land use tools and what often amounts to substantial sources of funding in the post-disaster environment in order to target smarter land use choices that account for natural hazards risk.

The underlying conditions affecting the adoption or modification of land use practices following Hurricanes Fran and Floyd are instructive as the coastal storms dramatically exposed the vulnerability of coastal and riverine communities. While large-scale relocation efforts were undertaken in a number of communities, the degree to which participating local governments adopted stronger land use measures to address future development in these same flood-prone locales remains highly varied. In some cases, communities sought to clear the floodplain and limit future development in adjacent areas whereas other local officials coordinated the buyout of willing participants while allowing new development to occur in similarly vulnerable locations elsewhere, relying on elevation requirements found in their local flood ordinances to mitigate future risk.

In the case of North Carolina's barrier island communities, no homeowners accepted the state's offer to purchase their highly vulnerable properties. This is a powerful reminder to North Carolina policy-makers, U.S. officials and others, that many barrier island property owners remain unlikely to adopt a voluntary strategy tied to retreat from the oceanfront. The actions taken by coastal residents in the Mississippi (USA) case study provides an interesting contrast in which dramatically higher post-disaster reconstruction standards and insurance rates have led to both a migration of property owners inland and an overall stagnation in the rate of reconstruction of oceanfront properties.[22]

The United States still faces strong opposition among some members of Congress to implement a national climate change adaptation policy. The politicisation of climate change and the framing of the debate as one of excessive regulation impinging on economic growth due to proposed greenhouse gas reductions have limited productive dialogue about the second key element of climate change policy, namely adaptation. Yet many of those who oppose funding climate change adaptation efforts are more than likely to seek post-disaster aid for their constituents to rebuild communities in areas just impacted by a disaster. The tentative approach taken by FEMA should be altered to more aggressively link existing policies, programmes and funding to implement initiatives that address many of the climate change adaptation problems cited in this and other chapters. Important changes in national policy should include limiting funding that perpetuates future disasters or is awarded without requiring compliance with heightened standards that recognise the effect of a changing climate.

### *9.5.3 Capitalising on Issue Salience: Post-Disaster Funding, Policy Change, and Disaster Recovery*

Following major disasters in the United States, like Hurricane's Fran and Floyd, there are large sums of federal, state, private and nonprofit assistance provided to communities. Furthermore, major disasters can elicit significant changes in national policy (Rubin 2007; Birkland 1997, 2006) and modifications to programme rules, especially when challenged by powerful state leaders backed by pertinent disaster-based data,

[22] Hurricane Ike (2008), which struck the Galveston Bay region of Texas (USA), devastated Bolivar Peninsula and the homes located in the area. Following the storm, Galveston County applied for HMGP funds to purchase over 650 homes located on what is essentially a low-lying barrier island. Once complete, this will be one of the largest single acquisition projects in the history of the program (Aulds 2011). The identification of willing sellers contrasts dramatically with the North Carolina experience and merits further analysis to determine the factors that led to such high levels of participation.

as was the case in North Carolina (Smith 2011, pp. 297–298). Richard Sylves argues that in the U.S., history has shown that disaster policies are often reactionary and differ depending on the nature of the event and those in positions of power when it occurs (2008, p. 47).

Taking advantage of these conditions is significant as it offers an opportunity to develop policies and fund reconstruction efforts, implement projects identified in pre-event plans and undertake educational and capacity building initiatives that are often under-resourced. Therefore, the pre-event development and effective implementation of integrated hazard mitigation, disaster recovery and climate change adaptation plans and policies becomes critically important as they can help direct post-disaster resources when they become available. The development of these plans and policies should be crafted in a thoughtful and inclusive manner as disasters have also been shown to result in the inequitable displacement of people that are less powerful or have been excluded from decision-making processes, stimulate the investment in larger infrastructure projects and protective measures that may incentivise more growth in known hazard areas resulting in more brittle, rather than more resilient communities or disasters may trigger quick reactionary choices rather than a more deliberative process of planning for the future.

## 9.6 Summary and Conclusions

The recommendations in this chapter stem from the importance of linking the activities described in the State of North Carolina case study to an expanded effort that includes adapting to climate change-induced and exacerbated hazards. All of the activities discussed in the chapter are relevant to such an undertaking and highlight the complementary nature of natural hazards risk management and climate change adaptation. While this chapter has emphasised disaster recovery and the implementation of post-event hazard mitigation measures, a greater emphasis on pre-event planning for post-disaster recovery and the pre-event adoption of hazard mitigation measures that preempt or reduce future risk should be the ultimate goal of states and communities.

Rising sea levels, more intense hurricanes and more severe flooding are all real threats that will require revisiting our notion of risk and modifying our adaptation and hazard mitigation strategies accordingly. Specific actions should include: (1) changing risk assessment models and benefit/cost analysis tools to include the use of modified return periods tied to future weather and climate-dependent disasters; (2) mapping projected changes; (3) educating individuals, policy-makers, private sector investors and other stakeholders about the implications of these changes; (4) creating incentives through grants, loans or other means to encourage the relocation and resettlement of communities out of flood-prone areas, including those susceptible to sea-level rise; and (5) modifying and/or strengthening design parameters (e.g., building codes, flood ordinances, levee heights, land uses and public investments in infrastructure) plans (e.g., hazard mitigation, disaster recovery, comprehensive and adaptation) and policies (e.g., national, sub-national and local) to include climate change-induced natural hazards.

New and modified policies should emphasise the integration of climate change adaptation strategies into both pre-event hazard mitigation and post-disaster recovery initiatives. Experimental policies should be undertaken in pilot states and communities, ultimately serving as learning laboratories from which to test approaches, measure success (and failures) over time and build a coalition of supporters. This incremental approach could be used to inform a more comprehensive pre- and post-event national strategy that recognises the state and local needs uncovered through this more organic process, while documenting the value of pre-event planning and risk reduction as a central part of an emergent adaptation strategy. Considering the global nature of this issue, U.S. lessons should be shared with other nations and lessons drawn from other cases should be studied and adapted to U.S. conditions as appropriate.

As the U.S. begins to explore the development of a unified national climate change adaptation strategy, it should reflect on the long-standing problems with the current natural hazard risk management policy framework (or lack thereof), particularly the strategies adopted to deal with hazard mitigation and disaster recovery. At the same time, the U.S. climate change adaptation community has the potential to draw from these lessons in order to better coordinate the actions of Federal agencies, including the National Oceanic and Atmospheric Administration (NOAA) and the Environmental Protection Agency (EPA), while encouraging FEMA to integrate adaptation into the DMA and NDRF guidance, training materials, policies and funding programmes.

FEMA, an agency charged with assisting states and local governments manage natural hazards and disasters, has remained quiet on the topic of climate change adaptation, although this is slowly changing. For instance, FEMA has released a policy memo noting how it will address climate change adaptation (FEMA 2011). Specific actions include:

1. To enhance climate research, monitoring, and adaptation capabilities, FEMA will continue to establish partnerships with other agencies and organisations that possess climate science and climate change adaptation expertise.
2. FEMA will continue to study the impacts of climate change on the National Flood Insurance Programme (NFIP) and incorporate climate change considerations in the NFIP reform report.
3. FEMA will evaluate how climate change considerations can be incorporated into grant investment strategies with specific focus on infrastructure and evaluation methodologies or tools such as benefit/cost analysis.
4. FEMA will seek to understand how climate change will impact local communities and engage them in addressing those impacts.
5. FEMA will promote building standards and practices, both within FEMA programmes and in general, that consider the future impacts of climate change.
6. Through partnerships with the climate science community, FEMA will evaluate the potential impact climate change may have on existing risk data and the corresponding implications for Threat Hazard Identification Risk Assessment (THIRA) development and operational planning.
7. FEMA will continue to pursue a flexible, scalable, well equipped, and well trained workforce that is educated about the potential impacts of climate change.

These actions are slated to be operationalised in FEMA's Climate Change Action Implementation Plan, which was under development when this book went to press. While it is still too early to tell how significant this shift in policy is, or the degree to which these high-level statements will be integrated with existing natural hazards risk management policies, there is no doubt that hazard mitigation and disaster recovery programmes have much to contribute to the United States' ability to more effectively adapt to climate change.

Improving the ability to integrate hazard mitigation, disaster recovery and climate change adaptation necessarily benefits from a national strategy that recognises the strengths and weaknesses in existing federal policies and strives to modify them where needed to confront a series of similar challenges that are germane to natural hazards risk management and climate change adaptation. Key elements of this strategy should include:

1. Draw lessons from the successes and failures of natural hazards risk management practice and incorporate these lessons into the climate change adaptation strategy.
2. Place a greater emphasis on pre-event planning and adaptive capacity building initiatives, rather than the continued overemphasis on post-disaster monetary assistance.
3. Conduct an audit of existing federal programmes, policies and funding sources across the natural hazards risk management and climate change adaptation sectors, identify complementary and contradictory policies and use this information to help develop an integrated natural hazards risk management/climate change adaptation strategy that may help inform others who are in the process of developing a larger national climate change policy framework.
4. Maximise the use of existing natural hazards risk assessment tools to inform climate change adaptation strategies, while modifying these tools to more readily incorporate climate change-induced/exacerbated hazards.
5. Develop new methods to calculate changes in natural hazard return periods that are influenced by climate change and incorporate these findings into new maps, plans, policies and tools.
6. Encourage agencies and organisations in the natural hazards risk management and climate change community to co-support experimental efforts, pilot programmes and educational initiatives in both the pre- and post-disaster environment that clearly demonstrate the co-benefits of hazard mitigation, pre-event planning for post-disaster recovery and climate change adaptation.

NOAA, EPA, select professional associations and environmental groups are beginning to work with local governments to link natural hazards risk reduction and climate change adaptation issues. For instance, NOAA's Coastal Services Center and Sea Grant programmes fund research and provide data, analytical tools and training programmes to local governments and other users for this purpose. Specific programmes include the creation of the sea-level rise viewer, the provision of funding to conduct outreach efforts, the writing of technical reports and studies, the translation and dissemination of research findings to broad audiences and the sharing of lessons across communities through peer-to-peer exchanges. EPA has also begun to implement initiatives like the climate ready estuaries programme; sponsor educational seminars and workshops; and host post-disaster recovery efforts linking disaster recovery, smart growth and climate change adaptation. In both instances federal efforts are not tied to a national climate change adaptation policy. Indeed, NOAA's initial effort to consolidate federal programmes in a national climate change office was discontinued following opposition from some members of congress. Rather, both NOAA and EPA are working with willing partners including communities, non-profits, states, professional associations and others to further a small but growing climate change/natural hazards risk management coalition. This reality, while perhaps disconcerting, is in some ways similar to the findings of other case study chapters as varied partners have assumed leadership roles with or without a clear national climate change adaptation policy. This does not preclude the need for a national strategy, but instead, suggests that as a national climate change policy is developed in the U.S. it should recognise and support those organisations that have shown the foresight to take the lead.

## References


Adger, W. N., Lorenzoni, I., & O'Brien, K. L. (Eds.). (2009). *Adapting to climate change: Thresholds, values, governance*. New York: Cambridge University Press.

Aulds, T.J. (2011). Bolivar Peninsula: Buyout of storm properties one of largest in US. *The Daily News* (khou.com. Posted on April 5), 2011 (6:01 AM). http://www.khou.com/news/neighborhood-news/Bolivar-Peninsula-Buyout-of-storm-properties-one-of-largest-in-US-119240414.html. Accessed 4 Jan 2014.

Bachrach, P., & Baratz, M. (1963). Decisions and non-decisions: An analytical framework. *American Political Science Review, 57,* 632–642.

Barnes, J. (2001). *North Carolina's hurricane history* (3rd Edn.). Chapel Hill: The University of North Carolina Press.

Berke, P., Smith, G., Lyles, W. (2012). Planning for resiliency: Evaluation of state hazard mitigation plans under the disaster mitigation act. *Natural Hazards Review, 13*(2), 139–150.

Birkland, T. A. (1997). *After disaster, agenda setting, public policy, and focusing events*. Washington, D.C: Georgetown University Press.

Birkland, T. (2006). *Lessons of disaster: Policy change after catastrophic events*. Washington, D.C: Georgetown University Press.

Burby, R. (1998). *Cooperating with nature: Confronting natural hazards with land-use planning for sustainable communities*. Washington, D.C: Joseph Henry Press.

Burby, R. (2006). Hurricane Katrina and the paradoxes of government disaster policy: Bringing about wise government decisions for hazardous areas. *The Annals of the American Academy of Political and Social Science, 64*(May), 171–191.

Crew, C. (2013). North Carolina State hazard mitigation officer. *Personal communication* (April 25), 2013.

Colten, C. E. (2005). *An unnatural metropolis: Wresting New Orleans from nature*. Baton Rouge: Louisiana State University Press.

Delia, A. A. (2001) Population and economic changes in Eastern North Carolina before and after hurricane Floyd. In Maiolo J. R., Whitehead J. C., McGee M., King L., Johnson J., & Stone H. (Eds.), *Facing our future: Hurricane Floyd and recovery in the coastal plain* (pp. 199–206). Greenville: Coastal Carolina Press.

Federal Emergency Management Agency. (2011). *FEMA climate change adaptation policy statement*. Washington, D.C.: Office of Policy and Program Analysis (2011-OPPA-01).

Federal Emergency Management Agency. (2012). Innovative flood plain management, Kinston, North Carolina. FEMA mitigation case studies. Washington, D.C. www.fema.gov/library/viewRecord.do?id=1790. Accessed 25 April 2013.

Freudenberg, W., Gramling, R., Laska, S., & Erickson, K. (2009). *Catastrophe in the making: The engineering of Katrina and the disasters of tomorrow*. Washington, D.C: Island Press.

Hearn, P. (2004). *Hurricane Camille: Monster storm of the gulf coast*. Jackson: University Press of Mississippi.

Lyles, W., Berke, P., & Smith, G. (2013). Do planners matter? Examining factors driving implementation of land use planning approaches into hazard mitigation plans. *Journal of Environmental Planning and Management*. http://dx.doi.org/10.1080/09640568.2013.768973.

Mileti, D. (1999). *Disasters by design: A reassessment of natural hazards in the United States*. Washington D.C.: Joseph Henry Press.

Mileti, D., & Gailus, J. (2005). Sustainable development and hazards mitigation in the United States: Disasters by design revisited. *Mitigation and Adaptation Strategies for Global Change, 10*(3), 491–504.

National Research Council of the National Academies. (2006). *Facing hazards and disasters: Understanding human dimensions*. Washington, D.C.: The National Academies Press.

National Research Council of the National Academies. (2010). *Adapting to the Impacts of Climate Change*. Washington, D.C.: The National Academies Press.

Neumann, J. E., & Price, J. C. (2009). *Adapting to climate change. The public policy response: Public infrastructure*. Cambridge: Resources for the Future.

North Carolina Division of Emergency Management. (1999). Hazard mitigation successes in the State of North Carolina. Raleigh: North Carolina Division of Emergency Management.

North Carolina Division of Emergency Management. (2000). Hazard mitigation in North Carolina: Measuring success. Raleigh: North Carolina Division of Emergency Management.

North Carolina Division of Emergency Management. (2011). Division of emergency management, office of geospatial and technology. North Carolina sea level rise management study: Potential impacts, risk assessment and management strategies. (January 18th). Study Plan 3.2.

North Carolina Office of the Governor and the Office of State Budget and Management. (2009). North Carolina disaster recovery guide. http://www.osbm.state.nc.us/disaster/. Accessed 4 Jan 2014.

Olson, R. S., Olson, R. A., & Gawronski, V. T. (1999). *Some buildings just can't dance: Politics, life safety, and disaster*. Stamford: JAI Press.

Patterson, R. G. (1998). The third sector: Evolving partnerships in hazard mitigation. In Burby, R. (Ed.), *Cooperating with nature: Confronting natural hazards with land use planning for sustainable communities*. Washington, D.C.: Joseph Henry Press.

Pilkey, O. H., & Dixon, K. L. (1996). *The corps and the shore*. Washington, D.C.: Island Press.

Pilkey, O. H., Neal, W. J., Riggs, S. R., Webb, C. A., Bush, D. M., Pilkey, D. F., Bullock, J., & Cowan, B. A. (1998). *The North Carolina shore and its barrier islands: Restless ribbons of sand*. Durham: Duke University Press.

Reice, S. R. (2012). *The silver lining: The benefits of natural disasters*. Princeton: Princeton University Press.

Riggs, S. R. (2001). Anatomy of a flood. In Maiolo J. R., Whitehead J. C., McGee M., King L., Johnson J., & Stone H. (Eds.), *Facing our future: Hurricane Floyd and recovery in the coastal plain* (pp. 29–45). Greenville: Coastal Carolina Press.

Riggs, S. R., Dorthea V. A., Stephen J. C., & David J. M. (2011). *The battle for North Carolina's Coast: Evolutionary history, present crisis, and vision for the future*. Chapel Hill: University of North Carolina Press.

Rubin, C. (2007). *Emergency management: The American experience, 1900–2005. Fairfax:* Public Entity Risk Institute.

Smith, G. (2002). *The 21st century emergency manager*. Emmitsburg: FEMA Emergency Management Institute, Higher Education Project.

Smith, G. (2009a). Planning for sustainable and disaster resilient communities, Chap. 9. In Pine J. (Ed.), *Natural hazards analysis: Reducing the impact of disasters* (pp. 221–247). Boca Raton: CRC Press.

Smith, G. (2009b). Disaster resilient communities: A new hazards risk management framework, Chap. 10. In Pine J. (Ed.), *Natural hazards analysis: Reducing the impact of disasters* (pp. 249–267). Boca Raton: CRC Press.

Smith, G. (2010). Disaster recovery planning in the United States: Lessons for the Australasian audience. Special Issue of the Australasian Journal of Disaster and Trauma Studies. *Natural Hazards Planning in Australasia*. ISSN: 1174:4707, Volume 2010–1.

Smith, G. (2011). *Planning for post-disaster recovery: A review of the United States disaster assistance framework*. Washington, D.C: Island Press.

Smith, G., & Wenger, D. (2006). Sustainable disaster recovery: Operationalizing an existing framework. In Rodriguez H., Quarantelli E., & Dynes R. (Eds.), *Handbook of disaster research* (pp. 234–257). New York: Springer.

Smith, G., & Flatt, V. (2011). *Assessing the disaster recovery planning capacity of the state of North Carolina: Research brief*. Durham: Institute for Homeland Security Solutions.

Smith, J. B., Jason M. V., & Cromwell J. E. III. (2009). An architecture for Government action on adaptation to climate change. An editorial comment. *Climatic Change, 95,* 53–61. (Springer Science+Business Media).

Smith, G., Lyles, W., Berke, P. (2013). The Role of the State in Building Local Capacity and Commitment for Hazard Mitigation Planning. *International Journal of Mass Emergencies and Disasters, 31*(2), 178–203.

State Hazard Mitigation Advisory Group Meeting. (March 19, 2013). Raleigh, North Carolina.

State of North Carolina. (1997a). North Carolina disaster recovery task force report: Recommendations for action, (February 14).

State of North Carolina. (1997b). North Carolina Disaster recovery task force report: Recommendations for action. Progress Report, (September 3).

Sylves, R. (2008). *Disaster policy and politics: Emergency management and homeland security*. Washington, D.C.: CQ Press.

United States Government Accountability Office. (September 2008). *Report to the committee on homeland security and governmental affairs, U.S. senate. Disaster recovery: Past experiences offer insights for recovering from hurricanes Ike and Gustav and other recent natural disasters*. Washington, D.C.: Government Accounting Office, GAO-08-1120.

United States Government Accountability Office. (March 2010). *Report to congressional requesters. Disaster recovery: FEMA's long-term assistance was helpful to state and local governments but had some limitations*. Washington D.C.: Government Accountability Office, GAO-10-404.

White, G. (1945). *Human adjustment to floods. Research paper no. 29*. Chicago: Department of Geography, University of Chicago.

# Chapter 10
# The 2004 Manawatu Floods, New Zealand: Integrating Flood Risk Reduction and Climate Change Adaptation

**Bruce C. Glavovic**

**Abstract** Climate change exacerbates the flood risk facing many New Zealand communities. The February 2004 floods severely impacted the Manawatu region and reveal valuable lessons and shed light on barriers and opportunities, and priority actions, for reducing flood risk and building resilience and adaptive capacity in the face of climate change. The 2004 floods and subsequent actions highlight five lessons: (i) Disasters are 'focusing events' that reframe risk perceptions and create opportunities for reducing risk and mainstreaming climate change adaptation; but realising such opportunities is fraught. (ii) There is a need to move beyond reliance on structural flood protection works to overcome the 'safe development paradox,' address the manifold drivers of flood risk and anticipate escalating flood risk given climate change. (iii) Managing flood risk and adapting to climate change need to be framed as an integrated and adaptive process for building resilience and sustainability. (iv) Strategies to build flood resilient communities need to take into account wider societal trends and shocks that may have no apparent association with natural hazard risk. (v) The past focus on Readiness and Response needs to be complemented by sharper focus on Reduction and Recovery; and the permissive approach to land use decisions and reliance on mitigation measures need to be superseded by the avoidance imperative. Efforts to mainstream climate change adaptation face barriers and opportunities in three vital arenas: Understanding risk; the institutional setting; and professional practice. Three priority actions are identified: (i) Institutionalise a national legislative directive to reduce natural hazard risks and build resilience in the face of climate change. (ii) Develop a capacity building programme to enable local government to translate this directive into practical reality. (iii) Proactively explore opportunities to reduce risk and adapt to climate change in day-to-day local planning and decision-making. A transformative practice of deliberative governance is ultimately needed to institutionalise the lessons learned from the 2004 flood experience.

**Keywords** Manawatu · New Zealand · Flood risk · Climate change adaptation

B. C. Glavovic (✉)
Massey University, Palmerston North, New Zealand
e-mail: b.glavovic@massey.ac.nz

B. C. Glavovic, G. P. Smith (eds.), *Adapting to Climate Change,* Environmental Hazards,
DOI 10.1007/978-94-017-8631-7_10, 

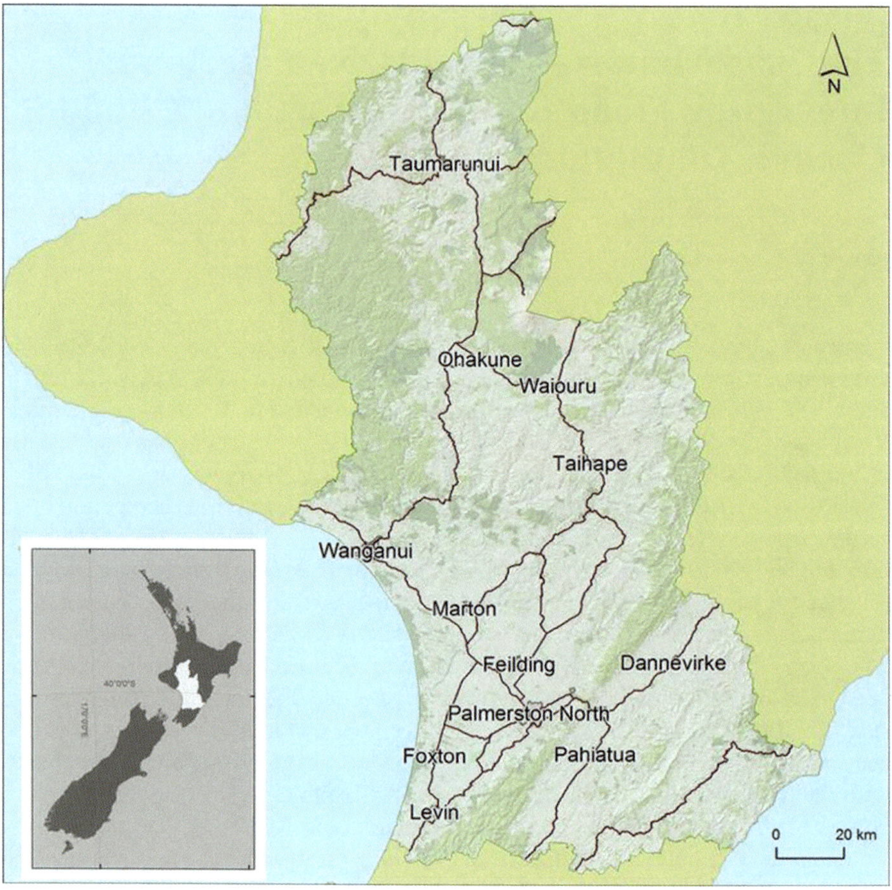


**Fig. 10.1** The Manawatu-Wanganui region, NZ. (Source: Prepared by Horizons Regional Council)

## 10.1 Introduction

The February 2004 floods caused widespread damage in the lower north and upper south islands of New Zealand (NZ). The predominantly rural Manawatu-Wanganui region was hardest hit (see Fig. 10.1). Rural livelihoods were disrupted for many months due to continued heavy rainfall and the time taken to repair damaged infrastructure, carry out farm repairs and restore livelihoods. For some the recovery process took considerably longer. The floods had a profound impact on risk perceptions, brought into sharp focus the issue of climate change and its impact on livelihoods and future flood risk and stimulated an array of activities that extended from the local to national scale. This chapter describes these experiences, explores the lessons learned and identifies barriers, opportunities and priority actions for building resilience and sustainability in the face of climate change.

## 10.2 Flood Risk and Climate Change: A Case Study of the 2004 Manawatu Floods

This section provides an overview of the case study setting and describes the 2004 flood event, its impacts and the subsequent response and recovery process.

### *10.2.1 The New Zealand and Manawatu-Wanganui Setting*

NZ comprises two large islands and numerous smaller ones in the southwest Pacific. Located on a plate tectonic boundary, the country is prone to a range of natural hazards, including geological, hydro-meteorological and biological perils (ODESC 2007). Flooding is the most frequently occurring natural hazard event and second most costly after earthquakes. An estimated two-thirds of the population live in areas prone to flooding, and there is widespread reliance on protection works such as stopbanks (or levees) and spillways (Waugh et al. 1997; Rouse 2012). Although NZ's exposure to climate change is expected to be modest, in part due to the moderating influence of the Southern Ocean, climate variability and extremes may have severe impacts on lynchpins of the economy—agriculture, forestry and tourism—and on the many communities located in areas prone to flooding and/or coastal hazards (Hennessy et al. 2007). Flood risk and climate change are consequently significant issues of national concern (MFE 2008a). The former is a long standing concern (Ericksen 1986). The latter is much more recent—mainly from the mid-2000s—and needs more focused attention in the future, especially with regard to the role of local government (Reisinger et al. 2011).

The Manawatu-Wanganui region has a population of about 220,000 people and makes up 8.1 % of NZ. It comprises two main catchments (the Whanganui and Manawatu), national parks, mountainous areas, farmland, the regional centres of Palmerston North (population circa. 85,000) and Wanganui (population circa. 40,000) and a number of small towns—with populations ranging from a few hundred to a few thousand—spread out in 10 districts. Flood risk is the top peril in the region (Manawatu-Wanganui CDEM Group 2009) but exposure to flooding and climate risk varies considerably. Whilst average annual rainfall in the region is not expected to vary much as a result of climate change, very heavy rainfall events are likely to become more frequent and increase flood risk four-fold by the end of the century[1]. The region is also likely to experience more frequent droughts. Some areas are likely to face increased problems with erosion, landslides and sedimentation in rivers as a result of the combination of drier conditions and more intense rainfall.

The regional council, Horizons Regional Council (HRC), is responsible for managing and monitoring the region's natural resources, including water and air quality, biodiversity and the coastal environment, as well as managing natural hazard risks.

[1] See http://www.mfe.govt.nz/issues/climate/about/climate-change-affect-regions/manawatu-whanganui.html.

**Fig. 10.2** The Motua sluice gates and spillway. (Source: Photograph by Massey University School of Aviation and sourced from Horizons Regional Council)

HRC oversees extensive river control and flood protection works, including 30 river and drainage schemes, 460 km of stopbanks, 700 km of drains, 20 pumping stations and 53 dams that together help to secure the safety and livelihoods of two thirds of the region's ratepayers. Most of the best farmland is on lowlands prone to flooding, and lowland settlements are reliant on protection works. The Lower Manawatu Scheme protects the city of Palmerston North and surrounding rural communities from flooding by the Manawatu River and its tributaries. The Moutoa sluice gates and spillway scheme, completed in 1962, allow floodwaters in the Manawatu River to bypass the meandering river course and cross low-lying farmland via a floodway that quickly releases floodwaters and thus prevents the build-up of floodwaters further up the river (Fig. 10.2).

### *10.2.2 The 2004 Manawatu Floods*

In mid-February 2004, the Manawatu-Wanganui region experienced one of the most destructive storms in over 100 years and perhaps the most intensive widespread rain since major deforestation in the twentieth century. What were the impacts of the 2004 floods and how did this experience shape risk perceptions and subsequent actions to address flood risk and climate change in NZ?

#### 10.2.2.1 The Flood Event and Impacts

The February 2004 storm affected about 70 % of the Manawatu-Wanganui region. Record or near record flood levels were experienced in all major rivers (see HRC 2004; MCDEM 2004; Fuller 2005) (see Fig. 10.3). Exceptionally heavy rainfall from the 13th to 16th of February had been preceded by unseasonably wet weather (5–6 times the average February rainfall) that saturated the ground and caused more rapid run-off than would otherwise have been the case. Rivers swelled quickly and flooding was widespread. The Moutoa sluice gates had to be opened to enable floodwaters to dissipate as quickly as possible once farmers in the floodway had been warned and stock moved.

Flood waters wreaked havoc in the region, damaging critical infrastructure and causing the closure of key regional roads; several trains were derailed; the main gas supply to the Hawkes Bay was disrupted for over a week due to a bridge collapse; and there was extensive damage to bridges (four destroyed and 21 damaged), water supplies, sewage treatment plants, flood-protection works, electricity and telecommunications. About 40,000 homes as well as businesses and schools were affected by power cuts, and water had to be trucked to some communities in the aftermath of the flooding.

If the stopbanks around Palmerston North had failed, it is likely that 1,000–2,000 houses would have suffered flood damage. The flood waters reached to within 900 mm of the top of the stopbanks. Flooding of this major regional centre was thus narrowly averted. But Scotts Ferry, Waitotara, Feilding, Tangimoana, Longburn, Whangaehu, Marton and Hunterville fared less well. The locally based army and airforce helped to evacuate over 2,300 people from the floodwaters and helped to avert loss of life and serious injuries. Overall, about 2,500 people were displaced; 1,000 people were homeless a week after the floods started; and about 400 homes were still not habitable three months after the floods (Fig. 10.4).

Rural communities were especially adversely impacted, with livelihoods severely disrupted by the floods and ongoing wet weather that prolonged adversity. At least 2,000 farms were flooded to some degree, about 800 severely (Ward 2005 in Smith et al. 2011). On-farm impacts ranged from destroyed livestock to broken fencing, inundation and destruction of crops and pastures, damage to buildings and homesteads, loss of plant and equipment, disruption of milking and dumping of about 10 million litres of milk, loss of feed, delays in re-establishing pastures, and loss of grazing from hill country landslips and erosion (HRC 2004).

Large areas of the region's hill country were subject to landslips: about 62,000 landslides were recorded (HRC 2004) across 8,000 $km^2$ of hill country (Hancox and Wright 2005) and caused major impacts on farm operations, roads and other infrastructure. An estimated 2,500 million t of soil and gravel went into waterways (equivalent to the annual discharge of the Manawatu River); with the lower reaches of the Manawatu and Rangitikei rivers experiencing severe silt loading, raising river beds and lowering flood protection levels (HRC 2004).

Overall, the economic impacts of the February floods on the lower North Island were in the range of NZD 300–400 million (HRC 2004; MCDEM 2004). HRC (2004) estimated that the cost of the storm included NZD 112 million in insurance claims;

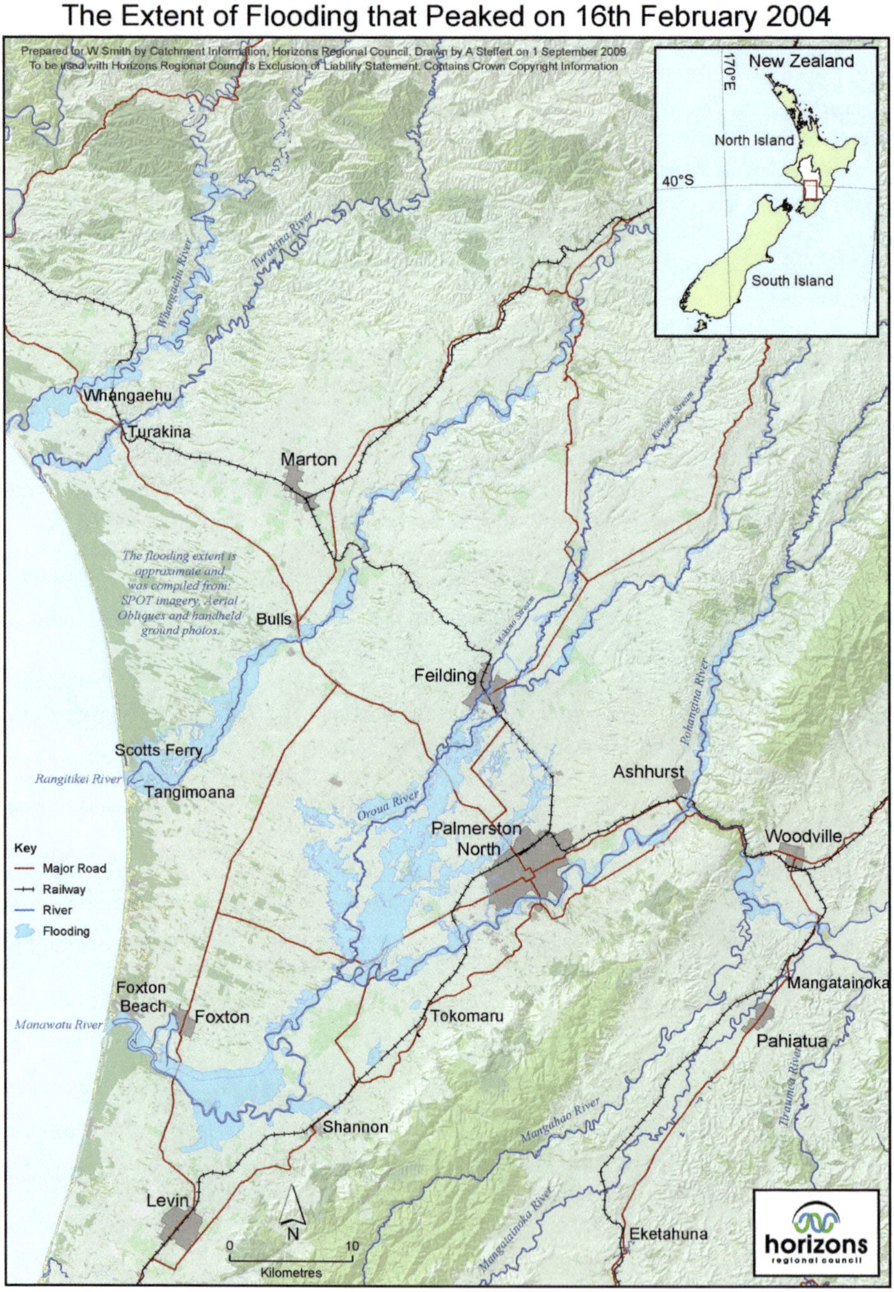


**Fig. 10.3** Peak flooding in the Manawatu in 2004. (Source: Prepared by Horizons Regional Council)

**Fig. 10.4** Flood impacts: Stock losses, infrastructure damage and flooding of rural and urban areas. (Source: Horizons Regional Council)

NZD 66 million in road repairs; costs to dairying of NZD 41.4 million; sheep, beef and deer NZD 66 million; crops NZD 24 million; and forestry NZD 29 million. The total cost to farming was in the order of NZD 159–180 million. The total damage to flood protection works and drainage schemes was estimated at NZD 19.6 million, with insurance covering only about NZD 9.2 million of the damage.

#### 10.2.2.2 Response and Recovery

The 2004 floods were the first major disaster in about two decades and the first emergency declared under the 2002 Civil Defence Emergency Management Act (CDEM Act). The region was the first in the country to have an active Civil Defence Emergency Management (CDEM) Group and Group Plan, and their preparations facilitated a timely and effective response according to independent reviews (Environmental Management Services 2004; Reid et al. 2004; Goodwin 2005). The Group found that the relationships that had been formed throughout the various agencies as part of the formation of the Group Plan were key, particularly those relationships that had been forged at the local level via the establishment of Emergency Management Committees (Ian Lowe 2013, personal communication). Many detailed recommendations were made to improve future readiness and response capacity, with particular attention focused on communications systems, flood prediction capacity, warning systems and a range of operational emergency management procedures. According to Reid et al. (2004), the clustered approach outlined in the CDEM Act for coordinating central and local government activities, as well as non-government and other organisations, was judged to have worked well. The principle of handling emergencies at the local level to draw on local knowledge, relational networks and local resources also worked well in practice. The local response was enabled by the timely mobilisation of support from central government, with the actions of Work & Income, Housing NZ and the Inland Revenue Department judged to have been especially timely and compassionate. The importance of early involvement of key providers of scientific and technical information, including the Meteorological Service, Crown Research Institutes and other providers, was also highlighted.

No formal reviews of the recovery experience have been published, leaving unanswered any questions about the effectiveness of recovery efforts and lessons learned. Hence the merit of this exploration of the impact of the 2004 floods on perceptions of and actions to deal with flood risk once the flood waters receded, especially given anticipated climate change impacts. Section 2.3 describes the institutional, legal and societal context within which the flood occurred and subsequent actions unfolded. Section 2.4 provides an overview of these actions and Section 3 distils lessons learned.

### *10.2.3 The Institutional, Legal and Societal Setting for Managing Flood and Climate Risks*

NZ is a representative democracy that has adopted a modified version of the Westminster political system with a single House of Representatives and independent judiciary. It has a two-tier government system, with responsibilities for natural hazards planning and risk management devolved by central government to the local level—administered by regional councils and territorial authorities (i.e., city and district councils). The NZ political economy has been shaped to a large extent by a neo-liberal agenda since the mid-1980s, with a strong emphasis on market-led growth, free trade and open markets, devolution, restructuring and downsizing of government, deregulation, privatisation of state-owned industries and more recently the sale of state assets. Government interventions that are construed as potentially inhibiting economic growth or impinging on private property rights are typically strongly resisted. NZ law recognises that private property rights are not sacrosanct and can be restricted in the public interest, including with respect to natural resource and hazard risk management. In practice, however, it is difficult for local government to constrain private property rights or development initiatives on the grounds of public safety, resilience and sustainability (Glavovic et al. 2010).

Flood risk management in NZ has evolved over time; shifting from historic ad hoc measures to a centralised approach in the last century and to a strongly devolved system from the 1990s (Waugh et al. 1997; Day 2005a) when major legislative and institutional reforms were introduced (Bührs and Bartlett 1993). The management of natural hazard risks is framed chiefly by the Resource Management Act (1991 as amended) (RMA), CDEM Act (2002), Local Government Act (2002 as amended) and Building Act (2004) (see Glavovic et al. 2010; Lee 2010). These and other legislative provisions focused on river control and protection works are central for managing flood and climate risk. In broad terms, territorial authorities and regional councils have the day to day responsibility for flood risk management in consultation with their local communities. Central government provides local government with legislative powers and directives as well as weather forecasts and warnings, funds natural hazards research, and assists local communities in preparing for, responding to and recovering from major events.

The CDEM Act, as set out in Part 3, aims to promote the sustainable management of hazards to protect people and property, encourage communities to investigate and communicate about hazard risks, plan and prepare for emergency response and to provide for coordinated and effective civil defence emergency management. The CDEM Act provides for an all hazards approach that is integrated across the '4Rs' of the hazard cycle—Reduction (equivalent to mitigation), Readiness (or preparedness), Response and Recovery. The Act emphasises the need for all actors—from individuals through to businesses, communities, local authorities, regional CDEM Groups, government agencies, emergency service providers and lifeline utilities—to understand the risks they face and take proactive steps to build resilience. Emergency response, community readiness, recovery planning and research into natural hazard risks are carried out chiefly under the CDEM Act and are implemented through provisions in the relevant Group Plan. Responsibility for Reduction has been assumed to be addressed primarily through land-use planning provisions in the RMA.

The RMA promotes sustainable management of natural resources and requires that regional councils control the use of land for the avoidance or mitigation of natural hazards, and that territorial authorities control the actual or potential effects of the use, development or protection of land, including for the purpose of avoiding or remedying natural hazards. Some councils are Unitary Authorities and have both regional and territorial functions. The RMA makes provision for government to declare National Policy Statements and National Environmental Standards but none have been promulgated for natural hazards in general or flood risk in particular. The RMA requires preparation of Regional Policy Statements and Regional Plans which set the framework within which District Plans are developed to articulate 'rules' for property development and land-use activities, including those in areas exposed to natural hazards. The RMA enables councils to prevent or restrict new development or development extensions in hazard-prone areas but this requires unambiguous plan objectives and policies that define non-complying activity status and, where appropriate, prohibited activities. Articulating clear objectives and policies to this effect requires robust information that is not always available (especially to less well-resourced councils) and is invariably subject to intensive scrutiny in the plan-making and consenting process. The onus of responsibility to demonstrate, for example, that potential economic benefits from a property development proposal are not overshadowed by natural hazard risks is borne by councils and local communities rather than development proponents. Consequently, council actions that are judged to impinge on private property rights or constrain economic development on the grounds of resilience, sustainability and public safety are likely to be challenged and end up in the Environment Court, exposing councils to potentially significant costs and reputational risk. Local politicians are understandably reluctant to expose their council's and communities to such costs and reputational risk, especially if in so doing their actions are seen to be 'anti-development' because of a low probability future natural hazard event. It is thus difficult for territorial authorities, especially smaller and less well-resourced councils, to counter proposals that promise immediate economic benefits even if resultant development exposes people to escalating natural hazard risks such as flooding in the long-term. Taking an 'avoidance' approach to natural hazard risks would be more feasible if lower level policy and planning provisions were compelled to do so by higher level policies or plans. Prior to 2004, there were no such requirements on local government, making it difficult to prohibit or even constrain development in areas exposed to natural hazards.

The Local Government Act is the primary law outlining local government responsibilities and functions and among other things requires territorial authorities to prepare a 10 year Long Term Plan that describes the outcomes that local communities seek to achieve, including the management of natural hazard risks. Provision is made for community engagement in defining these outcomes that are then planned and budgeted for through the Annual Plan process. Local authorities must consider the 'avoidance or mitigation of natural hazards' as a core service they provide to their communities according to section 11A of the Act.

The Building Act requires consideration of building standards in the face of natural hazard risks, including structural requirements and the characteristics of the land upon which buildings are located. Sections 71–74 of the Building Act require that territorial authorities refuse consent for building works proposed on land that is subject to natural hazards unless adequate mitigation measures are taken. Where exposure to a hazard cannot be minimised, the Act requires notices to be entered onto the title to record the risk they are exposed to. In addition to serving as a warning to potential purchasers, these notices can be used by insurance companies to determine the level of premium, or whether insurance will be provided at all. Territorial Authorities are also required to record information about hazards on Land Information Memoranda (LIM). But in some cases, Territorial Authorities have been pressurised to remove hazard information from LIM reports because of concerns about adverse impacts on property values.

A number of other laws influence natural hazard risk management through obligations and responsibilities for the provision of public services such as water supply, waste management and a range of emergency management and social services. Provisions specifically relevant for managing flood risk, as well as river control and drainage functions carried out by regional councils, are found in, among others, the Land Drainage Act (1908), the Soil Conservation and Rivers Control Act (1941) and the Local Government Act (1974). Also of note is the New Zealand Earthquake Commission (EQC)—a national insurance scheme for natural hazard events that impact residential homes, land and contents. Created under the Earthquake Commission Act 1993, the EQC provides insurance that would be difficult to obtain solely through the private market and insures against loss or damage from earthquakes, natural landslips, volcanic eruptions, hydrothermal activity and tsunami. It also insures residential land (within limits) against storm and flood damage; and from fire resulting from any of these perils. The EQC cover is government guaranteed.

Prior to 2004, local government was not required by law to address climate change other than in terms of the possibility of sea-level rise under the RMA NZ Coastal Policy Statement. An amendment to section 7 (i) of the RMA in 2004 requires all those exercising functions and powers under the RMA to 'have regard to climate change effects.'

The NZ legislative framework thus makes available to local government a variety of tools to manage flood risk including structural options, such as stopbanks, and non-structural options, such as catchment and riparian management measures. More generally applicable statutory provisions include legislative directives such as national policy statements and environmental standards, a variety of planning provisions such as CDEM Group Plans, Regional Policy Statements, Regional and District plans, Long Term Plans, Annual Plans, Asset Management Plans and Lifeline Plans, as well as making available to property owners and the public information about hazard exposure and sensitivity, warnings and evacuation plans and consent conditions. A range of non-statutory tools is also available, including guidelines developed by government agencies, crown research institutes and professional institutes as well as research, education, public awareness and advocacy. The extent to which these tools are effectively used in NZ varies from council to council (Glavovic et al. 2010; Lawrence and Quade 2011). Glavovic et al. (2010) demonstrate that the overarching legislative framework for natural hazards planning in NZ has a solid foundation and is imbued with good intentions. The challenge, however, is to translate these laudable intentions into practical reality in local communities. This challenge is significantly influenced by demographic, social, economic and political changes in recent decades that have reshaped resilience in NZ and that of rural communities in particular.

According to Smith et al. (2011), there has been a 'hollowing out' of rural resilience in NZ as a result of societal changes unrelated exposure to natural hazards and disaster risk. Such changes include the adoption of new farming practices; technological changes that among other things have reduced the demand for farm labour; a trend towards larger, corporate farming businesses that have consolidated smaller family-owned farms; a growth in small-scale 'hobby' farming or 'lifestylers;' increased mobility and a drift to urban centres; and a significant contraction of public services in rural areas, including post offices, policing, schools and other social services.

> The traditional view of rural New Zealand as primarily made up of farm households with a sense of common values and purpose and with a clear community structure centred on church, school or pub, has gone. Modern agriculture and the structural changes characteristic of rural societies in all Western industrial economies have introduced new elements to the rural environment unrelated to the farm industry and with values and interests that make community development in rural areas increasingly problematic. The 2004 flood clearly exposed the vulnerability of rural communities. It forced people to rethink what, or, more correctly, who, constitute their community. In certain cases, this revealed a complete breakdown of community (Smith et al. 2011, p. 550).

The prevailing institutional, legal and societal milieu, including the changing character of rural communities in the region, framed post-2004 endeavours to address flood risk and adapt to climate change.

### *10.2.4 Actions to Address Flood Risk and Climate Change Since the 2004 Floods*

The floods had a profound influence on public attitudes towards flood risk, and galvanised action on a number of fronts at local, regional and national levels. In the Manawatu, the reframing of risk perceptions was expressed in the prominent place of flooding in the political campaigns of several candidates in the local body elections in October 2004 and again in 2007. A number of candidates who took a strong stand on the need to reduce flood risk were elected to local and regional councils. Raised public awareness and political will to reduce flood risk was translated into significantly increased community investment in measures to understand and manage flood risk. This section briefly outlines some of the main initiatives that took place, including improved levels of flood protection; innovations in integrated approaches to natural resource and hazard risk management and policy, planning and development control; investment in research, monitoring, flood forecasting and warning systems; and legislative reform and guidance to address natural hazard risks, and support programmes to build community resilience.

#### 10.2.4.1 Flood Protection Works

The 2004 floods revealed that the protection provided by the Lower Manawatu Scheme was less than the intended 0.2 % Annual Exceedance Probability (AEP) (1 in 500 years) protection through the Palmerston North City reach principally because the flood carrying efficiency of the Manawatu River was less than predicted and previous flood frequency analyses needed to be reassessed. In addition, the flood carrying capacity of the rural reach downstream of Palmerston North had been reduced by silt and gravel accumulation. A major review and upgrade of the Lower Manawatu Scheme and the region's river management and flood protection works was imperative (Peter Blackwood 2013, personal communication). By late 2005, HRC developed a NZD 40 million programme of work to improve the Lower Manawatu Scheme. In addition, NZD 5 million was allocated to improve protection works on the lower Rangitikei, giving Scotts Ferry 1 % AEP (1 in 100 years) and Tangimoana 2 % AEP (1 in 50 years) protection; NZD 11.5 million was earmarked for the City Reach programme to upgrade the stopbanks around Palmerston North; and five new river management schemes for the region were developed. In short, since 2004, the HRC has improved the management schemes on the lower reaches of every major river in the region. In keeping with the requirement for local communities to shoulder the cost of flood risk management measures, this comprehensive

**Fig. 10.5** Stopbanks and rock revetments, Fitzherbert Bridge, Palmerston North. (Source: Bruce C. Glavovic)

upgrade of regional flood protection works required a review of the rating system to cover the costs of improvements in proportion with the benefits received. After extensive public consultation, the revised rating scheme was introduced in 2009; the first overhaul of the rating system since the 1950s.

Palmerston North opted to increase the level of protection to the revised design flow for 0.2% AEP (1 in 500 years) protection; a safeguard against floods 12.5% greater than the most severe flood on record. The City Reach programme was completed in 2013 with stopbanks having been raised to the desired level, rock revetments put in place in localities prone to riverine erosion, some sections of the river realigned and extensive riparian management and planting. About 3,500 houses would have been flooded if a 0.2% AEP flood event had occurred prior to these measures being put in place (Fig. 10.5).

In Wanganui, a community with a high proportion of retirees and residents on fixed incomes, there is understandable resistance to rates increases. The community rejected the rates increase necessary to upgrade stopbanks safeguarding some neighbourhoods vulnerable to flooding. Concerns were raised about the adequacy of the community consultation process carried out by the HRC and it is likely that the issue of flood protection will be explored further with at-risk residents at some point in the future (Max Benseman 2013, personal communication). New protection works have, however, been completed to safeguard the industrial precinct of the city. Consequently, there is considerable variability in the level of protection provided to different parts of Wanganui. The Balgownie compartment is protected to the 0.5% AEP (1 in 200 year) standard but Anzac Parade/Kowhai Park, including 57 houses, is protected to only about 2% AEP (1 in 50 years). The HRC has made provision in its 2013–2014 Annual Plan for works that will provide a uniform standard of 2% AEP protection without freeboard for Anzac Parade/Kowhai Park, at a cost of NZD 1.02 million, and planning and design work has commenced to this end. According to the Annual Plan, this upgrade will be undertaken in the current year with the Wanganui ratepayers being levied a one-off charge in 2014–2015 for 70%

of the cost of the works with the balance to be funded by the general ratepayer[2]. It is noteworthy that increasing attention has been focused on how to embrace the city's river heritage and strengthen the river connection. Evidence of attempts to learn to live with flood risk is indicated, for example, by the design and layout of the Visitor Information Centre on the riverfront. Opened in November 2010, the Centre is designed to accommodate flooding and the layout enables floodwater and debris to be swept out with minimal damage to the structural integrity and operations of the Centre. There has also been some discussion about managed retreat to reduce future flood risk in exposed areas of Wanganui but there is no known locality in NZ where such an approach has been successfully applied on the scale envisaged in Wanganui (Allan Cook 2013, personal communication). As important as protection works are for at-risk communities, the 2004 floods underscored that such works are not a panacea for managing flood risk.

#### 10.2.4.2 Sustainable Land Use Initiative

Community leaders recognised that new approaches needed to be explored to address flood risk in the region, especially given the expected increase in the frequency and magnitude of storms due to climate change. One issue underscored by the 2004 floods was the need to stop practices that accelerate hill country erosion—a pervasive problem with some 273,000 ha of the region categorised as highly erodible (Dymond and Shepherd 2006). Local government representatives, mayors, community leaders, representatives of the farming community, non-governmental organisations (NGOs), rural consultants and Crown Research Institutes met to explore ways to address hill country erosion and achieve multiple goals, including reducing flood risk, conserving soil and promoting more sustainable land use and improving water quality. A Governance Group, made up of HRC councillors and staff, district council mayor and chief executive, representatives of Federated Farmers and Farm Forestry and farm leaders, was established to define the purpose and general content of what became known as the Sustainable Land Use Initiative (SLUI), and was responsible for developing and overseeing the implementation process.

The SLUI aims to change land use practices that accelerate erosion by encouraging farmers to retire land, afforest eroding land with suitable production tree species, control pests, review and/or upgrade tracks and stream-crossings and re-establish indigenous vegetation. Participation in SLUI is voluntary. The SLUI pays for the development of a Whole Farm Plan that identifies opportunities for sustainable land use and improving overall farm business. The plan includes an assessment of land, water, living heritage and farm production resources; environmental services and threats to production; recommends best practices; benchmarks existing farm business; and prepares long-term business and work plans. The information about natural resources is retained by the HRC while business information remains the property of farmers. Each plan is tailor-made for particular farms so that locality-

[2] See http://www.horizons.govt.nz/assets/publications/about-us-publications/annual-plan/Annual-Plan-2013-14-FINAL-V4-Final-Changes-from-25-June-2013.pdf.

specific recommendations can be made. The SLUI endeavours to work alongside farmers to determine how best to implement each plan that can receive funding from central government. In addition to the Whole Farm Plans, the SLUI involves public awareness, education and advice programmes, regulations, monitoring and financial incentives. SLUI addresses issues at a catchment scale but focuses planning and action at a farm scale to generate practical benefits for landowners, the region and beyond. The aim was to develop about 1,500 Whole Farm Plans over a ten year period to significantly reduce hill country erosion by 2015.

SLUI provides benefits to farmers, local communities, the region and nation and efforts have been made to share the costs in proportion to benefits. Intensive lobbying of central government raised the profile of the SLUI and led to the Ministry of Agriculture establishing in 2007 a contestable fund initially called the Sustainable Land Management (Hill Country Erosion) Framework. In December 2007 some NZD 6.6 million over four years was secured from the fund for the SLUI. Individual farmers contribute and regional ratepayers are levied a SLUI Uniform Annual Charge to cover a portion of the costs. Others stakeholders are interested in participating in the programme, including forestry and carbon sequestration companies, and could provide a fourth income stream.

A concerted effort has been made to raise awareness about hill country erosion and garner support for SLUI via television, radio, print media, websites, meetings and other means to raise public awareness and encourage participation by landowners. By the end of 2011, 369 Whole Farm Plans had been prepared for an area of more than 280,000 ha or 28 % of the region's hill country, and over 85 % of these are active plans that have work underway. More than 9,260 ha of works have been completed, the bulk of which is afforestation. More than five million trees had been planted and over 280 km of fencing erected to protect erosion prone land[3].

It is noteworthy that Central Government commitment to continued funding of SLUI related activities appears to be waivering[4] and this calls into question the extent to which the merits of SLUI are appreciated and supported beyond the region. Many challenges have been faced over time as the SLUI seeks to engage stakeholders with diverse interests to bring about systemic changes in land use practices in the hill country (Ian Lowe 2013, personal communication; Greg Carlyon 2013, personal communication). This effort has taken place in the context of significant reformulation of local government policies and plans relevant to natural hazard risk management, and the HRC One Plan in particular.

#### 10.2.4.3 The Horizons Regional Council One Plan

The 2004 floods refocused HRC's attention on flood risk. The HRC developed an innovative regional policy and planning approach to guide natural resource man-

[3] See http://www.mpi.govt.nz/environment-natural-resources/funding-programmes/slm-hill-country-erosion-programme/slmhce-project-sustainable-land-use-initiative.

[4] See http://www.horizons.govt.nz/assets/Uploads/Events/Regional_Council_Meeting/2013-06-25_100000/RC25062013AGNAT.pdf.

agement referred to as the One Plan. The One Plan integrates into a single coherent plan a Regional Policy Statement and six separate plans, focusing on four pivotal regional issues: declining water quality, increasing demand for water, unsustainable hill country land use and threatened native habitats. Natural hazard risk management is a separate chapter in the Regional Policy Statement of the One Plan[5]. The formulation and institutionalisation of the One Plan has, however, been a protracted and contested process—essentially because the One Plan proposes a set of rules and imposes direct costs on farming activities that adversely impact sustainable natural resource management. After eight years of contestation and policy refinement, key elements of the One Plan were upheld in an Environment Court ruling in September 2012— despite criticism from organised interests in primary production such as Fonterra and Federated Farmers. There was, however, little overt resistance to the introduction of significant new provisions to address natural hazard risks and flood risk in particular.

Prior to 2004, the HRC did not explicitly advocate risk avoidance. In the absence of national or regional directives to avoid development in flood-prone areas, territorial authorities in the region, like those elsewhere in NZ at the time, generally adopted a permissive approach to activities in localities exposed to flooding and relied on structural measures to mitigate flood risk (Matthew Mackay 2013, personal communication). The 2004 flood experience demonstrated unequivocally the imperative to avoid putting people in harm's way and adopt a 'hazard avoidance' approach. The One Plan states: "Flood hazard avoidance must be preferred to flood hazard mitigation" (Policy 10–2)[6]. The previous standard for delineating areas prone to flooding has been raised from 1 % AEP (1 in 100 years) to 0.5 % AEP (1 in 200 years) to take into account the likely effects of climate change over a postulated 50 year life of a residential or commercial building. The One Plan also stipulates an avoidance approach to new activities or structures in areas likely to be inundated by a 0.5 % AEP, whilst recognising that some activities may be necessary on production land or out of functional necessity. No additional avoidance or mitigation measures are required within the area of Palmerston North protected to a 0.2 % AEP (1 in 500 years) level. The One Plan also states that the "Regional Council and Territorial Authorities must take a precautionary approach when assessing the effects of climate change on the scale and frequency of natural hazards with regard to among other things activities adjacent to rivers and flood mitigation activities" (Policy 10–6).

Overall, the One Plan advocates an avoidance and precautionary approach because it is better to keep people and associated infrastructure out of harms' way, especially given climate change. The Natural Hazards Chapter (Regional Policy Statement) does not contain rules regarding natural hazards but it does require that

[5] See http://www.horizons.govt.nz/assets/publications/about-us-publications/one-plan-publications-and-reports/proposed-one-plan/Chapter10_Hazards.pdf.

[6] In the United States, the term 'mitigation' connotes the elimination or reduction of natural hazard risk. In NZ, 'mitigation' typically refers to ameliorative measures to reduce risk, such as elevating a building above a particular flood level, whereas 'avoidance' means keeping people and the things they value out of harms' way.

territorial authorities develop not only policies and methods to manage the effects of natural hazards, but also rules. This in turn requires territorial authorities to give effect to the Regional Policy Statement via their respective District Plans. The requirement for District Plans to comply with the One Plan provides territorial authorities with the legal wherewithal to take a stronger stance against putting people in harms' way—even in the absence of a national legislative directive. But variable exposure to flood risk and the distinctive needs of local communities in the region present a challenge to the HRC as it seeks to implement a consistent avoidance and precautionary approach and yet maintain flexibility to support territorial authorities and their local communities as they address their particular circumstances. Prior to 2004, information about differential flood risk across the region was poorly developed and this compelled the HRC to improve its information platform.

#### 10.2.4.4 Investment in Research, Monitoring, Flood Forecasting and Warning Systems

Prior to the 2004 floods, HRC relied to a large extent on manually produced forecasts based on only 77 sites for recording rainfall and river flow (Jeff Watson 2013, personal communication). To improve the information base, the number of recording sites was almost doubled by 2010 and significant investment was made to build capability to forecast floods using mathematical rainfall and runoff models for the region's major flood risk catchments (Metservice provision of hourly rainfalls for the next two days is a key component of this forecasting capability). These flood forecasts are readily accessible via the HRC website and are available to a range of stakeholders including emergency managers, HRC engineers, emergency services, farmers and regional businesses and homeowners. Providing reliable flood forecasts and warnings has required improvements to the recording installations at numerous sites in the region. Multiple communications paths have been put in place from all key recording sites and Business Continuity Planning has enabled the hydrology telemetry to be relocated off-site with minimal loss of capability. Those living in areas prone to flooding can subscribe to an HRC warning service via an Interactive Voice Response System which sends out telephonic warnings to subscribers once their indicated warning level is reached. The number of subscribers has doubled since the 2004 floods, and now includes about 860 contacts (Jeff Watson 2013, personal communication).

Attention has also been focused on modelling and mapping floodplains to show areas prone to flooding under different scenarios to highlight flood depth, velocity and hazard potential. This investment has enabled the HRC to provide improved flood inundation information to district councils, emergency services, insurance companies, developers, farmers and the general public (Ian Lowe 2013, personal communication). Prior to 2004, providing such information was carried out by a procedure that was somewhat reliant on staff memories, photographic evidence and engineering designs. Post-2004 investment has enabled more sophisticated hydraulic modelling of floodplains to be undertaken for various AEP scenarios that enhances the accuracy of flood hazard advice and enables climate change scenarios to be taken into account. The intention is to make this inundation mapping publicly accessible via the HRC website to raise public awareness and inform local planning and development control decisions. The HRC also supplies information and advice (free of charge) to individuals, landowners, territorial authorities and developers regarding natural hazards in terms of land development planning advice.

#### 10.2.4.5 Local Planning and Development Decisions in Flood-Prone Localities

The influence of the 2004 floods on local development decisions is exemplified by proposed 'greenfield' development in the Te Matai precinct of Palmerston North. Prior to the 2004 floods, this area had been identified in the city's Urban Growth Strategy as a priority area for future suburban development—despite exposure to flood risk and high value agricultural soils. In the aftermath of the floods, and with the election of local politicians averse to flood risk, this area was deemed unsuitable for future growth. Subsequent to that decision, the City Reach programme provided 0.2% AEP (1 in 500 years) protection for the city and the new stopbanks now include a portion of the Te Matai precinct, thus potentially enabling an expansion of the city within the flood protected area. Subsequent decisions about priority areas for future urban growth have continued to explore alternatives in preference to this Te Matai area. If Te Matai land owners within the flood protected area were, however, to apply for a private plan change, to rezone their land use from agriculture to residential use, it may be difficult for the city council to counter such an application even though it wants to retain such land for productive purposes and avoid flood risk. It would be paradoxical if this land were to be developed in the future given the evolution of planning thinking and practice since 2004. If such a plan change were to be approved, it would constitute a classic example of the 'safe development paradox' in which protection against moderate natural hazards risk leads inexorably to development intensification and exposure to catastrophic risk if an event exceeds design standards (Burby 1998, 2006).

To complicate matters further, there have been proposals in the past to build a second bridge across the Manawatu River, as an extension of Te Matai Road, and to open up residential property development opportunities on the opposite bank of the river. The proposed bridge and associated development have not materialised for a variety of reasons, including the global financial crisis, fluctuations in the housing market and issues to do with responsibilities and financing of the road and bridge works. It is, however, not inconceivable that this proposal could be mooted again at some future date. If the bridge were built, there would be compelling reasons to use the road approach to the bridge as a foundation for additional flood protection works. The outcome would be two stopbanks within a few hundred metres of each other; each potentially providing 0.2% AEP flood protection. Future observers would wonder how such a situation could possibly arise in a country with a robust natural hazards planning framework that requires local level plans to comply with regional provisions that prescribe a risk avoidance approach; and where there is strong local political and community support for this policy approach. This evolving situation is indicative of the complexity, 'messiness' and unpredictability of local planning and development decisions that are shaped by an array of local and external drivers, some of which have nothing to do with local risk exposure and risk management preferences (Jonathan Ferguson-Pye 2013, personal communication). Local communities face difficult decisions as they seek to manage natural hazard risks and reconcile economic development opportunities, public safety, resilience and sustainability. The One Plan enables territorial authorities to adopt a risk avoidance approach. But, even with such a policy imperative in place, local communities in low-lying flood-prone localities face a conundrum if they have to choose between allowing property development that promises immediate economic gain and rejecting development to avoid exposure to a low probability but potentially devastating future event. The Palmerston North City Council continues to allow subdivision of existing lots and in-filling in areas prone to flooding in an event that exceeds design standards when it could restrict such development intensification to areas with no flood risk. If in-filling in areas that will be inundated in an over-design event is permitted, regular cost-benefit studies should be undertaken to determine at what point the stopbanks should be raised to cater for the greater loss that will occur when they are overtopped (Jeff Watson 2013, personal communication).

Not all communities in the Manawatu-Wanganui region want to or are able to resist proposals for new property development in localities prone to flooding. Several smaller communities in the region occupy land that is almost entirely within the 0.5 % AEP (1 in 200 years) flood zone and future expansion or development intensification is not desirable in the face of climate change; nor would it be consistent with the One Plan. Some communities have experienced repeated flooding in the past and face escalating risk in the future yet they cannot afford suitable structural protection and are reluctant to relocate. The HRC thus faces a conundrum in applying a consistent avoidance and precautionary approach while supporting communities that cannot meet the One Plan provisions.

The One Plan provides a strong mandate to avoid new activities or intensified development in areas prone to flood risk because territorial authorities must give effect to this higher level policy in their local planning and development decisions. But not all regions in the country have this regional policy requirement and, in the absence of a national legislative directive to avoid flood risk, territorial authorities are hard pressed to deal with escalating climate-driven flood risk.

#### 10.2.4.6 Legislative Directives and Guidance on Flood Risk and Climate Change

Central government is able to issue legislative directives to foster national consistency in addressing flood risk through, for example, a National Policy Statement or National Environmental Standard under the RMA. But, to date, the Government has opted not to provide such a directive. In the absence thereof, local government determines flood risk management approaches in keeping with regional and local community priorities; and the challenges alluded to above confront at-risk communities.

In 2001, a partnership involving the Ministry for the Environment (MfE), Ministry of Civil Defence and Emergency Management and a range of regional councils and territorial authorities published *Floodplain Management Planning Guidelines: Current Thinking and Practice in New Zealand* (Berghan and Westlake 2001) primarily to assist council staff in floodplain management planning and river and flood hazard management. The 2004 floods sharpened the focus of all levels and relevant agencies of government, other stakeholders such as professional institutes and researchers on understanding and addressing flood risk (see e.g., Day 2005a; Ericksen 2005a, b; Goodwin 2005). A draft *NZ Protocol on Managing Flood Risk* was published in 2005 (Day 2005b) based on inputs from local and central government and the Institute of Professional Engineers NZ. This protocol among other things informed an extensive two year review of flood risk management and river control that was led by the MfE, which released a report *Meeting the Challenges of Future Flooding in New Zealand* in 2008 (MfE 2008a). The draft *NZ Protocol on Managing Flood Risk* was intended to provide an overarching framework for further research that would help to develop implementation guidelines and modules on various flood issues. But this research was not funded and the work was halted (Rouse 2012). Standards NZ issued a voluntary process standard for managing flood risk in 2008 (Standards New Zealand 2008). The standard was based on the draft flood protocol and offers a best practice approach to managing flood risk for local and central government, communities, property developers and a range of professionals.

The MfE (2008a) review found that the institutional framework for managing flood risk was not fundamentally flawed but that significant matters needed to be addressed (see also Day 2005a). Central government had focused attention on the Response and Recovery stages of the flood hazard cycle but needed to focus more attention on Reduction by providing clear directives, information, guidance and assistance including resources to enable local government to more effectively assess and manage flood risk in the face of climate change. The review found that flood risks in NZ cannot be gauged accurately and neither can climate change impacts be predicted on levels of flood risk; and there is no way to carry out meaningful comparative analysis. Flood risk management practices were found to be variable around the country, with smaller at-risk communities less able to afford flood risk management. The review recognised that managing national flood risk was different from local flood risk and that the prevailing institutional arrangements needed to be amended to better align national and local flood risk management efforts. In particular, past reliance on protection works and the focus on Response and Recovery needed to be changed so that future decisions foster an avoidance strategy. It argued that in the absence of improvements to the institutional architecture of flood risk management, future generations would likely become more vulnerable to flooding, experience greater losses and require escalating expenditure on Response and Recovery efforts. Given ongoing pressure to develop land in flood-prone areas, and the lack of sound information, the imperative for change was clear. The Environment

Minister appointed an independent Board of Inquiry in August 2008 to consider a proposed National Policy Statement for flood risk management. MfE prepared a draft National Policy Statement. Government decided, however, that the costs and benefits of this approach needed to be investigated further. The initiative had stalled by the end of 2009 and the issue of flood risk has not been pursued further under the National-led government that assumed office in November 2008.

As recently as 2011, Local Government NZ (LGNZ 2011) argued that the prevailing Response-driven approach to flooding imposes significant national costs on government, local authorities, communities and citizens, and that the country and central government would benefit from a National Policy Statement on flood risk management. The costs of floodplain development are externalised through insurance, EQC cover and/or central government relief. Development proposals in flood-prone areas that promise economic benefits are difficult to counter even when public safety is potentially compromised. Efforts to counter such proposals invariably end up exposing local government and their communities to substantial litigation costs. The Environment Court tends to apply a precautionary approach in dealing with disputes over coastal hazards but is less able to do so for flood risk in the absence of a national policy prescription. A National Policy Statement on flood risk management would help to reduce the externalised costs of flood events and litigation costs that arise in countering development proposals in flood-prone areas.

In a study of the state of research on flood risk management, Rouse (2012) found that barriers to flood risk management are similar to those in managing coastal erosion and include the lack of a clear national directive; poor articulation of national and regional interests in local decision-making; inadequate long-term planning; development interests prevailing over community interests; and resourcing and information gaps in councils, especially smaller less well-resourced councils.

In parallel with concerns about flood risk, and the need to provide local government with national guidance and where appropriate legally binding directives, attention was also focused on climate change adaptation. In 2004–2005, the MfE commissioned the development of guidance that was later published as 2nd edition guidance on *Climate Change Effects and Impacts Assessment: A Guidance Manual for Local Government in NZ* (MfE 2008b), *Coastal Hazards and Climate Change: A Guide for Local Government in NZ* (MfE 2008c) and *Preparing for Climate Change: A Guide for Local Government in NZ* (MfE 2008d). MfE also focused attention on the intersection of flood risk and climate change, namely *Tools for Estimating the Effects of Climate Change on Flood Flow: A Guidance Manual for Local Government in New Zealand* (MfE 2010a), initially published in 2005, and a summary document *Preparing for Future Flooding* (MfE 2010b). The guidance aims to provide local government staff responsible for matters such as river and asset management with practical tools to manage flood risk in the face of climate change. There is, however, no legislative directive for how local government should deal with climate risk—other than take it into account in local decision-making. How local government deals with climate risk is thus up to each community. Adopting a phased or managed retreat from areas exposed to repeat flooding, for example, will depend on local community choices about how to reconcile this choice with compensation for affected landowners, community cohesion and council liabilities and responsibilities (Quade and Lawrence 2011; Lawrence and Quade 2011; Reisinger et al. 2011). Guidance documents are very helpful to local government and their communities, but they are insufficient for enabling communities to resolve this vexing challenge in the absence of legislative directives and other means of enabling support.

The Government developed a range of options to support rural communities in coping with adverse climatic events and assist with risk reduction and post-disaster recovery; recognising the increasing importance of this imperative in the face of climate change. For example, an 'On-farm Adverse Events Recovery Framework' was developed and revised in the light of the 2004 flood experience and other events; and additional resources budgeted[7]. It was recognised that individuals need to take responsibility for the risks they face but that when a disaster occurs, the Government will need to provide additional resources to help the impacted communities and regional economy recover quickly. This imperative has been borne out in the 2010–2011 Canterbury earthquakes.

#### 10.2.4.7 Canterbury Earthquakes and Legislative Reforms to Address Natural Hazard Risks

The 2010–2011 Canterbury earthquake series has recalibrated NZ risk perceptions (McClure et al. 2011a, b) and natural hazards risk is now a matter of national concern, with particular attention focused on ensuring that buildings meet appropriate seismic safety standards and that liquefaction prone areas are identified and taken into account in local planning and decision-making.

In October 2011, the Government established an independent Technical Advisory Group (TAG) to review the principles in sections 6 and 7 of the RMA which list "matters of national importance" and "other matters" that have a substantial role in shaping and directing how the RMA's purpose is given effect through planning and decision-making. The TAG was to take into account lessons learned from two decades of RMA practice, contemporary values and priorities including the Government's environmental and economic objectives and legislative reforms and the Canterbury earthquakes. The TAG (2012) review has been submitted to government and included among other things the following recommendations: section 6 of the RMA should require decision-makers to recognise and provide for natural hazard risks; a combined regional and district natural hazards plan should be prepared; CDEM Group management plans should be considered in preparing RMA regional policy statements and plans; regional councils should have the lead function in managing all the effects of natural hazards and territorial authorities should retain their current functions relating to natural hazards; and councils should be required to make natural hazards information available to all other councils in their region irrespective

[7] See http://www.mpi.govt.nz/news-resources/news/budget-strengthens-rural-communities-in-crises.

of information sharing constraints. It was recommended that section 106 should be amended to (i) reflect the full risk (likelihood and impact) of a hazard event rather than merely the likelihood of an event, (ii) include liquefaction and lateral spreading, (iii) ensure that resource consents[8] must be refused if there will be a significant increase in the risk associated with any natural hazard, and (iv) that consideration be given to including land use consents issued by regional councils. It was also recommended that Government should promulgate a National Policy Statement or National Environmental Standard on managing natural hazards.

The TAG recommendations were underscored by the findings of the Canterbury Earthquakes Royal Commission of Inquiry (2012) which also recognised that there is a need to improve the prevailing approach to natural hazard risk management and recommended that earthquakes and liquefaction be addressed at the level of RMA principles; that central government provide additional guidance to local government on earthquakes and liquefaction through better use of existing natural hazards planning tools and where appropriate develop new tools; and amend consent requirements so that earthquakes and liquefaction are more effectively addressed in practice.

In February 2013, the Government issued a Discussion Document (MfE 2013) that takes into account the TAG and Commission recommendations and made a number of suggestions for public consideration and feedback, including: Natural hazards could be added as a matter in the principles of RMA; section 106 could be amended to ensure that natural hazards are appropriately considered in subdivision and other land-use consent decisions; and the full risk of natural hazards—defined as the likelihood and magnitude of impacts—should be taken into account in these decisions. Government argues that such provisions would improve the effectiveness of consenting decisions and increase resilience. Importantly, the proposed amendments are not limited to earthquake risk and seek to ensure that the RMA promotes more integrated and consistent consideration of all natural hazard risks in local planning and resource consent decisions. Feedback on the Discussion Document was being reviewed at the time this chapter was written. It is, however, noteworthy that the Government's Discussion Document did not advocate a number of the TAG recommendations, including the need to ensure that the definition of natural hazards is consistently used in law or to promulgate a National Policy Statement or National Environmental Standard for managing natural hazard risks. Local Government New Zealand, which represents the national interests of councils and champions best practice in the local government sector, submitted a comprehensive commentary on the Government's Discussion Document (LGNZ 2013) and recommended with respect to natural hazard management that: natural hazards be added as a matter in the principles of the RMA; that section 106 be amended to ensure that all natural hazards can be appropriately considered in subdivision and other land-use consent decisions; and that magnitude and likelihood be taken into account in considering natural hazard risks. It also supported the TAG (2012) recommendation to strengthen linkages between and better integrate RMA and CDEM Act provisions relating to

[8] Any activities that are not permitted by the RMA, or by a rule in a city or district plan, require a 'resource consent' before they are carried out.

natural hazards risk management. It also recommends that the amendments take into account all natural hazards, explicitly noting the need to include flooding. It recommended that inconsistent use of the term 'natural hazards' in the RMA and CDEM Act be remedied. LGNZ (2013, p. 29) states that:

> "The proposed changes do not go far enough in prioritising avoidance over mitigation … [and] … do not require decisions (sic) makers to take account of risks to the community that endure (usually in perpetuity) regardless of whether the applicant accepted the risks at the time an application was made."

LGNZ (2013) also urges the Government to work with local government to further develop the proposals for better managing natural hazard risks.

The framing of natural hazard risk as outlined in the Government Discussion Document (MfE 2013) as well as the TAG (2012) report and LGNZ (2013) submission, amongst others, warrants more focused attention in the light of contemporary risk scholarship. The recommendation to amend the RMA to ensure that natural hazards risk is viewed as a combination of the likelihood and impact of an event, and not merely likelihood, is necessary but not sufficient. Leading risk scholars point out that not all natural hazard risks can be reduced to a likelihood and impact calculus or simple 'measurable uncertainty.' There is consequently a compelling need to deepen and extend the discourse about natural hazard risks in NZ and this is discussed further below.

What then are the main lessons learned from the 2004 Manawatu flood experience and subsequent efforts to address flood risk in the face of climate change?

## 10.3 Lessons Learned

Five key lessons are drawn from the preceding analysis.

First, disasters are 'focusing events' (Birkland 1997) that can reframe risk perceptions and create opportunities for reducing risk and mainstreaming climate change adaptation; but realising such opportunities is fraught. Despite being the most frequently occurring natural hazard event in NZ, flooding is typically treated with a degree of complacency. The 2004 flood experience shows that there is a window of opportunity after a major event for local communities to reduce long-term flood risk. But this window can close after a few years and this poses a serious problem for communities that face escalating flood risk due to climate change. The 2004 flood was a 'focusing event' that raised public awareness and concern that was translated into political will and local actions that aim to keep new development away from potential flooding. Adopting and sustaining such local commitment is, however, challenging in the absence of enabling directives at higher policy levels. The 2004 floods catalysed the regional institutionalisation of a hazard avoidance approach in the HRC One Plan that takes into account climate change effects. This regional 'focusing event' led to a number of local and regional actions to reduce flood risk; and reinforced the need for national guidance on flood and climate risk. But it did not compel national legislative or institutional reform to address this coupled challenge. The Canterbury earthquakes have, however, been a national 'focusing series of events' that have underscored the need for institutional and legislative reform to improve natural hazards planning and risk management in NZ. It remains to be seen whether the outcome of the government reform process will achieve this goal. The government proposes to elevate natural hazard risk to a matter of principle in the RMA, but has not indicated a willingness to adopt the TAG (2012) recommendation to introduce a legislative directive about how local government should address natural hazard risks. This recommendation together with several others by the TAG (2012) and LGNZ (2013) outlined above needs to be addressed to ensure better natural hazards risk management in NZ.

Second, there is a need to move beyond reliance on structural flood protection works to overcome the 'safe development paradox,' address the manifold drivers of flood risk and anticipate escalating flood risk given climate change. Historic choices to locate settlements on floodplains have compelled the construction of flood protection works and there is strong reliance thereon. In some localities, such as Palmerston North, the result is a 'safe development' paradox. The City Reach programme provides 0.2 % AEP (1 in 500 years) protection and so the residual risk is arguably minimal. However, the impact will be disastrous if, and arguably when, a flood event exceeds the City Reach design standards, especially with continued in-fill of residential dwellings behind the stopbanks. Other communities in the region must live with much higher levels of residual risk. The 2004 flood experience thus underscores the importance of having structural flood protection in place for communities already in perilous locations. This experience also shows that actions such as unsustainable land use practices in the hill country can exacerbate flood risk downstream and that non-structural measures are needed to complement structural protection. The SLUI demonstrates that adopting sustainable land use practices can facilitate flood risk reduction and enhance climate change adaptation prospects. Some of the actions taken by the Wanganui community, such as the design and layout of the Information Centre, are indicative of opting to 'learn to live with flood risk' rather than depend on structural protection. The 2004 floods thus demonstrate that a flood risk avoidance strategy is imperative for 'greenfield' development but that structural works are necessary for communities in low-lying areas and need to be complemented by non-structural measures. Vexing choices, however, have to be confronted in the case of at-risk communities that cannot afford suitable structural works and face exposure to escalating risk and/or repeat events.

Third, managing flood risk and adapting to climate change need to be framed as an integrated and adaptive process for building resilience and sustainability. The One Plan provides a policy framework that requires territorial authorities to reconcile long-term issues of public safety, resilience and sustainability with more immediate economic benefits without initiating or intensifying activities in areas exposed to high flood risk. The HRC adopted an integrated approach to flood risk and climate change by increasing the level of protection from 1 % AEP (1 in 100 years) to 0.5 % AEP (1 in 200 years) because of anticipated climate change driven increases in flood risk. The SLUI is predicated on and reveals the tangible co-benefits of regional initiatives that integrate conservation, flood risk reduction, climate change

**Fig. 10.6** Living at risk. (Source: Photograph by Manawatu District Council and sourced from Horizons Regional Council)

adaptation and sustainable livelihoods. Building and sustaining the partnerships and resourcing required to achieve the goals of SLUI is, however, difficult to achieve in practice.

Fourth, strategies to build flood resilient communities need to take into account wider societal trends and shocks that may have no apparent association with natural hazard risks. Many farming communities worked together incredibly well in response to the 2004 floods (John Keay 2013, personal communication). But the floods revealed that the resilience of rural communities in NZ has been 'hollowed out' over time due to systemic economic, social, demographic and institutional changes. Understanding and addressing the vulnerability exposed by the floods poses a significant challenge for the future. The legislative focus of the CDEM Act, for example, is on resilience rather than vulnerability and the two concepts are not merely flip sides of the same coin (e.g., Turner 2010). To date, most natural hazards research in NZ has focused on better understanding the factors that shape resilience. Much remains to be done to better understand and address the root causes and drivers of social vulnerability in NZ (Glavovic 2010; Glavovic et al. 2010).

Finally, the past focus on Readiness and Response needs to be complemented by sharper focus on Reduction and Recovery; and the permissive approach to land use decisions and reliance on mitigation measures need to be superseded by the avoidance imperative. The CDEM Act provisions were severely tested by the 2004 floods and, in general, proved to be effective in enabling a well-orchestrated response. Post-disaster recovery is a more complex challenge. Little attention has been focused on lessons learned from the 2004 flood recovery experience. The devastation of the Canterbury earthquakes has compelled more focused attention on Recovery and extensive research is now underway to understand and learn from this experience. Distilling and institutionalising the lessons learned is a challenge for the future. With regard to Reduction, formulating the One Plan has been contested and protracted but it demonstrates that institutionalising an avoidance and precautionary approach is achievable even when widely divergent local risk profiles, public perceptions and community characteristics have to be taken into account (Fig. 10.6).

What then does the 2004 flood experience, and lessons learned, reveal about barriers and opportunities for mainstreaming climate change adaptation into public policy and local planning and decision-making processes?

## 10.4 Barriers and Opportunities for Mainstreaming Climate Change Adaptation

Attention is focused on three vital arenas: understanding risk, the institutional milieu and professional practice.

### *10.4.1 Understanding Risk*

It is constructive to explore briefly the nature of risk because this concept is fundamental to the conceptualisation and practice of natural hazards planning and has important implications for adapting to climate change. This brief exploration will underscore the need to deepen and extend the discourse on risk in NZ and highlight the merits of framing natural hazards planning and climate change adaptation as an integral part of a deliberative governance approach for building community resilience and adaptive capacity.

Traditionally, risk has been defined as the probability (or likelihood) and consequences (or impacts) of a hazard event (after Knight 1921). This framing of risk dominates contemporary understanding about natural hazards risk in NZ, as reflected in the TAG (2012) review and government Discussion Document (MfE 2013) amongst others. Preoccupation with likelihood is obviously insufficient. Much can be gained by understanding both probability and consequences when making natural hazard planning decisions. But not all natural hazard risks can be reduced to simple 'measurable uncertainty' and especially not risks that are entangled with the 'super-wicked problem' of climate change (Levin et al. 2012 after Rittel and Weber 1973). Risk problems exhibit different characteristics according to varying levels of knowledge and/or competing knowledge claims; and understanding these characteristics is essential for assessing and managing risk (Stirling 2003, 2010; Renn 2008; Aven and Renn 2009, 2010; Assmuth 2011). Renn and colleagues (e.g., Klinke and Renn 2002, 2010, 2012; Klinke et al. 2006; Renn 2008; Renn et al. 2011) make a distinction between *complexity* (i.e., the difficulty of identifying and quantifying causal relationships between potential causal agents and particular observed effects), *scientific uncertainty* (i.e., the difficulty of predicting the occurrence of events and/or their consequences because of inadequate scientific knowledge) and *socio-political ambiguity* (i.e., different legitimate but divergent views about the same risk phenomena and their circumstances). Risk problems typically have varying combinations of these different characteristics and this has important implications for assessing and managing risk. When risk problems exhibit low levels of complexity, uncertainty and ambiguity they can be assessed using traditional quantitative risk assessment approaches and be routinely managed by regulators with minimal public involvement and deliberation. Risks characterised by high complexity but low levels of uncertainty and ambiguity need to be assessed through expert deliberation. Risk problems that have high levels of scientific uncertainty but low ambiguity require additional research to reduce uncertainty. If uncertainty cannot be reduced by research, precautionary and resilience-building strategies may be appropriate. Risks characterised by high levels of ambiguity require stakeholder and public deliberation to build a shared understanding of different values, beliefs and perspectives about the problem; and a common understanding of the risk problem (when evidence is controversial) and its acceptability and tolerability (when values are divergent). The difficult task of judging societal acceptability or tolerability needs to be undertaken in evaluating the risk. If risks are judged to be acceptable, risk can be managed by private actors. Government regulation is invariably required to deal with risk deemed to be intolerable. If the risk is tolerable, the appropriate management response needs to be decided on the basis of the risk characterisation. When complexity is high, expert-driven risk-informed management is appropriate. When uncertainty is high, stakeholders need to be engaged in a process of reflection and precaution-based management. When ambiguity is high, inclusive participatory discourse and deliberative management approaches are appropriate. Clearly, there is no 'one size fits all' for how to engage scientists, stakeholders, decision-makers and the public in assessing and managing tolerable risk—it depends on the characterisation of the risk problem. A contingent approach to risk assessment and management is thus necessary for dealing with the array of natural hazard risks facing NZ communities in this era of climate change. The prevailing framing of risk in NZ as 'measurable uncertainty' needs to be deepened and broadened to better understand and address natural hazard risks and to more effectively mainstream climate change adaptation into public policy and local planning and decision-making.

### *10.4.2 The Institutional Milieu*

This section portrays institutional barriers and opportunities for mainstreaming climate change adaptation as a triad of dilemmas that need to be resolved through natural hazards planning and public decision-making processes. The term 'institution' is used broadly to mean the formal and informal structures, processes and mechanisms that enable people to make collective decisions, e.g., religious institutions, the law, government, civil society or research. Institutions are made up of individual and organisational 'actors.' In this context, 'governance' refers to the steering activities of interacting government, civil society and private sector actors who make social choices (see Kooiman 2003), e.g., to build resilience, adaptive capacity and sustainability. The 2004 flood experience highlights three interwoven dilemmas.

First, the *uncertainty dilemma* needs to be confronted. Climate change involves varying levels of uncertainty or levels of knowledge and/ or competing knowledge claims—from 'known knowns' to 'known unknowns,' 'unknown unknowns' and 'unknowable unknowns'—and hence the imperative to adopt a contingent approach to assessing and managing natural hazard risks. Communities need to learn to live with uncertainty and even embrace change and surprise notwithstanding a desire for predictability and 'the familiar.' Building institutional adaptive capacity and resilience is vital for navigating the uncertainties of climate change—a challenging prospect when redundancy, diversity and keeping options open may appear inefficient and costly. A 'focusing event' like the 2004 floods can help to overcome the institutional barriers and inertia that are otherwise commonplace in seeking to build resilience and adaptive capacity at the local level.

Second, policy and planning processes face the *scale dilemma* in both geographic and temporal domains. The drivers of climate change, and measures to mitigate greenhouse gas emissions, are global in scope but impacts and adaptation occur at the local level (Adger et al. 2009). The 2004 floods reveal the complex local-national interconnections that arise in efforts to institutionalise flood risk reduction and adapt to climate change when community, regional and national interests are not easily aligned. Similarly, confounding short- and long-term considerations in addressing coupled flood and climate risk were exposed. For example, investing in improvements to regional flood protection works in anticipation of escalating flood risk due to climate change imposes significant short-term costs for uncertain deferred benefits.

Third, the uncertainty and scale dilemmas are compounded by the *responsibility dilemma* that bedevils efforts to mainstream climate change adaptation. Reflecting on the creation, bearing and sharing of flood risk in Australia, Handmer (2008) notes that private sector property development interests create the risk that is borne mainly by the public sector, small businesses and households. Moreover, reliance on Readiness, Response and Recovery does little to avoid exposure or reduce social vulnerability. The 2004 floods underscore this assessment and the merits of keeping people out of harms' way. However, as the narrative of the Te Matai precinct of Palmerston North reveals, giving effect to an avoidance strategy is far from simple and straightforward. Reconciling contending interests, rights and responsibilities lies at the heart of the climate change adaptation imperative (Adger et al. 2009). Resolving the responsibility dilemma raises vexed issues such as imposing national legally binding directives versus maintaining independent local decision-making autonomy; and intra- and inter-generational equity issues and responsibilities.

### *10.4.3 Professional Practice*

The foregoing barriers and opportunities also have implications for professional practice. The 2004 flood experience demonstrated the importance of building bridges between otherwise potentially balkanised professional domains—emergency management, land use planning, engineering and asset management, community development, the law, etc. Such balkanisation can occur within and between organisations at different scales and across sectors; and is reinforced when legislative coherence is under-developed (e.g., between the RMA and CDEM Act). Moreover, it extends into the arenas of science, policy and practice with the climate change science-policy interface being especially fraught (e.g., Jones et al. 1999). The process of developing the Manawatu-Wanganui CDEM Group and Group Plan helped to facilitate an effective flood response but the constellation of actors involved in recovery broadened and became more diffuse over time. It is difficult to sustain 'communities of professional interest' over time in a post-disaster situation. The 2004 floods and more especially the Canterbury earthquakes have brought to the fore the critical importance of developing cadres of professionals who are literate about natural hazard risks. These events have also stimulated new avenues of research and praxis that are likely to be built upon by the Government's plans to invest in the recently defined National Science Challenges[9], one of which is explicitly focused on building resilience in the face of natural hazards.

## 10.5 Priority Actions

Three priority actions are identified to overcome the obstacles and unlock the opportunities explored above to reduce natural hazard risks and facilitate effective climate change adaptation; each of which is discussed next:

First, institutionalise a national legislative directive to reduce natural hazard risks and build community resilience in the face of climate change. There is a long-standing compelling need for the NZ Government to provide local government with a clear legislative directive to reduce natural hazard risks in general and flood and climate risk in particular. Professional institutes, academics, researchers and other parties have recently recommended the adoption of a National Policy Statement and/ or National Environmental Standard to this end (e.g., Glavovic et al 2010; IPENZ 2012; NZPI 2012; Rouse 2012; Saunders and Beban 2012; TAG 2012; LGNZ 2013). The Government has initiated a reform process that includes a focus on natural hazards. But, to date, there has been no indication that the Government intends to introduce a legislative directive to manage natural hazard risks. This analysis of the 2004 floods and post-event actions underscores the imperative to institutionalise a national legislative directive that among other things determines how best to: (i) broaden and deepen understanding about risk; (ii) ensure legislative consistency in how key concepts such as natural hazards and risk are defined; (iii) coordinate different statutory planning processes; (iv) adopt a consistent approach to natural hazards planning and risk management in local government planning and decision-making processes; (v) prioritise avoidance over mitigation; (vi) shift from reliance on structural protection works, such as stop-banks, towards a combination

[9] See http://www.mbie.govt.nz/what-we-do/national-science-challenges.

of approaches, including non-structural measures, and enable communities to take into account climate change effects and chart adaptation pathways that help to resolve the triad of institutional dilemmas highlighted above; (vii) address the needs of communities exposed to high risk and where appropriate facilitate retreat in the face of escalating risk, e.g., communities who face repeat flood events but have minimal capacity to afford structural protection; (viii) enable local government to define hazard zones that take into account climate-driven escalating risks such as flooding; and (ix) take into account divergent community interests and needs.

Second, develop a capacity building programme to enable local government and communities to translate the above directive into practical reality. A national legislative directive on natural hazard risk management is necessary but not sufficient to ensure risk reduction and resilience at the local level. It needs to be supported by a capacity building programme that includes support measures for local government to translate such a directive into practical reality. It is beyond the scope of this chapter to outline the details of such a programme. But, among other things, it should include: (i) Appropriate resourcing (including financial, human and technical resources) to enable implementation of the directive. (ii) A locality-relevant public awareness campaign to build understanding about the nature of and linkages between natural hazard risks, resilience and climate change. (iii) A continuing professional development programme for those who shape community exposure to natural hazards and the associated risks, including community leaders, elected officials, property developers and professionals such as planners, architects, engineers, lawyers and community development specialists. The 2004 flood experience and post-flood actions demonstrate that reducing natural hazard risks and building resilience is complex and contested. Decision-makers must deal with issues that are characterised by uncertainty, changing circumstances and contradictory requirements that are not easily recognised let alone resolved. Change can be sudden or gradual; and surprise is inevitable. Promising solutions for prevailing problems can generate paradoxical outcomes (viz. the Te Matai precinct narrative) and/ or lead to more intractable problems downstream. The current framing of natural hazard risks as a combination of probability and consequence is better than merely focusing on the likelihood of an event. But analysing and addressing all risk problems as if they are 'simple' is misleading and can result in perverse outcomes including social amplification or irresponsible discounting of risk, and unwise public decisions that perpetuate controversy, compound conflict and legitimacy problems, and paradoxically escalate disaster risk (Renn et al. 2011; Klinke and Renn 2012). Conventional risk analysis and management approaches that work well for addressing 'simple risk' are ill-suited for addressing 'wicked problems.' Conventional approaches need to be complemented by more deliberative, open and flexible modalities of risk governance, public policy and planning in keeping with the complexity, uncertainty and ambiguity that characterise contemporary community issues. Until recently, however, scholars and educators in NZ, like elsewhere, have focused little attention on overcoming barriers and unlocking opportunities to reduce disaster risk and build resilience in the face of the 'super-wicked problem' of climate change. Continuing professional development opportunities are needed to better understand these issues and develop the necessary competencies and skills of decision-makers at local, regional and national scales.

Third, proactively explore opportunities to reduce risk and adapt to climate change in day-to-day local planning and decision-making. The SLUI clearly demonstrates the benefit of multi-objective initiatives that generate benefits for diverse stakeholders at different scales. There are many opportunities to integrate risk reduction, climate change adaptation and resilience into sector-specific and locality-specific initiatives and day-to-day decision-making processes. For example, considerable attention has been focused in recent years on improving water quality in the Manawatu watershed through, among other things, the Manawatu River Leaders' Forum[10]. Such initiatives open up opportunities to address flood risk in the face of climate change, and integrate emergency management, land-use planning, asset management and community development, and can be framed as an opportunity for collaborative and adaptive capacity building (van den Belt et al. 2013).

In conclusion, institutionalising the lessons learned from the 2004 flood experience, overcoming the barriers, unlocking the opportunities, and initiating the priority actions outlined above, necessitates a paradigm shift in prevailing thinking and practice. Natural hazards planning and climate change adaptation need to be conceptualised as a transformative practice of deliberative governance. The term 'governance' is not synonymous with government. Governance refers to the steering activities of interacting government, civil society and private sector actors as they make social choices through various formal and informal institutions, actor networks and practices (Kooiman 2003). These steering activities include norms, taboos, laws, policies and practices used by different social groups at varying scales to collectively address the array of natural hazard risks that communities face. Why a transformative practice of deliberative governance? First, notwithstanding a sound institutional framework, and the laudable efforts of many, the 2004 floods and more recent Canterbury earthquake series reveal systemic failures in how natural hazard risks are managed in NZ. Prevailing approaches need to be transformed so that an avoidance strategy is prioritised over mitigation and that opportunities are created that enable communities to better understand the natural hazard risks they face and choose pathways to reduce those risks. Second, deliberation is essential for addressing the 'super-wicked problem' of climate change and the specific barriers, opportunities and priority actions discussed above. Deliberation is discussion and debate that stimulates reasonable and well-informed views that can evolve through constructive dialogue, information sharing and participant exchange (Chambers 2003). It involves non-coercive communicative interactions that stimulate reflection on societal values, preferences and interests (Dryzek 2000). Public deliberation has considerable transformative potential for better understanding and addressing complex contemporary societal issues including risk reduction, resilience and sustainability (Fischer 2000, 2003; Baber and Bartlett 2005; Dryzek 2011; Glavovic 2013a, b). A transformative practice of deliberative

[10] See http://www.horizons.govt.nz/managing-environment/resource-management/water/manawatu-river-leaders-accord.

governance will help local communities better understand natural hazard risks and strengthen resilience and adaptive capacity in the face escalating disaster risk.

## 10.6 Conclusion

The February 2004 floods had a profound impact on the Manawatu-Wanganui region and stimulated a wide range of post-flood actions from the local to national level. These endeavours reveal five lessons for addressing coupled flood and climate risk through community resilience and adaptive capacity. First, disasters are 'focusing events' that reframe risk perceptions and create opportunities for reducing risk and mainstreaming climate change adaptation. Realising such opportunities is, however, fraught. Second, there is a need to move beyond reliance on structural flood protection works to overcome the 'safe development paradox,' address the manifold drivers of flood risk and anticipate escalating flood risk given climate change. Third, managing flood risk and adapting to climate change need to be framed as an integrated and adaptive process for building resilience and sustainability. Fourth, strategies to build flood resilient communities need to take into account wider societal trends and shocks that may have no apparent association with natural hazard risk. Fifth, the past focus on Readiness and Response needs to be complemented by sharper focus on Reduction and Recovery. Furthermore, the permissive approach to land use decisions and reliance on mitigation measures needs to be superseded by the avoidance imperative.

Barriers and opportunities to mainstream climate change adaptation into public policy and local planning and decision-making processes are explored in the context of three crucial arenas:

**Understanding risk** There has been a shift from a preoccupation with the likelihood of an event to a focus on probability and consequences or 'measurable uncertainty.' But some natural hazard risks exhibit high levels of complexity, uncertainty and/ or ambiguity that are not amenable to quantification. A contingent approach to risk assessment and management is needed, with the roles and responsibilities of scientists, stakeholders, decision-makers and the public best determined on the basis of the risk problem. The NZ discourse about risk needs to be deepened to better understand and address natural hazard risks, especially given the 'super-wicked problem' of climate change.

**The institutional milieu** A triad of dilemmas need to be addressed to overcome institutional barriers and unlock opportunities for reducing risk and strengthening resilience and adaptive capacity. First, in the face of climate change-driven escalating flood risk, communities need to maintain diversity and redundancy, despite yearning for predictability and the merits of efficiency and cost-saving measures—the *uncertainty dilemma*. A 'focusing event' can be the catalyst for institutionalising risk reduction, resilience and adaptive capacity. Second, divergent interests at different geographic and temporal scales pervade endeavours to build resilience and adaptive capacity—the *scale dilemma*. Third, reconciling contending interests, rights and responsibilities lies at the heart of natural hazards planning, risk management and climate change adaptation—the *responsibility dilemma*.

**Professional practice** Recent disasters in NZ have refocused attention on natural hazard risks, vulnerability and resilience and opened up new opportunities to bridge hitherto balkanised professional domains and compartmentalised ways of working; engage and sustain a vitally important cadre of professionals who are literate about natural hazard risks and resilience; and stimulate new avenues of research and scholarship.

Finally, three priority actions are identified: First, institutionalise a national legislative directive to reduce natural hazard risks and build resilience in the face of climate change. Second, develop a capacity building programme to enable local government to translate this directive into practical reality. Third, proactively explore opportunities to reduce risk and adapt to climate change in day-to-day local planning and decision-making. Ultimately, however, natural hazards planning and climate change adaptation need to be framed as a transformative practice of deliberative governance in order to institutionalise the lessons learned from the 2004 flood experience and subsequent events, including emerging lessons from the Canterbury earthquake recovery experience.

**Acknowledgements** This chapter draws on research conducted by the author over a number of years, insights gained from supervising several Planning student projects and many discussions with local professionals. EQC support of this research through *inter alia* a biennial grant is gratefully acknowledged. The views expressed in this chapter are, however, solely those of the author and do not in any way represent those of the EQC or any other organisation or individual. I sincerely appreciate the opportunity to explore these issues with local professionals over the years, as well as invaluable feedback on an earlier version of this chapter, including: Jeff Watson (HRC), Ian Lowe (HRC), Allan Cook (HRC), Peter Blackwood (HRC), Jonathan Ferguson-Pye and Matthew Mackay (Palmerston North City Council), Max Benseman (Wanganui District Council), Frances Sullivan (LGNZ), and Greg Carlyon (The Catalyst Group). I would also like to thank John Keay, a dairy farmer in the Rangitikei district, for generously sharing with my students and I his personal experiences during and after the 2004 floods which have provided us with deep insights into the lived reality of life in 'the deluge.' I also appreciate valuable feedback from Gavin Smith on a draft of this chapter. Lastly, I would like to thank Andrew Steffert (HRC) for his time and willingness to source photographs and maps from the HRC archives—you have helped to make visible the scale and impact of the 2004 floods.

## References

Adger, W. N., Lorenzoni, I., & O'Brien, K. L. (Eds.). (2009). *Adapting to climate change: Thresholds, values, governance*. Cambridge: Cambridge University Press.

Assmuth, T. (2011). Policy and science implications of the framing and qualities of uncertainty in risks: Toxic and beneficial fish from the Baltic Sea. *Ambio, 40*, 158–169.

Aven, T., & Renn, O. (2009). The role of quantitative risk assessments for characterising risk and uncertainty and delineating appropriate risk management options, with special emphasis on terrorism. *Risk Analysis, 29*(4), 587–600.

Aven, T., & Renn, O. (2010). *Risk management and governance*. Heidelberg: Springer.
Baber, W. F., & Bartlett, R. V. (2005). *Deliberative environmental politics: Democracy and ecological rationality*. Cambridge: MIT Press.
Benseman, M. (2013). Emergency manager, Wanganui District Council. *Personal Communication.*
Berghan, T., & Westlake, S. (2001). Floodplain management planning guidelines: Current thinking and practice in New Zealand, Wellington.
Birkland, T. A. (1997). *After disaster: Agenda setting, public policy, and focusing events*. Washington: Georgetown University Press.
Blackwood, P. (2013). Manager: Investigations and design, Horizons Regional Council. *Personal Communication.*
Bührs, T., & Bartlett, R. V. (1993). *Environmental policy in New Zealand*. Oxford: Oxford University Press.
Burby, R. J. (Ed.). (1998). Confronting natural hazards with land-use planning for sustainable communities. Washington, D.C.: Joseph Henry Press.
Burby, R. J. (2006). Hurricane Katrina and the paradoxes of government disaster policy: Bringing about wise governmental decisions for hazardous areas. *Annals of the American Academy of Political and Social Science, 604,* 171–191.
Canterbury Earthquakes Royal Commission. (2012). Final report volume 7, roles and responsibilities. http://canterbury.royalcommission.govt.nz/Final-Report-Volume-Seven-Contents. Accessed 26 July 2013.
Carlyon, G. (2013). Director: The Catalyst Group. *Personal Communication.*
Chambers, S. (2003). Deliberative democratic theory. *Annual Review of Political Science, 6,* 307–326.
Cook, A. (2013). Group manager: Operations, Horizons Regional Council. *Personal Communication.*
Day, T. J. (2005a). *Managing flood risk: The case for change*. Christchurch: Centre for Advanced Engineering.
Day, T. J. (2005b). *Managing flood risk: Draft flood management protocol*. Christchurch: Centre for Advanced Engineering.
Dryzek, J. S. (2000). *Deliberative democracy and beyond. Liberalism, critics, contestations*. Oxford: Oxford University Press.
Dryzek, J. S. (2011). *Foundations and frontiers of deliberative governance*. Oxford: Oxford University Press.
Dymond, J., & Shepherd, J. (2006). Highly erodable land in the Manawatu-Wanganui region, landcare research contract report: 0607/027.
Environmental Management Services Limited. (2004). *Review of Flood Response 2004*. A report prepared for the Horizons Regional Council.
Ericksen, N. J. (1986). Creating flood disasters? New Zealand's need for a new approach to urban flood hazard, *Water & Soil*, Wellington: Miscellaneous Publication No. 77.
Ericksen, N. J. (2005a). Hang-ups in flood hazard planning (Part 1). *Planning Quarterly*, (September 2005), *158*, 24–28.
Ericksen, N. J. (2005b). The hang-up in flood hazard planning (Part 2). *Planning Quarterly*, (December 2005), *159*, 28–32.
Ferguson-Pie, J. (2013). Senior policy planner, Palmerston North City council, *Personal Communication.*
Fischer, F. (2000). *Citizens, experts and the environment: The politics of local knowledge*. Durham: Duke University Press.
Fischer, F. (2003). *Reframing public policy: Discursive politics and deliberative practices*. Oxford: Oxford University Press.
Fuller, I. (2005). February floods in the lower North Island, 2004: Catastrophe-causes and consequences. *New Zealand Geographer, 61,* 40–50.

Glavovic, B. C. (2010). Future challenges and opportunities for natural hazards planning in Australasia, *Australasian Journal of Disaster and Trauma Studies*. http://www.massey.ac.nz/~trauma/issues/2010-1/glavovic2.htm. Accessed 26 July 2013.

Glavovic, B. C. (2013a). The coastal innovation paradox. *Sustainability, 5,* 912–933.

Glavovic, B. C. (2013b). The coastal innovation imperative. *Sustainability, 5,* 934–954.

Glavovic, B. C., Saunders, W., & Becker, J. (2010). Land-use planning for natural hazards in New Zealand: The setting, barriers, 'burning issues' and priority actions. *Natural Hazards, 54*(3), 679–706.

Goodwin, E. (2005). Lessons from mother nature. *Planning Quarterly*, September, 2–4.

Hancox, G. T., & Wright, K. (2005). Analysis of landsliding caused by the February 2004 rainstorms in the Wanganui-Manawatu hill country, southern North Island, New Zealand, Institute of Geological & Nuclear Sciences Science Report 2005/11.

Handmer, J. (2008). Risk creation, bearing and sharing on Australian floodplains. *International Journal of Water Resources Development, 24*(4), 527–540.

Hennessy, K., Fitzharris, B., Bates, B. C., Harvey, N., Howden, S. M., Hughes, L., Salinger, J., & Warrick, R. (2007). Australia and New Zealand. In M. L. Parry, O. F. Canziani, J. P. Palutikof, P. J. van der Linden, & C. E. Hanson (Eds.), *Climate change 2007: Impacts, adaptation and vulnerability*, contribution of working group II to the fourth assessment report of the intergovernmental panel on climate change (pp. 507–540). Cambridge: Cambridge University Press.

Horizons Regional Council. (2004). *Storm: Civil emergency storm and flood report, February 2004*. Palmerston North: Horizons Regional Council.

Institute of Professional Engineers New Zealand. (2012). *A safer New Zealand: Reducing our exposure to natural hazards*. Wellington: IPENZ.

Jones, S. A., Fischhoff, B., & Lach, D. (1999). Evaluating the science-policy interface for climate change research. *Climatic Change, 43,* 581–599.

Keay, J. (2013). Farmer, Rangitikei district. *Personal Communication.*

Klinke, A., & Renn, O. (2002). A new approach to risk evaluation and management: Risk-based, precaution-based, and discourse-based strategies. *Risk Analysis, 22*(6), 1071–1094.

Klinke, A., & Renn, O. (2010). Risk governance: Contemporary and future challenges. In J. Eriksson, M. Gilek, & C. Ruden (Eds.), *Regulating chemical risks: European and global challenges* (pp. 9–27). Berlin: Springer.

Klinke, A., & Renn, O. (2012). Adaptive and integrative governance on risk and uncertainty. *Journal of Risk Research, 15*(3), 273–292.

Klinke, A., Dreyer, M., Renn, O., Stirling, A., & van Zwanenberg, P. (2006). Precautionary risk regulation in European governance. *Journal of Risk Research, 9*(4), 373–392.

Knight, F. (1921). *Uncertainty and profit*. New York: Houghton Mifflin.

Kooiman, J. (2003). *Governing as governance*. London: Sage.

Lawrence, J., & Quade, D. (2011). *Perspectives on flood-risk management and climate change-Implications for local government decision making*. Wellington: New Zealand Climate Change Research Institute Report 07.

Lee, B-Y. (2010). Working together, building capacity-A case study of civil defence emergency management in New Zealand. *Journal of Disaster Research, 5*(5), 565–576.

Levin, K., Cashore, B., Bernstein, S., & Auld, G. (2012). Overcoming the tragedy of super wicked problems: Constraining our future selves to ameliorate global climate change. *Policy Science, 45,* 123–152.

Local Government New Zealand. (2011). *The local government case for a national policy statement for flood risk*. Wellington: Local Government New Zealand.

Local Government New Zealand. (2013). *Local government New Zealand in the matter of the discussion document: Improving our resource management system, Submission*. Wellington: Local Government New Zealand.

Lowe, I. (2013). Manager: Emergency management, Horizons Regional Council. *Personal Communication.*

Mackay, M. (2013). Policy planner, Palmerston North City council, *Personal Communication.*

Manawatu-Wanganui CDEM Group. (2009). *Civil defence emergency management group plan 2009-2014*. Palmerston North: Manawatu-Wanganui CDEM Group.

McClure, J., Wills, C., Johnston, D., & Recker, C. (2011a). How the 2010 Canterbury (Darfield) earthquake affected earthquake risk perception: Comparing citizens inside and outside the region. *Australasian Journal of Disaster and Trauma Studies*, 2011-2012. http://www.massey.ac.nz/~trauma/issues/2011-2/AJDTS_2011-2_McClure.pdf. Accessed 26 July 2013.

McClure, J., Wills, C., Johnston, D., & Recker, C. (2011b). New Zealander's judgements of earthquake risk before and after the Canterbury earthquake: Do they relate to preparedness? *New Zealand Journal of Psychology, 40*(4), 7–11.

Ministry for the Environment. (2008a). *Meeting the challenges of future flooding in New Zealand*. Wellington: Ministry for the Environment and the Flood Risk Management and River Control Review Steering Group.

Ministry for the Environment. (2008b). *Climate change effects and impacts assessment: A guidance manual for local government in New Zealand* (2nd ed.). Wellington: Ministry for the Environment.

Ministry for the Environment. (2008c). *Hazards and climate change: A guidance manual for local government in New Zealand* (2nd ed.). Wellington: Ministry for the Environment.

Ministry for the Environment. (2008d). *Preparing for climate change: A guide for local government in New Zealand*. Wellington: Ministry for the Environment.

Ministry for the Environment. (2010a). *Tools for estimating the effects of climate change on flood flow: A guidance manual for local government in New Zealand*. Wellington: Ministry for the Environment.

Ministry for the Environment. (2010b). *Preparing for future flooding*. Wellington: Ministry for the Environment.

Ministry for the Environment. (2013). *Improving our resource management system. A discussion document*. Wellington: Ministry for the Environment.

Ministry of Civil Defence and Emergency Management. (2004). The floods of February 2004. *Impact, 17,* 1–10.

NZPI (2012). NZPI Submission on TAG Report, Auckland: NZPI. http://www.planning.org.nz/Story?Action=View&Story_id=2455&Highlight1=Submission onTAG&Highlight2=Submission onTAG&Highlight3=Submission onTAG. Accessed 26 July 2013.

ODESC. (2007). *National hazardscape report*. Wellington: ODESC, Department of Prime Minister and Cabinet.

Quade, D., & Lawrence, J. (2011). Vulnerability and adaptation to increased flood risk with climate change-Hutt Valley household survey. *New Zealand climate change research institute report 06*. Wellington: NZCCRI.

Reid, P., Brundson, D., Fitzharris, P., & Oughton, D. (2004). *Review of the February 2004 flood event*. Wellington: MCDEM.

Reisinger, A., Wratt, D., Allan, S., & Larsen, H. (2011). The role of local government in adapting to climate change: Lessons from New Zealand. In J. D. Ford & L. Berrang-Ford (Eds.), *Climate change adaptation in developed nations: From theory to practice* (pp. 303–319). New York: Springer.

Renn, O. (2008). *Risk governance: Coping with uncertainty in a complex world*. London: Earthscan.

Renn, O., Klinke, A., & van Asselt, M. (2011). Coping with complexity, uncertainty and ambiguity in risk governance: A synthesis. *Ambio, 40,* 231–246.

Rittel, H. W. J., & Weber, M. M. (1973). Dilemmas in a general theory of planning. *Policy Sciences, 4,* 155–169.

Rouse, H. (2012). Flood risk management research in New Zealand: Where are we, and where are we going? GNS Science Report 2012/04.

Saunders, W. S. A., & Beban, J. S. (2012). Putting R(isk) in the RMA: Technical advisory group recommendations on the resource management act 1991 and implications for natural hazards planning. *GNS Science Miscellaneous Series 48*. Wellington: GNS Science.

Smith, W., Davies-Colley, C., Mackay, A., & Bankoff, G. (2011). Social impact of the 2004 Manawatu floods and the 'hollowing out' of rural New Zealand. *Disasters, 35*(3), 340–353.
Standards New Zealand. (2008). *Managing flood risk—A Process Standard, NZS 9401:2008*. Wellington: Standards New Zealand.
Stirling, A. (2003). Uncertainty and precaution: Some instrumental implications from the social sciences. In F. Berkhout, M. Leach, & I. Scoones (Eds.), *Negotiating change* (pp. 33–76). London: Edward Elgar.
Stirling, A. (2010). Keep it complex. *Nature*, *468*, 1029–1031.
Technical Advisory Group. (2012). Report to the Minister for the Environment's Resource Management Act 1991 principles Technical Advisory Group. http://www.mfe.govt.nz/publications/rma/tag-rma-section6-7/index.html. Accessed 26 July 2013.
Turner, B. L. (2010). Vulnerability and resilience: Coalescing or paralleling approaches for sustainability science. *Global Environmental Change, 20,* 570–576.
van den Belt, M., Schiele, H., & Forgie, V. (2013). Integrated freshwater solutions—A New Zealand application of mediated modeling. *Journal of the American Water Resources Association, 49*(3), 1–12.
Ward, C. (2005). The floods of 2004— looking back, *RM Update*, 16 (February), p. 5. http://www.maf.govt.nz/mafnet/publications/rmupdate/rm16/rm-february-2005.pdf. Accessed 26 July 2013.
Watson, J. (2013). Manager: Resource data, Horizons Regional Council, *Personal Communication*.
Waugh, J., Freestone, H., & Lew, D. (1997). Historic floods and droughts in New Zealand. In M. P. Mosley & C. P. Pearson (Eds.), *Floods and droughts: The New Zealand experience*. Wellington: New Zealand Hydrological Society.

# Chapter 11
# Learning From Analyses of Policy Frames and Informal Institutions in the Fire Management Sector of Victoria, Australia

**Karyn Bosomworth, John Handmer and Stephen Dovers**

**Abstract** If public policy sectors dealing with natural hazards are to play their part in climate change adaptation, the sectors must themselves be adaptive in their policies and larger governance contexts. Facilitating adaptive governance requires collaboration among many parties in the complex policy domain of natural hazard planning. To benefit from such collaborative processes as well as inquiries, reviews and experience, public sectors need to adopt a reflexive learning approach. Reflexive learning involves explicit consideration of current and alternate policy frames and informal institutions that structure a sector's governance arrangements, policy options and practices. The case for reflexive learning in enabling an adaptive governance is supported by lessons from a range of literatures. Therefore, this chapter does not discuss lessons from a particular bushfire event. Rather, it argues that for lessons to be learnt from natural hazards and adaptation planning, public policy sectors need a capacity to reflect upon and possibly change the policy frames and informal institutions that structure their current approaches. The chapter will argue this by discussing a study of policy frames and informal institutions of the fire management sector in Victoria, Australia, with a particular focus on the perspectives of middle or 'street level' bureaucrats.

**Keywords** Reflexive learning · Institutions · Policy frames · Public policy · Fire management

## 11.1 The Setting

South-eastern Australia has a reputation for being particularly fire-prone. Communities of this geographic region live in landscapes that have complex relationships with fire. Ecosystem health influences human well-being. Circularly, the health and thereby

K. Bosomworth (✉)
RMIT University, Melbourne, Australia
e-mail: karyn.bosomworth@rmit.edu.au

J. Handmer
RMIT University, Melbourne, Australia

S. Dovers
Australian National University (ANU), Canberra, Australia

B. C. Glavovic, G. P. Smith (eds.), *Adapting to Climate Change,* Environmental Hazards,
DOI 10.1007/978-94-017-8631-7_11, © Springer Science+Business Media Dordrecht 2014 

vulnerabilities of those systems depends on how we interact with and manage them, including how we interact with fire as both a tool and threat. Ecosystem diversity and vulnerabilities are shaped through suppression, exclusion and use of fire. Different land uses can fragment natural systems and thereby complicate or remove fire regimes from a landscape. The need to restore and 'reconnect' ecosystems also has implications for bushfire risk. Yet while fire plays a key role in the function of many Australian ecosystems, bushfires can present a real and grave threat to both social and ecological communities.

Even without climate change, fire management[1] is already a multi-stakeholder, multi-variable, multi-scale policy problem that has multiple partial 'solutions' and inevitable residual risk (Gill 2005). It is a complex policy problem not only because fire is a complex biophysical phenomenon (Cary et al. 2003), but because it is also a complex *social* phenomenon (Bradstock and Gill 2001; Gillen 2005). A holistic view of the governance of fire management reveals a diverse array of interacting policy issues that influence social and ecological vulnerabilities to bushfire. Those issues include: emergency management; public land and alpine management; water management; biodiversity conservation; forestry; tourism; health; education; community safety, engagement and development; fire suppression and prevention planning; land use planning; building standards; agriculture; and major infrastructure management including electricity, water, roads and rail. As a necessary consequence of the extent of skills and knowledge required in managing these issues, there is an array of organisations and agencies beyond fire management agencies involved in the fire management policy sector. Considering the potential implications of climate change for each of these issues, it becomes readily apparent that social-ecological vulnerabilities to bushfire are likely to change. This has potential to increase the complexity of fire management and drive an imperative for an adaptive approach to its governance.

## 11.2 Climate Change and Fire Management

A burgeoning body of research repeatedly highlights that climate change will affect fire weather. Increases in fire danger imply potentially higher fire intensities, longer fire seasons and more 'total fire ban' days (Hennessy et al. 2006; Lucas et al. 2007). Fire suppression effectiveness is limited to fires below about 3,500 kW/m intensity, which compares unfavourably with wildfire intensities that exceed 100,000 kW/m on 'blow up' days such as Ash Wednesday 1983 (Incoll 1994) and likely February 7th 2009. If we are to see more 'campaign[2]' fires such as those experienced in Victoria's Alpine regions in 2006/2007, personal and employer costs associated with volunteers including reliance or pressure on the good will of volunteers and their employers will increase (Hennessy et al. 2006). More very high and extreme

[1] Fire management refers to landscape or vegetation-based fires (i.e., not building fires).

[2] Campaign is defined here as a fire that remains burning for days or weeks.

fire danger days might also raise the chance of house loss. Analysis has shown that the majority of house loss in Australia's bushfires has occurred on days where the Forest Fire Danger Index (FFDI) exceeded 100 (Blanchi et al. 2004) and little house loss has occurred on days where the FFDI did not exceed 50 (Blanchi et al. 2010). While other important factors contribute to the degree of house loss, such as preparedness and actions on the day, any increases in fire weather conditions that mimic those seen in Victoria in February 2009 and Canberra in 2003 may well magnify the probability of house loss.

Looking beyond these biophysical projections, climate change also has the potential to alter underlying drivers of vulnerabilities to bushfire and fire regimes. Bosomworth and Handmer (2008) proposed several issues that may contribute to increased social-ecological vulnerability to a changing bushfire threat. These issues include implications for incomes from rural industries, water quality and availability, and their combination. For instance, there may be a negative impact on rural incomes, if climates in areas vulnerable to climate extremes become more hostile to farming (Lynn 2003; Mendelsohn et al. 2007). It is not clear that this will happen in Australia, but it is clear that climate change will likely have its most dramatic impacts on regions already stressed by high temperatures and low precipitation (Bosomworth and Handmer 2008). In south-east Australia, water availability may be reduced (Jones and Durack 2005) and may force consideration of alternate sources for suppression by agencies and householders. Water quality and yield may be affected if the 'protective' forests of our catchments burn and then suffer post-fire floods (Tryhorn et al. 2008). Post-fire floods, such as those seen in Gippsland in Victoria in 2006, may diminish recovery capacity of communities. An Insurance Australia Group paper (Coleman Undated) argues that there is potential for market erosion of insurance driven by the combined effect of increasingly severe climatic events and underlying socio-economic trends that can diminish investment viability and stress insurers, reinsurers, and banks to the point of impaired profitability and even insolvency. They suggest that communities will then have to deal with rising or unavailable premiums and the hardships of the events. When such issues combine to create 'double exposures' (O'Brien and Leichenko 2000), there may be an exodus from parts of rural Australia with implications for local economies through reduced demand for goods and services and an already decreasing volunteer base from an even further reduced population.

Bosomworth and Handmer (2008) also suggested that climate change has the potential to alter realistic expectations surrounding bushfire impacts, preparedness and suppression capabilities, and this has implications for communications and dialogue surrounding fire management. If the frequency of extreme fire danger increases, then translation of seemingly simple but highly sophisticated concepts that underpin programs such as 'stay or go' may become more difficult if not impractical, with no diminution of 'the tension between the ideology and practical need of asking those at risk to take more responsibility for managing their own risks' (Handmer and Tibbits 2005). Such challenges are only exacerbated in the context of tourism.

Explication and discussions of fire management complexities, in particular the limits of risk reduction (not elimination), have the potential to become increasingly demanding. When bushfires occur, 'blame' often gets placed on the hazard or the emergency managers rather than on the conditions of vulnerability that have resulted from a range of issues. These include lack of appropriate funding and resources for land management and community engagement, certain economic policies, poor private land management and inappropriate land use planning (Schipper and Pelling 2006; Handmer and Dovers 2008; Pelling and Dill 2009). Without open, explicit discussion of these issues, such 'blame games' will only escalate with a rise in major bushfires.

Finally, if a changing climate increases fire frequency ecosystem dynamics are likely to be altered, changing vegetation types, fuel loads and fuel types (Cary 2002; Mouillot et al. 2002). Precisely how this may manifest however, is not clear (Goldammer and Price 1998; Dyer et al. 2002; Sardans and Penuelas 2007). Concurrently, stressed human populations may place further burdens on ecosystem services through strategies to mitigate and adapt to climate change, including responses to an increasing fire threat. Actions to adapt to and mitigate the impacts of climate change will have other implications for fire management. The expanding markets in commercial forestry or native vegetation 'protection' for carbon sequestration will bring a concomitant need for protection from destructive fires, as well as consideration of how they may affect fire risk and regimes. Creation of habitat corridors to facilitate the adaptive capacity of ecosystems for example, raises similar challenges. These land use changes will produce valuable and additional assets in the landscape and likely pressures for protection of these, as well as existing assets.

Of greatest concern however, is that many of the most severe impacts won't stem from the influence of climate change on fire weather per se, but rather from synergistic interactions between climate change and existing pressures and trends that affect social-ecological vulnerabilities to fire (Thiele 2000; Thomas et al. 2004). For most public sectors, such a picture complicates assessment of impacts and development of policies to manage those impacts (UNFCCC 2004). Adapting to climate change requires a more sophisticated understanding of the underlying drivers of vulnerabilities and an adaptive approach to governance and policies. Without an adaptive approach to governance and policy, climate change will only serve to exacerbate the challenges of managing the complex social-ecological issues involved in fire management, and limit our capacity to adapt to climate change.

In the spirit of this argument, the remainder of this chapter will not discuss lessons from a particular bushfire event. Rather it will argue, through examples, that for sectors like fire management to learn lessons from planning, experience, inquiries and reviews (i.e., to be adaptive), they need a capacity to reflect upon and possibly change the policy frames and informal institutions that structure their current approaches to governance and policies. Moreover, consideration needs to be given to a range of alternate frames and ways of addressing, changing or working with a sector's informal institutions. Without such reflection, the disaster risk reduction (DRR) field may become better at things that maintain barriers to improved DRR and reduce our adaptive capacity. To explain this argument, the next section defines frames and institutions, and discusses their influence on the potential for learning and adaptation—drawing on lessons from a range of literatures.

## 11.3 Learning Lessons for Adaptation

This study drew on lessons from literatures surrounding public policy and administration, and organisational learning, as well as DRR and climate change adaptation (CCA). A central lesson from across those literatures is that for policy sectors to learn and adapt, there is a need for reflection upon the frames and institutions that structure a sectors' governance, policies and practices. In taking lessons from a broad range of literatures, this section will explain the concepts of policy frames and informal institutions as key influences on the ability for public sectors to learn from experience and to consider perspectives and impacts not yet experienced.

Underlying frames or ideas guide definition/s of a policy sector's fundamental goal, the 'problem' it is seeking to 'solve'. Policy frames are ideas structured around what is causing the problem or preventing attainment of a sector's fundamental goal, and thereby, what particular solutions or policy choices are appropriate. The notion of frames is often attributed to Goffman (1974) who denoted them as schemata of interpretation that enable people to locate, perceive, identify and label occurrences within their life space and the world at large (p21). Rein and Schön (1991, p. 263) define a frame as a perspective from which an amorphous, ill-defined problematic situation can be made sense of and acted upon. It is more a way of thinking about a problem or subject than it is an assemblage of facts (Fischer 2003, p. 103). "Like a window, we see the world through frames that determine our perspective while limiting our view to only a part of a complex world around us" (Creed et al. 2002, p. 36).

Consequently, different frames direct attention to different aspects of a situation, telling different stories about what is going on, and how the situation should be managed (Fischer 2003; Dewulf et al. 2004; de Boer et al. 2010, p. 464). For example, the framing of drivers of vulnerability influence the nature of policies and planning aimed at addressing vulnerability to various hazards (Blaikie et al. 1994; Turner et al. 2003; Birkmann et al. 2009; Handmer and Dovers 2008). This is most readily apparent in approaches that seek to manage the hazard alone, without consideration of underlying and complex drivers of vulnerabilities. One of the readily cited examples is the construction of floodwalls for flood management (Handmer and Dovers 2008). Others may frame vulnerability as an issue of exposure to the hazard. This framing typically underpins land use planning controls such as those that seek to prohibit development or at least, define development types within areas classified as bushfire prone. Yet others may turn to Blaikie et al.'s (1994) framing of vulnerability as being driven by various combinations of factors in our social, political and economic environments (as distinct from the natural environment alone), because these factors also structure our lives. While research regarding frames and framing is found throughout the social sciences (Druckman 2001, p. 226) and is currently finding increasing attention in the CCA field, in the natural hazards field there are very few authors who have directly considered the implications and influences of framing for hazards planning, policies and governance.

Policy frames have their greatest impact on public sector policies and governance when they are institutionalised—becoming part of the cultural-cognitive environment of a sector. Institutions include both formal and informal 'rules'[3]. They are different than frames because there is a collective or cultural sense of obligation to follow 'the rules', whether explicit, implicit or even unconscious. Because informal institutions have an obligatory dimension (Scharpf 2000; Helmke and Levitsky 2003), they can be every bit as important in shaping actors' behaviour as formally agreed procedures (High et al. 2005; Lowndes 2010, p. 68). They are often based in assumed ideas of appropriate behaviour rather than calculated returns expected from alternative options (March and Olsen 1989, p. 22). As norms and routines that guide social behaviour (Scott 2001; Ostrom 2005), informal institutions give rise to stable, recurring patterns of behaviour and thus structure collective actions (Balzer in Lauth 2000, p. 23; Lowndes 2010, p. 61). They are a form of systemic memory without which every issue, whether threat or opportunity, would have to be considered afresh and no action contemplated until basic decisions about proceeding were resolved (Considine 2005, p. 87). There is much discussion in the literature as to whether actors actively maintain institutions because of their material interests, or out of habit. Actors might also (unconsciously) follow an institution because they conceive no alternative, or they regard the alternatives they can imagine as unrealistic (Powell and DiMaggio 1991, p. 11) because they perceive those alternatives to be socially or culturally unacceptable. In public administration, institutions and their institutionalised frames are rarely (explicitly) explored because they are so embedded in the inherited assumptions of the public sector (Adams 2004, p. 32). Appreciating a sector's informal institutions provides further insight into processes of public administration and its adaptive capacity.

Frames and institutions are subtly different but closely interrelated concepts. Just as Blyth (2008) posits that ideas are the blueprints behind institutions, Scott (2001) describes frames as the *cultural-cognitive pillars* of institutions. What people think about the consequences of policies derives from their vision of a good society (Kahan and Braman 2006, p. 148). Epstein (2006) argues that a state observes collective norms or institutions because they are part of how it sees itself contributing to 'good government'. Similarly for a policy sector, its actors—bureaucrats, researchers, politicians, communities—may collectively observe its informal institutions because they are part of how the sector 'frames itself' as contributing to good government or the public good. Consequently, the impact of policy frames depends on the degree or 'thickness' of their institutionalisation (Riker 1980 in Lenschow and Zito 1998, p. 420)—to what degree a frame or frames is enacted through a sector's 'rules in use'.

Together, policy frames and informal institutions represent very strong influences upon the potential for public sectors to realise the depth of any lessons that may be learnt in both DRR and CCA planning. Institutions and institutionalised frames, create expectations of conformity and restrict the attention span of actors (governments, bureaucracies, businesses *and* communities) to approaches previously found to be appropriate (Koch and Hauknes 2005, p. 34). In doing so, institutions can

[3] Here institutions are defined as the formal and informal 'rules in use'. It does not refer to organisations.

create path-dependent actions and decisions, and resist new ideas because practices and structures are taken for granted, not questioned or not compared against alternatives (Powell and DiMaggio 1991, p. 192; Considine 2005, p. 105). However, learning and adaptation require a capacity for departing from the path-deterministic potential of institutions and institutionalised frames. Ignoring the structuring influences of frames and informal institutions ignores the administrative, cultural and often political realities that new ideas or approaches (such as adaptation) will necessarily have to interact and potentially challenge. As Macgregor (2006) argues:

> Without consideration of how the focus of a policy sector is (and can be) framed, 'symptoms' are more likely to be addressed rather than causes, and worse, actions may be misdirected because of a poor understanding of root causes.

### *11.3.1 Purposive Reflection to Learn Lessons From Natural Hazards Planning*

Without purposive reflection on frames and institutions, the success of mainstreaming any new concept or idea will be greatly dependent upon resonance of that idea with a sector's dominant or institutionalised frames. Past policies and frameworks have an important role in not only defining and problematising current issues (Parsons 1995, p. 230), but in providing channels through which current political process must flow (Fenna 2004, p. 136). Consequently, even analyses, inquiries and other lessons will occur within the dominant frame rather than through a multitude of frames that could provide a broader, more adaptive range of policy and practice.

Post-disaster inquiries, reviews and studies are often portrayed as 'windows of opportunity' for learning. Schmidt (2008, p. 307) argues that the ideas that underpin a public sector or policy generally sit in the background as assumptions or ideas that are rarely contested, *except* in times of crises. In natural hazards planning, such 'crises' are very often associated with a major hazard event. Allison and Halperin (1972) and more recently, Boin and t Hart (2003) and Boin et al. (2009), argue that crises generate framing contests to interpret events, causes, responsibilities and lessons in ways that suit the political purposes and visions of future policy directions of government actors and their critics. Explicit questioning of the fundamentals of a public sector is difficult in crises (and post-event responses to them), because managers under intense public scrutiny—as managers dealing with the impacts of natural hazards often are—may be inclined to adopt a 'fortress' mentality to shut out criticism (Weber and Khademian 2008, p. 342) or simply to cope with the stress. Similarly to fundamentally questioning a sector's raison d'etre, adaptive approaches to policy and administration may be viewed as a threat to existing programs and management (to the dominant frame), rather than as an opportunity for improvement (Folke et al. 2005). This may be particularly so where key actors have hard-won, long-term investments (financial, temporal and reputational) in current problem frames (Connor and Dovers 2004). Consider for example, the reputational and financial investment in long-lived infrastructure such as levees or aerial fire-fighting equipment. However, adaptation demands purposive reflection on a sector's frames and institutions not just in times of crises, but in the course of improving its efforts, let alone in proactively adapting to climatic changes and taking advantage of the proverbial 'window of opportunity'.

From a reflexive learning perspective, a diversity of views are a strength where different conceptions of issues give rise to contention and debate that ideally would enrich our understanding of policy issues and options. In adapting to climate change, consideration of a multitude of options would provide the public sector with robust[4] and adaptive policy suites and governance[5]. For this kind of reflexive learning, sector's need to move beyond 'groupthink', they *need* heterogeneity more than uniformity, disagreement more than consensus, creativity more than predictability and frankness more than conformity (t Hart and Kroon 1997, p. 103). Analyses of policy frames and informal institutions can not only aid reflection upon how the sector's policies may be constrained (or not) but can also support consideration of how to approach adoption of new ideas and approaches. A reflexive learning approach could generate different visions about central issues (Blyth 1997, p. 246; de Boer et al. 2010, p. 1), identify options towards addressing a sector's underlying assumptions, and identify different governance, policy and practice options. The following section will discuss results from research that examined policy frames and informal institutions in the fire management sector of Victoria, Australia. Also discussed are the implications for the sector's reflexive learning capacity.

## 11.4 Barriers and Opportunities

This section attempts to make the preceding theoretical argument practical by discussing some of the policy frames and informal institutions identified in a study of the fire management sector of Victoria, Australia. The research findings drew from the sector's policy practitioners and bureaucrats. As some of the policy experts within political systems, bureaucrats play particular roles in the development and shaping of understandings of policy issues and alternatives (March and Olsen 1989, p. 18)—of how the sector's issues are framed. Middle and street-level bureaucrats work at intersections between strategic and on-ground actions. They have key roles in co-ordinating a diversity of stakeholder inputs, facilitating negotiation between different values, objectives and demands and they make (or at least support) policy development and implementation choices. They have substantial discretion, can control resources, exercise power and are active participants in the preparation, formulation, implementation and enforcement of public policy (Olsen 2008, p. 26).

[4] A robust policy strategy is one that performs well when compared with the alternatives across a wide range of plausible futures, a strategy that need not be the optimal one in any future; but will, however, yield satisfactory outcomes in both easy-to-envision futures and hard-to-anticipate contingencies (Popper et al. 2005).

[5] Some authors, such as Rosenthal (1991) and Ostrom (2008), have argued for 'polycentric' approaches that appear to have reflexive, problem-solving potentialities because they can include multi-agency and departmental checks and balances, and informed and interested citizens and public officials.

Through their work practices, bureaucrats or policy practitioners reinforce the policy frames and institutions that structure a sector's governance approach and policy options. Yet outside of the policy analysis and public administration fields, the role of bureaucrats in developing policies and programs tends to escape analysis (Hajer and Wagenaar 2003, p. 88; Miller 2004; Schmidt 2008). Taking the perspective of a sector's bureaucrats provides greater insight into the often contested and always negotiated nature of policy processes, rather than an examination of policy documents alone. Consequently, the analysis presented here centres on the way in which bureaucrats in Victoria's fire management sector frame the fundamental goal of fire management, and on informal institutions that seemingly influence the options these people felt they had in enacting or realising their framing.

Two themes were common among the ways these bureaucrats framed fire management. The first was the idea of fire as a natural and inevitable landscape phenomenon. Equally ubiquitous was the image of 'good fire' (referring to ecological or fuel reducing, and usually deliberately lit) and 'bad fire' (a bushfire that threatens people and assets). However, two distinct 'master-frames' were identified. A master-frame is a metanarrative (McAdam 1994 in Koenig 2005) or a more enduring cultural theme (Gamson 1988 in *ibid*). In this study, the overwhelming majority of policy practitioners framed the fundamental goal of fire management using a sustainability master-frame[6]. A handful of participants[7] used an emergency or risk management master-frame. Characterisations of governance and policy purpose and options differentiated these two master-frames. Both sustainability and emergency management are globally popular storylines or cultural themes. Typical of policy frames, they each direct attention to different aspects of the situation and tell a different story about what is going on and what should be done (Dewulf et al. 2007, p. 53; Fletcher 2009; Isendahl et al. 2009; Spence and Pidgeon 2010).

## *11.4.1 A Sustainability Frame (Humans as Part of Nature)*

The majority of research participants framed fire management as an ongoing need to balance social and ecological objectives in order to achieve sustainability. It conceived humans as a part of nature, rather than nature as something from which to protect humans. Threaded throughout the storyline of this frame was the idea that because humans are a part of nature, how we choose to live with or against it influences not only our vulnerability to fire but our overall well-being. The language of this frame was accommodating in its rhetoric, with the repeated idea of finding an idealised balance. Biodiversity and natural resource management goals were viewed as legitimate as managing the very real threat that bushfires can pose to life

[6] It is necessary to stress that presentation of two master-frames should not be taken to suggest absolute delineation between the conceptions. Some of the arguments underpinning the master-frames drew upon similar normative orientations. Nonetheless, distinctions between the underlying sub-frames were enough to impose an analytically useful, yet artificial, structure.

[7] It is necessary to qualify that the authors can only talk about those who participated in the interviews or survey.

and assets, and that these two central drivers of fire management policy required negotiation, debate and 'balance'. However, those employing a sustainability framing expressed little certitude that the 'balance' was attainable. In this master-frame, one of the major barriers to achieving a sustainable way of living with fire was that many of the drivers underlying the sector's policy context are not readily identifiable, that many are unknown. Therefore, there are no sole, simple or permanent 'solutions' to achieving a sustainable balance.

> Achieving a balance between fire-safe and resilient human communities and sustainable and resilient ecosystems...between short-term and long-term objectives...a bit like the Holy Grail...I don't know how we'd do it, I really don't, but I think that's actually what management is about.

Through a sustainability frame, the choice of 'solution' (policy option) is a value judgement and a question of trade-offs, and the inherently political nature of such trade-offs contributes to the sector's complexity. Several respondents also suggested that the sector's complexities are exacerbated *because* people want a simple solution.

> The struggle in all of this—it isn't a scientific question. It's a how much do we value this part over this part over this part. Where do we juggle, how do we make the decisions that juggle at different points on the landscape?
> A lot of the issues and dilemmas that we're going to be facing in the future are going to be about what we value, not what we know.

The moral claims of this frame surrounded both social and ecological justice. Social justice included human security conceived as an ability to live safely with fire through vulnerability reduction, supported by an understanding of fire's role in ecosystem dynamics and an overall improved appreciation for 'the environment'. Ecological justice related both to fire's role in many ecosystems but also to broader objectives of land and native vegetation conservation. This frame's storyline was predicated on assumptions of diversity of views, values and knowledge within communities, and of complexities inherent in working with this diversity. For several participants, the inherent struggle within the sector is a broader reflection of our struggle to understand and live with our environment and landscapes, including how we have changed them.

> We are not living in a landscape the same as when Europeans turned up. The bush is no longer 'natural' because we have to put out lightning strikes to protect life and property. The presence of 'man' has affected the type of landscape we live in forever and we have to understand that better.

The narrative of the sustainability framing was one of living with an environment in which fire is one of a number of natural, albeit potentially life threatening, facets. Within the sustainability frame, the knowledge needed to achieve that goal is considered limited and the goal is perhaps unattainable. Parallels between the underlying rhetoric of much of this frame and the well-known 'harmony with nature' theme (Jenkins-Smith and Sabatier 1993; Koenig 2006) were obvious. Mirroring that globally popular storyline, the sustainability frame arguably reflects broader societal shifts towards sustainability or 'ecological modernisation' (Dovers and Handmer 1993, p. 205; Eder 1996; Dovers 2004).

### *11.4.2 An Emergency/Risk Management Frame (Defending Ourselves From Nature)*

The second but less espoused master-frame was of fire management as an emergency or risk management challenge. The basic operating assumption of this frame was that human safety is paramount to all else, and that because fire can threaten that safety, any actions to defend humans from that threat are morally superior to any other issue[8]. Fire was described as something to be battled, prevented, suppressed or controlled, automatically portraying fire as a negative force; as an enemy to be subdued (Wuerthner 2006, p. 60). A language of risk management permeated this framing, focussing attention on the risk of human safety being impacted by fire. In this frame, 'the causes' of bushfire risk were conceived as identifiable and quantifiable—mostly based on a physical location in the landscape. Therefore management of bushfire risk was depicted as equally identifiable and a matter of rational choice.

This emergency management master-frame seemed based on narrower ideas of disaster management than those expressed in many key DRR documents, such as the Hyogo Framework. Although there were expressions surrounding collective or community-based approaches to managing bushfire risk, those employing an emergency management frame used a defensive language that was most concerned with hazard management essentially through fuel reduction or suppression activities. These participants argued that with more funding and resources agencies could ensure 'the community' is aware of bushfire risk (and accepts some responsibility for dealing with that risk) and 'assets' could be protected through suppression and fuel management. Agencies were consistently portrayed as being readily able to define risk—based on exposure to hazard—because of their knowledge and understanding of fire behaviour, fire suppression and fuel management. It was because of this supposed central knowledge that the agencies—the authorities—were also framed as being 'in control' in the face of the threat of fire.

> It's an anathema to our thinking in this organisation that anybody could die or we could lose anything.
> In the good old days we were the shining yellow knights on the big red charger and "don't worry your pretty little head, we'll look after you".

Yet within this frame, there was an argument that people cannot expect to be defended by the 'authorities'; that communities must take responsibility for defending themselves. While perhaps a reflection of the reality of limited resources, this argument continued the theme that if communities simply understood the 'logical knowledge' that the agencies have, they would gain an understanding of how to defend themselves from a bushfire and prepare 'appropriately' by managing fuel.

---

[8] In Maslow's hierarchy of human needs, safety is a fundamental requirement (Aucoin 2006, p. 72).

In this frame, complexity was engendered by the number of organisations and thereby issues that require consideration, which removed the focus from the safety objective. Consequently, while arguing for integrated approaches, the purpose of collective approaches was generally about achieving effective suppression, fuel management and community engagement to encourage and accept fuel reduction activities[9]. Assumptions and predispositions toward this rationale were evident in technocratic interpretations of risk management that peppered many of the interviews and surveys.

> It is not rocket science to understand that our fire management planning needs to be risk based; we need to plan collectively to manage and to develop treatment options to treat a risk, and it needs to be tenure blind ('It' being fuel).

The concept of emergency or risk management seemed to serve as discursive glue that many of these bureaucrats rallied around, even if when pressed it turns out they have different understandings of the terms (Lees 2004, p. 102). While an emergency management master-frame may reflect a particular bureaucratic rationality that tends toward order and process, it may also reflect broader societal notions (and thereby demands of government) that with enough of 'the right' information, the 'problem' can be solved[10].

### *11.4.3 Informal Institutions and Their Influence on the Sector's Capacity for Reflexive Learning*

Reflexive learning requires that governance and policy processes use multiple frames or perspectives. Part of this research sought to identify the degree to which this may be occurring in the fire management sector of Victoria. An institutional analysis suggested that while most participants framed fire management in sustainability terms, the sector's informal institutions corral options into the emergency management frame. In other words, the emergency management frame is highly institutionalised and the idea or frame of sustainability has only a slight bearing on current policy options. A key indication was that while most bureaucrats framed the sector in terms of sustainability, when asked about dealing with the implications of climate change for the sector, the majority utilised risk management language. This is perhaps unsurprising as risk management is already a strong discourse within the DRR arena, and is increasingly adopted in the adaptation field. Risk management provides a very powerful discourse for policy practitioners and researchers in conceptualising a structured and clear management approach to the complex challenges of DRR and CCA. The notion of managing the risk of 'extreme events' is increasingly advocated in both arenas.

An institutional analysis provided the second indication that the emergency management frame is highly institutionalised within this fire management sector. First-

[9] The language of disaster risk reduction was not expressed throughout any of this data.

[10] Fire management is not alone in this belief. Much of the climate change literature argues for production of 'better' science driven by the idea that this information would encourage action.

ly, the analysis suggested that participating bureaucrats share a sense of obligation to be seen to responding or acting and that those actions must be infallible.

> The job that's been done in the last 15 years or so isn't perfect by a long way, but it's getting tougher. The political lack of forgiveness, I mean they're meant to get it perfect or else, so we tend not to take the calculated risks that are inherent in properly managing fire regimes, because of fear of legal sanction.

Perceived as constraints on the policy options available to the sector, these 'rules' were often attributed to a broader societal expectation that agencies can control all bushfires and the ideal of evidenced-based policy that is clearly part of the rational model of public administration. The presence of these overarching institutions provided insights into the presence of several other informal institutions.

One of the strongest institutions was that of scientism (Dupre 2001): the idea that science is the ultimate, neutral arbiter, because it can provide 'the solution'. Irrespective of the master-frame employed, most participants argued that science would support their position. The implication of this institution or 'rule' for the sector's adaptive capacity is twofold. First, it raises the question of which (and whose) science is used to inform policy options, such as whether options are limited to those with a technical or engineered basis. Without reflection, the scientific discipline used to inform fire management's policy options may be quite narrow in scope. This is not to discount the essential role the sciences play in DRR, fire management and CCA, but rather to highlight the second implication. The second implication is that science in general can inform our choices—indicating the implications of various courses of action—but it cannot make those choices because many responses to public policy issues are about value judgements.

The institutional analysis also suggested that within this sector, knowledge of fire fighting and fuel management is bestowed greater legitimacy than other bodies of knowledge.

> They are 'old' fire management. Policy people, community engagement, business improvement didn't exist 20 years ago. So "they're fire management and they know how to put fires out", so they're the only ones who know anything. I know that's a bit harsh, but that's a perception that we get.
>
> Fire related views are accepted and pushed by people who have a particular perception and often don't come from a fire background. They don't understand the fire business from a suppression nor prevention perspective they are more involved in policy and budget related tasks and just follow policy when it comes to the nitty gritty….
>
> I was there for 11 years—they still had difficulty in accepting my expertise because of my background. I wasn't a forester.

The institution or 'rule' implied that knowledge that may challenge this 'conventional wisdom' is either illegitimate or even 'taboo'.

> It seems like it's a parallel with not being able to stand up and say there are physical limits to suppression. To be able to say basically that fuel reduction burning is hard for all these reasons…Because there is no right answer. It is based on a number of different people's objectives and quite different values…There's no permission to have that conversation.

The parallels with the emergency management frame are obvious—fire-fighting and hazard reduction are central to managing bushfire emergencies—and indicated again an institutionalisation of that frame. Each of these institutions contains an assumption that factors contributing to the sector's challenges and decision options are readily calculable. This parallels the implicit depiction within many risk management approaches of hazards as amenable to conventional procedures of calculation, management and control within the capacities of established institutions (Pidgeon and Butler 2009, p. 679). Such institutions often exclude many social, ecological and political considerations because these types of underlying factors or consequences are conceived to be indescribable, unamenable to use of probability calculations or cost-benefit analyses, and lie outside prevailing scientific risk-knowledge (Jasanoff 1993; Wynne 2002, p. 469). This is anathema to the concept of reflexive learning, to seeking out a multitude of frames.

### *11.4.4 Barriers to Learning*

While language such as threat, hazard and danger validly leads to prioritisation of strategies focussed on risk and harm minimisation, a singular focus on threat preparation and emergency response tends to exclude consideration of other values, choices and struggles inherent to both fire management and adaptation separately, let alone in combination. These value choices and struggles include how, why and where we live in landscapes, as well as to which risks we give attention and with what (and whose) approaches. Under the dominance of a hazard management rubric, policies and practice are more likely to be instrumentalist in nature, ignoring underlying drivers of vulnerability and resilience in social-ecological systems and limiting options to physical, technical or engineered approaches. This has direct implications for community-resilience, as well as adaptive capacity. On one hand, there is an argument that if communities only understood the requisite, supposedly obvious and rational knowledge that the authorities have, then they would take actions to protect themselves from a bushfire hazard. On the other hand, the institutionalised notions of credible knowledge and skills, reinforces notions of expertise, which may actually be disempowering for community members. Consider the conceptual differences between needing specialist fire trucks and aerial equipment to suppress a bushfire and preparing to defend ones home from the same bushfire with mops and buckets.

In the context of adaptation, an emergency management frame likely limits the perceptual scope of policy and management options to paradigms that imply the calculability of risk to exclusion of other potentially incalculable, complex factors. It also tends toward an implied 'singular' solution—fuel management, community education, more fire suppression equipment. Yet singular solutions by definition, distinguish definitive positions around which stakeholders immediately gravitate, often precluding collaboration, innovation and creativity (Coffman and Umemoto 2010, p. 599). Furthermore, such panaceas are fallacies in complex, dynamic social-ecological systems (Meinzen-Dick 2007; Ostrom et al. 2007). A difficulty of bounding 'legitimate' knowledge to that which is quantifiable not only creates perceptual tensions between agencies and communities, and between policy areas, but traction for adopting adaptive approaches to governance and policy may be especially difficult. This is particularly so where the knowledge required is *not* quantifiable, immediately demonstrable nor even perhaps known. Climate change associated uncertainties are unlikely to be assuaged by a 'requisite' knowledge that is solely drawn from readily quantifiable data. Climate change will undoubtedly increase the need for us to make value judgments. As stated earlier, science can help inform such judgements, but it cannot make choices for us.

Recourse to rationalist, technocratic approaches is salutary for efforts aimed at evolving the concept of vulnerability reduction in the fire management sector, let alone in CCA. It is even more salutary for any effort to move the sector from an approach that implies a mythical panacea, toward one of reflexive learning (in this, fire management is most certainly not alone). This is not to discount the imperative of emergency management. Rather, this study's results suggest that despite use of the phrase 'adaptive management' in several policy documents, Victoria's fire management sector may not yet have shifted *culturally* to embrace the concept of adaptive management—a concept focused on reducing underlying drivers of vulnerabilities to natural hazards.

### *11.4.5 Opportunities for Learning*

The presence of at least two master-frames, each with their own logics and foci, suggests that this sector has at least two perspectives from which to explore a broader range of options for policy and practice in confronting a changing climate. From a reflexive learning perspective, differences between these master-frames can be viewed as strengths rather than conflicts. Overlaps between the frames can serve as a basis for discussion aimed at transforming circular debates between the various perspectives into an expanded and therefore, potentially more adaptive suite of policies and practice. For example, in the face of the complexities and uncertainties of climate change, an overlap in framing of land use planning as a major driver of risk presents itself as an obvious focal point for the sort of discussion that could actively make explicit the underlying policy frames.

## 11.5 Future Action

If sectors are to be adaptive in their policies *and* governance, they must give attention to facets beyond formal governance and institutional arrangements. The research presented here indicates a least two steps a public sector might take to building its own adaptive capacity in order to support the adaptive capacities and sustainability of our communities: deliberate reflection on underpinning policy frames and informal institutions, and addressing (or beginning to discuss) them in terms of adaptation.

Sectors could undertake a kind of deliberative reflexive practice to identify additional frames and encourage an expansion of the frames within their purview. For example, Victoria's fire management sector could build upon some initial work with ideas of cultural burning and learning from indigenous knowledge and relationship with fire. It could also build upon some of its more innovative programs based in principles of community development. Reflexive practice may at least identify some key areas of frame convergence.

This reflection must include the frames that direct research, which goes to inform governance and policies. Much of the fire management research literature also follows an institutionalised framing of fire management as an issue of risk or emergency management. For example, although there is an important growing body of social research, the fire management literature is still dominated by work examining emergency response planning and implementation, and on fire behaviour. Very little of the fire management literature addresses the policy sector as a whole (some of the notable exceptions being Dombeck et al. 2004; Gill 2005; Lavorel et al. 2006). In Australia, there is scant fire related research literature that takes a DRR perspective and considers underlying drivers of vulnerability. Where it does exist, such work often concentrates on a singular policy approach, such as 'community engagement'. This fact alone suggests that for research to encourage adaptation in the fire management sector, the research community itself needs to conduct more holistic work—expanding their own often disciplinarily defined frames. In the context of CCA, most of the published literature has, to date, focussed on climate science, the likely increase in weather conducive to bushfires and the narrowing of windows of opportunity for prescribed burning. There is a nascent body of work exploring the implications of climate change for biodiversity through its interaction with fire regimes. While research that considers social-ecological adaptation to climate change in the context of fire management is virtually non-existent. As has been repeatedly espoused in much of the literature, there is much to be gained in an exchange and mutual learning between the fields of DRR and CCA.

Another more difficult, but arguably necessary action DRR sectors might take to build their and their communities' adaptive capacity is to begin to address their informal institutional landscape. While reflexive learning asks us to expand our purview, to actively consider and perhaps include a more diverse array of frames, it may be challenging to introduce different frames that challenge or do not 'fit' with the sector's current informal institutional landscape. For example, the current study bespoke an institutional landscape underpinned by broader societal ideas of control over nature and the well-worn rational model of public administration. Consequently, attention is given to approaches and frameworks perceived to have a degree of certainty or control, including an emergency management frame focussed on controlling 'the hazard' through suppression or fuel management. Exploring the sector's issues from a polar-opposite frame that (hypothetically) suggests fuel reduction contributes nothing to overall community safety or resilience would likely be ignored and probably scorned within this sector.

A key lesson from this research is perhaps an old one. There is no single answer or solution to the societal challenges of DRR, let alone adapting to climate change. Certainly, the fire management sector has a suite of policy options within its portfolio. It would be encouraging to see this suite presented and analysed as a whole more often. Perhaps even the idea of 'solution' is misleading when we are living in a changing climate. Besides encouraging a war-like narrative that conveys a battle for control over nature, bolstering an emergency management capacity is not the end of the adaptation story—nor is it the beginning. It does not encourage us to question where and how we live, and how that may have to continually adapt and change into the future. It certainly ignores the social construction of risk and vulnerabilities. Addressing institutionalised notions of control over nature will likely require a society-wide discussion about the limits of control and the fallacy of the idea of certainty—something that is becoming increasingly important as our climate changes. Given that DRR could encompass a social-ecological approach to living sustainably in a landscape of numerous natural hazards, this field may have the emotive power at least to initiate such a conversation.

Using language that draws from and builds upon shared ideas or frames, such a conversation could begin to expand the range of existing policy perspectives. Information presented in forms that affirm rather than denigrate individuals' policy frames and beliefs are less likely to be resisted (Kahan and Braman 2006, p. 165). As such, identification of areas of frame convergence or overlap is pertinent. Human safety should be an equally explicit and central part of the discussion as ecology. The concept of a sustainable future could combine risk management efforts with those aimed at vulnerability reduction and ecological conservation. Connection of different frames into a jointly meaningful story can generate motivation and commitment for collective action and, provided participants are able to deal with the variety, confers the potential for crafting innovative solutions (Dewulf et al. 2011, p. 52). Visioning of a sustainable future might include increased adaptive capacity and reduced vulnerabilities of our social and ecological systems to inappropriate fire regimes, including bushfires.

Finally, in policy sectors such as fire management, where hazard events are touted as 'windows of opportunity' for learning, such events will only provide opportunities for reflexive learning and adaptation when the learning process is genuinely girded by attempts to appreciate the underlying drivers of vulnerabilities and the frames and institutions that guide overarching objectives. As the findings presented in this chapter have indicated, without a frame reflective basis, so-called 'windows of opportunity' may simply serve to reinforce existing policy paradigms that have yet to embrace the nexus between natural hazards planning and climate change adaptation.

## References

Adams, D. (2004). Usable knowledge in public policy. *Australian Journal of Public Administration, 63*(1), 29–42.

Allison, G. T., & Halperin, M. H. (1972). Bureaucratic politics: A paradigm and some policy implications. *World Politics* 24 Supplement: Theory and policy in international relations(Spring), 40–79.

AuCoin, L. (2006). Don't get hosed: How political framing influences fire policy. In G. Wuerthner (Ed.), *The wildfire reader: A century of failed forest policy.* Washington: Island Press.

Birkmann, J., Tetzlaff, G., & Zentel, K.-O. (2009). Addressing the challenge: Recommendations and quality criteria for linking disaster risk reduction and adaptation to climate change. In

J. Birkmann, G. Tetzlaff, & K.-O. Zentel (Eds.), *Dkkv publication series 38* (p. 60). Bonn: German Committee for Disaster Reduction (DKKV).
Blaikie, P., Cannon, T., Davis, I., & Wisner, B. (1994). *At risk—natural hazards, people's vulnerability, and disasters*. London: Routledge.
Blanchi, R., Leonard, J., & Maughan, D. (2004). *Towards new information tools for understanding bushfire risk at the urban interface*. Adelaide: Bushfire 2004.
Blanchi, R., Lucas, C., Leonard, J., & Finkele, K. (2010). Meteorological conditions and wildfire-related house loss in Australia. *International Journal of Wildland Fire, 19,* 914–926.
Blyth, M. (1997). "Any more bright ideas?" The ideational turn of comparative political economy. *Comparative Politics, 29*(2), 229–250.
Blyth, M. (2008). *The secret life of institutions: On the role of ideas in evolving economic systems entretien avec Mark Blyth*. Revue de la régulation capitalisme, institutions, pouvoirs, Revues. Org.
Boin, A., & t Hart, P. (2003). Public leadership in times of crisis: Mission impossible?. *Public Administration Review, 63*(5), 544–553.
Boin, A., t Hart, P., & McConnell, A. (2009). Crisis exploitation: Political and policy impacts of framing contests. *Journal of European Public Policy, 16*(1), 81–106.
Bosomworth, K., & Handmer, J. (2008). Climate change and community bushfire resilience. In K. Haynes & J. Handmer (Eds.), *Community bushfire safety* (pp. 175–183). Collingwood: CSIRO.
Bradstock, R., & Gill, A. M. (2001). Living with fire and biodiversity at the urban edge: In search of a sustainable solution to the protection problem in southern Australia, In, Cary, G. 2004, Climate change and bushfire incidence. Proceedings of the New South Wales Clean Air Forum 2004.
Cary, G. (2002). Importance of a changing climate for fire regimes in Australia. In R. Bradstock, J. E. Williams, & A. M. Gill, *Flammable Australia: Fire regimes and biodiversity of a continent* (pp. 26–49). Melbourne: Cambridge University Press.
Cary, G., Lindenmayer, D. B., & Dovers, S. (Eds.). (2003). *Australia burning: Fire ecology, policy and management issues*. Melbourne: CSIRO Publishing.
Coleman, T. (Undated). *The impact of climate change on insurance against catastrophes*. Report. Insurance Group Australia.
Coffman, M., & Umemoto, K. (2010). The triple-bottom-line: Framing of trade-offs in sustainability planning practice. *Environment, Development and Sustainability, 12,* 597–610.
Connor, R., & Dovers, S. (2004). *Institutional change for sustainable development*. Cheltenham: Elgar.
Considine, M. (2005). *Making public policy. Institutions, actors, strategies*. Cambridge: Polity Press.
Creed, W. E. D., Langstraat, J. A., & Scully, M. A. (2002). A picture of the frame: Frame analysis as technique and as politics. *Organisational Research Methods, 5*(1), 34–55.
de Boer, J., Wardekker, J. A., & van der Sluijs, J. P. (2010). Frame-based guide to situated decision-making on climate change. *Global Environmental Change*. doi:10.1016/jgloenvcha.2010.03.003.
Dewulf, A., Craps, M., & Dercon, G. (2004). How issues get framed and reframed when different communities meet: A multi-level analysis of a collaborative soil conservation initiative in the Ecuadorian Andes. *Journal of Community & Applied Social Psychology, 14,* 177–192.
Dewulf, A., François, G., Pahl-Wostl, C., & Taillieu, T. (2007). A framing approach to cross-disciplinary research collaboration: Experiences from a large-scale research project on adaptive water manageent. *Ecology and Society, 12*(2), 14.
Dewulf, A., Mancero, M., Cardenas, G., & Sucozhanay, D. (2011). Fragmentation and connection of frames in collaborative water governance: A case study of river catchment management in southern Ecuador. *International Review of Administrative Sciences, 77*(1), 50–75.
Dombeck, M. P., Williams, J. E., & Wood, C. A. (2004). Wildfire policy and public lands: Integrating scientific understanding with social concerns across landscapes. *Conservation Biology, 18*(4), 883–889.
Dovers, S. (2004). Sustainability and disaster management. *The Australian Journal of Emergency Management, 19*(1), 21–25.
Dovers, S., & Handmer, J. (1993). Contradictions in sustainability. *Environmental Conservation, 20*(3), 217–222.

Druckman, J. N. (2001). The implications of framing effects for citizen competence. *Political Behaviour, 23*(3), 225–256.

Dupre, J. (2001). *Human nature and the limits of science*. New York: Oxford University Press.

Dyer, R., Jacklyn, P., Partridge, I., Russell-Smith, J., & Williams, R. J. (Eds.). (2002). *Savanna burning: Understanding and using fire in northern Australia*. Darwin: Tropical Savannas CRC.

Eder, K. (1996). The institutionalisation of environmentalism: Ecological discourse and the second transformation of the public sphere. (Ch: 9) In S. Lash, B. Szerszynski, & B. Wynne (Eds.), *Risk, environment and modernity*. London: Sage Publications.

Epstein, C. (2006). The making of global environmental norms: Endangered species protection. *Global Environmental Politics, 6*(2), 32–54.

Fenna, A. (2004). Australian public policy. Frenchs Forest: Pearson Education Australia.

Fischer, F. (2003). *Reframing public policy: Discursive politics and deliberative practices*. Oxford: Oxford University Press.

Fletcher, A. L. (2009). Clearing the air: The contribution of frame analysis to understanding climate policy in the United States. *Environmental Politics, 18*(5), 800–816.

Folke, C., Hahn, T., Olsson, P., & Norberg, J. (2005). Adaptive governance of social-ecological systems. *Annual Review of Environmental Resources, 30*, 441–473.

Gamson, W. A. (1988). Political discourse and collective action. *International Social Movement Research, 1*, 219–246.

Gill, A. M. (2005). Landscape fires as social disasters: An overview of 'the bushfire problem'. *Environmental Hazards, 6*, 65–80.

Gillen, M. (2005). Urban vulnerability in Sydney: Policy and institutional ambiguities in bushfire protection. *Urban policy and research, 23*(4), 465–476.

Goffman, E. (1974). *Frame analysis: An essay on the organization of experience*. Boston: Northeastern University Press.

Goldammer, J. G., & Price, C. (1998). Potential impacts of climate change on fire regimes in the tropics based on MAGICC and a GISS GCM-derived lightning model. *Climatic Change, 39*, 273–296.

Hajer, M. A., & Wagenaar, H. (Eds.). (2003). *Deliberative policy analysis: Understanding governance in the network society. Theories of institutional design. Cambridge: Cambridge* University Press.

Handmer, J., & Dovers, S. (2008). *Handbook of disasters and emergency policies and institutions*. London: Earthscan.

Handmer, J., & Tibbits, A. (2005). Is staying at home the safest option during bushfires? Historical evidence for an Australian approach. *Environmental Hazards, 6*, 81–91.

t Hart, P., & Kroon, M. B. R. (1997). Groupthink in government: Pathologies of small-group decision making. In J. L. Garnett & A. Kouzmin (Eds.), *Handbook of administrative communication*. New York: Dekker.

Helmke, G., & Levitsky, S. (2003). *Informal institutions and comparative politics: A research agenda*, Kellogg Institute for International Studies, University of Notre Dame. Working Paper #307.

Hennessy, K., Lucas, C., Nicholls, N., Bathols, J., Suppiah, R., & Ricketts, J. (2006). *Climate change impacts on fire-weather in south-east australia*. Report. CSIRO Marine and Atmospheric Research, Bushfire CRC and Australia Bureau of Meterology.

High, C., Pelling, M., & Nemes, G. (2005). *Understanding informal institutions: Networks and communities in rural development*. In *Transition in Agriculture, Agricultural Economics in Transition II, Institute of Economics, Hungarian Academy of Sciences*, 28–29 Oct 2005, Budapest, Hungary.

Incoll, R. (1994). *Asset protection in a fire-prone environment*. Fire and Biodiversity: The effects and effectiveness of fire management, Footscray, Melbourne, Australia.

Isendahl, N., Dewulf, A., Brugnach, M., François, G., Möllenkamp, S., & Pahl-Wostl, C. (2009). Assessing framing of uncertainties in water management practice. *Water Resource Management, 23*, 3191–3205.

Jasanoff, S. (1993). Bridging the two cultures of risk analysis. *Risk Analysis, 13*(2), 123–129.

Jenkins-Smith, H. C., & Sabatier, P. A. (1993). The dynamics of policy oriented learning. (Ch: 3) In P. A. Sabatier & H. C. Jenkins-Smith (Eds.), *Policy change and learning. An advocacy coalition approach* (pp. 41–56). Boulder: Westview Press.

Jones, R., & Durack, P. (2005). *Estimating the impacts of climate change on Victoria's runoff using a hydrological sensitivity model*. Report. Research Report for the Victorian Greenhouse Unit, Department of Sustainability and Environment, Melbourne.

Kahan, D. M., & Braman, D. (2006). Cultural cognition and public policy. *Yale Law and Policy Review, 24,* 147–170.

Koch, P., & Hauknes, J. (2005). *On innovation in the public sector—today and beyond*. Report. D20. Publin. Oslo.

Koenig, T. (2005). Identification and measurement of frames. Research Methods. http://www.ccsr.ac.uk/methods/publications/frameanalysis/measurement.html. Accessed 15 Apr 2012.

Koenig, T. (2006). Compounding mixed-methods problems in frame analysis through comparative research. *Qualitative Research, 2006,* 6–61.

Lauth, H.-J. (2000). Informal institutions and democracy. *Democratization, 7*(4), 21–50.

Lavorel, S., Flannigan, M. D., Lambin, E. F., & Scholes, M. C. (2006). Vulnerability of land systems to fire: Interactions among humans, climate, atmosphere, and ecosystems. *Mitigation and Adaptation Strategies for Global Change, 12*(1), 33–53.

Lees, L. (2004). Urban geography: Discourse analysis and urban research. *Progress in Human Geography, 28*(1), 101–107.

Lenschow, A., & Zito, A. R. (1998). Blurring or shifting of policy frames? Institutionalisation of the economic-environmental policy linkage in the European community. *Governance: An International Journal of Policy and Administration, 11*(4), 415–441.

Lowndes, V. (2010). The institutional approach. (Ch: 3) In D. Marsh & G. Stoker (Eds.), *Theory and methods in political science*. Hampshire: Macmillan.

Lucas, C., Hennessy, K., Mills, G., & Bathols, J. (2007). *Bushfire weather in southeast Australia: Recent trends and projected climate change impacts*. Report. Bushfire CRC, Australian Bureau of Meteorology and CSIRO Marine and Atmospheric Research.

Lynn, K. (2003). Wildfire and rural poverty: Disastrous connections. *Natural Hazards Observer, 29*(2), 10–11.

MacGregor, D. G. (2006). The future of fire in environmental management. *Futures, 38*(4), 505–518.

March, J. G., & Olsen, J. P. (1989). *Rediscovering institutions: The organizational basis of politics*. New York: The Free Press.

Meinzen-Dick, R. (2007). Beyond panaceas in water institutions. *PNAS—Proceedings of the National Academy of Science, 104*(39), 15200–15205.

Mendelsohn, R., Basist, A., Kurukulasuriya, P., & Dinar, A. (2007). Climate and rural income. *Climatic Change, 81,* 101–118.

Miller, L. L. (2004). Rethinking bureaucrats in the policy process: Criminal justice agents and the national crime agenda. *The Policy Studies Journal, 32*(4), 569–588.

Mouillot, F., Rambal, S., & Joffre, R. (2002). Simulating climate change impacts on fire frequency and vegetation dynamics in a mediterranean ecosystem. *Global Change Biology, 8,* 423–437.

O'Brien, K. L., & Leichenko, R. M. (2000). Double exposure: Assessing the impacts of climate change within the context of economic globalization. *Global Environmental Change, 10,* 221–232.

Olsen, J. P. (2008). The ups and downs of bureaucratic organization. *Annual Review of Political Science, 11,* 13–37.

Ostrom, E. (Ed.). (2005). *Understanding insitutional diversity*. New Jersey: Princeton University Press.

Ostrom, E., Janssen, M. A., & Anderies, J. M. (2007). Going beyond panceas. *PNAS—Proceedings of the National Academy of Science, 104*(39), 15176–15178.

Parsons, W. (1995). *Public policy: An introduction to the theory and practice of policy analysis*. Cheltenham: Elgar.

Pelling, M., & Dill, K. (2009). Disaster politics: Tipping points for change in the adaptation of sociopolitical regimes. *Progress in Human Geography*, 1–17. DOI:10.1177/0309132509105004.
Pidgeon, N., & Butler, C. (2009). Risk analysis and climate change. *Environmental Politics, 18*(5), 670–688.
Popper, S. W., Lempert, R. J., & Bankes, S. C. (2005). Shaping the future. *Scientific American, March 28.*
Powell, W. W., & DiMaggio, P. J. (Eds.). (1991). *The new institutionalism in organizational analysis*. Chicago: The University of Chicago Press.
Rein, M., & Schon, D. (1991). Frame-reflective policy discourse.(Ch: 12) In P. Wagner, C. H. Weiss, B. Wittrock, & H. Wollman (Eds.), *Social sciences and modern states* (pp. 262–289). New York: Cambridge University Press.
Sardans, J., & Penuelas, J. (2007). Drought changes phosphorus and potassium accumulation patterns in an evergreen mediterranean forest. *Functional Ecology, 21*(2), 191–201.
Scharpf, F. W. (2000). Institutions in comparative policy research. *Comparative Political Studies, 33*(6–7), 762–790.
Schipper, L., & Pelling, M. (2006). Disaster risk, climate change and international development: Scope for, and challenges to, integration. *Disasters, 30*(1), 19–38.
Schmidt, V. A. (2008). Discursive institutionalism: The explanatory power of ideas and discourse. *Annual Review of Political Science, 11*(1), 303–326.
Scott, W. R. (2001). *Institutions and organizations*. London: Sage Publications.
Spence, A., & Pidgeon, N. (2010). Framing and communicating climate change: The effects of distance and outcome frame manipulations. *Global Environmental Change, 20*(4), 656–667.
Thiele, L. P. (2000). Limiting risks: Environmental ethics as a policy primer. *Policy Studies Journal, 28*(3), 540–557.
Thomas, C. D., Cameron, A., Green, R. E., Bakkenes, M., Beaumont, L. J., Collingham, Y. C., Erasmus, B. F. N., Ferreira de Siqueira, M., Grainger, A., Hannah, L., Hughes, L., Huntley, B., van Jaarsveld, A. S., Midgley, G., Miles, L., Ortega-Huerta, M. A., Peterson, A. T., Phillips, O. L., & Williams, S. E. (2004). Extinction risk from climate change. *Nature (letters to), 427*(8 January), 145–148.
Tryhorn, L., Lynch, A., Adbramson, R., & Parkyn, K. (2008). On the meteorological mechanisms driving postfire flash floods: A case study. *Monthly Weather Review, 136*(5), 1778–1791.
Turner, B. L., Kasperson, R. E., Matson, P. A., McCarthy, J. J., Corell, R. W., Christensen, L., Eckley, N., Kasperson, J. X., Luers, A. L., Martello, M. L., Polsky, C., Pulsipher, A., & Schiller, A. (2003). A framework for vulnerability analysis in sustainability science. *Proceedings of the National Academy of Sciences (PNAS), 100*(14), 8074–8079.
UNFCCC. (2004). Application of methods and tools for assessing impacts and vulnerability, and developing adaptation responses. Background paper, United Nations framework convention on climate change 24.
Weber, E. P., & Khademian, A. M. (2008). Wicked problems, knowledge challenges, and collaborative capacity builders in network settings. *Public Administration Review, 68*(2), 334–349. (March/April)
Wuerthner, G. (2006). Incendiary language: How words affect perception. In G. Wuerthner (Ed.), *The wildfire reader: A century of failed forst policy* (p. 54). Sausalito: Island Press.
Wynne, B. (2002). Risk and environment as legitimatory discourses of technology: Reflexivity inside out? *Current Sociology, 50*(3), 459–477.

# Chapter 12
# Recovering from the 2004 Indian Ocean Tsunami: Lessons for Climate Change Response

**Ahana Lakshmi, Purvaja R. and Ramesh R.**

**Abstract** The impact, relief and recovery processes after the devastation caused by the 2004 Indian Ocean tsunami offer a number of lessons for building the resilience of the coast and coastal communities to overcome challenges associated with climate change. This chapter focuses on the response and recovery undertaken in the coastal state of Tamil Nadu, the worst affected state on the Indian mainland. These lessons are useful for facing climate change-related challenges, especially since India's long coastline is subject to a variety of hazards such as cyclones and storm surges apart from sea-level rise which, on an annual basis, cause extensive damage to coastal areas, and which are expected to increase in intensity in the future. Thus the reduction of vulnerability has to be done in a planned and phased manner. Mainstreaming disaster risk reduction into development activities requires conducting a needs assessment to inform capacity building and livelihood enhancement initiatives and improving the horizontal and vertical coordination of activities that are likely to enable adaptation to various impacts due to climate change along the coast. Integrated coastal management offers opportunities to achieve these objectives.

**Keywords** Tsunami · Tamil Nadu · Vulnerability · Resilience · Risk reduction

## 12.1 The 2004 Indian Ocean Tsunami

### *12.1.1 Introduction*

On 26th December, 2004, the Great Sumatra Earthquake ($M_w$9.3) generated a tsunami that devastated long stretches of the shoreline of a dozen countries in the Indian Ocean (Lay et al. 2005) including Indonesia, Thailand, Sri Lanka and India. In India, the tsunami affected the mainland states of Tamil Nadu, Andhra Pradesh and Kerala and the union territory of Puducherry. Being close to the epicentre of the

---

Ramesh R. (✉) · A. Lakshmi · Purvaja R.
National Centre for Sustainable Coastal Management,
Anna University Campus, Chennai 600 025, India
e-mail: rramesh_au@yahoo.com

B. C. Glavovic, G. P. Smith (eds.), *Adapting to Climate Change,* Environmental Hazards,
DOI 10.1007/978-94-017-8631-7_12, 

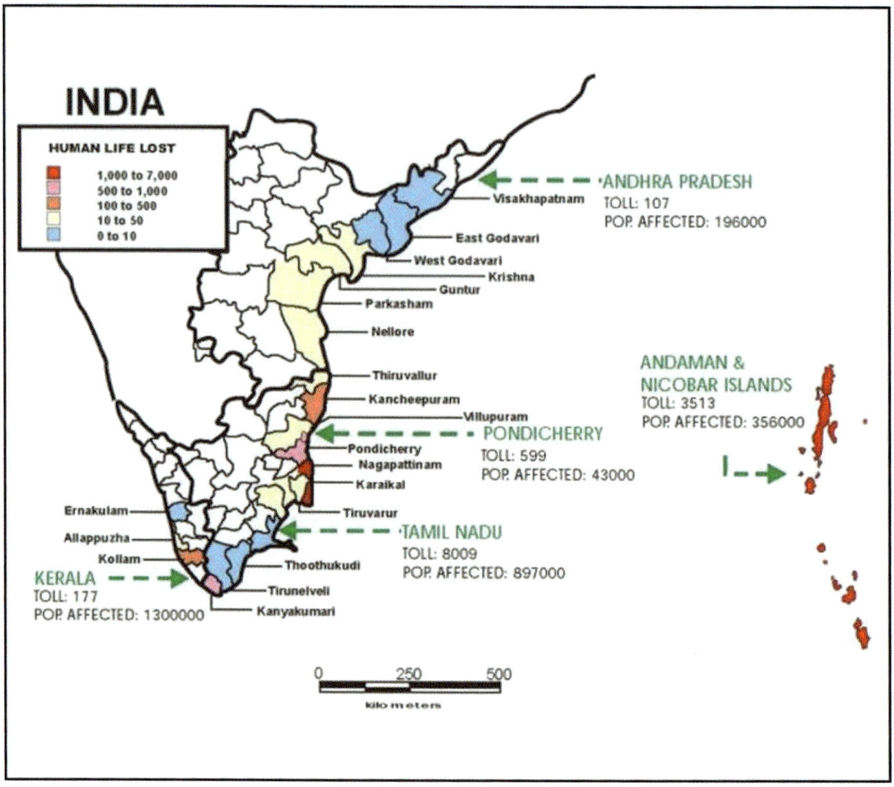


**Fig. 12.1** Impact of Tsunami on India. (Source: UNTRS 2005)

earthquake, the union territory of Andaman and Nicobar Islands were affected by both the earthquake and tsunami. The extent of devastation is shown in Fig. 12.1.

This chapter examines the impact of the tsunami on the south Indian state of Tamil Nadu and the recovery process that followed in order to see what lessons can be distilled from the response to the disaster that could help inform climate change adaptation processes.

### *12.1.2 The State of Tamil Nadu—An Overview*

Tamil Nadu is one of India's 28 states, covering an area of 130,058 km$^2$. Population of the state according to the 2011 census was 72 million, up from 62.4 million in 2001 (Census of India 2011). The 1076 km coastline is divided into thirteen districts (Fig. 12.2). All the coastal districts border the Bay of Bengal and the southernmost district of Kanyakumari includes a short stretch of coastline adjacent to the Arabian Sea. The state is the eleventh largest in India and also the most urbanised. It has

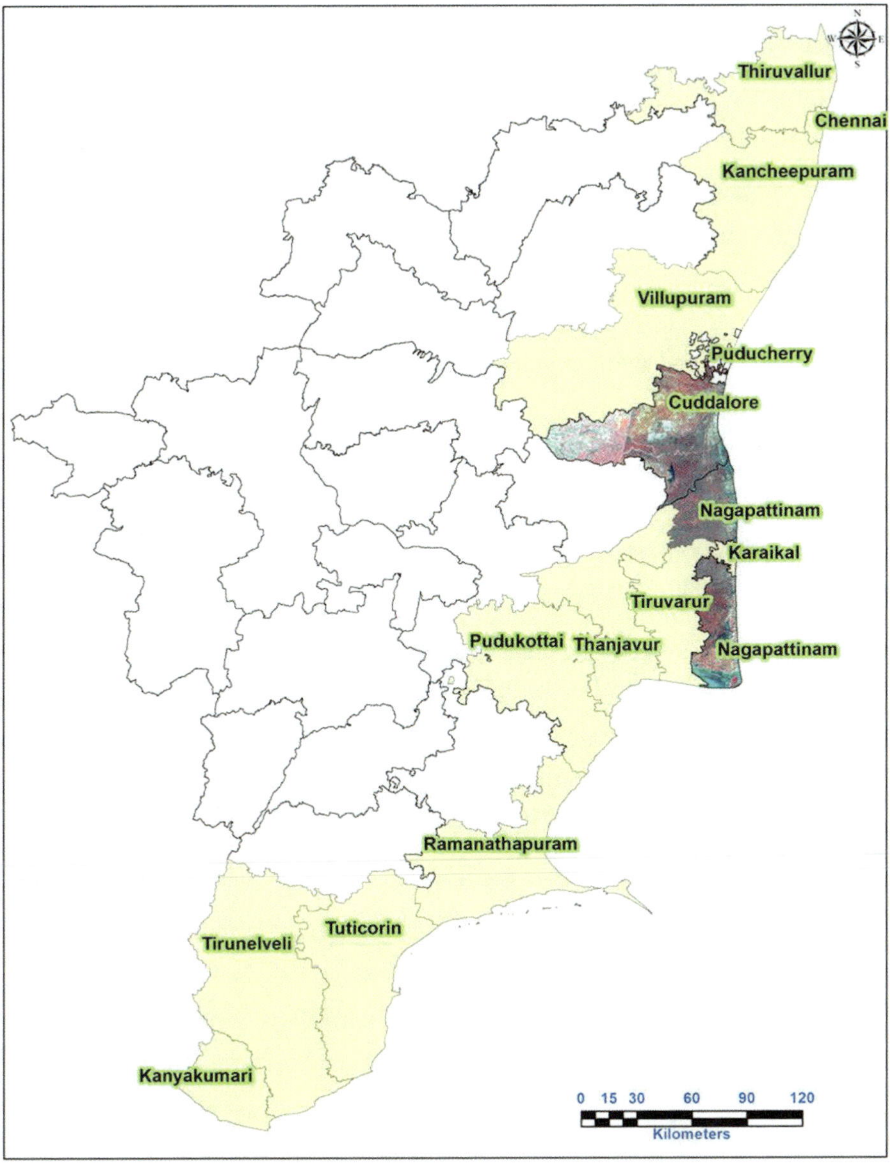


**Fig. 12.2** Map of the coastal districts of Tamil Nadu. Cuddalore and Nagapattinam are highlighted using satellite imagery.

a good rail and road network as well as a well-developed communication system ensuring connectivity between all parts of the state.

Along the coast, the dominant livelihood is based on marine fishing. The state has a long history of fishing conducted by those living within 591 fishing villages. In 2004–2005, Tamil Nadu had a population of about 0.737 million fishermen, of which 0.280 million were actively engaged in this activity. There were 12,000 mechanised fishing boats and 50,700 traditional crafts registered, of which 20,000 had outboard motors (DoF 2004). Fishing is largely a caste-based occupation and is not related to the mainstream agrarian system.

The Hindu Pattinavar community is located along the Bay of Bengal or Coromandel Coast, the Christian Parava community along the Gulf of Mannar and the Christian Mukkuva Community along the Arabian Sea coast (FIMSUL 2011a). Other communities (e.g. agriculturalists, dalits) may have their own traditional governance systems but they are not quite as powerful or involved in the activities of their constituents (Babu et al. 2008). The evolution of traditional management systems has depended on the resource, the environment in which the resource existed and the interactions between people to extract these resources (Kurien 1995).

Fishing communities live close to the shoreline in dwellings that include mud houses with roofs of coconut/palm thatch, tiled houses and well-built Reinforced Cement Concrete (RCC) structures. In Nagapattinam district, for example, of the 16,609 houses of the fishers, 9,827 houses were thatched, 1,861 were tiled and 1,675 were terraced (Fisheries Department Census 2000). Less than 1 % of the houses were rented.

A large number of the fishers use catamarans, fibre glass boats or larger plank boats while there is a growing number that rely on mechanised boats and trawlers. Most fishing villages have their own landing site which is usually a strip of the sandy shore, where the craft are beached. Trawlers and mechanised boats are berthed in fishing harbours. At the time of the tsunami disaster, fishing hamlets, also called *kuppams* in many places, were distinct entities mostly controlled by traditional leadership in the Hindu dominated areas and by the Church in the Christian dominated areas (Gomathy 2006).

### *12.1.3 Impact of the Tsunami in Tamil Nadu*

The coast of Tamil Nadu was severely affected with all its thirteen coastal districts suffering various degrees of devastation. Table 12.1 gives an overview of the impact of the Tsunami on the case study area.

### *12.1.4 Tsunami Run Up and Extent of Inundation*

Among the thirteen districts affected by the tsunami, the Cuddalore-Nagapattinam stretch of Tamil Nadu's coast experienced the worst impacts of the tsunami surge and inundation. In the Cuddalore District, run-up ranged from 2.5 to 3.3 m with inundation distances extending between 330 m and 1,680 m. At Nagapattinam, the run up ranged between 0.86 m and 5.7 m while the inundation reached a maximum of 3 km (Subramanian 2006). Detailed analysis of the geophysical

**Table 12.1** Impact of 2004 tsunami on Tamil Nadu. (Source: GoTN 2008)

| | |
|---|---|
| Districts affected | 13 |
| Villages affected | 238 |
| Hamlets affected | 418 |
| Lives lost | 7,997 |
| Missing | 846 |
| Injured | 3,625 |
| Children orphaned | 242 |
| Houses damaged | 118,000 |
| People evacuated | 0.49 million |
| Population affected | 1.078 million |
| Number of families whose immediate livelihood was affected | 0.3 million |
| Damage assessment—infrastructure & housing | US$ 880 million |

**Table 12.2** Impact of the 2004 tsunami on Cuddalore and Nagapattinam districts of Tamil Nadu. (Source: GoTN 2005)

| | Cuddalore | Nagapattinam |
|---|---|---|
| Villages affected | 8 | 38 |
| Hamlets affected | 43 | 73 |
| Lives lost | 610 | 6,065 |
| Injured | 259 | 2,375 |
| Children orphaned | 12 | 179 |
| Houses damaged | 2,872 | 19,630 |
| People evacuated | 61,054 | 196,184 |
| Population affected | 99,704 | 196,184 |

data showed that the structure of the underlying basement, the morphology and the land–ocean tectonics were the main factors influencing the run-up heights in the case of the Nagapattinam-Cuddalore shelf. The fault controlled basement structure, and a straight coastline with a narrow and gentle shelf fostered a rapid transgression of the surge inundating the coastal area (Murthy et al. 2006). The impact of the tsunami on Cuddalore and Nagapattinam districts is summarised in Table 12.2.

The Cuddalore-Nagapattinam district falls within the 'high damage risk zone' with respect to wind and cyclones according to the Vulnerability Atlas prepared by the Building Materials Technology Promotion Council (BMTPC 2008). Cyclone Thane, the strongest tropical cyclone of 2011 within the Northern Indian Ocean region, affected large areas of Cuddalore district as it made landfall between Cuddalore and Puducherry on December 30, 2011 (IMD 2011). Nagapattinam district has also been affected in the past by cyclones and floods as it is located at the terminus of the Cauvery delta. This region is also situated in the low elevation coastal zone (LECZ) and comprises a high population density (Fig. 12.3) (McGranahan et al. 2007) making it highly vulnerable to sea-level rise. It is therefore useful to examine how the recovery process after the tsunami was carried out and what lessons can be drawn from this experience and applied to climate change adaptation.

**Fig. 12.3** Population density within and outside of the 10 m Low Elevation Coastal Zone. (Source: Centre for International Earth Science Information Network (CIESIN) 2007)

## 12.2 Response to the 2004 Tsunami

### *12.2.1 Introduction*

While Indian communities along the coast as well as government agencies at federal and provincial levels are aware of and have standard operating procedures to respond to disasters caused by cyclones and floods, they were largely unprepared to deal with the effects of the tsunami. Its impact was unexpected and swift. There was no warning unlike in the case of cyclones and floods which are preceded by rainy/stormy weather (the day was sunny), and hence it was unforeseen. In a matter of minutes, large swathes of the coast were devastated by the tsunami. Yet the response from the state government as well as the civil society, after initial paralysis, was rapid.

The role of the mass media, especially television, was important as the photographs and accounts sent from the disaster site gave graphic details of the devastation which resulted in a deluge of humanitarian response from all over India as well as the globe. Tamil Nadu's Chief Minister immediately sent handpicked senior government officers to oversee relief measures. Area teams were formed that included personnel from various departments such as public health, public works, electricity board, water and drainage (Babu et al. 2008). In the following sections, an attempt is made to examine some of the key issues in the post-tsunami scenario as a response to 'building back better' by examining the impacts and responses in Nagapattinam district as well as Cuddalore district which was less affected by the tsunami but significantly impacted by Cyclone Thane in 2011.

### *12.2.2 Coordination and Information Dissemination*

Nagapattinam is a narrow district whose entire coastline of about 190 km was affected by the tsunami. The extent of devastation overwhelmed the district's capacity to respond (Babu et al. 2008), including area hospitals who struggled to contend with over 6,000 deaths. There was a break-down in regular communication systems and as a result, emergency short wave systems were activated. Teams were rushed from the neighbouring district with relief supplies. Humanitarian aid, including people and supplies started pouring in as soon as the extent of devastation was shown in the mass media. Prior to the tsunami, only a few NGOs worked in the district. Afterwards over 500 NGOs and volunteer groups arrived in Nagapattinam alone (Fig. 12.4).

Among the many teams that arrived to help in relief efforts, was a team from Gujarat which had played an important role in the recovery after the 2001 earthquake in Bhuj, Gujarat. They emphasised the need to coordinate the relief as well as the long-term recovery process, which led to the establishment of a coordination centre by prominent NGOs including SIFFS (South Indian Federation of Fishermen Societies)

**Fig. 12.4** NGO Coordination Centre, Nagapattinam. (Source: Photograph by Annie George (NCRC))

and SNEHA (Social Needs, Education and Human Awareness) and supported by the district administration. Establishment of the NGO Coordination and Resource Centre, Nagapattinam (NCRC, www.ncrc.in) fostered appropriate, equitable and sustainable post disaster response during relief and recovery (Babu et al. 2008). It should be noted that in the analysis of the response to the tsunami (GoTN 2008), the first topic focuses on coordination and information-sharing.

In a disaster situation, coordination and information sharing happen at multiple levels responding to needs associated with different phases of recovery (Lakshmi and Bau 2007). In the post-tsunami scenario, while there was horizontal coordination between the various line departments (fisheries, agriculture, health, revenue etc.), there was also vertical coordination (state—district—village) present. In Nagapattinam, the Village Information Centres (VICs), designed for two-way transmission of information, communicated the needs of the communities to the government and the government's policies to the beneficiaries. The coordination was facilitated by the NCRC. At the state level, the Tsunami Rehabilitation Information Network (TRINet), conceived in March 2005 as a network of resource centres, provided a neutral platform for discussions and learning, while the Tamil Nadu Tsunami Resource Centre (TNTRC), established much later and supported by the state government and a consortium of international NGOs and donor organisations, helped with state level coordination as well as capacity building.

The use of appropriate technology for communication and information sharing also played an important role. When the telephone system broke down, radio networks were used. Once cellular communication systems were back on stream, mobile phones and laptops were provided to all key personnel to facilitate rapid communication. Websites were set up, such as www.tsunami2004-india.org for SIFFS that quickly provided an access point for information ranging from prices of engines for fishing boats to reports of sector coordination meetings. Soon after, websites for NCRC (www.ncrc.in) and TRINet (www.trinet.in) were created. Cuddalore district, led by a dynamic and IT savvy District Collector (the chief administrator of the district), created a detailed website that gave hamlet-wide listing of the deceased, compensation provided and other details of immediate importance. Subsequently, the state government began to host information on the tsunami rehabilitation programme on http://www.tn.gov.in/tsunami/ which still continues to be regularly updated. With an information overload in the media, TRINet began a weekly digest of tsunami recovery-related news that helped NGOs and Civil Society groups as well as government officers stay informed about what was happening in the different districts as well as the state, national and international levels. NCRC and TRINet websites also carried reports of the sector-group meetings which were not available in the mainstream media.

### *12.2.3 Livelihoods and Shelter—Improving Resilience*

Immediately after the relief phase came the recovery phase. In the recovery phase, two major activities included the provision of housing and restoration of livelihoods, which were closely related. As mentioned earlier, the tsunami's major impact was on the fishing community whose houses are located closest to the shore. Dwelling units as well as livelihood-related assets such as boats, nets and vessels (used by women to carry fish) were washed away or destroyed. The priority was to restore livelihoods as well as move the people out of relief camps which were located in local schools and community halls into temporary shelters before the damaged houses were restored or new houses built. Given the close knit nature of fishing communities there was a need to house families from the same hamlet together. Fishing craft, mainly Fibre-Reinforced Plastic (FRP) boats, were donated in large numbers to fishermen to enable them to get back to work as quickly as possible. The traditional panchayat played a major role in providing lists of beneficiaries and negotiating with NGOs and donor organisations to get the best deal (Gomathy 2006).

**Shelter reconstruction** The housing-reconstruction issue was very complex. In the immediate aftermath of a major disaster, which could recur at any time in the future, it was believed that moving fisher communities inland was the best solution. This was hotly contested and widely debated because some families preferred to move inland and others preferred to stay close to the shore. During the reconstruction process, the government provided the land (acquired by buying land from private owners as well as allocating government owned land) while the NGOs, supported by donor organisations, built the houses. The government also decided the minimum area (plot size and plinth area) of the houses and the amenities to be provided apart from insisting that the house was registered in the name of the husband and the wife. Multi-hazard resilient houses became the norm during the rebuilding process and a 10 year insurance policy was made compulsory thus ensuring a mechanism for risk transfer. Guidelines for reconstruction were provided and the widespread adoption of hazard-resistant construction practices throughout the construction programme was a key achievement (Fig. 12.5).

The shelter reconstruction programme was largely NGO/Donor-driven. Some of these organisations ensured participation of the beneficiaries to include the selection

**Fig. 12.5** New houses for those affected by the tsunami, Cuddalore district. (Source: Photograph by Ahana Lakshmi)

of the plot of land in which the housing unit was built (GoTN 2008; Lakshmi and Babu 2008). In addition to the houses, other amenities such as community halls, schools and fishing related infrastructure (e.g., auction sheds) were also constructed using NGO funding. Information about the status of housing in the different districts and the resource centres were regularly updated which enabled transparency in the reconstruction programme. The establishment of a shelter advisory group also helped to ensure that good construction practices were adhered to. The success of the programme in Nagapattinam district suggests that a similar programme at the state level would carry similar benefits (UN 2008).

**Restoring livelihoods** Restoring the livelihoods of those in the study area requires understanding the gender-based roles assumed by men and women. In the fisheries sector, the men fish and the women take care of post-harvest activities such as processing and retailing. With fishing becoming more capital intensive, the role of women is steadily diminishing (Salagrama 2006). The increasing importance of access to sufficient capital is coupled with circumstances that make it increasingly difficult to obtain safe access to reliable sources of credit. As a result, the level of social vulnerability among fishing communities has increased over time. In an effort to address these concerns, a large number of self-help groups were established to provide revolving funds as well as a variety of training programmes. Livelihood diversification schemes were promoted. Alternate livelihoods included tailoring, handicrafts, fashion designing, manufacture of incense sticks, fish pickling and even driving autorickshaws used for transporting fish.

Some of these programmes were carried out by NGOs while others were part of internationally funded schemes such as the Japan Fund for Poverty Reduction and the Asian Development Bank's Tsunami Emergency Assistance Project (TEAP). While some of these ventures survived due to external support, most of the programmes that were related to traditional livelihood activities (especially in the case of fisherwomen) became self-sustaining (Babu et al. 2008). In other cases, some equipment such as solar fish dryers provided to fisherwomen were not successful

(UNTRS/FAO 2008). The failure of many ventures was related to lack of market linkages or the need to travel long distances to work. Policies that encouraged participation of women in public activities and empowered them through education and finance initiatives helped to better articulate existing problems (Lakshmi 2007). Discussions with women showed that they had become more articulate and aware about their rights resulting in a positive change in their attitudes and outlook:

> Before the tsunami we did not come out of our houses. But now there is a great change. We have the confidence to come out, talk and even fight for our rights. (http://www.lwsi.org/html/cs-gender_justice.html)

In addition to fisheries and allied livelihoods, the tsunami also affected agriculture. The initial understanding was that only a narrow width of the coastline was affected; this was supported by graphic visuals of damaged fishing boats and houses located in fishing communities. However, the tsunami wave was found to have travelled inland through backwaters and creeks inundating fields and resulting in the salination of cropland and surface water sources. It was expected that recovery would take at least three years. Various approaches were tried to hasten recovery, including organic agriculture which was found to speed the process while fostering more sustainable practices post-disaster (NCRC 2006). Thus, in a disaster recovery framework, it is necessary to examine the various livelihoods found in an area and use multiple approaches to assist in their recovery.

### *12.2.4 Disaster Preparedness*

Maintaining a strong pre-event level of preparedness plays a crucial role in reducing the impact of natural disasters. Efforts to accomplish this aim are the responsibility of the United Nations Development Programme (UNDP)-Government of India Disaster Risk Management (DRM) Programme. Two months before the tsunami struck, villagers of Samiyarpettai in Cuddalore district had been trained in search, rescue and prevention from drowning. The pre-event training helped to limit the death toll from the tsunami to 24. In the adjoining village of Pudupettai, a little less than 2 km away, the death toll was as high as 92 (UNDP 2005). This sent a clear message about capacity building for disaster preparedness, including the realisation that members of the local community are the first responders following a disaster.

In the immediate aftermath of the tsunami, a number of NGOs worked with local communities to build their capacity and prepare vulnerability maps. A study was carried out in 2008 in Cuddalore and Nagapattinam looking at the post-tsunami changes in the community in terms of disaster resilience (Babu et al. 2008). The study found that in six of the ten villages studied in Cuddalore and Nagapattinam, NGOs had organised a participatory vulnerability mapping process and shared the results with the communities. In Cuddalore, the vulnerability maps and the associated evacuation routes were painted on the panchayat walls so that they were easy for people to see. However, there was no standardised approach in the mapping process nor were the communities trained in how to interpret or use the maps. The study reported that among 20% of those who had claimed continued "lack of confidence" in facing future disasters, respondents cited the lack of training as a major concern. After the 2004 tsunami, a number of training programmes have been organised by the state government at the district level. The Anna Institute of Management, which is the state government's training division, regularly organises training programmes on disaster management for government officials at various levels and is beginning to initiate a focused effort tied to the report's findings. The government has also introduced disaster related information as part of the school curriculum (from Grade 8 onwards) to ensure that information about various hazards, vulnerability and disaster preparedness are taught to students.

### *12.2.5 Coastal Protection Measures*

The tsunami affected a narrow band along the coast. This same area is also subject to other, more frequently occurring, natural hazards such as coastal flooding, tidal inundation due to cyclones and storm surges as well as shoreline erosion. In addition to affecting fishing hamlets, the tsunami impacted agricultural fields inland as the water rushed through canals into backwater areas. In South Poigainallur village, tsunami waves broke through the drain provided for agricultural water and destroyed part of the man-made dune. The loss of life in this village was limited because of the presence of mangroves which had been protected by the local people (GoTN 2005; Hedao 2006). Similar findings were also reported in other villages in Cuddalore district (Kathiresan and Rajendran 2005). It was also found that villages shielded by mangroves or casuarina plantations suffered less damage as the vegetation attenuated tsunami waves and protected shorelines (Danielsen et al. 2005). Mangrove plantations as well as coastal shelterbelts (usually casuarina plantations) have also been found to serve as wind and wave-breakers during cyclones (e.g. Das and Vincent 2009). Hence, the maintenance of mangrove plantations as well as casuarina plantations as coastal shelterbelts were promoted extensively following the tsunami (MSSRF 2008). The forest department achieved its target of 700 ha of mangrove plantation and 2000 ha of shelterbelt plantation in the coastal areas of Tamil Nadu by the end July 2008 (GoTN 2008) (Fig. 12.6).

### *12.2.6 Reducing Vulnerability*

The tsunami disaster was seen as providing a window of opportunity to reduce the vulnerability of people in coastal areas. Some of the highlights of the process are shown in Fig. 12.7. Efforts to achieve vulnerability reduction included physical, social, economic and institutional components. For example, physical vulnerability has been addressed by providing multi-hazard resilient houses as well as better infrastructure such as roads and water supply. In the case of social vulnerability, the focus was on building equity, inclusive housing, education and empowerment of women and mainstreaming of the

**Fig. 12.6** Developing a coastal shelterbelt. Casuarina plantation (*left*), Mangrove (*right*). (Source: Photograph (*left*) by Ahana Lakshmi; Photograph (*right*) by Purvaja, R.)

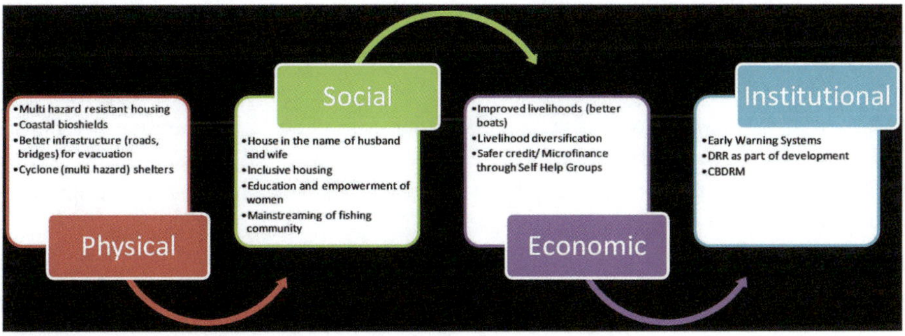


**Fig. 12.7** Reducing vulnerabilities in the post-tsunami context.

fishing community. Economic vulnerabilities were addressed by improving as well as diversifying livelihoods and providing safe credit through self-help groups. Linking disaster risk reduction with development activities, early warning systems (including ensuring last-mile connectivity) and community based disaster risk management programmes were used to overcome institutional vulnerabilities.

## 12.3 Climate Change in India: Learning from the Tsunami Experience

### *12.3.1 Climate Change Impact on the Indian Coast*

Tsunamis are not related to climate change, but its impacts are akin to the impacts of cyclones, a frequent hazard in many parts of coastal India. According to the Climate Change Vulnerability Index (CCVI) released by the global risk advisory firm

Maplecroft in 2010, India is ranked second among all nations in terms of its vulnerability to the impacts of climate change. The index was based on 42 social, economic and environmental factors used to assess national vulnerabilities across three core areas (Maplecroft 2010). An assessment carried out by a network of institutions (INCCA 2010) indicated that the most significant impacts would be due to increased intensity of cyclonic disturbances and sea-level rise. The Cuddalore-Nagapattinam region is also highly vulnerable to cyclones, the most recent examples being Cyclone Nisha in November 2008 and Cyclone Thane in December 2011, which created surge heights between 1 and 1.5 m (Saxena et al. 2012).

The INCCA study indicated that while the number of rainy days is likely to decrease by 1–5 days along the east coast, the intensity of rainfall is likely to increase from 1–4 mm/day. The number of cyclonic storms may decrease but the intensity is likely to increase (Niyas et al. 2009; Ramesh Kumar and Shankar 2010). This means that storm surges and flooding due to intense events are likely to increase over time, especially in the southern part of the east coast where tidal ranges are low (Unnikrishnan et al. 2010). The INCCA study (INCCA 2010) estimated the inundation of coastal areas due to sea-level rise at two locations along the east coast (Nagapattinam, Tamil Nadu and Paradip, Odisha). The study showed that the inundation area would be around 4.2 $km^2$ for a 1 m rise in sea level. Apart from being in the LECZ (Fig. 12.3), the Cuddalore-Nagapattinam coast is also located at the terminus of the Cauvery delta where agriculture is highly vulnerable to the impacts of climate change.

### *12.3.2 Lessons from the Tsunami Recovery Process*

Key impacts of climate change include:

- Sea-level rise
- Higher intensity of cyclones
- Higher intensity of rainfall

The following results are expected:

- Rising sea levels will inundate coastal areas, turning land and water saline
- Rising sea levels and more intense cyclones will increase the saline intrusion of coastal aquifers
- More intense cyclones and rising sea levels will increase the erosion of coastal shorelines
- Rising sea levels, more intense cyclones and extreme rainfall events will increase the impact on coastal infrastructure
- Rising sea levels, more intense cyclones and extreme rainfall events will increase the impact on coastal economies

While some extreme events, such as cyclones, are more predictable and well-developed early warning systems have been created to alert the public, other hazards such as sea-level rise and saline intrusion create another set of problems. Hence, while the overall scheme would be to reduce vulnerability at all levels, the response mechanisms are likely to vary across hazard types. Based on the results described earlier in this chapter, there are several important lessons regarding specific activities that can be used to build capacity to adapt and cope with climate change. These include the following:

#### 12.3.2.1 Early Warning Systems for Predictable Hazards

- *Responding to cyclones*: The Indian Meteorological Department (IMD) has a network of meteorological observatories covering the entire coastline. The more recent creation of the National Cyclone Risk Mitigation Project (NCRMP) further strengthens the early warning systems. For instance, Doppler weather radars have been installed at key meteorological offices (including Chennai in Tamil Nadu). Proof that such systems work was obtained in 2011 when Cyclone Thane crossed the coast of Tamil Nadu at Cuddalore. Once the cyclone came within the range of the Doppler radar, better estimation of intensity and landfall location was made available (Kurian 2012) and immediately, extensive warnings were provided over the IMD website (www.imd.gov.in), television and radio, people were evacuated well in advance of the storm and the loss of lives was minimised. The IMD also has a well-developed system for tracking weather patterns and providing alerts for heavy rainfall.
- *Responding to tsunami*: Following the 2004 tsunami, a tsunami warning system has been developed for the Indian Ocean. A series of bottom pressure recorders placed in the Bay of Bengal and the Arabian Sea can detect changes in the sea level and this is conveyed to the Indian Tsunami Early Warning Centre (ITEWC) located in the Indian National Centre for Ocean Information System (INCOIS) which runs models to predict the height of the tsunami as well as identifying the coastline that could be affected. Since October 2011, this has been functioning as the regional tsunami advisory service.

#### 12.3.2.2 Vulnerability Reduction

- *Physical vulnerability reduction*: While implementing the tsunami rehabilitation programme, it was found that a large number of families, though not directly affected by the tsunami, were living in vulnerable areas or in poor quality housing. As a result, the state government decided to undertake a second phase of housing reconstruction by building disaster resilient houses for vulnerable communities complemented by construction of cyclone shelters, delineation of evacuation routes, development of early warning systems and establishment of better connectivity between roads and bridges (GoTN 2008).
- *Livelihood diversification*: Addressing social vulnerability requires efforts to reduce both physical vulnerability by providing safe shelter and improving income security. Coastal communities, especially those that are highly dependent on a fisheries-based economy, are being encouraged to take up alternate or diversified

livelihoods so that they have a steady source of income. Post-tsunami experience in livelihood enhancement and diversification activities has shown that it is essential to perform a needs assessment before designing a programme, and based on the findings, the programme should reflect the assets and needs of the population. Implementing the programme requires sufficient training, the provision of long-term funding, and a focus on local empowerment objectives.

- *Demarcation of the hazard line*. The Coastal Regulation Zone Notification 1991 issued under the Environment (Protection) Act of 1986, was designed to restrict activities in a strip of land 500 m from the high tide line. A committee was set up by the Ministry of Environment and Forests in 2004 to look into its revision taking into account current threats to the coast including climate change-related impacts. In 2011, when the CRZ Notification was reissued, it included the demarcation of a composite hazard line to delineate the boundaries of the coastal zone in mainland India (MoEF 2011). Delineation of the composite hazard line is now established based on historic erosion rates or extreme water levels rather than adopting arbitrary distances which do not truly represent the threat from erosion or coastal flooding (Saxena et al. 2012). The hazard line is mapped by the Survey of India taking into account tide, waves, sea-level rise and shoreline changes. The area between the hazard line and the 500 m contour (CRZ line) falls within the coastal regulation notification area thereby restricting certain activities (MoEF 2011). A hazard line has been demarcated for Cuddalore district (Saxena et al. 2012) on a pilot basis.
- *Promotion of coastal bioshields*. Since it has been proven that mangroves and other coastal plantations help in reducing the impact of cyclones and tsunamis on human settlements, shelterbelt plantations have been extensively promoted. In some places, these are linked with livelihood activities such as crab fattening in mangrove areas (MSSRF 2008). There is also legal protection available through the CRZ by designating 'ecologically sensitive areas' which include mangroves and sand dunes.

## 12.4 Post-Tsunami Rebuilding and Subsequent Disasters

### *12.4.1 Cyclone Thane*

Cyclone Thane made landfall between Cuddalore and Puducherry on December 30, 2011. It caused large scale devastation despite the fact that it was described as a Category 1 cyclone as the maximum sustained wind speeds reached only about 140–150 kmph which is supposed to cause minimal damage. The Indian Meteorological Department helped to predict the storm's landfall and surge heights, which impacted an area that was much larger than the narrow coastal belt affected by the 2004 tsunami. In addition to waves, strong winds pulled down trees and electrical poles resulting in severe damages to the area's power infrastructure. A number of useful lessons were learned as a result of this disaster (George and Lakshmi 2012).

1. *Effectiveness of EWS*: The early warning system provided continuous information about the cyclone which was aired over radio and television stations. However, as a precautionary measure to prevent accidental electrocution, electricity was shut down 24 hours before the cyclone struck which meant that people could not watch the TV. Other traditional modes of communication were also used to supplement mass media channels including the ringing of temple/church bells and messages on PA systems. Modern communication techniques such as mass text messaging were not used however, and should be utilised in the future. It is also important to recognise who delivers the message. For instance, providing information through the FM channel was a good idea, but since these channels are largely for entertainment, it is not clear if the messages were taken seriously. In some places, people were urged to move to safety, yet there were no follow-on instructions on the safe places to move to or any organised attempt to shift them to other locales as new information became available.
2. *Experience helps*: Compared to Puducherry and Cuddalore, fewer boats were lost in Nagapattinam district as fishermen dragged their boats as far inland as possible, having experienced considerable loss during the 2004 tsunami. On the other hand, in Puducherry, the impact of the tsunami was lower and here as well as in Cuddalore, people were more complacent because cyclone warnings were often announced only to see them veer off the coast without causing significant impacts.
3. *Disaster resilient housing*: People living in disaster resilient housing experienced less wind-related damages than other residents but they too had to face issues of flooding. Disaster resilient housing should give greater consideration to the role of site design (including proper drainage) and the physical proximity of housing to areas vulnerable to flooding, including current cyclonic and rainfall events as well as projected rising sea levels.
4. *Impact on industries*: Cyclones and associated hazards such as strong winds and heavy rain can impact industries as well—Cuddalore and Puducherry have a large number of medium and small scale industries as well as large industries, compared to Nagapattinam. Unlike the tsunami which affected largely the fishing industry on the coast, heavy rain and wind associated with cyclones caused impacts further inland, affecting other activities such as industries. Infrastructure such as roads and power lines were affected due to falling trees and debris. Thus, climate change related disasters would require the inclusion of non-coastal industries and those responsible for the construction and maintenance of infrastructure in the development of preparedness as well as recovery strategies.
5. *Impact on agriculture*: Plantation crops such as coconut and Jackfruit were extensively damaged in Cyclone Thane. This kind of damage was not seen in the aftermath of the 2004 tsunami. A conclusion therefore is that damage assessment protocols and rehabilitation packages have to be created for multiple situations. In the case of agriculture, those most likely to be affected are the highly vulnerable small and marginal farmers. Their resilience needs to be improved to cope with the variety of impacts expected in a changing climate scenario.

### 12.4.2 Tsunami Warning

On April 11, 2012, an earthquake struck close to the origin of the 2004 earthquake off Indonesia. According to the USGS, the 8.7 magnitude earthquake was tsunamigenic. Media reports that a tsunami was expected to reach the Nicobars increased concern among people along the coast of Tamil Nadu. Phone lines became overwhelmed and the information conveyed in the media focused only on the repeated reports that an earthquake had occurred. In Chennai, the Marina beach was evacuated and officers of the government went to hamlets located close to the shoreline to alert the local populace (Anon 2012). The region faces acute power crisis and as part of their management strategy, there was a routine power cut in Nagapattinam and other places which had experienced the brunt of the 2004 tsunami. Not being able to switch on the radio or TV caused significant anxiety among residents as few had clear information on the situation. Websites remained inaccessible or were not updated on a regular basis. Ultimately, a tsunami did not materialise as the earthquake was a strike-slip fault and did not result in a vertical displacement of the seafloor.

Some of the lessons learned from this case include (Lakshmi and George 2012):

- Connectivity was bolstered by the use of text messaging, taking advantage of the extensive penetration of the cellular network.
- Communities were alert and ready to evacuate on short notice, including some who were able to bring their valuables with them.
- Families in 'safe houses' constructed after the 2004 tsunami felt confident about staying, saying they would relocate to the roof if the wave was large.
- Long power cuts hindered the ability of people to receive updated information from television channels.
- The state administration set up a control room manned by officers with prior experience in handling disasters.

## 12.5 Barriers and Opportunities

As described in this chapter, disaster risk reduction (DRR) as well as preparedness is becoming a more readily accepted practice linking disaster management and climate change adaptation under the umbrella of sustainable development. However, despite the fact that there is supposedly a paradigm shift in the approach to disaster management (of being proactive rather than reactive), there is still a long way to go towards achieving it as even if there are policies for proactive DRR, the mind-set of the people continues to be relief-centric. One of the problems is that people have become accustomed to being provided with compensation for damages due to disasters. This was particularly evident in the aftermath of the 2004 tsunami and became an expectation after Cyclone Thane. This is a major barrier that needs to be overcome.

A major breakthrough achieved in the aftermath of the tsunami was the improved communication and interaction between communities and government. This is an opportunity that needs to be capitalised on to ensure better governance. Communities are the long term players in adaptation at the local level and hence they should become partners in governance because in the long term it is the intangible relationships that are sustained over time that are among the most important. These horizontal relationships should, however, be coupled with strong vertical linkages thereby improving not only the identification of local needs and the maximisation of indigenous knowledge and resources, but also take advantage of the varied governmental programmes that can support locally-grounded approaches to creating more adaptive communities.

Vulnerability reduction through a variety of welfare schemes are already available and efforts have been made to ensure that socially vulnerable groups such as the elderly, widows, deserted women and the physically and mentally challenged are included in various social security nets. Implementation of such schemes needs to be innovative and opportunistic. An example of such efforts includes the use of the National Rural Employment Guarantee Scheme (www.nrega.nic.in) to clear debris after Cyclone Thane by the Government of Tamil Nadu. Using this social security mechanism can ensure "cash-for-work" following disasters. More important, it takes care of the asset-less people like labourers who are not eligible for compensation based on assets lost. As the decision-making and disbursements are implemented through the Panchayati Raj Institutions (PRI), this represents a move towards decentralising disaster management (George and Lakshmi 2012). Barriers to proper functioning of such schemes can be overcome by increased levels of community involvement and flexible policies are able to adjust to identified local needs.

The United Nations General Assembly designated the 1990s as the International Decade for Natural Disaster Reduction (IDNDR), with the aim to decrease the loss of life, property destruction and social and economic disruption caused by natural disasters. Since India is an agrarian country, the focus has been on floods and droughts and it was the agriculture ministry that dealt with disasters. Disaster response in India has traditionally been reactive and relief-centric. The turning point in the approach to disasters was probably the 1999 Supercyclone Odisha (Orissa). The 2001 Bhuj (Gujarat) earthquake and the 2004 tsunami disaster accelerated the efforts towards a paradigm shift from being reactive to proactive, with focus on early warning systems, disaster resilient technologies, capacity building and risk mitigation strategies.

On 23 December 2005, the Government of India enacted the Disaster Management Act, which envisaged the creation of the National Disaster Management Authority (NDMA 2008), headed by the Prime Minister, and State Disaster Management Authorities (SDMAs) led by respective Chief Ministers. The purpose of the NDMA was to spearhead and implement a holistic and integrated approach to Disaster Management in India. NDMA mandated the policies, plans and guidelines for Disaster Management to ensure timely and effective response to disasters. A National Disaster Management Policy has been prepared. There has also been a paradigm shift in the approach to disaster management, from reactive response to proactive preparedness. The NDMA has also led to a series of guidelines ranging from management of tsunamis, cyclones, earthquakes and other natural hazards to guidelines for minimum standards for relief and the role of NGOs in disaster situations.

In recent years, coastal India has experienced extensive development. Settlements as well as other anthropogenic activities currently occupy more than 58 % of the coastline (Lakshmi et al. 2012). Ports, power plants, special economic zones and industrial zones are growing rapidly. The Environmental Impact Assessment (EIA) Notification, 2006, outlines the process for assessing, mitigating and managing environmental impacts. The CRZ Notification of 2011 provides for creating a buffer zone between the sea and human activities on land to reduce vulnerability from coastal hazards. The notification also states that no development activities may be permitted in ecologically sensitive areas such as sand dunes and mangroves which perform important ecosystem services as they have been rated very high in their potential to reduce the impact of waves generated by high tides, storm surges and even tsunamis.

## 12.6 Conclusions—The Way Forward

This chapter examines the response to the 2004 Indian Ocean tsunami by the coastal state of Tamil Nadu to see what lessons may help coastal communities adapt to climate change. As far as coastal India is concerned, communities have to be prepared for creeping disasters like sea-level rise (and its various impacts on the shoreline) as well as extreme events such as cyclones. Three specific recommendations for building community resilience, adaptive capacity and sustainability include:

1. *Mainstream disaster risk reduction.* Disaster risk reduction must be seen as an inherent part of development schemes. Reduction of vulnerability includes reducing physical vulnerability as well as socio-economic vulnerability. Reduction in physical vulnerability can be done in multiple ways, including the identification of vulnerable coastal areas through the mapping of hazard zones, the relocation of at-risk communities to safer areas, providing multi-hazard resilient housing and associated infrastructure and more effectively managing bioshield buffers. The local community is a key source for information on local vulnerability as well as mitigation and management options. Hence, a bottom-up participatory approach in building action plans for development activities at the local level must be put in place. Operationalising this approach requires the application of integrated coastal management principles. Integrated coastal management is defined as a continuous and dynamic process by which decisions are made for the sustainable use, development and protection of coastal and marine areas and resources. The main purpose of ICM (or ICZM—Integrated Coastal Zone Management) is to overcome fragmentation due to sectoral management approaches and uncoordinated local jurisdictions (Cicin-Sain and Knecht 1998; Kay and Alder 1999). In order to be effective and sustained over time, communities must take ownership of the plans while the government pro-

vides the requisite support such as adequate financing, capacity building and technology transfer.

2. *Conduct local needs assessment.* Establishing a baseline of local needs helps to target proposed capacity building efforts and assess whether the right approaches are being undertaken. This is especially important in terms of livelihood diversification (FIMSUL 2011b). Coastal fishing communities are innovative and are fairly clear about what is likely to succeed when it comes to livelihood diversification programmes. Key elements of success include placing emphasis on local employment opportunities related to their existing livelihoods. Providing appropriate market and credit linkages apart from required infrastructure is also needed. Improved economic status as well as working towards improved social status by providing appropriate incentives to vulnerable populations as well as addressing poverty will ensure inclusive adaptive capacity building initiatives.
3. *Enhance coordination.* The most important lesson is the need for improved horizontal and vertical coordination after a disaster strikes as well as during ongoing development activities. The government has a variety of welfare schemes and various policies that are designed to reduce vulnerability. However, there is often extensive overlap between schemes and some policies actually work against each other, resulting in an increase in vulnerability. Important changes require improving levels of inter-departmental coordination as well as coordination at multiple administrative levels. In both cases, coordination should be supported by relevant information that is constantly updated. This will enable both communities and government departments to make informed decisions, increase institutional flexibility and improve the ability of larger governance systems to adapt to a changing climate while also more effectively managing the causes and effects of disasters.

**Acknowledgments** The authors thank V. Vivekanandan, Annie George and Satish Babu for many useful discussions.

## References

Anon. (12 April 2012). How official machinery reacted to the tremors. *The Hindu*, Chennai.

Babu, S., George, A., Lakshmi, A., Katticaren, R., & Sukumaran, S. (2008). Building local capacities for disaster response and risk reduction. An Oxfam-BEDROC study. BEDROC. www.bedroc.in.

BMTPC. (2008). *Vulnerability atlas for India*. Building Materials Technology Promotion Council, New Delhi, 2008.

Census of India. (2011). Provisional population totals: Tamil Nadu. http://www.censusindia.gov.in/2011-prov-results/prov_data_products_tamilnadu.html. Accessed 26 Oct 2012.

Cicin-Sain, B., & Knecht, R. W. (1998). *Integrated coastal and ocean management—Concepts and practices*. Washington DC: Island Press.

CIESIN. (2007). Columbia University. Centre for International Earth Science Information Network. http://preventionweb.net/go/7689. Accessed 24 Oct 2012.

Danielsen, F., Sørensen, M. K., Olwig, M. F., Selvam, V., Parish, F., Burgess, N. D., Hiraishi, T., Karunagaran, V. M., Rasmussen, M. S., Hansen, L. B., Quarto, A., & Suryadiputra, N. (2005, October 28). The Asian tsunami: A protective role for coastal vegetation. *Science*, *310*, 643.
Das, S., & Vincent, J. R. (2009, May 5). Mangroves protected villages and reduced death toll during Indian super cyclone. *Proceedings of National Academy of Sciences*, *106*(18), 7357–7360. doi:10.1073/pnas.0810440106. (Published online before print April 20, 2009)
DoF (Department of Fisheries). (2004). Fisheries department policy note 2004–2005. Animal husbandry and fisheries department, government of Tamil Nadu. http://www.tn.gov.in/policynotes/archives/policy2004-05/fish2004-05.htm. Accessed 26 Oct 2012.
DoF (Department of Fisheries). (2005). *Census of Marine Fisherfolk 2000*. Department of Fisheries, Government of Tamil Nadu.
FIMSUL, Vivekanandan, V., & Kasim, H. M. (2011a). Fisheries management options for Tamil Nadu & Puducherry. A report prepared for the fisheries management for sustainable livelihoods (FIMSUL) Project, undertaken by the UN FAO in association with the World Bank, the Government of Tamil Nadu and the Government of Puducherry. Report No. FIMSUL/R20. FAO/UTF/IND/180/IND. New Delhi, Chennai and Puducherry, India.
FIMSUL. (2011b). Fisheries management for sustainable livelihoods (FIMSUL) Project in Tamil Nadu and Puducherry, India. FINAL REPORT. Delhi: Food and Agriculture Organisation of the United Nations. FAO Office, Lodi Estate, Delhi, India. FAO/UTF/IND/180/IND. (Vol. I—Key Findings and Recommendations; and Vol. II—Summary of FIMSUL Process and Outcomes under each Work Package).
George, A., & Lakshmi, A. (2012, February). Cyclone Thane: Disaster preparedness and response. TRINet Newsletter. The Resource and Information Network for the Coast. www.trinet.in.
Gomathy, N. B. (2006, January 18–19). The role of traditional panchayats in coastal fishing communities in Tamil Nadu, with special reference to their role in mediating tsunami relief and rehabilitation. Proceedings of the Regional Workshop on Post-Tsunami Rehabilitation of Fishing Communities and Fisheries-based livelihoods. Chennai, India. International Collective in Support of Fishworkers (ICSF) (pp. 211–244).
GoTN. (2005, December). *Tiding over tsunami—Part I*. State Relief Commissioner, Revenue Administration, Disaster Management and Mitigation Department, Government of Tamil Nadu.
GoTN. (2008, October). *Tiding over tsunami—Part II*. State Relief Commissioner, Revenue Administration, Disaster Management and Mitigation Department, Government of Tamil Nadu.
GoTN. (2011). Action taken on World Bank funded ETRP (Emergency Tsunami Reconstruction Project (as on 21–12–2011) and Action Taken on ADB (Asian Development Bank) funded TEAP (Tsunami Emergency Action Project). http://www.tn.gov.in/tsunami. Accessed 23 Oct 2012.
Hedao, P. (2006, June 12). "Natural ecosystems along the Coromandel coast". Presentation at the TRINet Workshop on "Post Tsunami Coastal Protection Projects. Chennai. Tsunami Rehab Information Network. www.trinet.in. Accessed 15 June 2006.
IMD. (2011). Very severe cyclonic storm "THANE" over the Bay of Bengal, (25–31 December, 2011): A report. http://www.imd.gov.in/section/nhac/dynamic/THANE.pdf. Accessed 15 Oct 2012.
INCCA. (2010, November). *Climate change and India: A 4 by 4 assessment. Indian network for climate change assessment*. Ministry of Environment and Forests, Government of India.
Kathiresan, K., & Rajendran, N. (2005). Coastal mangrove forests mitigated tsunami. *Estuarine Coastal and Shelf Science*, *65*, 601–606.
Kay, R., & Alder, J. (1999). *Coastal planning and management*. E & FN Spon. Routledge: London.
Kurian, V. (4 January 2012). IMD claims its radars chased 'Thane' to the wire. *The Hindu Business Line*.
Kurien, J. (1995). The Kerala model: Its central tendency and the outlier. *Social Scientist, 23*, (1–3, Jan–Mar), 70–90.
Lakshmi, A. (2007, March). "Agents of change". TRINet Newsletter: Tsunami rehab information network. www.trinet.in. Accessed 1 Mar 2007.
Lakshmi, A., & Bau, M. (2007). Project management needs in disaster situations: Lessons learnt from the Boxing day Tsunami. Originally published as a part of 2007 PMI (Project Management Institute) Global Congress Proceedings—Hong Kong.

Lakshmi, A., & Babu, S. (2010). Community based reconstruction experiences after 2004 Tsunami. In R. Shaw & R. R. Krishnamurthy (Eds.), *Communities and Coastal Zone Management* (pp 11–27). Research Publishing: Singapore.

Lakshmi, A., & George, A. (2012, May). Changing responses to natural hazards. TRINet Newsletter: The resource and information network for the coast. www.trinet.in. Accessed 1 May 2012.

Lakshmi, A., Schiavina, A., Banerjee, P., Reddy, A., Mandeen, S., Rodriguez, S., & Apte, D. (2012, October). The challenged coast of India. A report prepared by PondyCAN (Pondicherry Citizen Action Network) in collaboration with BNHS (Bombay Natural History Society) and TISS (Tata Institute for Social Sciences).

Lay, T., Kanamori, H., Ammon, C., Nettles, M., Ward, S., Aster, R., Beck, S., Bilek, S., Brudzinski, M., Butler, R., DeShon, H., Ekström, G., Satake, K., & Sipkin, S. (2005). The great Sumatra–Andaman earthquake of 26 December 2004. *Science*, *308*, 1127–1133. doi:10.1126/science.1112250.

Maplecroft. (2010). Big economies of the future—Bangladesh, India, Philippines, Vietnam and Pakistan—Most at risk from climate change. http://maplecroft.com/about/news/ccvi.html. Accessed 26 Oct 2012

McGranahan, G., Balk, D., & Anderson, B. (2007). Low Elevation Coastal Zone (LECZ) Urban-Rural Population Estimates, Global Rural-Urban Mapping Project (GRUMP), Alpha Version. Palisades: NASA Socioeconomic Data and Applications Center (SEDAC). http://sedac.ciesin.columbia.edu/data/set/lecz-low-elevation-coastal-zone. Accessed 26 Oct 2012.

MoEF. (2011). The coastal regulation zone notification, Ministry of Environment and Forests, Government of India.

MSSRF. (2008). *Ecological rehabilitation of the coastal areas of Pichavaram: Establishing community-based Mangrove and other coastal vegetation as bioshield.* Annual Report of the M. S. Swaminathan Research Foundation, 2007–2008. www.mssrf.org. Accessed 26 Oct 2012.

Murthy, K. S. R., Subrahmanyam, A. S., Murty, G. P. S., Sarma, K. V. L. N. S., Subrahmanyam, V., MohanaRao, K., Suneetha Rani, P., Anuradha, A., Adilakshmi, B., & Sri Devi, T. (2006, June 10) Factors guiding tsunami surge at the Nagapattinam–Cuddalore shelf, Tamil Nadu, east coast of India. *Current Science*, *90*(11), 1535–1538.

NCRC. (2006). Report of national workshop on disaster preparedness in agriculture, NGO coordination and resource centre, Nagapattinam, August 3–4, 2006.

NDMA (2008). National disaster management guidelines: Management of cyclones. National Disaster Management Agency. www.ndma.gov.in. Accessed 24 Oct 2012.

Niyas et al. (2009). Climate Profile of India. Contribution to the Indian Network for Climate Change Assessment (National Communication-II). Ministry of Environment and Forests, Government of India. (Quoted in S. D. Attri and A. Tyagi, 2010)

Ramesh Kumar, M. R., & Sankar, S. (2010, December). Impact of global warming on cyclonic storms over north Indian Ocean. *Indian Journal of Marine Sciences*, *39*(4), 516–520.

Salagrama, V. (2006). Trends in poverty and livelihoods in coastal fishing communities of Orissa State, India. FAO Fisheries Technical Paper 490. Food and Agriculture Organization of the United Nations, Rome, 2006a.

Saxena, S., Purvaja, R., Suganya, G., Divya, M. & Ramesh, R. (2012). Coastal hazard mapping in the Cuddalore region, South India. *Natural Hazards*, Published Online 13 September 2012. doi:10.1007/s11069-012-0362-7.

Subramanian, B. R. (2006). Inundation of seawater due to Indian Ocean Tsunami along the Indian Coast. In G. V. Rajamanickam, B. R. Subramaniyan, M. Baba, R. Ramesh, L. Elango, & M. Prithviraj (Eds.), *26th December 2004 tsunami: Causes, effects, remedial measures pre and post tsunami disaster management: A geoscientific perspective* (pp 10–44). New Delhi: New Academic Publishers.

UN. (2008). *Disaster recovery: A focus on good practices. UN Tsunami Response—India.* Compiled by Dave Storey and Anupama Pai.

UNDP. (2005). Forearmed for surviving natural disasters. Good Practice 2005-2004-18.4648364550. UNDP Asia-Pacific.

Unnikrishnan, A. S., Manimurali, M., & Ramesh Kumar, M. R. (2010, November 16). Sea-level changes along the Indian coast—Impacts & vulnerability. Presentation made at the National

Workshop 16th Nov 2010. http://moef.nic.in/downloads/others/Sea-level_Unnikrishnan_Manimurali%20and%20Kumar.pdf. Accessed 19 Sept 2011.
UNTRS. (2005). Tsunami damages in India: United Nations team for tsunami recovery support. http://reliefweb.int/sites/reliefweb.int/files/resources/3F6E3FF46EA8E9B48525717A00525387-undp_TS_ind300406.pdf. Accessed 4 Feb 2009.
UNTRS/FAO. (2008). Setting directions for sustainable fisheries and coastal livelihoods in the post-tsunami context, India. A report of the interventions by the Food and Agriculture Organisation of the United Nations (FAO) under the United Nations team for Tsunami Recovery Support.

# Chapter 13
# Coastal Hazards Planning: The 2009 Tsunami and Lessons Learned for Climate Change Adaptation in Samoa

**Namouta Poutasi, Michele Daly, Jude Kohlhase and Filomena Nelson**

**Abstract** This chapter describes a coastal hazards planning project in Samoa which reduces the vulnerability of coastal communities and strengthens institutional and community capability to adapt to and manage climate-related disasters. Key elements included respect for *fa'a Samoa* and *fa'a matai*, culturally appropriate processes and plans and integration of planning and disaster management frameworks.

**Keywords** Coastal hazards · Coastal infrastructure · Coastal planning · Samoa · Community resilience

## 13.1 Introduction

Samoa is a small island developing state in the south-west Pacific and is subject to a range of climate-related hazards such as storm surge, cyclones and landslips, as well as other coastal hazards such as tsunami. Approximately 80% of the 403 km coastline is 'sensitive' or 'highly sensitive' to erosion, flooding or landslip (Gibb 2001).

Between 70 and 80% of the population of 181,081 people (World Bank 2010) live on or near the coast (100% live within 100 km of the coast) with the majority living on the island of Upolu. Most of the country's physical and social infrastructure

M. Daly (✉)
Risk and Society Department, GNS Science Ltd, Avalon, Lower Hutt, New Zealand
e-mail: M.Daly@gns.cri.nz

N. Poutasi
Integrity Professionals, Auckland, New Zealand

J. Kohlhase
Planning and Urban Management Agency, Ministry of Natural Resources and Environment Apia, Samoa

F. Nelson
Disaster Management Office, Ministry of Natural Resources and Environment

B. C. Glavovic, G. P. Smith (eds.), *Adapting to Climate Change,* Environmental Hazards,
DOI 10.1007/978-94-017-8631-7_13, © Springer Science+Business Media Dordrecht 2014

**Fig. 13.1** The villages of Taelefaga and Maasina, Fagaloa Bay, north coast of Upolu. Most villages are located along the coast with associated roads, power, schools, health clinics and other infrastructure vulnerable to coastal hazards. (Source: Photograph Michele Daly)

is located along the coast (Fig. 13.1), indicating a need for improving development policy within the context of climate-related and other coastal challenges.

Samoa's agricultural and fishing-based economy has traditionally depended on development aid and family remittances from overseas, conforming to many aspects of Bertram and Watters' (1985) Migration, Remittances, Aid and Bureaucracy (MIRAB) economy for small islands. Tourism is an expanding sector accounting for 25% of GDP. In 2007, 122,000 tourists visited the islands.

Samoa's cultural context is also an important factor when prioritising and selecting climate change adaptation policies. The traditional model of community decision-making is by consensus under the leadership of the *matai* (chief). The authority of a village matai and customary land ownership rights are respected and negotiated settlements between the government and village matai that can take a long time to develop. There is a commitment to supporting village based participatory consultation processes which are inclusive of women and untitled youth. Raising awareness of climate change and other development concerns through village based consultation is seen as an effective and sustainable way of supporting the traditional decision-making model.

> Individual tropical cyclones…are unpredictable; so most South Pacific islands are exposed to some degree of risk every year and must be always prepared. (Jenwick 2000)

Between 1980 and 2000 it was estimated that Samoa (and areas within 100 km$^2$ of the island nation) experienced an average of 1.3 cyclones per year (Jenwick 2000). Samoa suffered two major cyclones in two consecutive years (Ofa 1990 and Val 1991). Cyclone Ofa was reported to be one of the strongest storms to affect Samoa in the twentieth century leaving seven people dead and causing at least US$ 180 million in damage to multiple countries (Ready and Woodcock 1992). Damage to dwellings, businesses and social facilities was extreme and generally caused by either large waves which inundated the coastal zone or high winds. Sea inundation and salt water poisoning severely affected the subsistence economy of many of the coastal

**Fig. 13.2** Example of typical damage to houses along the south coast of Upolu as a result of the September 2009 tsunami (photo taken March 2010). (Source: Photograph Michele Daly)

villages. Government buildings suffered considerable damage during the cyclone, including the Apia Observatory, which resulted in the loss of important historical meteorological and scientific records. Coastal roads, in particular, sustained significant damage from the force of the breaking waves overtopping them (Rearic 1990). Cyclone Val (a Category 5 cyclone) caused 13 deaths, destroyed half of Samoa's coconut trees and severely stripped 90% of Samoa's land. Both cyclones Val and Ofa led to international funding and aid of approximately US$ 500 million to assist with recovery efforts.

In September 2009, Samoa experienced a devastating tsunami resulting in the loss of 143 lives, 310 people seriously injured and an estimated 5,274 people (1:50 Samoans) directly affected. Over 3,000 Samoans lost their homes (Fig. 13.2). Twenty southern Upolu coastal villages were destroyed or severely affected. The total value of the disaster-related damages caused by the tsunami is estimated at US$ 124.04 million, more than 22% of Samoa's GDP (World Bank 2009, 2010).

Changes in the scale and impact of these types of disasters is expected to worsen in Samoa because the increases in sea-level and average sea surface temperatures will increase the intensity and damage from major storms. Other potential impacts are linked to weather patterns tied to El Niño-driven Southern Oscillation (ENSO) events, leading to agriculture impacts associated with lower rainfall totals (World Bank 2010). The economic implications are not thought to be significant (0.6–1.3% of the present value of GDP to 2050) if sound development policies are adopted to minimise the impact of existing and predicted increased weather risks and other natural hazards (World Bank 2010). These include updating and implementing stronger building design standards to reflect the more common occurrence of high intensity winds, developing more effective warning systems and adopting coastal land development policies and plans that address climate change-induced sea-level rise, higher storm surge inundation levels and areas prone to landslips. Extensive community consultation has been undertaken over the past 5 to 10 years to raise awareness of disaster management and to identify community priorities for sustain-

**Fig. 13.3** Apia road flooding in 2006. (Source: Photograph Beca International Consultants Ltd.)

able development and climate change adaptation. A cornerstone to climate change adaptation practice has been Samoa's approach to the development of its land use planning framework.

Land use planning in Samoa is in its relative infancy, with the Planning and Urban Management Act being passed in 2004 (PUMA Act 2004). Prior to the passage of the PUMA Act (2004), planning-related functions were implemented by various government agencies, often in isolation and sometimes in conflict with one another. In the absence of a planning framework, urban development in Apia was largely unregulated and has created a haphazard urban environment (Taule'alo 2000). R. Willcox, former Public Works Department Town Planner (1972), describes the area along Beach Road: '*workshops and small factories are located next to residential, commercial and office buildings where the noise, fumes and traffic necessary to their operations is a nuisance to their neighbours*'.

Inadequate drainage leading to flooding, poor sewerage disposal and uncontrolled urban growth continued in 2000 (Taule'alo 2000). Flooding resulting from inadequate drainage continues into more recent times (Fig. 13.3). The absence of an institutional planning structure has been identified as a main cause of urban problems in small island states, with improvements to the planning delivery system and inclusion of environmental considerations seen as key challenges (Connell and Lea 1992).

Much of the previous academic work on land use planning in Samoa (as opposed to only describing land use) was related to tourism (e.g. Pearce 2000; Twining-Ward and Butler 2002).

The purpose of the PUMA Act (2004) is to establish a Planning and Urban Management Agency and to implement a framework for planning the use, development, management and protection of land in the present and long-term interests of all Samoans and for related purposes. The Act is modelled on New Zealand's legislation with the intention to progressively introduce statutory land use plans, while representing the Samoan socio-political context. The Act provides a planning framework to direct and coordinate the development, use and protection of Samoa's natural and physical resources and to address the infrastructure and vulnerability issues of the past. The Act has a number of objectives which are currently being implemented. Objective 8(c) focuses on the urban structure and form for the development of Apia, the capital and largest city in Samoa. In late December 2011 and early 2012, the Government of Samoa was developing an Apia Spatial Plan (a long term vision for Apia that will guide the development of the city) and undertaking flood vulnerability and adaptation assessments for the city.

In order to reduce the vulnerability of Samoa's coastal infrastructure to climate change-related hazards within the context of wider development challenges such as urban planning (e.g. Jones and Lea 2008), the Government of Samoa undertook a significant project in partnership with communities which sought to assess the vulnerability of the country's infrastructure to climate-related hazards and to develop a land-use and disaster management framework.

The Samoa Infrastructure and Asset Management (SIAM) project provided a unique opportunity to explore how these national frameworks could be integrated into policy and practice, including how implementation could occur at a local level through the development of a Coastal Infrastructure Management Strategy (CIM Strategy) and Coastal Infrastructure Management Plans (CIM Plans). The national policy framework is intended to provide consistent messages at all levels and better support across national, district and village levels.

The CIM Plans provide non-statutory guidance for future planning in Samoa in a number of ways including providing a framework for local village land use planning. Land is predominantly communally owned by extended families and the allocation of various land uses is traditionally controlled at a village level by the village council (*fono a matai*). The CIM Plans provide information in support of village decision-making concerning land use. The CIM Plans also provide a framework for the PUMA Act 2004 to process the required development consents which the village must apply for to undertake certain activities such as coastal works. The list of identified mitigation actions in the CIM Plans, such as relocating infrastructure, repairs to community assets and coastal works, provide a framework for further public and private investment allowing funding to be directed toward projects that have already been identified as having a community benefit. In the future, the CIM Plans will assist in providing a framework to introduce statutory land use plans under the PUMA Act (2004). Statutory plans will likely be developed slowly over the next 10 years, initially focusing on developed areas such as Apia where more land is held in private ownership. In rural areas, a statutory approach will need to take into account the predominant communal land tenure and village-style decision-making. This aspect is discussed in more detail later in the chapter.

The CIM Plans were used after the 2009 tsunami to identify broader development goals for the villages affected. They provided a base map of infrastructure (pre-tsunami), coastal hazard zones and a documented list of issues, concerns and possible solutions for each village. These were taken into consideration when planning for the recovery of the affected villages, and in some cases became the *de facto* community recovery plan (World Bank 2010). Added to the CIM Plan information was subsequent information from later village consultations on wider development issues including food security and ideas for economic self-sufficiency (e.g. micro-enterprise options). Incorporating all of this information enabled the recovery process to build on a solid information base and provided greater certainty around the recovery options for the affected villages. A number of villages had already identified relocation away from the coast as a potentially desirable solution to reduce the impact of weather related hazards. Some villages have relocated in entirety (*Saleapaga*) while for others tourist operations remain on the coast with villagers' homes located on higher ground in the surrounding hills (*Lalomanu*). The concept of relocation has been implemented in some villages which were directly affected by the 2009 or past tsunami. Relocation (as recommended in the CIM Plans for certain villages) is a long term goal that requires a substantial amount of planning and coordination. Initial activities which assist with relocation such as sealing inland work roads and providing services along these roads have been identified, prioritised and put onto a work programme database ready for funding to be allocated when available. Lack of funding to implement activities other than core business continues to be an issue.

## 13.2 Lessons Learned

To draw on the lessons of the CIM planning experience, additional background about its development is required. This section describes the planning process in more detail and identifies key successes, challenges and lessons learned.

The SIAM project was a World Bank funded initiative that was led locally by the Planning and Urban Management Agency (PUMA) and National Disaster Management Office (NDMO) of the Samoa Ministry of Natural Resources and Environment (MNRE). The Government of Samoa commissioned New Zealand based consultancy Beca International Consultants Ltd to help coordinate the project, which included the provision of training and support for government staff and local partners to undertake much of the work.

Two of the major integrated streams of work were:

1. The development of a national level policy for the management of coastal infrastructure (CIM Strategy) and local implementation plans (CIM Plans) for all Samoa's coastline.
2. The development of a national level disaster management framework which was operationalised through the Samoa Disaster and Emergency Management Act (2007) and the Samoa National Disaster and Emergency Management Plan (2007). This framework covered disaster management, institutional strengthening of the National Disaster Management Office and engagement of private sector response agencies in disaster risk management linked to the wider context of climate and development endeavours.

The workstreams were supported by State of the Environment reporting which involved surveying all of Samoa's 403 km coastline and mapping the extent and condition of natural environments (e.g., landforms, mangroves and lagoons), identifying natural resources (e.g., aggregate and offshore sand resources) and mapping coastal hazards (coastal inundation, erosion and landslip). The latter workstream took an all hazards approach and addressed tsunami risk reduction in addition to dealing with climate-related hazards.

The national level CIM Strategy which drove the planning process has the following central vision: "*Resilience: Coastal Infrastructure & Communities resilient to natural hazards.*" The Strategy provides a clear direction for coastal hazard and environmental information gathering and monitoring, education and awareness raising, use and management of resources and for undertaking coastal defences and works. The CIM Plans were designed to serve as one of the key methods of implementation (Beca 2001).

The CIM Plans describe the existing environment and identify and assess the resilience of existing infrastructure against coastal hazards as well as identify potential solutions to reduce exposure. Infrastructure in the context of the CIM Plans includes not only important utility and network infrastructure (roads, power, water, telecommunications) but also important village buildings and other social infrastructure (bathing pools, schools, committee houses, health clinics, etc). The CIM Plans assess resilience by identifying key infrastructure, associated problems and potential solutions and assign actions to implement the identified measures. They also include a disaster response plan for the village and district. This guidance enhances the ability of stakeholders to be adaptive and responsive while enabling an informed recovery process after a hazard event (i.e., to be resilient according to Paton 2006).

The CIM Plans were developed through extensive traditional village and repeated district *fono* (meetings) spanning seven years, involving 7,000 people and 60 government staff. The consultation process is described in more detail below (see also Daly et al. 2010).

Many of the solutions identified in the CIM Plans were developed by the villagers themselves. Research shows that community resilience can be increased when communities are empowered to discuss hazards and risks and to participate in collective problem solving (Paton 2008; Mercer 2010).

Actions to address the solutions identified were assigned to various government departments and local villages to implement. The CIM Plans were formally adopted by village representatives, the Chief Executive Officer (CEO) and the Minister of the Ministry of Natural Resources and Environment, thus demonstrating the Government's commitment. This cemented the partnership amongst the key participants to implement each Plan's provisions while signifying an acknowledgement by each party of the roles and responsibilities of the others. The CIM Plan actions requiring government assistance have also been tabulated and are in the process of being prioritised and implemented. These are important steps for evaluating the sustainability and effectiveness of the project over the long term.

Community preparedness, response and immediate recovery measures for cyclones were discussed with the villages and also incorporated into the CIM Plan. The villagers did not see any separation between risk reduction and response and therefore sought a conversation about all aspects of dealing with cyclones and other climate-related challenges in the context of development. In particular, they understood how the sustainable management of their natural resources, such as offshore sand, was important to mitigate cyclone effects and how dwellings built on reclaimed land might increase their vulnerability to climate-related hazards. All of these concepts (land use planning, natural resource management and disaster risk management) were integrated and treated as a seamless topic for policy and practice at the village level, demonstrating the successful integration of climate and disaster topics into development approaches. The need for integration only came to light after an earlier pilot stage was undertaken. Partly as a result of the seamless conversations sought by villagers on all aspects of the disaster cycle (risk reduction, readiness, response and recovery), these topics were introduced into subsequent CIM Plans. A growing body of work at the time also recognised that sustainable development should include disaster risk management as a component of climate-related development policy and practice (e.g. Lewis 1999; Shea 2001, 2003; Douglas 2006).

To demonstrate the commitment to the process by the government and the intention to use the CIM Plans to guide land-use and development planning in Samoa, the completion of a CIM Plan by a village was a prerequisite for access to government funds for village mitigation projects.

The CIM Plans were completed in two stages. Stage 1 (pilot) (2000–2002) covered 15 districts and 92 villages, with approximately 2,000 people directly consulted, resulting in the development of 15 CIM Plans. The pilot enabled a number of improved aspects to be introduced in Stage 2, including enhanced maps (with better visual aids), ensuring that a cross section of the community (including women and untitled youth) are consulted and extending the CIM Plans to include a response plan component (as discussed above). The national disaster management framework workstream was also added in Stage 2.

Stage 2 (2005–2007) covered 26 districts and 191 villages, with approximately 5,000 people directly consulted (Figs. 13.4 and 13.5). An additional 26 CIM Plans were developed and at the conclusion of the project, the entire country was covered by CIM Plans. In addition, 42 organisations participated in developing the national disaster management framework, including the development of individual agency response plans and participation in three simulations. These simulations were designed to test the national response framework and provided an opportunity for agencies to test their own agency response plans and their role in the national response arrangements. The simulation scenarios were based on a cyclone event, but a pandemic scenario was also included as this was perceived to be a real threat at the time.

In order to undertake such extensive consultation, a large number of personnel were drawn from the consulting firm and the Ministries of the Government of Samoa. Staff training was an important part of the project and involved more than 60 Government of Samoa personnel spread across several government ministries. Training was delivered in preparatory workshop sessions where standard team protocols, information sheets and data recording schedules were discussed. This was supplemented by practical training on-site during the consultation when new staff were added to the outreach teams. The Government of Samoa now has a substantial capacity to maintain and update the CIM Plans as well as to engage with communities on development issues.

**Fig. 13.4** Villagers looking over aerial photo based hazard maps during CIM plan consultation. (Source: Photograph Beca International Consultants Ltd.)

**Fig. 13.5** Youth looking over aerial photo based hazard maps during CIM plan consultation. (Source: Photograph Beca International Consultants Ltd.)

As a result of the concurrent CIM Plan development process, climate (and other) risk reduction policies were integrated across disaster management and environmental management frameworks. The importance of community-developed response plans (e.g., Boura 1998), which were a component of the CIM Plans, was recognised in both the legislation and National Disaster and Emergency Management Plan.

### *13.2.1 What Worked Well?*

Reflecting on the CIM Plan development process and subsequent implementation, a number of factors have been identified as making an important contribution to the process and provide important learning opportunities for those implementing land-use planning projects in other small island developing nations. These include:

#### 13.2.1.1 Inclusive Consultation and Respect for Fa'a Samoa & Fa'a Matai

An important element driving the success of the process has been the extent and style of consultation used in the villages. Approximately 7,000 people were engaged in developing the 41 District CIM Plans, representing around 4.8% of the country's adult population.

Until recently, as evidenced by PUMA, consultation by the government with local villages was uncommon. Previously, infrastructure was generally built without direct involvement or comment by landowners. The CIM Plan project involved a systematic and coordinated approach using the village *pulunu'u* (mayor) to coordinate village meetings. The project team adopted a conventional New Zealand style consultation process adapted to reflect the particular social and political structure in Samoa. These included the concepts of *fa'a matai* and *fa'a Samoa* which loosely translated are "the way of the *matai* (chief)" and "the way of Samoa" (Hooper 1998; Watson 2007). Both concepts relate to the traditional model of community decision-making by consensus under the leadership of the *matai*.

The impact of the consultation was evident in the evaluation of those villages in which Stage 1 of the CIM Plan project was undertaken. Three years later, 39% of all villagers surveyed were aware of the CIM Strategy and 18% were aware of the CIM Plans.

The planning process emphasised numerous opportunities for village residents to have meaningful and sustained input, including: (1) village meetings (2) additional district meetings held to confirm issues raised and discuss broader district issues, (3) after draft plans were prepared and (4) the final meeting where all representatives signed the document.

The use of participatory planning processes allowed for the incorporation of traditional village coping mechanisms which proved particularly important during major decisions such as the relocation of the village of *Falealupo,* which occurred in response to a tsunami in the early 1990s. By engaging the villages throughout the process of creating the action plans, community-level ownership and support for the CIM plans were achieved.

Another creative addition to the consultation methodology instigated in Stage 2 was the departure from the traditional meeting involving *matai*, high chiefs and orators to include the involvement of women and *ali'i* (untitled youth) (Fig. 13.5). While this approach follows the principles of gender inclusion articulated in the climate, disasters and development literature (Enarson and Morrow 1998; Sweetman 2009), it was highly unusual to include the latter groups in formal village consultation because Samoan culture has a hierarchy where everyone has a clearly defined position and role in society. To facilitate discussions (because women and youths are not typically allowed to speak if they do not have a *matai* title), smaller groupings of *matai*, orators and mixed groups of women and/or youths were created.

Additionally, a village representative committee was established with a *matai*, an untitled youth, a women's representative and the *pulunu'u* which attended district meetings. Meetings were conducted in Samoan, so at least a basic understanding of the language was required. District level meetings were also held to address cross-boundary risks and management issues. It was important that the underlying theme

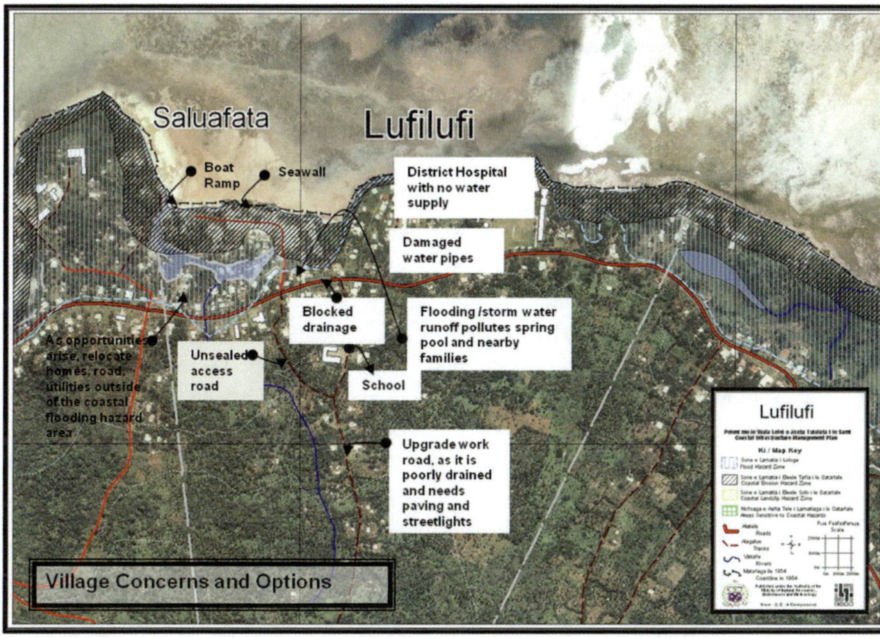


**Fig. 13.6** An example of a hazard map highlighting concerns and options in Lufilufi Village (north coast of Upolu). All the CIM Plans are available online at the Samoan Ministry of Natural Resources and Environment's website http://www.mnre.gov.ws. (Source: Beca International Consultants Ltd.)

of community participation, which is fundamental to the PUMA legislation and which will ultimately be a major determining factor in the success of CIM Plans, was secured in an appropriately Samoan manner. The CIM Strategy and Plans provide a substantial basis for this to occur.

All village consultations are now coordinated by the Ministry of Women, Community and Social Development and are required to follow the format developed with adaptations as required. Increasingly junior and intermediate staff from agencies undertaking the consultations are involved which helps build capacity. Villagers are increasingly open to having more junior members of staff (both men and women) involved although senior members of the team are still required as a matter of respect for the formal oratory and consultation with the *matai*.

#### 13.2.1.2 Culturally Appropriate Processes and Plans

Much consideration was given to the use of materials taken to and left in the villages. Rather than traditional maps (with a two-dimensional perspective), simple aerial photo-based hazard maps were developed in a way that laypeople could understand them, including being developed in both Samoan and English while being shown at a large scale with landmarks clearly identified (Fig. 13.6) (e.g., Haynes et al. 2007; Maceda et al. 2009). The villagers were able to verify buildings' locations, discuss the hazard zones and speak of their experience and knowledge of past flood and storm events. It is essential that the knowledge of indigenous communities is a primary source of information for climate and development activities (Mackinson and Nottestad 1998; Shaw et al. 2008) as well as natural hazards mitigation (Berke et al. 2011). Being able to discuss past events and to have this knowledge and experience directly recorded in the CIM Plan and on the hazard maps empowers the villagers involved and contributes to building community resilience (Paton 2007, 2008).

The plans are written in both Samoan and English. The most effective means of ensuring the CIM plans are implemented and used at both a local and national level was to make them easy-to-read and readily accessible. The CIM Plans' content and layout were specifically developed to be user friendly. They consist of a large portrait flip book with substantial numbers of illustrative photos, plans, charts and lists of actions (e.g. Fig. 13.6 and Table 13.1). The CIM Plans for all of Samoa include approximately 1,200 Samoan language plans (since each village received four copies) plus additional hard and electronic copies in both Samoan and English for the Government of Samoa. Because paper is rare in villages, at each Stage 2 meeting, eight detailed maps of the villages and district were printed and laminated, and then left with the villages. The CIM Plans are also now available on the MNRE website at http://www.mnre.gov.ws.

#### 13.2.1.3 National Approach: National Framework was Top Down (strategy) and Bottom Up

A key factor influencing the level of success in adaptation planning was securing a high level government commitment early in the process and fostering a coordinated national approach by the Government of Samoa to manage coastal infrastructure. What makes the CIM Plan process innovative is the combination of the top down national framework approach of the CIM Strategy, with implementation achieved through the grassroots/local (bottom up) approach of the CIM Plans. This process involved consensus decision-making transcending all governance levels—village, district and national. This approach helped to avert the conflicts that arose in other development projects in Samoa. (For example, Cox and Elmqvist (1997) describe, conflicts due to the introduction of reserves that were owned, controlled and managed by villagers based on the imposition of western conservation paradigms and power structures on local Samoans which were incompatible with traditional concepts of conservation and human dignity. Furthermore, local Samoan's decision-making powers were removed. The grassroots approach taken in the CIM Plan process created local buy-in which is essential in ensuring the ongoing success and implementation of the project and is considered a critical element to any climate change adaptation initiative.

The continuity of critical Ministry of Natural Resources and Environment leadership during the development of the CIM Plans has also been an important factor in their success. This combined with a lack of a statutory land-use planning framework in place at the time allowed for an innovative approach to take root. As mentioned

**Table 13.1** Lufilufi Village (north coast of Upolu) CIM Plan solutions

| Infrastructure | Best Solutions and Other Solutions Proposed | Other Benefits | Implementation Guidelines |
|---|---|---|---|
| Village houses, village historical assets (malae i Mulinu'u) and other infrastructure within the coastal hazard zones | To continue to consider building foundations at a level that takes into account the CEHZ and the CFHZ in the vicinity of buildings<br>Relocate buildings outside CEHZ and the CFHZ when replaced or ensure investment within the hazard zones is considered in relation to the potential for damage from coastal erosion and flooding<br>Village council to control sand mining<br>Plant appropriate plants in wetlands and coastal areas<br>Provide for drainage around the malae<br>Upgrade and raise concrete wall protection for Fuioa Spring Pool<br>Regularly clean the pool to clear seaweed growing in the pool<br>Enlarge or add more culverts at the pool outlet<br>*Responsibility: MNRE/MWCSD/MWTI/Village/Families* | More resilient to natural hazards<br>Better use of economic resources<br>Improved protection and resilience<br>Improved sustainability of natural resources | Family decision when building to provide long-term protection from flooding<br>Village/Family decision to relocate in long term<br>Village/Families to recognize the role that vegetation plays in stabilizing coastal areas<br>Coastal areas to be planted as soon as possible with suitable species<br>MNRE to advise on appropriate tree species and where possible, provide seedlings<br>Village to implement rules on sand mining and reclamations<br>Village to work with MWTI to upgrade and protect pool<br>Programme and budget works |
| Village work roads | Upgrade work roads to include paving and construction of drainages<br>Provide more streetlights and ensure that they are working at all times<br>Extend water and electricity utilities inland along work roads<br>*Responsibility: MWTI/MWCSD/Village/Families* | | Investigate access road upgrades with MWTI, undertake environmental impact assessment if necessary<br>Investigate drainage systems needed<br>Programme and budget works |
| Local electricity and telephone lines | Realign low lying lines at the village site<br>Replace damaged electricity poles<br>Provide underground electricity lines in the long term<br>*Responsibility: EPC/SamoaTel/Village* | | Investigate relocation and undergrounding of electricity and telephone lines and poles<br>Programme and budget works |

previously, there were a number of issues associated with Samoa's coastal island vulnerability and past haphazard plans which needed to be addressed through the introduction of a planning framework. The CIM Plans provided a sound starting point for understanding the extent of existing coastal hazards, the ability to highlight and address existing issues and to plan for future development.

#### 13.2.1.4 Integrating CIM Plan and Disaster Management Frameworks

Another successful element of the CIM Plans was the integrated approach across hazard mitigation and preparedness. A key enabling factor was the co-location of the planning agency (PUMA) and the Disaster Management Office (DMO) within the Ministry of Natural Resources and Environment. Other related functions within this Ministry include the Meteorological Office (responsible for hazard monitoring, warnings and climate change science), mapping and environmental management. The CIM plans are underpinned by physical hazard and climate science and a hazard map forms the basis of each CIM Plan. It is important that up-to-date and robust scientific information is relevant and used to support decision-making. The collection of information, such as coastline monitoring, hazard event frequency and severity and climate change science, is prioritised as part of the Ministry's overall programme. Many of the scientific and GIS mapping staff were involved in the village consultations highlighting for them the relevance and importance of the work they do. The inclusion of the planning agency within a Ministry responsible for these other functions, enables science and technical information to be more effectively integrated into planning and disaster management functions.

In addition, villagers saw risk reduction (mitigation), preparedness, response and recovery as a seamless discussion. Planning needed to take this into account and facilitators needed to be prepared to discuss a range of issues from coastline protection and enhancement options, to the provision of warnings and response actions in the event of an approaching cyclone. Often planning practice seeks to focus on one aspect (e.g. response or contingency planning), however, the villagers saw a seamless continuum across these four phases. Being able to discuss all phases provides for a more meaningful discussion between the facilitators and villagers and assists in the building of trust between the parties. The ability to connect cause and effect (hazards and consequences) with options to reduce and mitigate risk, including simple actions to take if a disaster were to occur, all promote resilience (Paton 2006; Becker et al. 2011).

#### 13.2.1.5 The Partnership Principle and Strengthened Agency Relationships

One of the key successes of this project included recognising up front the importance of partnerships and meaningful consultation (e.g., also Chambers 1980, 2002; Cooke and Kothari 2001; Kumar 2002). A partnership approach identifies what the government 'can and will do' and what villages and private agencies 'can and will do' in creating mutually supportive roles. This approach creates better ownership of climate (and other) risks by the community along with a shared responsibility for managing those risks (e.g. Ballinger et al. 2002; Valency 2007).

Implementation was meaningful for the villages in that, at the end of the process, each village had a prioritised list of hazard mitigation and preparedness projects which they could work on in partnership with the government. A CIM Plan is required to secure access to a government small projects fund and is used to support land use decision-making. A CIM Plan database of villages, infrastructure at risk and project-based solutions enables the Government to prioritise financial support.

The development and implementation of the CIM Plans required greater inter-government agency communication and coordination which has resulted in strengthened relationships. The project directly involved over 40 Government of Samoa staff including 30 staff from MNRE, Ministry of Women, Community and Social Development (10 staff). In addition, Government staff are involved in the implementation of actions such as the Ministry of Works, Transport and Infrastructure (MWTI).

The CIM Plans have been pivotal in establishing a new way of writing plans, using a collaborative, consensus-building, multi-level governance approach, blending traditional decision-making systems with contemporary ones in the Pacific. The process used to develop the plans is understood to be unique in the Pacific; usually coastal hazard planning occurs through either a top down or bottom up/grass roots approach. In this instance, the development of the CIM Plans created a combination of both.

#### 13.2.1.6 CIM Plans were Recognised as a Key Climate Change Adaptation Strategy Early and are Linked to the Development Consenting Process

The CIM Strategy and Plans are viewed, particularly by the Ministry of Natural Resources and Environment, as a key climate change adaptation initiative and linked to climate adaptation in practice (Daly and Bannock, 2010). For instance, the CIM Plans (which contain hazard zone maps showing areas of potential flooding, erosion and landslips) are being referred to during the development consenting process in order to ensure new developments are not located in previously identified hazard zones. The information is publically available and therefore is able to be used by developers as well as the Government in determining the best location for development, including new infrastructure.

#### 13.2.1.7 Benefits of Incremental Approach and Initial Development of a Pilot Study

Splitting the project into two stages proved to be a valuable exercise. This enabled the pilot stage to focus on those areas with the greatest identified need, for funding to be thoroughly sorted out and to fine tune the methodology as much as possible prior to expanding the project across the entire country. Examples of improvements in the second phase included consultation separately with *matai,* women and untitled youth and the improved practice of breaking the meeting into smaller groups to obtain more comprehensive feedback. Further, holding short intensive consultation rounds (on average 2–3 weeks) maximised staff time and reduced overhead costs (accommodation, transport etc). Conversely, there was a heavy time requirement from the Planning and Urban Management Agency staff, which often resulted in postponing of non-urgent work and the use of a minimal number of staff.

## 13.3 Project Challenges

The following are some of the challenges and barriers faced during the project which at times looked to threaten the longer term sustainability and ongoing implementation of the programme.

### *13.3.1 Limited National Resources for Implementation and Consultation*

Limited national resources were assigned to implement the CIM plans, including both staff and funds to undertake improvement projects in the villages. A fund was available for small projects which villages could apply to, however this fund wasn't utilised fully. An unclear application process and lack of a prioritised list of works hampered the utilisation of the fund. A recent (2010/2011) project funded by the World Bank is looking at how the actions identified by the villages in the CIM Plans will be prioritised and implemented which should help keep the programme moving. This project will also provide a much needed evaluation phase.

Another future challenge will be any government project involving consultation. The successful consultation undertaken for Stage 2 of the CIM Plan development set a high standard which will be difficult to meet without significant support from donor agencies. The costs associated with consultation on this scale (including payments to the village for catering as well as gifts) along with materials such as map and plan printing costs are significant and are not able to be met from the Ministry's operational budget.

The logistics of managing this scale of consultation were also challenging, but was possible with careful planning and coordination. For Stage 2 of the project, two teams worked at the same time in neighbouring districts in order to undertake the consultation in a consolidated manner, reducing costs of accommodation and travel, but increasing workload on government staff. These short, intensive work periods worked well in some respects in providing a period of consolidated focus away from the distractions of day-to-day work activities and provided invaluable experience for staff. However, these short, intensive consultation periods involved a high proportion of the Planning Urban Management Agency's staff time which meant all non-urgent work either had to be postponed or additional time was required to catch up. The Agency also had to operate on minimal staffing levels during these periods.

### *13.3.2 Changing Community*

The urbanisation of Samoa's townships decreases 'in built' resilience to natural hazards. This is particularly apparent in *Apia* (*Upolu*) and *Salelologa* (*Savai'i*) where land tenure changes from traditional *aiga* (extended family) ownership to private ownership reduces the joint responsibility to maintain infrastructure, such as cleaning out culverts and planting and maintaining sand banks. The westernisation of housing styles and preference of location (e.g., by the sea) also proves problematic in that western houses are harder to rebuild. Traditional houses, or *fales* can be dismantled and re-erected more easily, thereby increasing part of the resilience equation, namely the speed at which recovery occurs. Retaining 'in built' or indigenous island resilience to natural hazards in the face of increasing urbanisation and westernisation is important and the CIM Plans provided an opportunity to recognise this knowledge and incorporate it in the action plans where possible.

### *13.3.3 Changing Planning Environment*

One of the most problematic challenges in Stage 2 of the project was the transition to a more statutory planning environment with the introduction of the 2004 PUMA Act (PUMA 2004). The CIM Plans were required to transition from a non-statutory instrument to become Sustainable Management Plans (SMP) under the new Act.

Table 13.2 outlines the merits of converting the CIM Plans to SMPs. This issue was hotly debated as it draws parallels to conflicts raised previously regarding the imposition of western paradigms on indigenous cultures (Cox and Elmqvist 1997) which could be viewed as being incompatible with traditional concepts.

Following extensive discussions, it was agreed that the existing and current CIM Plans would not be converted into SMPs under the Planning and Urban Management Act 2004.This decision was shaped by a number of factors. First, the process of formulating an SMP was as yet untested. A pilot project to prepare an SMP for an Apia peri-urban location is currently being pursued with support from a donor agency. This initial trial project will be able to determine the extent of work required to develop and implement an SMP. Second, the CIM Plan process is based around a collaborative partnership approach. The SMP process has significant statutory obligations and introduces a new formalised legal process between the Government and villages. This means a considerable change in the way these two groups interact and will likely be difficult to implement. Third, the resources required from MNRE to develop and implement SMPs were largely unknown. They were, however, expected to exceed available capacity, in particular when other existing commitments to support CIM Plan preparation and other matters were taken into consideration. Fourth, the CIM Plans could be capable of delivering similar outcomes to those

**Table 13.2** The reasons for and against converting CIM Plans to SM Plans

| Advantages of converting the CIM Plans to SM Plans | Reasons against converting the CIM plans into SM Plans |
|---|---|
| No legal/technical reasons why the existing and future CIM Plans could not be converted into SM Plans | A philosophical position that wishes to retain the strong partnership ethic that underpins the basic CIM Plan concept. The strength of the CIM Plans are that they are developed through tradition *fa'a Samoa* and *fa'a Matai* or consensus decision-making mechanisms |
| CIM Plans meet most of the SM Plan requirements currently and could be converted simply | There are alternative mechanisms within PUMA 2004 which may be more appropriate to managing rural development in Samoa |
| The SM Plans provide a clear and enforceable direction for land use planning through zoning techniques | The scale of environmental issues to be addressed does not warrant the application of a rigorous statutory framework at this stage in the process |
| | The resource implications for PUMA and the community are significant |
| | A staged/phased approach to the implementations of a statutory land use planning and regulatory framework for Samoa may be more appropriate with a focus on development control and land use planning in the urban area before considering rural issues in detail |

expected from SMPs without the need for extensive statutory procedures to be undertaken. Fifth, many of the issues that would be addressed in a village/district level SMP can be handled more efficiently at a national level via policy statements or through existing legislation. This also reduces the risk of alienating existing villages and provides consistency in planning across the country. Finally, there is always the ability to convert CIM Plans to SMPs in the future.

### *13.3.4 Reviewing the CIM Plan Programme*

A review of the implementation of the CIM Strategy and Plans was undertaken in 2010–2011 (Daly and Bannock 2010; Roberts et al. 2011). The review concluded that there was good knowledge of the Strategy and Plans by those directly involved in its application within MNRE, while less awareness was found in other Ministries. Enhancing the effectiveness of the CIM and associated plans requires a much greater degree of integration of the CIM Strategy's key principle of increasing resilience into the policy objectives for all Divisions within the Ministry (not just the PUM Agency). The CIM Plans are the primary implementation tool for the Strategy and these should be supported in legislation alongside the partnership approach fostered with the joint signing of each CIM Plan by the Minister, the CEO and the Village representatives.

An initiative adopted after the creation of the CIM Plans includes the development of Village Sustainable Development Plans which include issues such as food security and micro-enterprises. This required another extensive consultation process, which built on the CIM Plan process. This additional information complements that captured by the CIM Plans. The potential for greater integration of these Plans with the CIM Plans to provide a "Ridge to Reef" approach to sustainable land management should be explored. The two Plans could be combined into one village plan that promotes the ownership of land management issues as being a village responsibility. These plans could serve as a vehicle to incorporate coastal infrastructure management, climate change adaptation, sustainable land management, biodiversity, renewable energy, natural hazards, emergency and disaster management, forestry, marine and agricultural development.

The Apia urban area faces quite different development pressures when compared to the rural areas and would benefit from an integrated approach to land management that reflects these different pressures. The Government of Samoa is currently looking at a planning concept for parts of Apia, however, a more comprehensive Apia Urban Area Plan could be developed that addresses the development of Apia as the main business and employment centre for Samoa and should take into account coastal infrastructure management, climate change adaptation, natural hazards (such as flooding, landslips and erosion), economic development, transportation, infrastructure development and potential for growth.

### *13.3.5 Barriers and Opportunities*

Two of the most significant barriers to mainstreaming the Samoa CIM Strategy and CIM Plan into planning and decision-making processes are funding and lack of knowledge. The CIM Plan consultation process uncovered a lack of awareness among stakeholders regarding the actions influencing climate change risk. For example, in the village of *Fagamalo* one fishing technique included blowing up coral reefs with dynamite. Coral reefs are among the first lines of defence against tsunamis, so the maintenance and enhancement of these ecological features is essential. In Apia during heavy rainfall, downtown often floods due to built-over or blocked culverts. Both examples highlight the greater need for education at all levels. Further, a number of government buildings, schools and the *Savai'i* hospital are all built in hazard prone areas. Prior to the SIAM project there was limited information regarding suitable building locations. The CIM Plans contain aerial photographs which show flood, erosion and landslide prone areas and should be used to better inform land use choices.

MNRE/PUMA use the CIM Plans in assessing development consent applications. However, as is common in most countries, political influence can be great, and in one particular example, a seawall has been constructed in a low CIM Plan priority area. This is another example of lack of knowledge and education regarding the CIM Plan process at high levels.

The CIM Strategy and Plans appear to have worked well for Samoa due to the culturally appropriate, combined (top down and bottom up) partnership process, and their integration with disaster management frameworks. However, they required a significant financial investment by a donor agency to establish and additional support is required for an expanded implementation programme. A plan review process has documented good ongoing use of the Plans as a primary land-use planning and climate adaptation tool and the Government of Samoa is committed to their ongoing implementation. Given the current use of the CIM Plans and their potential for even greater utility once implementation is increased, the investment has arguably been worth it.

It is difficult to weigh the benefits of hazard mapping in combination with the application of a land use planning framework against the cost of such an initiative. But in these circumstances, if the CIM Plans have raised awareness regarding potential high hazard risk locations and if the implementation of these plans assists in reducing village vulnerability and improving resilience to natural hazards they could arguably be considered a success. Funding for the implementation of the CIM Plan actions will continue to be a significant barrier to mainstreaming climate change adaptation measures into planning and decision-making. As noted previously, the expectations created by the significant level of consultation and reciprocation involved in developing the CIM Plans will be difficult for Government to match in other projects.

Implementation of the CIM Plans provides an opportunity to avoid exposure to and recover from climate change impacts. In order to reduce and recover from potential impacts on people and property, the CIM Plans should continue to be implemented by the government, private sector and the public. Actions include building new government buildings, key infrastructure and houses outside of the hazard risk zones and/or building high foundations, encouraging people to move away from the coast by sealing inland work roads and providing infrastructure such as power, water and telephone connections along them. Relocating key road routes (particularly the lifeline routes to the hospitals) inland represents another important option. Regularly clearing culverts, vegetating sand banks and maintaining coral reefs, are actions that can and are being implemented by both the Government and local villages. The CIM Plans provide a guide as to what key actions need to be undertaken by each village and district to avoid exposure to and recover from climate change impacts.

The CIM Plans are being used as a reference in other consultations, particularly an expanded disaster risk management programme which is currently focussed on tsunami. While tsunami was not specifically included in the CIM Plans, the assets at risk from climate related and tsunami hazards are similar and the environmental concerns and mitigation actions identified are relevant to reducing tsunami risk as well as climate related risk. The facilitator training in the tsunami inundation modelling and evacuation map project includes awareness of the CIM Plans in the villages targeted. The resulting tsunami inundation map is expected to be added to the suite of hazard maps developed for the CIM Plans.

The CIM Plans included a response plan component, seeking to address mitigation, preparedness and response in one process. An opportunity exists to expand this to other hazards and adding the additional information captured to updated and expanded CIM Plans.

Improving public, government and private sector education and knowledge about the impact of land use activities (including housing) on increasing vulnerability to climate change impacts is needed. For example, the impacts that sand mining and wetland drainage have on removing natural defence mechanisms for storms and tsunami, and the poor suitability of reclaimed land for building dwellings and hospitals should be incorporated into ongoing and future training programmes. This is currently underway with the implementation of the PUMA Act which requires consent from the agency for any land use activity.

## 13.4 Future Action

The following are three practical recommendations for building community resilience, adaptive capacity and sustainability. These recommendations apply to Samoa and potentially other Pacific nations.

### *13.4.1 Improve Knowledge and Education Regarding Environmental Management, Natural Resource Management and Natural Hazards Management at All Levels*

A starting point would be to continue to develop the base of knowledge and capacity of Government staff. Training could then be filtered through to key stakeholders (e.g., infrastructure providers and developers) and villagers. Another way to improve environmental knowledge, build resilience and adaptive capacity is through implementation of the CIM Plan actions which can occur at all levels. This needs momentum and, as mentioned previously, funding continues to be a challenge. A champion of status from each village would be ideal in assisting with the implementation of the plans.

There are a number of development programmes already occurring in villages and there is a need to link and coordinate these where possible to link knowledge awareness initiatives. Improving knowledge and education at all levels is critical to improving community resilience, adaptive capacity and sustainability.

### *13.4.2 Continue to Integrate Planning—By Combining a National Strategic Framework and Grass Roots/Bottom Up Planning Process*

Villages need to be involved in the identification of natural hazard threats and developing ways to address them. In Samoa, building community resilience through village involvement encourages buy-in and ownership of the CIM Plans and the implementation of actions. It is further recommended that integrated planning incorporate both a national framework and that local village plan implementation is a core component of any Pacific project.

### *13.4.3 Develop Effective Government and Village Partnerships*

In the Samoa CIM Plan, project partnerships were encouraged through the project development process. Specific elements of the process include agreement through sign off by the CEO of MNRE and village representatives, the identification of incentives (e.g., access to resources), the development of agreed actions and the creation of clear roles and responsibilities. Another key aspect of developing partnerships through integrated planning is the respect provided for tradition and traditional decision-making processes. Examples of a village's traditional adaptive techniques can be seen through the relocation of a village inland and the existing high building levels of some *fales* (houses) above flood prone levels.

## 13.5 Where to From Here?

As mentioned previously the Government of Samoa is developing the Apia Spatial Plan and undertaking flood vulnerability and adaptation assessments in the capital city. It is anticipated that the Apia Spatial Plan will lead to or form the basis of the first SMP for Samoa. The degree to which the spatial plan can provide lessons for others and leads to similar efforts across the country merits close attention as Samoa faces a number of serious challenges tied to natural hazards, including those influenced by a changing climate.

## References

Ballinger, R. C., Taussik, J., & Potts, J. (2002). Managing coastal risk: Making the shared coastal responsibility work. Summary Report to the LGA Coastal Special Interest Group. London: Local Government Association.

Beca. (2001). Coastal infrastructure asset management project. Government of Samoa.

Berke, P., Cooper, J., Salvesen, D., Spurlock, D., & Rausch, C. (2011). Building capacity for disaster resiliency in six disadvantaged communities. *Sustainability, 3*(1), 1–20.

Becker, J. S., Johnston, D. M., Daly, M. C., Paton, D., Mamula-Seadon, L., Petersen, J., Hughes, M. E., & Williams, S. (2011). *Building community resilience to disasters: A practical guide for the emergency management sector* (p. 40). Lower Hutt: GNS Science. GNS Science report 2011/09.

Bertram, G., & Watters, R. F. (1985). The MIRAB economy in South Pacific Microstates. *Pacific Viewpoint, 26*(3), 497–519.

Boura, J. (1998). Community fireguard: Creating partnerships with the community to minimise the impact of bushfire. *Australian Journal of Emergency Management, 13*(3), 59–64.

Chambers, R. (1980). The small farmer is a professional. *Ceres,* March-April, 19–23.
Chambers, R. (2002). *Participatory workshops: A sourcebook of 21 sets of ideas and activities*. Brighton: Institute of Development Studies.
Connell, J., & Lea, J. P. (1992). "My country will not be there". Global warming, development and the planning response in small island states'. *Cities, 9*(4), 295–309.
Cooke, B., & Kothari, K. (Eds.). (2001). *Participation: The new tyranny?* London: Zed Books.
Cox, P. A., & Elmqvist, T. (1997). Ecocolonialism and indigenous-controlled rainforest preserves in Samoa. *Ambio, 26*(2): 84–89.
Daly, M., & Bannock, S. (2010) Economics of climate change adaptation in Samoa: Summary of in-country visit march 2010. Client report for the World Bank, June 2010. Beca International Ltd. p. 40.
Daly, M., Becker, J., Parkes, B., Johnston, D., & Paton, D. (2009). Defining and measuring community resilience to natural disasters a case study from Auckland. *Tephra, 22,* 15–20.
Daly, M., Poutasi, N., Nelson, F., & Kohlase, J. (2010). Reducing the Climate Vulnerability of Coastal Communities in Samoa. *Journal of International Development, 22,* 265–281.
Douglas, C. H. (2006). Small island states and territories: Sustainable development issues and strategies-challenges for changing islands in a changing world. *Sustainable Development, 14*:75–80. doi:10.1002/sd.297.
Enarson, E., & Morrow, B. H. (Eds.). (1998). *The gendered terrain of disaster: Through women's eyes*. Connecticut: Praeger Publishers.
Gibb, J. (2001). Samoa Sensitivity Index Database, Beca, Auckland, New Zealand.
Haynes, K., Barclay, J., & Pidgeon, N. (2007). Volcanic hazard communication using maps: An evaluation of their effectiveness. *Bulletin of Volcanology, 70*(2), 123–138.
Hooper, A. (1998). Pacific islands stakeholder participation in development: W Samoa. Pacific Islands Discussion Paper Series Number 3, East Asia And Pacific Region, Papua New Guinea And Pacific Islands Country Management Unit, World Bank: Washington, DC.
Jenwick, J. (2000). Media release: Different patterns for tropical cyclones in the Pacific. NIWA. http://www.niwa.co.nz/news-and-publications/news/all/2000/2000-11-27–1. Retrieved 27 November 2000.
Jones, P., & Lea, J. P. (2008). What has happened to urban reform in the island Pacific? Some lessons from Kiribati and Samoa. *Pacific Affairs, 80*(3), 473–491.
Kumar, S. (2002). *Methods for community participation: A complete guide for practitioners*. Warwickshire: ITDG Publishing.
Lewis, J. (1999). *Development in disaster-prone places: Studies of vulnerability*. London: Intermediate Technology Publications.
Maceda, E., Gaillard, J.-C., Stasiak, E., Le Masson, V., & Le Berre, I. (2009). Experimental use of participatory 3-dimensional models in island community-based disaster risk management. *Shima: The International Journal of Research into Island Culture, 3*(1), 46–58.
Mackinson, S., & Nottestad, L. (1998). Combining local and scientific knowledge. *Reviews in Fish Biology and Fisheries, 8,* 481–490.
Mercer, J. (2010). Disaster risk reduction or climate change adaptation: Are we reinventing the wheel? *Journal of International Development, 22,* 247–264.
Paton, D. (2006). Disaster resilience: Integrating individual, community, institutional and environmental perspectives. In D. Paton & D. Johnston (Eds.), *Disaster resilience: An integrated approach*. Springfield: Thomas.
Paton, D. (2007). Measuring and monitoring resilience in Auckland. GNS Science Report 2007/18. Wellington: GNS Science.
Paton, D. (2008). Modelling societal resilience to pandemic hazards in Auckland. GNS Science Report 2008/13. Wellington: GNS Science.
Pearce, D. G. (2000). Tourism plan reviews: methodological considerations and issues from Samoa. *Tourism Management, 21*(2), 191–203.
Planning and Urban Management (PUMA) Act, 2004 (amended in 2005). Retrieved from http://www.mnre.gov.ws/index/php/mandate.

Ready, S., & Woodcock, F. (1992). The South Pacific and Southeast Indian Ocean Tropical Cyclone Season 1989-1990. *Australian Meteorological Magazine* (Bureau of Meteorology Australia)*, 40,* 111–121.

Roberts, G., Wells-Green, P., Frentz, K., & Daly, M. (2011). SIAM-2, Components C4.01 and C4.0102 environmental, risk and resource management: Update and prioritisation report. Client report prepared for the Government of Samoa. Beca International Consultants Ltd.

Rearic, D. M. (1990). Survey of Cyclone Ofa damage to the Northern Coast of Upolu, Western Samoa. Sopac technical report 104. SOPAC (South Pacific Applied Geoscience Commission): Suva.

Shaw, R., Uy, N., & Baumwoll, J. (Eds.). (2008). *Indigenous knowledge for disaster risk reduction:* Good practices and lessons learned from experiences in the Asia-Pacific Region. United Nations International Strategy for Disaster Reduction: Bangkok.

Shea, E. (2001). *Pacific island regional assessment of the consequences of climate change and variability (The Pacific Assessment)*. Honolulu: East-West Center.

Shea, E. (2003). *Living with a climate in transition: Pacific communities plan for today and tomorrow*. Honolulu: East-West Center.

Sweetman, C. (Ed.). (2009). Climate changes and climate justice. Special issue of *Gender and Development, 17*(1), 1–135.

Taule'alo, T. (2000). Urban planning in Samoa-issues for decision -making. *Samoan Environment Forum, 2001*(2), 3–9.

Twining-Ward, L., & Butler, R. (2002). Implementing STD on a small island: Development and use of sustainable tourism development indicators in Samoa. *Journal of Sustainable Tourism, 10*(5), 363–387.

Valency, R. A. (2007). *Building disaster resilient communities: Good practices and lessons learned*. Geneva: UNISDR (United Nations International Strategy for Disaster Reduction).

Watson, B. E. (2007). Reconceptualising disasters: Lessons from the Samoan experience. Master of Arts in Geography dissertation, University of Canterbury, Christchurch.

World Bank. (2009). Samoa: Post-disaster needs assessment following the earthquake and Tsunami of 29th September 2009. Government of Samoa, December 2009.

World Bank. (2010). Economics of adaptation to climate change: Samoa Case Study (p. 58).

# Chapter 14
# Disaster Recovery in Coastal Mississippi, USA: Lesson Drawing from Hurricanes Camille and Katrina

**Gavin P. Smith**

**Abstract** Hurricane Katrina is often thought of as an extreme event that struck New Orleans, Louisiana (USA). Less attention has been placed on the effects of this devastating storm on coastal Mississippi (USA) where more than 70,000 homes were damaged or destroyed by a storm surge that exceeded 30 feet in height at its peak. The State of Mississippi has a long hurricane history, including Hurricane Camille, which made landfall a few miles from where Katrina collided with a number of small and mid-sized communities 36 years later. This chapter will describe the lessons from both of these events, including how they can inform future climate change adaptation efforts in small coastal communities across the United States and elsewhere that may not possess the capacity to effectively deal with disasters, including those created or exacerbated by a changing climate. Hurricanes Camille and Katrina serve to highlight three important themes: (1) large-scale disasters in the U.S. trigger considerable amounts of post-disaster assistance and attention; (2) the manner in which risk is communicated and tied to new policies and reconstruction standards can play a significant role in shaping recovery outcomes, including the adoption of hazard mitigation measures, the stagnation of housing reconstruction and more intensive commercial development; and (3) the creation of state-led recovery organisations can play a key role in coordinating disaster aid across a diffuse network of resource providers.

**Keywords** Disaster recovery · Risk communication · State recovery office · Housing resettlement · Financial assistance

## 14.1 Introduction/Setting

This chapter describes the actions taken by the State of Mississippi, United States (U.S.) to recover from two major hurricanes and uses this information to inform potential climate change adaptation strategies in the U.S. and other nations. The two

---

G. P. Smith (✉)
University of North Carolina at Chapel Hill, North Carolina, USA
e-mail: gavin_smith@unc.edu

B. C. Glavovic, G. P. Smith (eds.), *Adapting to Climate Change,* Environmental Hazards, DOI 10.1007/978-94-017-8631-7_14, © Springer Science+Business Media Dordrecht 2014

hurricanes-Camille (1969) and Katrina (2005) caused major damage due to storm surge and high winds; two effects that a growing number of scientists believe will worsen in the U.S. given the projected increase in the intensity of hurricanes tied to rising sea surface temperatures (Emanuel 2005), although this is still subject to debate (National Research Council 2010b, p. 49). The policies adopted and institutions created to deal with the aftermath of these storms reflect both similarities and differences when compared to those advanced following the North Carolina (U.S.) case study. The composition of the larger policy milieu at the time of the disasters merits attention as Mississippi, like North Carolina, was led by a governor who was closely aligned with the political leadership in Washington, D.C. and the Mississippi state legislature at the time Katrina struck. The State of Mississippi effectively exploited this reality, and like in North Carolina, political influence coupled with good post-disaster information resulted in the more effective acquisition of post-disaster assistance when compared to adjacent states facing the same disaster, particularly Louisiana (Smith 2011), which is described in Chap. 15.

The effects of Hurricanes Camille and Katrina are discussed first in order to provide a historical context and to help frame the actions taken by the state following these two events and the lessons that follow. Embedded in these lessons are three important themes: (1) large-scale disasters in the U.S. trigger considerable amounts of post-disaster assistance and attention; (2) the manner in which risk is communicated and tied to new policies and reconstruction standards can play a significant role in shaping recovery outcomes, including the adoption of hazard mitigation measures, the stagnation of housing reconstruction, and more intensive commercial development; and (3) the creation of state-led recovery organisations can play a key role in coordinating disaster aid across a diffuse network of resource providers. Factors influencing resource distribution, policy change, and ultimately recovery outcomes include a number of political, institutional, social, and economic factors.

### 14.1.1 The Historical Context of Hurricanes in Mississippi

In coastal Mississippi, Hurricane Katrina's destructive force was closely tied to its storm surge, which exceeded 30 feet in some areas, resulting in the loss of 238 people (Smith 2012). Over 70,000 homes were severely damaged or destroyed, while an additional 1,60,000 homes were impacted. One year after the storm, over 97,000 Mississippians were living in over 36,000 Federal Emergency Management Agency (FEMA)-provided temporary housing units. Five months later, FEMA temporary housing levels peaked at 38,000 units. Hurricane force winds were felt more than 150 miles inland and over 80 % of the state's three million residents were left without power. The storm generated more than 45 million cubic yards of debris in the state, exceeding the previous record following Hurricane Andrew, a Category 5 storm which struck the Florida coast in 1992 (State of Mississippi 2006, 2007, 2008).

In any other U.S. disaster, these figures would have received greater attention were it not for an even more significant loss of life and damages sustained in nearby

New Orleans, Louisiana. While the failed levee system caused major urban flooding in New Orleans, the coastal storm surge that struck Mississippi may, in fact, provide a more likely scenario facing communities located along the U.S. Gulf Coast and Eastern Seaboard. However, a number of factors make Mississippi particularly vulnerable when compared to other coastal states, including a shallow continental shelf that extends into the Gulf of Mexico, a low-lying shoreline with limited elevation gain, and geographic positioning on the Gulf Coast that is susceptible to land falling hurricanes (FEMA 2006, pp. 1–2).[1] Hurricane Katrina, a large Category 5 hurricane prior to landfall, drove substantial amounts of water into the shallow Gulf waters. When the storm came onshore as a Category 3 hurricane the storm surge more closely approximated that associated with a Category 5 storm.[2]

Hurricane Camille, which struck the Mississippi coast on August 17, 1969 within 10 miles of where Katrina made landfall, is one of only three Category 5 hurricanes to reach the U.S. mainland (Fig. 14.1). Top winds exceeded 200 miles per hour and the barometric pressure of 26.85 in. (a key measure of a storm's intensity) remains among the lowest on record. The storm, while relatively small compared to Katrina, produced a 24 ft. storm surge, which devastated coastal communities in Mississippi. One hundred and seventy two people were killed in the state, including 41 that were never found. Property damage exceeded US$ 11 billion in 2004 dollars (Hearn 2004, p. ix). After making landfall in Mississippi, the storm traveled through the southeastern states of Tennessee, Kentucky, West Virginia, and Virginia, causing flash flooding and landslides that killed over 100 people in Virginia and 347 people in total (Hearn 2004, p. ix).

Prior to Camille, coastal communities in Mississippi were relatively small, comprised of a mix of hotels, marinas and souvenir shops, as well as permanent and secondary homes, the latter often owned by those residing in nearby New Orleans. The steady breeze off the Gulf Coast provided an alternative to the stifling summer

[1] Mississippi has been struck by a number of hurricanes including the Hurricane of 1893, the hurricane of 1947, Audrey (1957), Ethel (1960), Betsy (1965), Bob and Fredrick (1979), Elena (1985) Georges (1998), Lili (2002), Ivan (2004), Dennis (2005), Katrina (2005) and Gustav (2008).

[2] The underestimation of hurricane-induced storm surge caused a number of Mississippians to stay in vulnerable coastal communities rather than evacuate. This has led the National Hurricane Center, National Oceanic and Atmospheric Administration (NOAA) and other federal agencies to consider revising the principal means used to measure the intensity of hurricanes. The Saffir-Simpson scale, which evaluates hurricane intensity, is based on wind speed and yet does not effectively incorporate storm surge into this equation. The storm's track and location at landfall played an important role in the resulting storm surge as the upper right hand quadrant of the storm, which includes the greatest wind and storm surge (due to the off-shore winds associated with its counter-clockwise rotation in the Northern hemisphere), struck along the Mississippi/Louisiana border. As a result, the water driven by the storm collected in the wedge between the "boot" of Louisiana and the Mississippi shoreline, and was eventually pushed on shore. The improved modeling of future climate change-induced storm intensity, a more accurate assessment of expected storm surge, and the effects of rising sea levels on hazard vulnerability in areas prone to coastal storms represent the frontiers of climate change science. The interpretation and dissemination of model results, including the development of more effective short- and long-term risk communication messaging, and the incorporation of these findings into hazard mitigation and climate change adaptation plans and policies are also equally important and challenging.

**Fig. 14.1** The tugboat dubbed the S.S. Camille represents an iconic reminder of the 1969 storm. While the tugboat remained in place after Hurricane Katrina, more than 70,000 homes were damaged or destroyed along the Mississippi Gulf Coast. (Source: Photograph by Gavin Smith)

heat found in the crescent city. Over time, Mississippi Gulf Coast towns became recognised as de-facto suburbs of New Orleans as residents made the 50 mile commute along Interstate 10 (Brinkley 2006, p. 83). In many ways coastal Mississippi has more in common with New Orleans than the rest of the state, including its deep Catholic roots, architectural heritage and ties to the original French settler of New Orleans, Sieur Di Bienville, who first landed in what is now Biloxi, Mississippi.

The intensive development of permanent homes along the Mississippi coast began in earnest during the 1970s, a time of hurricane inactivity along the Gulf Coast (Brinkley 2006, p. 77). During the 1990s, even more rapid growth occurred, in large part due to the effects of casino development, which had previously been disallowed. Between 1995 and 2000 more than 90,000 individuals moved into the state's three coastal counties (Hearn 2004).

## 14.2 Drawing Lessons from Mississippi's Hurricane Experience

The Mississippi experience clearly demonstrates how differing disasters and the political, economic, social and environmental conditions present at the time they struck influence the policy and institutional choices made by various stakeholders.

It also reveals how states can play a role in altering recovery trajectories. While the state of Mississippi developed a range of post-disaster "plans," following hurricanes Camille and Katrina, the degree to which policy recommendations were adopted in the aftermath of these events as well as the translation of lessons from one event to another will be explored, including the merits of developing pre-event plans for post-disaster recovery. Embedded within this discussion is the challenge inherent in advancing a risk communication strategy that permeates evacuation behavior, the type of pre- and post-disaster development and reconstruction activities employed, and the adoption of varied hazard mitigation measures.

The Mississippi case study also dramatises the problems associated with taking a reactive approach, including the limited degree to which lessons were learned from previous events and applied to later disasters. The ability to translate the lessons from an event that was at the time the worst disaster in the state's history to a later, even more destructive disaster is instructive. The findings also show that a more effective post-disaster response to an event is possible given the development of strong state institutions.

The State of Mississippi has suffered the effects of two of the most devastating storms (Hurricanes Camille and Katrina) to ever strike the U.S. mainland. These storms serve as important benchmarks from which to assess how the state recovered from these extreme events while providing clues about how the state may draw on these experiences to address climate change adaptation, particularly the threat of sea-level rise and the possible intensification of coastal storms. A number of environmental and socio-political factors combine to make the state's coastal communities extremely vulnerable, including low lying coastal property, high rates of coastal erosion, and high levels of social vulnerability (National Research Council 2010a, pp. 82–83). Vulnerability is also changing over time, due to the inherent dynamism of the coastal environment, the acceleration of hazard processes, and the development of coastal communities in a manner that makes them less able to adapt to these conditions.

### *14.2.1 Recovery Following Hurricane Camille: Missed Opportunities that Presaged the Next Major Disaster*

In 1968, one year prior to the arrival of Hurricane Camille, the National Flood Insurance Program was passed by the U.S. Congress. In order to join the programme communities are required to adopt a Local Flood Damage Prevention Ordinance which establishes a set of recognised building standards in the floodplain. These standards are intended to reduce exposure to a 100-year flood event and those of a lesser magnitude. Once a community joins the programme, property owners are able to purchase federally subsidised flood insurance.[3] At the time Hurricane

[3] A consequence of the NFIP has been to encourage development in known flood hazard areas as flood insurance rates are not actuarially sound. Rather, they are subsidised by the federal government in order to encourage greater participation. In 2012, the passage of the Biggert Waters Act requires flood insurance policies to account for flood risk through new insurance rates that more accurately monetise the risk. This policy change, once implemented over time, has the potential

Camille struck in 1969, no coastal communities in Mississippi had joined the programme. As a result, NFIP construction standards for structures located in areas prone to coastal storm surge (i.e., elevated on a pier foundation) were not in place, which greatly affected the level of damages sustained by coastal property owners (Godschalk et al. 1989, p. 53).

Prior to Camille, communities along the Mississippi coast chose to adopt a range of construction codes and standards, and in some cases, none at all, even though the state has been struck by a number of major hurricanes, including Hurricane Betsy (1965), which caused over US$ 400 million in damages (Godschalk et al. 1989, p. 55). Some communities mistakenly believed that a 26 mile long seawall constructed after a hurricane in 1911 provided an adequate level of protection.[4] After being breached following a 1947 hurricane, an artificial beach was created to protect the seawall (Godschalk et al. 1989, p. 55).

In the mid to late 1960s, the coastal towns of Biloxi, Gulfport and Long Beach began to develop land use plans, created in consultation with the Gulf Regional Planning Commission and local contractors. However, none of the plans incorporated hazard mitigation, while their zoning ordinances did not include designated flood-prone areas or associated regulations restricting certain uses (Godschalk et al. 1989, p. 56).

### 14.2.2 Governor's Emergency Council

Several days after Hurricane Camille made landfall, the Governor created the Governor's Emergency Council (GEC) whose purpose was to assess local needs, coordinate the delivery of assistance and develop recommendations for action (Hearn 2004, p. 180). The Council was tasked with four principal objectives: (1) assess "… the factors that relate to the long-range development of the affected area …," (2) "explore in detail and in depth all available avenues of assistance, both public and private …," (3) "make specific recommendations of the most efficient

---

to dramatically alter the makeup of those who can afford to maintain flood insurance, particularly those who own property in high hazard areas along the coast. It also has the potential to change the nature of coastal development. One possible result may be the replacement of single-family permanent and rental housing with larger structures like condominiums and hotels that are less able to be relocated once built. The effects and implications of higher pre-event construction and post-disaster reconstruction costs are discussed later in this chapter.

[4] After the 1900 Galveston, Texas (U.S.) hurricane, which killed between 6,000 and 8,000 people, the city constructed a three mile long seawall and elevated a 500 block area behind it more than 15 ft. on fill material. More than 2,100 structures were elevated to accommodate the major engineering project (Smith 2011, p. 160). Following Hurricane Ike, which struck the Galveston Bay area in 2008, some have proposed the creation of the "Ike Dike" which would extend the seawall along the rest of Galveston Island and nearby Bolivar Peninsula and install large gates capable of limiting a hurricane-induced storm surge from entering Galveston Bay, thereby protecting the Houston Ship Channel and a number of bayside communities. The likelihood of funding the multi-billion dollar project seems unlikely, particularly given the large federal outlays following Hurricane Sandy.

and effective roles that should be played by local and state governments in co-operation with the Federal Government and private interests, to the end that the total resources of all might be mobilised swiftly and decisively to accomplish this objective," and (4) "recommend a comprehensive plan for the accomplishment of the maximum long range development of the area's recreational, cultural and economic life" (State of Mississippi 1970, p. 2).

#### 14.2.2.1 Post-Disaster Codes and Standards

Specific actions of the GEC included the establishment of a temporary building moratorium, the creation of an "emergency building code," and the formation of a nonprofit building inspection organisation (Hearn 2004, p. 181). These actions were not uniformly supported by local towns, as some did not enforce the moratorium, monitor permit compliance or implement suggested activities in the GEC report. This led to wide ranging redevelopment actions taken by homeowners and business owners as many local governments were more focused on repairing and rebuilding damaged infrastructure and public facilities rather than developing a comprehensive recovery plan that incorporated elements of hazard mitigation into the community's decision-making processes. The potential to learn lessons from Camille, particularly the importance of adopting more rigorous state, regional, or local land use policies and codes that recognised the inherent vulnerability of the coastal region were largely ignored (Pielke et al. 1999).

#### 14.2.2.2 Disaster Relief

At the National level, the President designated the GEC as the state recipient of federal assistance, which ultimately influenced future approaches to federal disaster relief in the U.S. (Godschalk et al. 1989, p. 58), including those adopted by Mississippi following Katrina. For instance, the GEC, working with the federal government, established relief centers in each town so residents and local business owners could better access information regarding varied recovery programmes. While the establishment of relief centers in disaster-stricken areas has become a standard operating practice at FEMA, local officials involved in Camille recovery efforts resented the council's policies which they believed were imposed on them without their meaningful input. This led to significant violations of code provisions, unsupervised repairs of homes and businesses, the waiving of permit requirements in some locations and the lifting of the building moratorium before new standards could be adopted (Hearn 2004, p. 181). Ultimately, the mix of federal, state and local policies developed after Camille resulted in more intensive development, including the rezoning of former residential areas for commercial use (Godschalk et al. 1989, pp. 56–57).

National efforts to improve the coordination and delivery of non-profit relief also emerged following Hurricane Camille. Among the most significant examples of this was the formation of the National Organisations Active in Disasters (NVOAD). The goals of NVOAD, a consortium of non-profit relief organisations, include the coordinated distribution of assistance; the sharing and dissemination of information; the coordination of policies and programmes; the provision of training, education, and outreach programmes; leadership development; and the adoption of hazard mitigation practices (NVOAD 2012).

The degree to which non-profits and other members of the larger disaster assistance network can play in disaster recovery was further expanded upon following Hurricane Katrina in Mississippi, in part, because of the more effective translational organisations created by state government, including the Governor's Commission on Recovery and Renewal (Governor's Commission) and later the Governor's Office of Recovery and Renewal. Among the most important tasks for both organisations was to translate new and evolving risk assessment information and encourage the adoption of higher reconstruction standards.

### 14.2.3 Disaster Recovery Following Hurricane Katrina: Adopting a New Approach to Risk Communication and Risk Reduction

Many of the issues associated with Hurricane Camille were revisited more than 35 years later following Hurricane Katrina. Key concerns centered on the destructive effects of Hurricane Katrina and the approach to take during recovery. This time, however, the state took a more active role in the communication of risk while simultaneously encouraging communities to adopt higher post-disaster recovery and reconstruction standards. For instance, the Governor established the Governor's Commission on Recovery, Rebuilding and Renewal and later the Governor's Office of Recovery and Renewal, two organisations tasked with the drafting and implementation of disaster recovery policy recommendations; the identification of disaster assistance needs that exceeded the resources provided by the federal government; and conducting education, outreach, and training initiatives focused on building local capacity (State of Mississippi 2006). The focus on design standards and less emphasis on land use measures, however, resulted in a number of secondary consequences discussed throughout the remainder of this chapter.

### 14.2.4 Governor's Commission on Recovery, Rebuilding, and Renewal

The Governor's Commission on Recovery, Rebuilding, and Renewal was charged with writing the report *After Katrina: Building Back Better than Ever*. The report, completed three months after Hurricane Katrina made landfall, contained 240 recommendations that spanned four categories: (1) Infrastructure, including land use, transportation, public services and housing; (2) economic development, including tourism and small businesses, agriculture, forestry, marine resources and defense and government contracting; (3) human services and non-governmental organisations; and (4) other considerations, including finance, long-term policy recommendations and a roadmap to greater accountability (Governor's Commission on Recovery, Rebuilding and Renewal 2005). Area committees were also created, including one representing each of the six most heavily impacted counties as well as one spanning the southwestern part of the state. More than 300 people volunteered to serve as members of these committees (Governor's Commission on Recovery, Rebuilding and Renewal 2005, p. i). During the drafting of the report, over 50 public meetings were held in an effort to identify issues and solicit input from coastal residents. The input was captured by committee members and incorporated into the final document.

#### 14.2.4.1 Mississippi Renewal Forum

In addition to the writing of the report and the development of recovery policies, the Governor's Commission sought to involve the professional design community in recovery. This was achieved through the hosting of the Mississippi Renewal Forum. The forum brought together over 200 design professionals, community officials, citizens and others to prepare design plans employing the principles of New Urbanism in 12 coastal towns located on the Mississippi Gulf Coast. New Urbanist designs emphasise compact urban form, walkable communities and mixed land use (Barnett et al. 2004; Duany et al. 2000).

The week long process, referred to by architects and other design professionals as a "charrette", represented one of the boldest attempts to directly involve design professionals in disaster recovery. The intensive process resulted in the development of form-based "plans" for each of the 12 coastal towns (see www.mississippirenewal.org). However, several issues hindered their effectiveness, including the limited degree to which hazard mitigation measures were incorporated into the final plans (particularly the role of land use), the timing of the event and the limited involvement of the public (Berke et al. 2009; Smith 2011, pp. 88–90). The charrettes were held three weeks after the storm. This may have been too early in the process to effectively engage members of the stricken communities, many of whom still did not have access to adequate housing, a number of schools remained closed, and residents were often overwhelmed with activities such as trying to understand the implications of new flood hazard determinations, proposed building codes and flood ordinances, filing insurance claims and identifying potential grants or loans to rebuild their homes, repair damaged infrastructure and reconstitute local government services.

Identifying the appropriate time for the delivery of varied types of assistance is among the great challenges in disaster recovery due to differing levels of absorptive capacity among those individuals and communities receiving that assistance (Smith 2011, pp. 16–22). Factors that can increase absorptive capacity include the presence of adequate staffing (including in-house personnel, on-call disaster reservists, pre-event contracts with private sector consultants, and resource sharing agreements with nearby communities), previous disaster recovery experience (and the institutionalisation of lessons learned), and the existence of an actionable plan that specifies roles and responsibilities of the larger assistance network. Furthermore, the post-disaster provision of technical assistance, particularly that which involves potential changes in existing settlement patterns, land use, and the adoption of new design standards, often results in highly contentious debates that benefit from lengthy dialogue.

In the case of the charrette process, many of the New Urbanists had preconceived ideas of what coastal communities should look like and when disagreements arose in some towns, the design professionals criticised opposition in the media leading to increased levels of conflict. Furthermore, many design professionals who attended the charrette initially failed to recognise the importance of rules associated with the National Flood Insurance Programme, including the value of elevating or relocating at-risk structures and updating outdated (pre-Katrina) Flood Insurance Rate Maps (FIRMs) that significantly underestimated flood elevations. The reanalysis of coastal Mississippi's floodplains after Hurricane Katrina resulted in dramatic changes in 100-year flood elevations; increasing flood heights by as much as 3–8 ft. (Federal Emergency Management Agency 2005a, b).

The New Urbanists strongly disagreed with the use of the new maps that were grounded in a more accurate assessment of flood and storm surge vulnerability, arguing that homes elevated to such extreme heights would limit interaction among neighbours, alter the past urban form that made the area so attractive and affect the areas sense of community. The failure to fully recognise the importance of designing communities in a way that better reflects the hazard risk prevalent in the area is symptomatic of many New Urbanist developments (Berke et al. 2009).

### *14.2.5 Governor's Office of Recovery and Renewal*

Following the hosting of the Mississippi Renewal Forum and the completion of the Governor's Commission report, *After Katrina: Building Back Better Than Ever*, the Governor created the Governor's Office of Recovery and Renewal. This group was tasked with four primary objectives: (1) identifying sources of recovery funding beyond those offered by FEMA, to include Congress, non-profits and foundations and private sector interests; (2) providing policy counsel to the Governor, state agency representatives and local officials; (3) identifying organisations best suited to implement the recommendations found in the Governor's Commission report; and (4) delivering training, education, and outreach efforts intended to assist coastal recovery and build local capacity (State of Mississippi 2005, p. 6). Four activities led by the Governor's Office of Recovery and Renewal exemplify these objectives. They include the development of the Mississippi Alternative Housing Programme, the hosting of disaster recovery workshops and regular meetings with local officials, explaining the implications of new advisory flood recovery maps and their implications for recovery and the hosting of the Governor's Recovery Expo.

#### 14.2.5.1 Mississippi Alternative Housing Programme

The Mississippi's Governor's Office of Recovery and Renewal, working with the Housing Committee of the Governor's Commission and several private sector design firms, developed a US$ 400 million proposal that sought funding to pay for the design, construction, placement and management of an improved emergency housing programme. The idea, which originally emerged from the Mississippi Renewal Forum, sought to demonstrate that better long-term housing alternatives could be created than those used by FEMA after major disasters.[5]

The State of Mississippi was awarded funding to design, construct, and deploy three housing types: the Park Model, the Mississippi Cottage, and the Green Mobile (Fig. 14.2). The Park Model, a 492 sq. ft. unit was intended to serve as a temporary home for a small family, typically located on a homeowner's lot while they rebuilt their home. The dwelling was built on a permanently attached wheeled undercarriage so that it could be maneuvered into tight spaces, quickly made operational and wheeled out of a community after a permanent home was rebuilt. The Mississippi Cottage included two and three bedroom models attached to a wheeled undercarriage that could be removed in order to set the unit on a permanent foundation if purchased by the tenant. Both the Park Model and Mississippi Cottages were intended to serve as reusable units if not purchased by the original occupant.[6] The Green Mobile included one and two bedroom units that were focused on energy efficiency, building on earlier work by a Mississippi State University Professor who originally developed the unit as an alternative to the mobile home for low-income, rural residents.

#### 14.2.5.2 Informational Workshops and Policy Briefings

The Governor's Office of Recovery and Renewal included staff in Jackson, Mississippi, (the state capital) as well as field staff located on the coast. A major role of the staff was to keep local officials, residents, non-profits and others aware of policy issues and uncover local needs that were not being met by traditional disaster recovery programmes. This information was discussed among staff and

[5] While the State of Mississippi developed the concept and initially presented the idea to members of Congress, the states of Louisiana and Alabama, two states that were also affected by Hurricane Katrina, were allowed to submit competing proposals.

[6] FEMA's current approach to emergency housing tends to view campers and mobile homes as disposable property after a disaster as they are either sold for scrap or to others once they are used to house disaster victims. The more substantial modular housing design used to construct the Park Model and Mississippi Cottages met International Residential Code standards which meant that they were able to better withstand high winds and flooding (through the use of flood-resistant materials). An additional benefit of using modular construction is that the units are more durable and could be used as emergency housing multiple times. Given the experimental nature of the Mississippi programme, the Park Models and Mississippi Cottages were sold to individuals for use as outbuildings, guest quarters and hunting cabins after their initial purpose was served.

**Fig. 14.2** The Mississippi Park Model, pictured here, has been sited and is being used to house a small family as they await the reconstruction of their damaged home. (Source: Photograph by Gavin Smith)

the Governor on a regular basis and attempts were made to identify appropriate programmes, modify policies or create new ones to meet these needs. One approach regularly used by staff was to host informational workshops whereby federal agencies, non-profits, foundations and private sector programmes were discussed with local officials and individuals affected by the disaster.

#### 14.2.5.3 Communicating Risk: The Post-Disaster Adoption of New Flood Insurance Rate Maps and Building Codes

One of the most contentious issues surrounding disaster recovery efforts in Mississippi was the uncertainty surrounding reconstruction standards on the coast. Hurricane Katrina's storm surge, recorded in excess of 30 ft. in some areas, devastated the communities located along its 80 mile-long coastline. It also exposed residents to what coastal scientists and storm surge modelers already knew; Mississippi's Gulf Coast communities are among the most vulnerable to the effects of hurricane-induced storm surge in the U.S. Prior to the storm, FEMA was in the midst of a restudy of the state's FIRMs. After Katrina struck, FEMA refocused their efforts to create what are called Advisory Base Flood Elevation (ABFE) maps. The maps, which provided a more accurate representation of coastal flooding based on the latest data and applied use of mapping technology, were intended to help communities make more informed permitting and land use decisions during the initial process of reconstruction while the final FIRM's were being developed.

Many coastal residents, community leaders and the New Urbanists brought in to assist with recovery were initially opposed to both the use of the ABFE maps and the adoption of new FIRMs citing the increased cost of reconstruction. In some cases, individual homeowners did, in fact, rebuild prior to the adoption of new standards. The Governor's Office of Recovery and Renewal regularly met with community officials and residents along the coast to discuss the importance of using the best available data to inform post-disaster reconstruction activities as well as adopting the Flood Insurance Rate Maps once vetted and finalised. Coastal communities ultimately came to the realisation that the advisory standards should be adopted during reconstruction as compliance with the final standards would be required in order to remain in the National Flood Insurance Programme.

The storm and its effects also reopened a dialogue that had remained largely dormant since Hurricane Camille struck in 1969. Prior to Hurricane Katrina, the State of Mississippi was one of the few coastal states that did not maintain a state-wide building code. The state, which, encouraged adoption, left the decision up to county and local officials. Eventually five of the six coastal counties chose to adopt a building code while Mississippi decided not to adopt a statewide standard.[7]

#### 14.2.5.4 Governor's Recovery Expo

Another approach used to convey information to a number of stakeholders involved the hosting of the Governor's Recovery Expo. This two-day event, attended by over 50,000 people, was designed to inform residents, builders and local officials about a number of issues surrounding disaster recovery, hazard mitigation, and risk communication. For instance, over 50 workshops were held, addressing varied building techniques being used during reconstruction (e.g., modular and panelised construction) the steps needed to comply with new building codes and standards (e.g., post-disaster ABFE's and new FIRM's), how to incorporate hazard mitigation measures into repair and reconstruction (e.g., wind retrofitting, flood-proofing facilities, elevated construction) and post- disaster grant-making, including eligibility issues.

In addition, more than 80 booths were set up, including those staffed by home improvement retailers, federal and state agencies, developers, non-profits and foundations. Mississippi's National Public Radio conducted their live radio show inside the Coliseum during the expo. The show helped to further advertise the event and offer a chance for public officials, including the Governor, mayors and federal and state agency officials to discuss pertinent policy issues associated with recovery. Located outside the facility, over 20 homebuilders set up modular housing units on the grounds in order to show prospective buyers replacement housing options.

## 14.3 Key Lessons from the Mississippi Experience

A number of important lessons have emerged from Mississippi that are directly relevant to climate change adaptation. The lessons are tied to the following activities, policies and programmes: (1) the remapping of the state's coastal floodplains and the adoption or modification of pre- and post-disaster codes and standards; (2)

[7] The State of Louisiana adopted a statewide building code after Hurricane Katrina.

the issues and challenges surrounding the migration of people from the immediate coastal fringe and the development of new or expanded inland communities; (3) the active involvement of the design community in recovery; (4) the creation of an innovative post-disaster emergency housing programme intended to serve as an alternative to the traditional federal approach; and (5) the creation, administration, and maintenance of state-level recovery organisations.

### *14.3.1 The Remapping of Mississippi's Coastal Floodplains: Implications for Risk Communications, Hazard Mitigation, Disaster Recovery and Climate Change Adaptation*

In Mississippi, Hurricane Katrina caused severe damage to coastal communities. FEMA developed Advisory Base Flood Elevation (ABFE) maps and later, new FIRM's to guide initial reconstruction efforts as well as provide the baseline information required to administer floodplain management activities over time. In both cases, FEMA used the latest technology and an improved assessment of coastal flood hazard risk. For instance, the ABFE maps were produced in a digital format and overlaid with aerial imagery of the Mississippi coast taken prior to Hurricane Katrina as well as mapped inundation levels collected after the storm. This allowed coastal communities to see the extent of storm surge-induced flooding relative to their property, public facilities and other landmarks, providing a powerful image of the storm's geographic reach and associated damages. It also provided context for coastal residents, property owners and local officials as they weighed the merits of adopting advisory maps to guide reconstruction efforts.

The remapping effort also required an extensive outreach campaign in order to convince local communities that the adoption of advisory maps was a good idea and indicative of sound policy making given the implications of failing to do so would result in rebuilding communities to their pre-event condition or in a manner that did not reflect the area's new depiction of flood hazard risk. The outreach efforts, conducted by Local Floodplain Administrators; FEMA; members of the Governor's Commission on Recovery, Rebuilding and Renewal; and the Governor's Office of Recovery and Renewal, paid off as all coastal communities adopted the ABFE's. The FIRMs, once completed by FEMA, were also adopted by coastal communities, thereby providing the fact base upon which to regulate development in the floodplain. The generation of new data and associated maps used for the purpose of informing communities during post-disaster reconstruction and later used as part of a community's Local Flood Damage Prevention Ordinances contrasts with that following Camille as local officials either ignored proposed standards or chose not to adopt a uniform code.

Several factors affected the decision of local officials to adopt the ABFEs and FIRMs which offer an important corollary to the Camille experience and can help inform future efforts to map and devise policies that address sea-level rise in Mississippi and elsewhere. These include: (1) the extensive outreach efforts conducted by the Governor's Commission, the Governor's Office of Recovery and Renewal, FEMA and local floodplain administrators; (2) the realisation among local officials that adherence to the ABFE maps was an important part of regulating reconstruction and the eventual FIRMs would codify building standards in the floodplain; (3) the adoption of the FIRMs was required to remain in the National Flood Insurance Programme (thereby allowing residents to maintain flood insurance); and (4) institutionalised local floodplain management programmes were already in place at the community level that would regulate compliance once the FIRMs were adopted.

The remapping of flood hazard areas in Mississippi highlights the importance of education and outreach initiatives to educate communities about the importance of adopting new standards tied to the latest understanding of risk and how that risk is subject to change over time. The remapping of hazard areas also benefits from an inclusive process that incorporates the findings into plans, uses this information to shape policies at the local level and inform individual decision-making, and serves to better educate the design community, some of whom may be reluctant to support more rigorous risk reduction standards if they seem to conflict with a more narrow definition of what comprises "good" design parameters. Ideally, these issues should serve as part of an interconnected vision achieved through the improved integration of community- and state-level hazard mitigation, disaster recovery and comprehensive land use planning.

The dynamism of hazard risk is perhaps nowhere greater than in coastal areas of the U.S. due to the natural migration of barrier islands toward the mainland, coastal erosion, subsidence and storm-related flooding. The climate change induced factors associated with rising sea levels and increased levels of storminess, heighten this growing vulnerability. Climate change-related effects have also added a degree of uncertainty that has led many states and communities in the U.S. to cite this ambiguity as a reason for inaction rather than a call to arms in support of management and governance strategies that accept uncertainty and develop plans that are able to adjust to new information as it becomes available (Hansen and Hoffman 2011, pp. 42–43). Indeed, Philip Berke argues in Chap. 8 of this book that a new planning approach is required that incorporates both robust policies that address low-regret and no-regret strategies that are implementable in the short run as well as contingent policies grounded in the best available information and scenarios collectively developed by experts and relevant stakeholders.

The ability to convey the often extreme vulnerability of coastal communities, map estimated future hazard risks based on an accepted degree of uncertainty, and incorporate this information into an integrated set of actionable policies is critically important as investments and development decisions made today influence growth in areas that are likely to become increasingly vulnerable over time. These policy options can be tied to the inherent dynamism of the coast regardless of climate change and hence the value of no-regrets policies that account for this variability. In some ways this is exemplified by the adoption of new flood hazard policies in the aftermath of Hurricane Katrina in Mississippi.

However, the same cannot be said for the adoption of contingent policies based on future hazard risk scenarios, including those tied to sea-level rise or increased storminess. Rather, new NFIP standards were tied to the latest (i.e., post-Katrina) understanding of the 100 year or 1 % chance annual flood. Ideally, contingent policies, integrated into improved hazard mitigation and disaster recovery plans, should recognise the changing nature of coastal hazard vulnerability, including our emerging understanding of sea-level rise (tied to an agreed upon set of future scenarios). Based on this understanding, the mix of policies may include, for instance, the relocation of flood-prone communities and abandoning supporting infrastructure while simultaneously adopting a disinvestment strategy in areas prone coastal flooding, erosion, storm surge and rising sea levels.

### *14.3.2 Secondary Policy Impacts, Migration of Coastal Residents and the Construction of New Inland Communities*

In coastal Mississippi, residents, community officials, and the state did not adopt a hazard mitigation strategy tied to large-scale buyouts of storm damaged properties like in the case of riverine communities in the North Carolina (U.S.) chapter. Rather, most hazard mitigation funds were used to retrofit (strengthen) critical public facilities (i.e., police and fire stations and emergency operation centers) and construct tornado safe rooms in schools and private residences throughout the state. Congressionally-appropriated Community Development Block Grant Funds were provided to residents to help offset some of the costs of complying with new reconstruction standards. For the most part, the use of federal and state funds to relocate those owning or renting property on the coast were not used for this purpose, nor was there widespread support among community officials or property owners to pursue this voluntary programme. At the local level, coastal communities did not consider the use of eminent domain or the application of other land use strategies post-disaster to regulate the location of development.

Rather, the state's hazard mitigation strategy encouraged local governments to adopt more stringent flood ordinances and building codes that stressed how structures were to be built in known flood-prone areas subject to coastal storm surge. Less emphasis was placed on where housing, infrastructure, and public facilities should be rebuilt following Hurricane Katrina or where new construction should be allowed in the future. Instead, the movement of homeowners and low income renters away from the waterfront has occurred on an ad hoc basis due to the secondary effects of the adoption of more rigorous codes and flood ordinance standards.

A number of organisations, including the U.S. Corps of Engineers, local environmental groups, and some attendees of the Mississippi Renewal Forum, suggested that the area devastated by coastal storm surge could serve as a large linear park, stretching along much of the state's 80 mile-long coastline. This idea was met by strong opposition from local property owners and elected officials citing a desire to return to their homes once repaired or reconstructed and the fear of lost tax revenues associated with businesses located along Highway 90, more commonly referred to as Beach Boulevard.

More than six years after the storm, the U.S. Corps of Engineers, through the Mississippi Coastal Improvement Project, has begun to purchase homes on the coast as the realities of the cost of reconstruction have begun to sink in. Like the eligibility requirements of FEMA's Hazard Mitigation Grant Programme (HMGP), discussed in the North Carolina (U.S.) chapter, properties purchased under the Corps of Engineers programme are converted to open space in perpetuity and are maintained by the local unit of government. As such, the property (and any structure that was once located on the parcel) no longer is taxed. As a result, local officials remain largely opposed to the purchase of individually owned residential lots in areas that provide future opportunities for more intensive development of casinos, shopping centers and condominiums. In a growing number of cases, individual property owners are selling their property to land speculators who plan to purchase and cluster properties in order to redevelop the land as a commercial venture. Of those that are participating in the Corps of Engineers buyout, most are located in areas adjacent to wetlands and along coastal rivers, not directly fronting the Gulf of Mexico.

The majority of those who have resettled in areas further inland have done so due to a number of other factors, including high reconstruction costs, the affordability of insurance and the lack of suitable replacement housing. The post-Katrina map determinations and their associated estimation of flood extent, has dramatically shaped the physical reconstruction of coastal Mississippi, particularly along the waterfront. The new flood hazard determinations and associated codes have significantly increased the cost of reconstruction due to the nature of new elevation standards. Once the new maps were completed, insurance companies and the state insurance commissioner significantly raised insurance premiums, further increasing the costs associated with maintaining property in these areas. As a result, eight years after the storm, tens of thousands of lots remain vacant, although many have been purchased by private investors, some of whom are waiting for an improvement in the economy and an ability to access the needed capital to build new commercial structures.[8]

In Mississippi, many low and middle-income residents whose homes were destroyed have chosen to relocate several miles inland, in largely rural and suburban communities. Developers have tried to meet this growing demand by constructing new neighbourhoods and expanding existing ones. The rapid growth has over-taxed existing school systems and other supporting infrastructure such as roads, water, and sewer capacity. The ability of communities to receive this rapid influx of new residents post-disaster has created a number of challenges, particularly among smaller, largely rural communities or counties that are less likely to employ a planner on

[8] Similar findings were reported in Miami, Dade County, Florida (U.S.) following Hurricane Andrew, although these houses tended to be occupied by low income Blacks (Girard and Peacock 2000), whereas in coastal Mississippi, the remaining vacant lots include a range of wealthy secondary and primary homeowners, middle-class residents and working class poor. In addition, the housing stock destroyed during Hurricane Andrew was largely the result of high winds whereas housing damages in Mississippi were due to hurricane-induced storm surge. This reality raises an important issue associated with the role of land use versus building codes across differing hazard perils, including those tied to climate change.

staff, or have at their disposal adequate land use policies to help guide and manage this growth. Welsh and Esnard (2009) argue that good pre-disaster recovery planning allows communities to address these and other important resettlement issues, including the identification of possible locales in which those individuals and communities may be resettled, referred to by Berke and Campanella (2006) as "sending zones".[9]

Decisions surrounding the large-scale relocation of vulnerable coastal communities will likely become more common in hurricane-prone coastal states. As water temperatures rise (the energy source from which hurricanes feed), their intensity is expected to increase and greater storm surges are expected. In addition, these warm waters may begin to extend farther north, thereby increasing the breadth of their likely tracks, making major metropolitan areas such as New York City more likely to feel the effects of these storms. New York's physical infrastructure, including its roads, tunnels, stormwater, and sewer (which use the same piping system) pump stations and airport are all highly vulnerable to storm surge-induced flooding. Reducing the likelihood of future disasters will necessitate large investments in protective measures and the selective relocation of at-risk infrastructures (Cullen 2010, pp. 237–241). These issues were being discussed among city officials as this book went to press as Hurricane Sandy (2012) severely damaged parts of New York City and a number of coastal communities in the State of New Jersey (U.S.).

Planning for these eventualities in locations like coastal Mississippi and major metropolitan areas like New York City will require the development of unique strategies that recognise local conditions. For instance, Mississippi has experienced multiple disasters over its history that destroyed tens of thousands of homes while New York City (until recently with the occurrence of Hurricane Sandy) has remained largely unscathed from past hurricanes and yet has made a massive investment in physical infrastructure in an area highly vulnerable to the effects of coastal storms and climate change. Both cases exemplify real design and planning challenges. Is the prevailing strategy among communities in Mississippi to reinvest in the reconstruction of lost housing stock, albeit in a manner that meets higher construction standards adequate or will this approach result in a mix of individual choices including the de facto abandonment of private property without an adequate caretaker that can afford to rebuild to these higher standards? In many cases, those that can afford to rebuild on coastal lots are not homeowners. Rather, they include those that seek to build hotels, casinos, condominiums and apartment complexes (Fig. 14.3).

The resulting development pattern is inherently less adaptable. This is due in large part to the difficulties associated with the movement of these large structures should they be threatened by coastal erosion or rising seas. Further, the development of large structures often has the effect of stimulating additional supporting infrastructure and growth around it, further limiting the physical relocation of these structures landward. As development of this type increases, options tend to become constrained and lean

---

[9] Hillsborough County, Florida (U.S.) has developed a pre-disaster recovery plan which designates potential inland areas where new development may occur in order to compensate for the loss of housing stock in storm surge-prone areas that is not redeveloped, but rather converted to open space.

**Fig. 14.3** The billboard advertises a new condominium slated for post-disaster redevelopment in an area that was recently devastated by a 25 ft. storm surge. (Source: Photograph by Gavin Smith)

towards the use of hardening structures like levees and seawalls which typically have the effect of encouraging even more development. Once the design parameters of these protective structures (which also tend to be massive and largely immovable) are exceeded, major disasters are the largely predictable result.

Do Mississippi communities return to initial discussions held shortly after Katrina about the possible mix of options, including the construction of a coastal linear park, the return of some coastal residential neighbourhoods and a more balanced approach to commercial development? Does New York City, as part of a modification to PlaNYC 2030 (created in 2007) following Hurricane Sandy, apply sea-level rise and hurricane-induced coastal flooding scenarios as part of a forward-looking assessment of risk and develop a mix of armoring and strategic retreat as suggested by Heidi Cullen in her book titled *The Weather of the Future: Heat Waves, Extreme Storms, and other Scenes from a Climate-Changed Planet* (2010, pp. 249–259)?[10] In both cases, the design community has an important role to play. Yet, as the Mississippi case study shows, the degree to which they are involved in developing potential solutions that are incorporated into actionable plans often remains underutilised.

[10] In some cases, past extreme events may be a prelude to the future and will require the exploration of strategies that a few years ago may have seemed unthinkable. For instance, while Hurricane Andrew represented the United States' most costly disaster to date when it struck in 1999, (estimated losses of US$ 28 billion), had it made landfall only 20 miles to the north, it would have struck the much more densely populated City of Miami, leading to much higher levels of damages (Morrow 2000, p. 6).

### 14.3.3 The Role of the Design Community in Disaster Recovery, Hazard Mitigation and Climate Change Adaptation

The design community plays an important role in disaster recovery and hazard mitigation. Yet the full value of the skills they possess often remain unrealised as design professionals are frequently unaware of existing pre-event actions, policies and plans that exist within the emergency management community and emergency management policy-makers often fail to encourage the involvement of design professionals in pre-event planning initiatives (Smith 2011, p. 113). This reality was clearly evident following Hurricane Katrina in Mississippi when architects sought to recreate communities as they existed prior to the storm even though a post-storm analysis resulted in dramatic new estimations of flood risk.

Furthermore, the involvement of the New Urbanists, like in many other events, occurred after the disaster struck. The importance of better incorporating design professionals like the New Urbanists into policy dialogues that educate them about the merits of incorporating hazard mitigation and the complementary aspects of climate change adaptation into their future housing designs, site planning and charrette processes is timely. Many design professionals are actively modifying their practices to incorporate energy efficient standards, use recycled materials, reduce sprawl and institute a range of other techniques to mitigate greenhouse gas emissions and advance sustainable development principles. Many of these same individuals and firms have yet to fully recognise their role in natural hazard risk reduction (including that exacerbated by climate change) and the closely associated concept of disaster resilience described throughout this book. The ability to incorporate these lessons is becoming increasingly important given the ubiquitous nature of master planned communities, including the spread of the New Urbanist approach to community design.

This suggests that through proper co-education and training of practitioners and policy-makers schooled in design, natural hazard risk reduction, disaster recovery policy and climate change adaptation, the New Urbanists and other design professionals could more effectively integrate architectural and site design practices with tested risk reduction techniques in a manner that respects existing and emerging hazard mitigation, disaster recovery and climate change adaptation goals and policies.[11] Specific means to achieve this goal include incorporating the often complementary aims of hazard mitigation, disaster recovery and climate change adaptation into the curricula of planning and design schools; conducting professional associa-

[11] A good example of the nexus between design and hazard risk reduction has been undertaken in coastal Mississippi by the Gulf Coast Community Design Studio (GCCDS), a Mississippi State University School of Art and Architecture-led effort. The GCCDS has focused their efforts on providing architectural services to those rebuilding their homes, including the creation of housing and site design plans. The GCCDS also provides landscape and planning assistance, educational opportunities and research findings to organisations and communities along the Mississippi Gulf Coast (GCCDS 2010). The GCCDS achieves these ends by working with non-profit organisations, local governments, universities, developers and other partners. For more information about the Gulf Coast Community Design Studio see www.gccds.org/.

tion workshops and seminars; and creating planning and design teams, schooled in hazard mitigation, disaster recovery and climate change adaptation, that can be deployed to communities before and after a disaster strikes to share their own experiences and lessons.

### *14.3.4 Post-Disaster Temporary and Transitional Housing*

In perhaps the most significant contribution of the design community following Hurricane Katrina, the New Urbanists brought to light the many problems associated with the provision of post-disaster emergency housing in the U.S. and offered a number of potential solutions that were eventually operationalised, constructed and deployed to the Mississippi coast by the state. The use of post-disaster temporary and transitional housing is common following major disasters in the U.S. and is highly relevant to future actions associated with climate change adaptation strategies, particularly the resettlement of large urban populations. The reluctance of most communities and states to develop robust pre-event disaster recovery plans that pinpoint housing-related challenges, craft appropriate policies based on sound information and public input and recognise those parties (including non-profit aid groups) responsible for the implementation of housing policies both before and after disasters, significantly hinders the ability of states and local governments to address what often amount to some of the most complex issues underlying post-disaster recovery (Welsh and Esnard 2009).

The New Urbanists and the State of Mississippi recognised the problematic living conditions experienced by those in FEMA-provided temporary housing and sought to develop a series of improved alternatives. While disasters can precipitate the coming together of government agencies, non-profit organisations, members of the private sector, university faculty, and others to address identified problems, they do not necessarily result in a sustained commitment to pre-event planning for post-disaster recovery that has the potential to put in place procedures that will enable this collection of stakeholders to be better prepared to deal with a largely predictable set of activities in the future. For instance, in the case of the Mississippi Alternative Housing Programme, the units, which were designed to be reusable, were not used for this purpose by FEMA. Instead, the federal agency has returned to the use of campers and mobile homes following more recent disasters. The reluctance to adopt designs that represent improved living conditions for those affected by disasters highlights the difficulties of overcoming federal institutional inertia and a reticence to embrace innovative ideas, particularly those created by state governments.

One important lesson derived from this process is the need to more actively involve all relevant stakeholders, both early in the process and over time. In the case of the Mississippi Alternative Housing Programme, the design and construction of the units were done in relative isolation and FEMA was not an active participant in this process. In some ways the post-Katrina contractual process hindered a close association between the State of Mississippi and the federal agency whose role is to manage emergency housing. Following the allotment of federal funds through a

Congressional appropriation, Mississippi, Louisiana and Alabama were responsible for the creation of their own temporary housing designs. While FEMA helped select the winning designs, the agency remained skeptical of the idea overall and believed the use of campers and mobile homes was still appropriate for temporarily housing disaster victims. Thus, it is not surprising that FEMA has not embraced the concept as they were never fully vested in it nor involved in the design process.

The Mississippi experience provides a good example of how state leadership led to the creation, adoption and implementation of an innovative solution to a complex problem after a disaster. Indeed, states in the U.S. are often faced with the creation of an organisation that can help local governments navigate the confusing array of federal programmes, identify gaps in assistance, and develop state-led strategies to confront these shortfalls. The degree to which these lessons are shared with others who may be tasked with the re-creation of similar organisations in future disasters is important as it remains unclear who at the federal level will lead the growing threat posed by climate change-induced hazards. In order to confront these emerging threats, one option to consider is the creation or modification of existing state organisations or consortia tasked with hazard mitigation, disaster recovery, and adaptation to climate change. Central to their success is the ability to create and sustain a network of organisations committed to this effort.

### *14.3.5 Creating and Sustaining a State Disaster Recovery Organisation Committed to Natural Hazards Risk Management and Related Elements of Climate Change Adaptation*

Mississippi created state-led organisations to address disaster recovery issues after both Hurricanes Camille and Katrina. In Mississippi, like other states in the U.S. there are two principal reasons for this phenomenon: (1) State emergency management agencies are largely unprepared to deal with the complexities of disaster recovery following major events, in large part because they have not effectively developed a sound pre-disaster recovery plan, and (2) States have not identified who will assume a coordinative role dealing with long-term disaster recovery operations (Smith 2011, pp. 43–49). Instead, state emergency management officials tend to focus on the administration of post-disaster FEMA programmes to the frequent exclusion of the much larger array of relevant stakeholders found in the private sector, quasi-governmental organisations, non-profits and emergent groups (Smith 2011, p. 49). The importance of these non-federal organisations is magnified in larger disasters or those events that are unique and/or lack the presence of a mature set of federal policies to deal with their effects. It stands to reason that hazards that are either the direct result of climate change or exacerbated by their effects will require broad collaborative networks like those that emerge following major disasters (Smith 2011, p. 225). It is also true that such organisational networks, unlike those that typically form to address disaster recovery, will benefit from an ability to sustain themselves over long periods of time.

Following Hurricane Camille, the State of Mississippi Governor's Emergency Commission was established to coordinate recovery efforts. In the end it was not as successful as the Governor's Commission on Recovery, Rebuilding and Renewal at developing a broad set of recommendations (that were tied to public input and an underlying understanding of extreme hazard vulnerability) and identifying appropriate organisations tasked with sustaining a commitment to action. Organisations created to assist with the implementation of the Governor's Commission Report included the Governor's Office of Recovery and Renewal, the Gulf Coast Business Alliance (a collection of coastal business officials initially led by the former Deputy Director of the Governor's Office of Recovery and Renewal) and the Renaissance Corporation (a quasi-governmental organisation tasked with financing affordable housing). Many of the original recommendations were assigned to commission members who implemented these actions after the report was completed. The Governor's Office of Recovery and Renewal remains operational more than eight years after Katrina and was used most recently to address the inter-organisational issues associated with the Deepwater Horizon Oil Spill which occurred in the Gulf of Mexico in 2010 and the Mississippi River Floods which struck the western part of the state in 2011.

## 14.4 Barriers and Opportunities: Recommendations for Action

Hurricane Katrina caused extreme, highly localised damages to communities along the 80 mile stretch of the Mississippi Gulf Coast. The level of devastation resulted in a widely held refrain that the event represented a unique opportunity to rebuild in a way that was better than what existed prior to the storm. This echoed similar promises among state and local officials following Hurricane Camille, which struck this same area 36 years earlier. Leaders also discussed how they would act on the lessons from Hurricane Camille, particularly the need to adopt more stringent codes and standards. There is little doubt that coastal communities better heeded the need to do so as evidenced by the unanimous adoption of new NFIP standards. Yet in some ways the adoption of these new standards reflects an understanding of the latest disaster and not necessarily the level of risk that is expected in the future. That is, the new flood insurance standards reflect design parameters associated with the next 100 year flood event. How will the advent of climate change and its effects, including sea-level rise and the intensification of coastal storms, be incorporated into new return periods that reflect this changing reality? Further, these standards focus on how structures are built and less on where they should be sited. Compounding the vulnerability of Mississippi's coastal communities are the future consequences of the policy choices made after Katrina that have led to indecision in some locales and the intensification of development in high hazard areas.

The results of this case study highlight the need to better understand the dynamic nature of coastal hazard risk and take steps following disasters to account for both episodic and long-term threats. The post-disaster environment presents a unique set of conditions that enable adroit governance networks to take advantage of a heightened political saliency surrounding natural hazards vulnerability and access to resources not available on a regular basis. The ability to use these resources to build an enhanced adaptive capacity and implement identified risk reduction projects and policies in the aftermath of a major event should represent an important part of a comprehensive adaptation strategy that is tied to the emerging use of scenario-based planning. The ability to act on these changing conditions in a way that injects a future oriented risk reduction strategy into the post-disaster dialogue is also fundamentally important.

### *14.4.1 Understanding the Changing Face of Risk: Disaster and the Alteration of Future Settlement Patterns*

In Mississippi, Hurricane Camille triggered a discussion about land use as a way to reshape communities and reduce future disaster losses. The conversation was short-lived and did not result in significant or enduring land use policies. The National Flood Insurance Programme, which was established one year prior to the arrival of Hurricane Camille, would ultimately play a major role in reconstruction practices and resettlement patterns following Hurricane Katrina. The post-Katrina adoption of new FIRMs has resulted in the de facto change in settlement patterns as coastal property owners are faced with several choices: rebuild to new higher standards that are very costly (due to the costs of construction and the maintenance of flood insurance); sell their property to others (often private commercial investors that plan to develop casinos, condominiums, and shopping centers); or hold on to the vacant lots, given the uncertainty of future development costs, access to reinvestment capital and potential profit margins derived from the sale of land over time. These choices, taken individually, may seem insignificant, yet when assessed relative to the thousands of property owners along the Mississippi coast who are facing these choices, the effects are substantial. This has resulted in uneven reconstruction patterns across the coast, including a mix of vacant lots that remain in the hands of individual property owners or private development interests, the construction of new commercial development and the creation of large new neighbourhoods inland.

The unfortunate reality remains that many of the resulting settlement patterns are not tied to underlying land use plans at the neighbourhood, municipal, and regional level that account for the changing face of risk, directly involve citizens in the development of such plans or link plan goals to the massive sums of post-disaster assistance that was made available to the State and local governments after Katrina. Early efforts to develop local plans based on New Urbanist principles were not widely adopted, nor were these measures incorporated into codes and standards that helped guide post-disaster redevelopment. After Hurricane Camille, some communities adopted new land use and zoning measures while others did not. The degree to which hazard mitigation was incorporated into community land use decisions varied after Hurricane Katrina as well although the widespread participation of coastal communities in the NFIP provided an established programme upon which to hang more rigorous post-disaster reconstruction standards.

In practice, a clear national strategy, undergirded by adequate training and capacity building initiatives, and the construction of diverse coalitions before and after disasters, remain unrealised. Such a strategy must confront the realities of a disinterested set of stakeholders before disasters and the challenges inherent in building a functioning coalition of diverse interests after a disaster when there is an overwhelming attention focused on the speed of recovery and returning to what was rather than an honest evaluation of current and future risk and developing a comprehensive strategy to address both exposure and differing levels of physical and social vulnerability. The inherent tension between the speed of recovery and the time required to engage in a more meaningful dialogue was evidenced by a number of examples. For instance, the New Urbanists, who arrived after Hurricane Katrina to assist communities develop post-disaster plans, initially rejected the incorporation of hazard mitigation measures into their designs, arguing that they would negatively affect the coastal region's sense of community. And their efforts to develop urban form-based plans in the throes of disaster recovery did not effectively incorporate issues associated with land use in high hazard areas. When plans did try and address this issue, property owners and local officials rejected limiting development in these areas. Instead, market forces drove land use and emerging settlement patterns, replacing residential construction with more intense land uses in many of the most hazardous locations.

In other cases, some homeowners with ready access to cash or non-profit assistance (who primarily targeted low income property owners) rebuilt their homes to their pre-event condition before the new more stringent regulations went into effect. In the latter example, non-profits, which are often able to provide disaster assistance more rapidly than government agencies can adopt new reconstruction policies, targeted low income residents for assistance and often repaired or reconstructed homes as they were before the storm, thereby perpetuating social vulnerability (Smith 2011, p. 18).

A brief discussion of socially vulnerable populations is warranted here, including the issues surrounding their disaster-induced displacement. Girard and Peacock (2000) argued in their assessment of housing relocation following Hurricane Andrew, that risk assessment and mapping should better account for differing levels of social vulnerability and this information should be used to target socially vulnerable populations through the provision of additional assistance (p. 204). These policies should also recognise the disproportionate and long-term impacts of coastal storms on socially vulnerable populations (Dash et al. 2007). In the case of the Mississippi Gulf Coast, the costs of reconstruction have been difficult to overcome for many low-income residents. The growing unaffordability of the U.S. coast is one of many problems that will likely increase in the future. At the same time, people are continuing to move to U.S. coastal areas in increasing numbers, often with a limited understanding of the natural hazards prevalent in these locales.

These and other factors described throughout this chapter portend a growing need to adopt a national adaptation strategy that recognises the unique nature of the coastal zone, including the inherent and emerging risks associated with living here. Yet, the U.S. still faces strong opposition among some members of Congress to implement a national climate change adaptation policy. Its politicisation and more recent framing of the debate as one of excessive regulation impinging on economic growth has limited productive dialogue, including the recognition that policies, programmes and funding are available to implement initiatives that address many of the problems cited in this and other chapters. One way to confront the apparent denial is to incorporate climate change adaptation measures into recognised hazard mitigation and disaster recovery programmes, policies, and associated funding mechanisms; document their effectiveness over time; share the findings with others, including those in the climate change community; and ensure that post-disaster hazards management activities are a key part of any national climate change adaptation strategy.[12]

### *14.4.2 Post-Disaster Funding, Policy Change and Disaster Recovery*

Following major disasters in the United States like Hurricane Katrina, there are massive sums of federal, state, private, and nonprofit assistance provided to communities. Furthermore, major disasters can often elicit significant changes in national policy and modifications to programme rules, especially when supported by powerful state leaders backed by pertinent disaster-based data (Smith 2011, p. 298). The ability to capitalise on these realities is significant as it offers an opportunity to develop policies and fund reconstruction efforts, implement projects identified in pre-event plans, and carry out educational and capacity building initiatives that are often under-resourced.

Disasters also provide an "opportunity" to displace people that are less powerful or have been excluded from decision-making processes, invest in larger infrastructure projects and protective measures that may incentivise more growth in known hazard areas, or it may serve as an event that results in reactionary choices rather than a more deliberative process of planning for the future (Barry 1997; Peacock et al. 2000; Freudenburg et al. 2009). The ability to tackle the complex issues associated with post-disaster recovery, including the collective understanding

[12] The National Oceanic and Atmospheric Administration's (NOAA) Sea Grant Programme has begun to initiate a series of initiatives focused on many of the recommendations for action suggested in this chapter For instance, the Mississippi-Alabama Sea Grant Consortium is working with a number of coastal communities in Mississippi. Specific activities include: engaging in outreach campaigns describing the implications of sea-level rise and increased storm surge, mapping projected sea-level rise, incorporating sea-level rise adaptation-related activities into local hazard mitigation plans and flood ordinances, inventorying a community's adaptive capacity and adopting higher elevation requirements in anticipation of rising sea levels. For more information on adaptation projects underway in coastal Mississippi, see the Mississippi-Alabama Sea Grant Consortium website at: masgc.org/coastalstorms/slr.

of local needs, the coordinated timing of disaster assistance and facilitating inter-organisational coordination, benefits from the adoption of a well-conceived, inclusively crafted disaster recovery process in advance of an event (Smith 2011).

## 14.5 Summary and Conclusions

The Mississippi experience provides a number of important lessons for coastal nations and communities. Key lessons include: (1) Develop sound pre- and post-disaster risk communication strategies; (2) Create post-disaster recovery policies that reflect the best understanding of current and projected future risk and maximise the injection of risk reduction measures into recovery and reconstruction processes; (3) Improve the pre-event involvement of the design community in recovery; (4) Develop the ability to sustain recovery organisations over time; (5) Recognise and act on the unintended consequences of adopting a hazard mitigation strategy focused on design standards and less on land use; and (6) Plan for the relocation of communities away from the coast, including those sites likely to receive them.

The issues raised here all point to a set of questions facing those who live on the coast. For instance, how do coastal communities balance the role of market forces with good pre- and post-disaster recovery planning that includes a long-term perspective tied to climate change adaptation? Given the rise in commercial development such as casinos, condominiums, and shopping centers following Hurricane Katrina, what does the future hold when assessed relative to the disasters of tomorrow? Other coastal communities across the U.S. will confront similar questions as storms and their destructive effects present unique challenges and opportunities to affect change during reconstruction that will directly impact their ability to adapt to sea-level rise.

The ultimate message, which spans each of the lessons described above, stems from the value of pre-event planning for post-disaster recovery and the corollary of failing to do so. Efforts to plan for recovery after a major disaster are always difficult and can lead to a number of negative consequences. Tough choices like abandoning high hazard areas are often hampered by the incessant pressure to make quick choices that facilitate the rebuilding of an area rather than asking should these areas be rebuilt in the first place, who should pay for it, and what are the future effects of pre- and post-disaster policies on settlement patterns in the immediate aftermath of the event? Quick actions may have unforeseen long-term consequences that once set in motion are difficult to alter at a later date. These lessons can and should be transferred to others facing similar conditions so that they can begin to plan now for what appears to be an increasingly uncertain future.

This means more than planning for recovery in areas at greatest risk of future disasters as we typically define them. It also means planning in areas likely to receive the influx of new residents, particularly in places adjacent to the coast as these locations will soon become the front lines of climate change. Tackling the challenges presented in this chapter prior to an extreme event is hard and requires a sincere and enduring commitment from a network of partners, including elected officials and others in positions of power within a community, region, state, and nation. Good pre-disaster recovery plans are not a panacea for what awaits many coastal communities, but rather they should serve as a procedural roadmap to assist them plan for the future and take well-conceived actions when a disaster does occur, including those steps that simultaneously address our growing understanding of the hazards management-climate change adaptation relationship.

## References

Barnett, J., Russell, J., & Greenburg, E. (2004). Codifying new urbanism: How to reform municipal land development regulations. Chicago, Illinois: American Planning Association. Planning Advisory Service Report Number 526.

Barry, J. (1997). *Rising tide: The great Mississippi flood of 1927 and how it changed America.* New York: Simon and Schuster.

Berke, P. R., & Campanella, T. J. (2006). Planning for postdisaster resiliency. *The Annals of the American Academy of Political and Social Science, 604*(1), 192–207.

Berke, P., Song, Y., & Stevens, M. (2009). Integrating hazard mitigation into new urban and conventional developments. *Journal of Planning Education and Research, 28*(4), 441–455.

Berke, P., Smith, G., & Lyles, W. (Pending). Planning for resiliency: An evaluation of coastal state hazard mitigation plans. *Natural Hazards Review*.

Brinkley, D. (2006). *The great deluge: Hurricane Katrina, New Orleans, and the Mississippi Gulf Coast*. New York: Harper Collins.

Cullen, H. (2010). *The weather of our future: Heat waves, extreme storms, and other scenes from a climate-changed planet*. New York: Harper Collins.

Dash, N., Morrow, B. H., Mainster, J., & Cunningham, L. (2007). Lasting effects of hurricane Andrew on a working class community. *Natural Hazards Review, 8*(1), 13–21.

Duany, A., Plater-Zyberk, E., & Speck, J. (2000). *Suburban nation: The rise of sprawl and the decline of the American dream*. New York: North Point Press.

Emanuel, K. (2005). The increasing destructiveness of tropical cyclones over the past 30 years. *Nature, 436*(7051), 686–688.

Federal Emergency Management Agency. (2005a). Reconstruction guidance using hurricane Katrina surge innundation and advisory base flood elevation maps. November. www.fema.gov/pdf/hazard/flood/recoverydata/katrina_reconstruction.pdf. Accessed Dec 15, 2005.

Federal Emergency Management Agency. (2005b). Hurricane Katrina surge inundation and advisory flood elevation maps summary of methods. November. www.fema.gov/pdf/hazard/flood/recoverydata/katrina/katrina_ms_methods.pdf. Accessed Dec 15, 2005.

Federal Emergency Management Agency. (2006). Summary Report on Building Performance: Hurricane Katrina. April.

Freudenburg, W. R., Gramling, R., Laska, S., & Erikson, K. T. (2009). *Catastrophe in the making: The engineering of Katrina and the disasters of tomorrow*. Washington, D.C.: Island Press.

Girard, C., & Peacock, W. G. (2000). Ethnicity and segregation: Post-hurricane relocation. In W. G. Peacock, B. H. Morrow, & H. Gladwin (Eds.), *Hurricane Andrew: Ethnicity, gender, and the sociology of disasters* (pp. 191–205). Miami: International Hurricane Center Laboratory for Social and Behavioral Research.

Godschalk, D. R., Brower, D. J., & Beatley, T. (1989). *Catastrophic coastal storms: Hazard mitigation and development management*. Durham: Duke University Press.

Governor's Commission on Recovery, Rebuilding and Renewal. (2006). *After Katrina: Building back better than ever*. December, 31.

Gulf Coast Community Design Studio. (2010). *Alternative construction research guide: Methods for building communities*. Starkville: Mississippi State University, College of Architecture Art + Design (Obtained from the Gulf Coast Community Design Studio website www.gccds.org on April 15, 2010).

Hansen, L. J., & Hoffman, J. R. (2011). *Climate savvy: Adapting conservation and resource management to a changing world*. Washington, D.C.: Island Press.

Hearn, P. D. (2004). *Hurricane Camille: Monster storm of the gulf coast*. Jackson: University Press of Mississippi.

Morrow, B. H. (2000). Disasters in the first person. In W. G. Peacock, B. H. Morrow, & H. Gladwin (Eds.), *Hurricane Andrew: Ethnicity, gender, and the sociology of disasters* (pp. 1–19). Miami: International Hurricane Center Laboratory for Social and Behavioral Research.

National Research Council of the National Academies. (2010a). *Adapting to the impacts of climate change*. Washington, D.C.: The National Academies Press.

National Research Council of the National Academies. (2010b). *Advancing the science of climate change*. Washington, D.C.: The National Academies Press.

National Volunteer Organizations Active in Disasters. (2012). www.nvoad.org/. Accessed 15 Feb 2012.

Peacock, W. G., Morrow, B. H., & Gladwin, H. (2000). *Hurricane Andrew: Ethnicity, gender, and the sociology of disasters*. Miami: International Hurricane Center Laboratory for Social and Behavioral Research.

Pielke, R. A., Jr., Simonpietri, C., & Oxelson, J. (July 1999). *Twenty years after hurricane Camille: Lessons learned, lessons lost*. Hurricane Camille Project Report.

Smith, G. (2011). *Planning for post-disaster recovery: A review of the United States disaster assistance framework*. Fairfax: Public Entity Risk Institute.

Smith, J. P. (2012). *Hurricane Katrina: The Mississippi story*. Jackson: University Press of Mississippi.

State of Mississippi. (1970). Final Report on Governor's Emergency Council Activities.

State of Mississippi. (2006). One year after hurricane Katrina: Progress report on recovery, rebuilding and renewal. Office of Governor Haley Barbour. August 29, 2006.

State of Mississippi. (2008). Three years after hurricane Katrina: Progress report on recovery, rebuilding and renewal. Office of Governor Haley Barbour. August 29, 2008.

Welsh, M. G., & Esnard, A.-M. (2009). Closing gaps in local housing recovery planning for disadvantaged displaced households. *Cityscape: A Journal of Policy Development and Research, 11*(3), 87–104.

# Chapter 15
# Waves of Adversity, Layers of Resilience: Floods, Hurricanes, Oil Spills and Climate Change in the Mississippi Delta

**Bruce C. Glavovic**

**Abstract** The Mississippi delta is a place of remarkable ecological, cultural and economic significance. Prevailing practices are, however, unsustainable; and climate change compounds disaster risk in the region. Delta communities need to build layers of resilience as a buffer against the waves of adversity they face. The historical context and distinctive social-ecological systems of this region are described and the relationship between resource use, disaster risk and resilience explored, with a focus on Hurricane Katrina and the BP-Deepwater Horizon oil spill. This exploration highlights four delta imperatives: (i) stem wetland loss and restore delta ecosystems to sustain coastal livelihoods and reduce disaster risk in the face of climate change; (ii) confront the 'safe development paradox'; (iii) address the drivers and root causes of social vulnerability that predispose marginalised groups and communities to disaster; and (iv) reframe governance thinking and practices that lead to environmental degradation and compound disaster risk. Barriers and opportunities are then discussed with respect to the human, physical, economic, social and natural capital needed to construct layers of resilience. A process of deliberative delta governance is recommended to foster community resilience, adaptive capacity and sustainability. Three priority actions are highlighted to translate this recommendation into practical reality: (i) articulate, share and celebrate delta narratives about overcoming adversity and building resilience; (ii) design and institutionalise inclusive processes of community disaster risk reduction and resilience planning; and (iii) sustain region-wide strategic collaborative planning processes to address the intractability of climate change that delta communities cannot resolve alone.

**Keywords** Mississippi delta · Katrina · BP oil spill · Climate change adaptation · Governance

---

B. C. Glavovic (✉)
School of People, Environment & Planning, Massey University,
Palmerston North, New Zealand
e-mail: B.Glavovic@massey.ac.nz

B. C. Glavovic, G. P. Smith (eds.), *Adapting to Climate Change*, Environmental Hazards,
DOI 10.1007/978-94-017-8631-7_15, 

## 15.1 Introduction

The Mississippi delta is an iconic location of immense strategic, ecological, economic and cultural value. It is a region that has experienced floods, hurricanes and technological disasters. It is a global hotspot for climate change impacts. Vast areas of the deltaic plain will be under water in coming decades because the rate of sea-level rise is much faster than the rate at which the slowly subsiding deltaic plain is being built up. To compound matters, many delta communities are poor and socially vulnerable. Communities living in this low-lying region need to build layers of resilience as a buffer against waves of adversity that will get progressively worse in the face of climate change. This chapter provides an overview of recent delta history, the distinctive social-ecological systems of this region and describes recent disaster experience, with a focus on Hurricane Katrina and the BP-Deepwater Horizon (BP-DWH) oil spill. It identifies lessons learned, explores opportunities and challenges for building resilience and suggests priority actions for adapting to climate change.

## 15.2 Life in the Mississippi Delta: Abundant Natural Resources, Disaster Risk and Resilience

The Mississippi delta is a cornucopia of natural resources that has attracted people for millennia. The river drains over 40 % of the lower 48 states of the United States (US) and discharges into the Gulf of Mexico via an ecosystem complex of incredible diversity, productivity and value. The delta is home to over 2.2 million people who have contributed manifold cultural riches to the nation, including distinctive Cajun and Creole cultures, cuisine, music and literature. The delta plays a pivotal role in the regional and national economy through, among other things, navigation and shipping, oil and gas, a range of petrochemical and other industries as well as fisheries, forestry, agriculture and flood protection, water supply, tourism and recreation. Delta ecosystems generate annual benefits in the order of US$ 13–47 billion and the minimum asset value of these ecosystems would be US$ 330 billion—1.3 trillion (at 3.5 % discount rate) if this natural capital were treated as an economic asset (Batker et al. 2010) (Fig. 15.1).

For over a century, the delta has been exploited as if it can provide a limitless supply of natural resources (Saikku 2005; Morris 2012). Resource over-exploitation and transformation of the lower reaches of the Mississippi river and surrounding wetlands now jeopardise the ability of the delta to sustain the provision of cherished ecosystem goods and services. About 4,870 $km^2$ of wetlands have been converted to open water since the 1930s and a further 4,530 $km^2$ could be lost over the next 50 years (Boesch et al. 1994; NRC 2006; LACPRA 2012). Barrier islands have shrunk and the coastline has retreated by nearly 50 km in places. Many of those exposed to floods, coastal storms or oil spills have limited access to the resources necessary

**Fig. 15.1** Wetlands of the Mississippi delta. (Source: photograph by Bruce Glavovic)

to cope with sudden shock events let alone the prospect of rising sea-level. Delta communities and those whose livelihoods depend on these ecosystems face escalating disaster risk; a prospect that is compounded by climate change. How did this predicament come to pass? What are the prospects for the future? And what can be learned from recent disasters to chart pathways towards resilience and sustainability? The next section provides a brief historical overview of the delta, with a focus on New Orleans, to highlight the intricate and complex interconnections between patterns of resource use, disaster risk, resilience and sustainability.

### *15.2.1 Waves of Adversity*

Delta communities face an array of hazards. Early settlers had to learn to live in this watery milieu and cope with regular riverine flooding. In more recent times, and especially since the mid-twentieth century, extensive protection works have been constructed to control flooding and facilitate physical development in the delta and New Orleans in particular (Colten 2000). Major floods were experienced in 1927, 1937 and 1973. Flooding was averted as recently as 2011 when spillways were opened to alleviate high flow rates past New Orleans. Between the 1850s and late 2000s, Louisiana was struck by 54 hurricanes and 52 tropical storms. Hurricanes have caused serious flooding in New Orleans no fewer than 38 times. Those living in the delta also face a host of technological hazards. The river between Baton Rouge and New Orleans is lined with more than 130 petrochemical plants and many other industries that together have led some to describe the area as 'Cancer Alley' (Allen 2003) or the 'Chemical Corridor' (Lerner 2005). In 2010, the BP-DWH oil spill caused the worst environmental disaster in US history. To compound matters, the combined impact of sea-level rise and deltaic subsidence are likely to result, conservatively, in the submergence by 2100 of all terrain less than one m above current sea level (Blum and Roberts 2009). Delta prospects are dismal because "significant drowning is inevitable" (Blum and Roberts 2012, p. 655). Understanding the nature of this constellation of risk and what can be done to reduce it needs to be located in the context of historic patterns of resource use, and the vulnerability and resilience of the social-ecological systems of the delta.

The deltaic plain has been extending seawards since the relative stabilisation of sea level about 6,000–7,000 years ago (Day et al. 2007). It has alternated between cycles of land loss followed by gain with the creation and retraction of deltaic lobes every 500–1,500 years (Törnqvist et al. 1996); forming six distinct lobes where the river has discharged into the Gulf as it meandered across the plain. Over time, the delta, comprising wetlands and marshes that trap sediments and form peaty soils, as well as distributaries, barrier islands and ridgelines, encompassed an area of about 25,000 $km^2$. The health and productivity of these ecosystems are dependent on the flow of river water and sediments to build up the land and 'feed' the delta. The flow patterns of the river including periodic pulses of floodwaters together with coastal storms have created a distinctive disturbance regime that sustains delta ecosystems and underpins their profound productivity and abundance of coastal and marine life, migratory birds and wildlife.

Human settlement in such a dynamic and high risk environment poses a significant challenge. Archaeological evidence shows occupation dating back 12,000 years. Early occupants and more recent Native Americans, including Natchez, Choctaw and Chickasaw tribes, thrived in the delta and used a range of strategies to cope with floods and storms, including making their camps on high ground, building substantial earthen mounds and migrating to and from the region on a seasonal basis. From the mid-1500s, the region was explored by the Spanish and then successive waves of settlers, including French from the seventeenth century, and more recently Arab, African, German, English, Irish, Scots-Irish, Jewish, Italian, Chinese, Mexican and southeast Asian peoples. European settlement was characterised by conflict, slavery and white colonial rule (Cobb 1992; Woods 1998). In the twentieth century, the region gained ill-repute for political wrangling and corruption, racial discrimination and waxing and waning economic fortunes (Parent 2004)—a fraught history that shapes contemporary disaster risk.

European settlers faced waves of adversity, including wars, pestilence like cholera, typhoid and yellow fever; extremes in temperature, humidity and rainfall; storms, and riverine and coastal flooding. New Orleans was founded by French settlers in 1718 on high ground on the banks of the Mississippi River near the present day French Quarter. It took more than 200 years "to wrest the city from nature" (Colten 2006a). By 1840, New Orleans was the third largest city in the US, with over 102,000 people, indicating its strategic import at that time. Through the course of the nineteenth century, backwater flooding from the river was more of a threat than Gulf storms because the wetlands acted as a protective barrier. By 1930, the New Orleans population had reached over 458,000 and it reached a peak of over 627,500 in 1960. Two main driving forces facilitated the growth of New Orleans in the twentieth century: the drainage of low-lying swampland and the construction of flood protection works. In 1893, city officials set out to drain the swamplands east of the river to stimulate economic development and improve public health prospects. The innovative Wood Screw Pump was designed by New Orleans native A.B. Wood in 1913 and installed soon thereafter to drain the swamps. The second major factor was the construction of an elaborate flood protection system, including levees, spillways and flood protection barriers. Extensive improvements to the levee system were stimulated by the 1927 Mississippi River flood—the most catastrophic river flood in US history (Barry 1997). As the floodwaters headed towards New Orleans, the distraught city elite successfully lobbied state and federal officials to breach levees downstream from the city so that levees protecting the city would not be compromised. The levees at Caernarvon were dynamited causing devastating flooding in St. Bernard and Plaquemines parishes. This self-serving action by the New Orleans elite lingers in the region's social memory prompting some to think that the Katrina levee failure was a deliberate act to protect elite interests (Dyson 2006; Lindahl 2012). In 1928, flood protection was declared a federal responsibility and the US Army Corps of Engineers (USACE) was designated responsible for flood protection works. Levee construction under federal direction ensued; as did other works such as spillways and barriers that effectively brought the river under control. The lower Mississippi River became an engineered system no longer prone to periodic shifts from one lobe to another. The incessant flooding and 'unsanitary conditions' of the previous century became a thing of the past.

Hurricanes have repeatedly devastated the region and stimulated federally supported flood protection works. In 1947 and again in 1965, hurricane-induced flood losses prompted the federal government to intensify investment in extensive protective works that, among other things, facilitated expansion of suburban development into low-lying swamplands east of downtown New Orleans. Hurricane Betsy ravaged communities in the region in 1965 when storm surges overtopped the levee system east of New Orleans and inundated St. Bernard Parish and large areas of Orleans Parish. Betsy killed 74 people and caused over US$ 1 billion in property damage in the state. The US Congress quickly passed flood protection legislation and soon remnant wetlands east of the city were being barricaded by levees. In 1968, Congress passed the National Flood Insurance Act to provide insurance against flood damage. In 1969, Hurricane Camille caused widespread damage in the region, killing 256 people. But with improved flood protection and insurance underpinned by the federal government, suburban development expanded rapidly into former low-lying swampland that began subsiding as peaty soils dried out. In the early twentieth century, nearly all city neighbourhoods and residents lived above sea level. By the late 1960s, up to half of the city's population lived below sea level as New Orleanians flocked to the new suburbs (Campanella 2006).

The Mississippi River has long been a major transport artery and trade route, with New Orleans being a strategic location from its inception. Harnessing this potential was a major driver for engineering interventions to control the river. The Port of New Orleans and Port of Southern Louisiana together now constitute one of the largest and busiest port systems in the world. Major works were undertaken in the first half of twentieth century to create an inner harbour Port complex. Federal funds were secured to connect these inner harbour facilities more directly to the Gulf of Mexico via a 122 km shipping canal called the Mississippi River Gulf Outlet (MRGO). Despite warnings of severe environmental, economic and societal impacts, construction of MRGO was approved and completed in 1965. The promises of economic boon were never realised; erosion of the marshes lining MRGO was rampant—up to 5 m per annum; usage by ocean-going traffic declined sharply despite significant annual maintenance costs; and, to compound matters, it was judged by experts to act as a funnel that amplified storm surges and played a key role in catastrophic levee failure when Hurricane Katrina struck in 2005 (Freudenburg et al. 2008; Shaffer et al. 2009).

Channelizing and leveeing of the river together with damming and flood control works have caused a drop of up to 50 % in sediment loads previously transported by the river (Blum and Roberts 2009). Water flow has been reduced and sediments are deposited in Gulf waters rather than building up the subsiding delta. To compound matters, from the 1930s onwards, oil and gas exploration and production intensified in the marshes and extended out into the Gulf, peaking in the 1960s through to the 1980s. Even in the absence of a major spill, oil and gas activities have had significant negative ecological impacts (Ko and Day 2004). The delta is now a labyrinth of navigation channels and pipelines that enable seawater to penetrate salt-intolerant ecosystems. The extraction of hydrocarbons reduces subsurface pressure and has led to much faster subsidence in areas near extraction activities than elsewhere (Day et al. 2007). To make matters worse, nutria—a beaver-sized rodent—were introduced in the 1930s for their fur but they proliferated and accelerated wetland loss because of their voracious appetite for marsh plants. Increased use of fertilisers in the catchment has raised nitrate levels fourfold and results in periodic 'dead zones' or hypoxia that can extend across large areas of the Gulf and have negative impacts on delta ecosystems. By the time Hurricane Katrina struck in 2005, an average of about 88 $km^2$ of delta marsh had been lost every year for five decades. Concern by local citizens, scientists, activists, the media, politicians and others grew from the 1960s, and led to increased focus on wetland restoration. The result was the creation of the Coastal Wetlands Planning, Protection and Restoration Act in 1990. Yet, despite growing concern and legislative and other efforts, the loss of delta wetlands was relentless right up to when Katrina made landfall in 2005.

Looking back over time, one could argue that laudable societal goals, such as public health and safety and economic prosperity, drove efforts to reduce flood and public health risks through levee construction and protective works, improve navigation and provide access to natural resources by channelizing the river and cutting through the swamps. Others argue, however, that exploitation and transformation of the delta, and the growth of New Orleans in particular, is the product of more nefarious forces. For example, Freudenburg et al. (2009) contend that the demise of public safety and environmental sustainability in the delta is the product of a 'growth machine': a cabal of self-interested property developers, business tycoons and public officials who garnered public funds to undertake projects of dubious social benefit that profited a few in the short-term but created spiralling disaster risk and environmental degradation. Regardless of viewpoint, the cumulative impact of delta practices over the last century has unquestionably compromised the biophysical processes that underpin and sustain life in the region and beyond. Engineered controls and alteration of naturally occurring river flows and fluxes have starved wetlands of water, sediments and nutrients and disrupted disturbance regimes and salinity balances that sustain delta ecosystems. Extractive and exploitive practices have compounded wetland loss and are key drivers of escalating disaster risk in the delta. In his environmental history of the lower Mississippi, Morris (2012) argues that disaster risk has deeper roots than inadequate scientific understanding or poor engineering; rather, it stems from the misguided belief that people can and should separate water from land. Paradoxically, this endeavor has destroyed vital delta ecosystems and increased exposure or physical vulnerability to flooding and storms. But exposure to physical perils is only one dimension of disaster risk; the other dimension is social vulnerability (Oliver-Smith and Hoffman 2002; Wisner et al. 2004).

There is a long-standing history of vulnerability in the delta that is driven by a combination of political corruption, racial discrimination and social inequity. The 'colourful' nature of delta politics can be traced back to the early 1800s when Jean Lafitte carried out pirate operations from the bayous of Louisiana. Louisianan politicians have an unenviable reputation for corruption (Parent 2004). Huey Long, Louisiana governor from 1927 until his assassination in 1935, was notorious for his flagrant theft and corruption. Illegal activities were continued by a slew of politicians, bureaucrats and business people, including subsequent Governors who received federal prison sentences: Leche (1936–1939; mail fraud, corruption and bribery) and Edwards (1972–1996; extortion, mail fraud and money laundering). Despite efforts to 'clean up' state and local politics, such practices have continued to the present. Nine term Louisiana Congressman William Jefferson (1991–2009) received a 13 year federal prison sentence for bribery in 2009.[1] At the start of 2013, the New Orleans Mayor at the time of Katrina, Ray Nagin (2002–2010), was indicted on federal charges of corruption.[2] This political culture significantly impacted post-Katrina recovery efforts (Jurkiewicz 2007a) as explained below.

Social vulnerability in the delta has been shaped by segregation and discrimination that can be traced back to slavery. Racial disparities are more pronounced in Louisiana than most other states. From the 1960s, flood protection and federally backed insurance enabled many, especially white, New Orleanians to move into

[1] See http://www.justice.gov/opa/pr/2009/November/09-crm-1231.html

[2] See http://edition.cnn.com/2013/01/18/justice/louisiana-former-mayor-indicted

newly created suburbs in former swamplands. Public housing was constructed for a growing urban population of predominantly African Americans. Employment opportunities were limited and many poor people lived in concentrated deprivation with little prospect of a decent education, meagre access to basic social services and scant hope for the future. In 2000, the city of New Orleans had the second highest concentration of poverty among large US cities, with poverty disproportionately concentrated amongst African Americans who made up 67 % of the city but 84 % of the population below the poverty line; many living in neighbourhoods with poverty in excess of 40 %. Virtually every socio-economic indicator of life in poor neighbourhoods in pre-Katrina New Orleans painted a dismal picture. African Americans were consequently disproportionately vulnerable to the ravages of Katrina; especially because many lived in low-lying areas and had limited access to private vehicles to evacuate before the hurricane struck the city (Colten 2006b; Laska and Morrow 2006; Campanella 2007).

The combination of physical exposure to an array of hazards and social vulnerability makes disaster inevitable. A synopsis of the Katrina and BP-DWH oil spill disasters reveals important lessons about disaster risk and natural hazards planning that provide vital insights for adapting to climate change.

### *15.2.2 Hurricane Katrina*

The Katrina tragedy has been recounted in a vast array of popular books, scientific analyses and government reviews too numerous to cite. This section briefly describes the event and recovery process to date to draw lessons for adapting to climate change (Fig. 15.2).

Hurricane Katrina severely impacted the entire Gulf Coast from Florida to Texas. Federal disaster declarations covered an area of about 233,000 km$^2$. It is the costliest 'natural disaster' in US history—total economic losses, including insured and uninsured property and flood damages amounting to about US$ 150–200 billion including about US$ 48.7 billion in private insured losses (King 2005). The most severe damage occurred along the Mississippi coast and in Louisiana, with catastrophic levee failure resulting in the flooding of approximately 80 % of New Orleans. Although about 80 % of the 1.3 million people living in the metropolitan New Orleans region evacuated ahead of Katrina, over 100,000 people were left in the city when Katrina struck. Those who remained behind were predominantly African American, poor and elderly or otherwise vulnerable. An estimated 1,836 people died as a direct or indirect result of the hurricane and associated flooding. There were many incredible stories of heroism, altruism and selflessness in the initial response. The shocking television images of people left stranded for days amidst flood waters without drinking water, food or shelter, however, exposed the wholesale failure of the government response (Schneider 2005; U.S. House of Representatives 2006; Jurkiewicz 2007b). Hurricanes Rita (Sept. 05) and then Wilma (Oct. 05) compounded devastation in the delta and made the 2005 hurricane season the worst in living memory.

**Fig. 15.2** 'Katrina was here'—near London Avenue Canal, New Orleans. (Source: photograph by Bruce Glavovic)

New Orleans was 'shut down' for six weeks after Katrina as the levees were repaired, flood waters pumped out and efforts made to restore basic services. The direct impact of Katrina was determined by property elevation and proximity to levee breaches. But disparities in income, race, class, gender and age shaped exposure and vulnerability to flooding and recovery prospects (Colten 2006b; Campanella 2007; Masozera et al. 2007; Finch et al. 2010). Communities in the Gulf coast faced a torrid time for years after the 2005 hurricane season. Many had to come to terms with the loss of loved ones and disruption to all aspects of life, including homes, jobs, access to critical infrastructure and public services including education and health care as well as the intangible losses of community identity and sense of place (Hawkins and Maurer 2011). Post-traumatic stress disorder and clinical depression became prevalent (Kessler et al. 2008). Many left behind the neighbourhoods they grew up in, familiar rituals and traditions and social networks, and had to start over in a new locality far from home. For those who remained behind, the post-Katrina devastation was a daily reminder of the disaster. They had to navigate a morass of red tape to secure government support and insurance payments to start the rebuild process. Economic and business continuity woes prevailed. Public infrastructure and social service provision in every sector proved challenging for years. Katrina laid bare the deep racial and class cleavages that had characterised New Orleans for decades (Dyson 2006). Nearly every facet of public and private life needed to be repaired or rebuilt, posing a monumental recovery challenge. The widely acknowledged failed response to Katrina continued well into recovery (Comfort et al. 2010). This recovery experience sheds valuable light on barriers and opportunities for adapting to climate change (Fig. 15.3).

A plethora of recovery efforts was initiated: from the federal level through to state, parish, neighbourhood and individual businesses and households (Burby 2006; Kates et al. 2006; Olshansky 2006; Nelson et al. 2007; Olshansky et al. 2008; Olshansky and Johnson 2010; Barrios 2011). FEMA and the state of Louisiana started the Long-Term Community Recovery Emergency Support Function (ESF-14) of the National Response Plan in October 2005. Local recovery efforts were fraught

**Fig. 15.3** Derelict house, Lower 9th Ward, New Orleans. (Source: photograph by Bruce Glavovic)

as reflected in the chaotic recovery efforts in New Orleans. Initial formal recovery planning for the city revealed the complex and contested nature of recovery and had a perverse long-term outcome. The Mayor appointed the Bring New Orleans Back Commission (BNOBC) in September 2005 which tasked the Urban Land Institute to prepare a recovery plan that was issued three months after the flood. The technically sound plan sought to reduce risk, prioritise redevelopment resources and provide services for the anticipated smaller population. It advocated selective rebuilding of less damaged properties and a buy-out of properties in the most damaged areas so that they could be converted to open space. It was not clear what would happen to the people—predominantly African Americans—who lived in the infamous 'green dot' zones that were identified as being too risky for rebuilding. Many feared that the political and economic elite wanted to profit out of the misfortunes of others and public opposition to the plan was strident. A follow-up BNOBC plan was also rejected. The Mayor faced a political firestorm and distanced himself from the BNOBC recommendations, rendering their work impotent. Alternative recovery plans were soon in train but subsequent efforts avoided the issue of disallowing redevelopment in areas exposed to high flood risk and relocating at-risk communities. In short, local politics dictated that the city be rebuilt according to the pre-Katrina 'footprint' regardless of flood risk, let alone escalating risk in the face of climate change.

The dismal failure of initial top-down planning efforts galvanised local citizens to wrest control of recovery planning from the 'technocrats.' The City initiated a neighbourhood recovery planning process—the Neighbourhood Rebuilding Plans—that was called the Lambert Plans. Extensive public engagement underpinned these plans. In addition, a number of independent neighbourhood plans were carried out with the support of non-profit groups and universities and various projects were supported by philanthropic and faith-based organisations. Funding could not be secured from FEMA for the Lambert Plans. The Louisiana Recovery Authority (LRA), established in October 2005 to coordinate rebuilding efforts and channel federal funding to local communities and the city, initiated the Unified New Orleans Plan (UNOP) process in the summer of 2006. Funded by foundation grants, the aim was to build upon and integrate previous neighbourhood plans and guide future reconstruction investment. The UNOP, however, created some confusion and raised questions about the legitimacy of the Lambert Plans. A consolidated city plan was prepared together with 16 district plans based on extensive public consultation and professional input. The city prepared a hazard mitigation plan to secure funds from the post-disaster Hazard Mitigation Grant Program (HMGP) because there were no funds to implement the UNOP. This hurriedly compiled plan did not dovetail well with the UNOP or other plans. Normally, federal assistance for hazard mitigation requires pre-event preparation of local hazard mitigation plans. Many Louisiana communities did not have such plans in place before Katrina struck and provision was made to allow post-event plans to be developed to access this assistance. In December 2006, as the UNOP was being finalised, the city established the Mayor's Office of Recovery Management (ORM) headed by charismatic but controversial Ed Blakely. ORM prepared an implementation plan drawing on UNOP recommendations and identified 17 target recovery areas across the city that were widely supported by the public. The LRA was responsible for disbursing special disaster funds from US Dept. of Housing and Urban Development (HUD) Community Development Block Grants (CDBG) for targeted recovery projects. The ORM's plan to allocate the CDBG funds to recovery projects—the Long-Term Community Recovery Plan—was approved by the LRA in June 2007. Blakely's oft-quoted promise to have "cranes in the sky" by September 2007 was not realised. Regardless of whether or not this was a literal or metaphorical reference, the gap between 'plans' and physical reconstruction reflects the challenge of turning recovery promises into reality.

In January 2008, a US District Court found the US Army Corps of Engineers (USACE) responsible for the levee failures, a decision upheld in a federal court in 2009, but the judgement was overturned on appeal in September 2012 on the grounds that the government cannot be sued for actions that an agency or government employee makes, or fails to make, if the function is discretionary.[3] Superficially, the flooding of New Orleans was the result of an engineering failure—there were systemic flaws in the design and construction of the levees as well questionable oversight and maintenance by widely discredited levee boards. Much of the USACE recovery effort focused on repairing the levee system to protect the

[3] See http://edition.cnn.com/2012/09/25/us/louisiana-katrina-lawsuit

**Fig. 15.4** Louisiana Speaks Public Meeting, New Orleans, 21 July, 2006. (Source: photograph by Bruce Glavovic)

city from a one percent storm surge event and ensure continued access to flood insurance within the levees. The federal government allocated US$ 14.6 billion to the USACE to strengthen the New Orleans Risk Reduction System based on levees, floodgates and storm surge barriers, including a 37 km long storm surge barrier eight meters high to prevent storm surge from the Gulf entering the City's inner harbour-port complex.[4] Despite opposition from USACE, MRGO was closed in 2009 as a result of growing public pressure and lack of evidence of the promised benefits (Shaffer et al. 2009). There was growing recognition that structural measures to protect against flooding and storms are not a panacea for reducing disaster risk and enabling long-term recovery for New Orleans and other delta communities. In parallel with the USACE efforts and New Orleans recovery planning, there were many other sectoral and region-wide recovery planning efforts, including the *Louisiana Speaks Regional Plan* (LRA 2007), *Louisiana Comprehensive Master Plan for a Sustainable Coast* (LACPRA 2007/2012), *USACE Louisiana Coastal Protection and Restoration Program* (NRC 2009) and parallel recovery efforts in neighbouring Mississippi and the Gulf region e.g., the Gulf of Mexico Alliance—a partnership between Gulf states that seeks to foster regional cooperation to promote the ecological and economic health of the Gulf of Mexico[5] (Fig. 15.4).

In November 2008, New Orleanians approved a change to the City Charter to prepare a Master Plan backed by the force of law that sought to remove politics from land-use choices by requiring that all zoning maps, land-use decisions, public projects and government actions conform to the Master Plan. There was vocal opposition to the proposed change, notably by some African American leaders, on the grounds that it could lead to the relocation of 'at-risk', predominantly African American neighbourhoods. Memories of the BNOB Commission's infamous 'green dot' plan were stirred—revealing persistent concerns about the role of race and class in the city's recovery, and mistrust and fear about the role planning could play in

[4] See http://www.mvn.usace.army.mil/Missions/HSDRRS?RiskReductionPlan.aspx

[5] See http://www.gulfofmexicoalliance.org/index.php

perpetuating inequitable development. The City of New Orleans' first ever Master Plan was adopted in 2010 with a legally enforceable Comprehensive Zoning Ordinance that was in draft form at the time of writing. Among other things, the Master Plan recognises the city's dependence on protective measures such as levees and the need for 1 in 500 year flood protection. It also recognises that the long-term sustainability of the city is dependent on wetland restoration. The draft Zoning Ordinance introduces a range of measures that are intended to help people learn to live with the reality of abundant water and flood risk by adopting measures such as passive surface and groundwater management strategies.[6]

Overall, some 20 recovery related planning processes were undertaken in New Orleans between 2005 and 2010. This array of plans ostensibly engaged the public in the crucial process of charting recovery pathways for communities in the city and beyond. Many people had, however, had their lives turned upside down by the 2005 hurricanes, were profoundly disappointed in the dismal government response and then had had to participate in a slew of official and unofficial planning processes that seemed to overlap and duplicate each other. At the same time, people were trying to get their lives back together and deal with frustratingly slow responses from government and insurance companies, the challenges of getting repairs and rebuilding done, kick-start businesses and communities as well as become familiar with and respond to USACE levee repair plans, state Road Home and Hazard Mitigation programmes, Small Business Administration support, changes to building codes and other planning processes for the schooling system, wetland restoration, public health and justice systems, etc. To compound matters, Hurricanes Gustav and Ike necessitated evacuation of the region in 2008 and caused a hiatus in recovery planning and implementation efforts. The global financial crisis further compounded recovery woes. Despite commitments from every level of government and manifold recovery plans and initiatives supported by the allocation of billions of dollars of recovery funding, the post-Katrina recovery process has been convoluted and plagued by controversy from the initial response through to the present, especially in New Orleans. In April 2010, the BP-DWH oil spill subjected delta communities to yet another disaster that reopened still raw wounds from the 2005 hurricane season.

### *15.2.3 BP-DWH Oil Spill*

The BP-DWH platform burst into flames on the 20th of April 2010, killing 11 workers and injuring another 17. The story of this disaster has been recounted in many books, analysed in detail by government agencies and appointed bodies including the President appointed National Commission on the BP-DWH oil spill and Offshore Drilling (National Commission 2011) and an array of scientific studies.[7] Within a month of the blowout, President Obama[8] said:

---

[6] See http://new.nola.gov/city-planning/master-plan/

[7] This section distils aspects of a study by Glavovic (2013a).

[8] Remarks by the President to the Nation on the BP oil spill, 15 June 2010, http://www.whitehouse.gov/the-press-office/remarks-president-nation-bp-oil-spill (Accessed 17 December, 2012).

> Already, this oil spill is the worst environmental disaster America has ever faced. And unlike an earthquake or a hurricane, it is not a single event that does its damage in a matter of minutes or days. The millions of gallons of oil that have spilled into the Gulf of Mexico are more like an epidemic, one that we will be fighting for months and even years.

The blowout occurred in about 1,500 m of water when an exploratory well was being drilled to a depth of about 4,000 m below the seabed in the Macondo Prospect some 66 km off the Louisiana coast in the US Exclusive Economic Zone. Containing a blowout at such depth is extraordinarily difficult and it took 87 days to do so; after about five million barrels of crude oil had been unleashed into the Gulf. Every effort was made to stem the flow of oil and contain damage, including removing it from the water, diluting and dispersing oil in less sensitive areas and keeping it on the surface away from sensitive areas. Over 6.8 million l of dispersant were used—an unprecedented amount—notwithstanding potential impacts on human health as well as wider environmental health issues (Solomon and Janssen 2010). Initially, some 200,000 km$^2$ of the US EEZ were closed to fishing with major impacts on commercial and recreational fishers as well as direct and indirect impacts on those whose livelihoods were reliant on Gulf ecosystems (Gill et al. 2012; Lee and Blanchard 2012). A preliminary estimate indicated that the loss of ecosystem services could range from US$ 34–670 billion (Costanza et al. 2010). But the full extent of the impacts of the oil spill disaster is difficult to assess with precision and will be revealed in years to come (National Commission 2011; Committee on the Effects of the Deepwater Horizon Mississippi Canyon-252 oil spill on Ecosystem Services in the Gulf of Mexico 2012; Gill et al. 2012; Silliman et al. 2012).

By mid-2012, media sources[9] calculated that BP would face costs in the order of US$ 38 billion—including some US$ 14 billion in response and clean-up costs, US$ 1 billion in restoration projects, US$ 9 billion in compensation pay-outs and US$ 7.8 billion to resolve outstanding claims. The company faces a slew of court cases. In November 2012, BP representatives were found guilty of criminal charges and agreed to pay US$ 4 billion to the US Justice Department and a further US$ 525 million to the US Securities and Exchange Commission for misleading investigators about the rate of oil flow into the Gulf. The company has been denied access to US government contracts because of a 'lack of business integrity'. Sprawling civil proceedings began in 2013 as the US Government and others sought as much as US$ 17 billion in civil damages under the Clean Water Act and other statutes.

The National Commission (2011, p. vii), issued in January 2011, provides valuable insights into the causes of this disaster:

- The explosion and subsequent sinking of the DWH well could have been prevented.
- Key parties made identifiable mistakes that caused the blowout and reveal systemic failures in risk management that raise questions about the safety culture of the entire industry.

[9] See http://www.guardian.co.uk/business/2012/jul/31/bp-deepwater-horizon-costs

- Oil and gas exploration and production in deep water pushes the boundaries of experience and creates risks for which neither the industry nor government have been adequately prepared.
- Regulatory reforms have been instituted since the disaster. But regulatory oversight of leasing, energy exploration and production reforms need to go much further to ensure human and environmental safety. Fundamental reform is needed in how regulatory agencies are structured and how they make decisions to ensure political autonomy, build technical expertise and enable full consideration of environmental concerns.
- Regulatory oversight per sê will not be sufficient. The oil and gas industry will be required to take its own, unilateral steps to improve safety throughout the industry.
- There is a significant lag between the technology, laws and regulations and practices for containing, responding to and cleaning up oil spills compared to the technology innovations that enable drilling into large, high-pressure reservoirs of oil and gas located far offshore and thousands of meters below the sea-surface and sea-bed. This gap needs to be closed, with active industry involvement and support rather than resistance.
- Much remains to be done to improve understanding about the environmental conditions in sensitive environments in deep Gulf waters, the region's coastal habitats and in areas proposed for more drilling, such as the Arctic; as well as the human and ecological impacts of oil spills.

Among other things, the BP-DWH oil spill disaster refocused attention on the imperative to halt the destruction of coastal wetlands and restore the valuable ecosystems of the delta. The scale and complexity of this undertaking was underscored by the intricacies of the oil spill response and recovery process. Massive financial investment and the coordinated involvement of all levels of government and key actors from the private sector, civil society and scientific community are required (Stokstad 2010; Barbier 2011; Bjorndal et al. 2011). Paradoxically, the BP-DWH oil spill disaster has been a source of funding for wetland restoration, with about US$ 2.5 billion already going to research and restoration as a result of the criminal proceedings and fine (Malakoff 2012).

What lessons can be learned from recent disaster experiences that will enable delta communities to build layers of resilience as a buffer against waves of adversity that will get progressively worse with climate change?

## 15.3 Institutionalising Lessons Learned from Recent Delta Disasters

Learning from past experience to build more adaptive and resilient social-ecological systems is crucially important. Colten and Sumpter (2009) point out that the tabulation of lessons is necessary but not sufficient for averting future disasters.

They remind us that pledges were made after Hurricane Betsy to ensure that nothing like that disaster ever occurred again in Louisiana. Yet, between that time and 2005, practices continued apace that further unravelled the resilience of the delta and its communities. What might be done to ensure that the lessons learned from recent disasters become part of the social memory of the delta and are institutionalised to build more resilient and adaptive practices and livelihoods? Moreover, in the turbulent world of climate change, how can delta communities learn to live with uncertainty and cope with surprise? This section will first provide an overview of the state of recovery and future prospects after the recent disasters, and the global recession that started in 2008. It will then identify key imperatives for the delta in the face of climate change.

### *15.3.1 The State of Recovery in and Future Prospects for the Delta*

There are diverse information sources and divergent views about the state of recovery in New Orleans and the delta region. Notwithstanding USACE improvements to the flood protection system, the 2005 hurricanes increased exposure to coastal storms by transforming about 260 km$^2$ of marsh into open water in southeast Louisiana[10]—an area that had been projected to be lost over a 50 year period. There have been significant improvements but social vulnerability persists.[11] Census data shows that in mid-2011, the New Orleans population was 74 % of its 2000 population of about 485,000 people; and the population levels of the wider metro region had reached 90 %. The city was the fastest growing large US city between 2010 and 2011. The metro region is more diverse than it was pre-Katrina, with the percentage of Latinos growing to 63 % between 2000 and 2011. The proportion of African Americans in the city declined from 67 % in 2000 to 59 % in 2011. There is evidence of improvements in various measures of employment and entrepreneurship, tax revenues, aspects of school performance and declining numbers of blighted residences. Other trends, however, reflect the struggle that many continue to face. The mainstay industries of the regional economy—oil and gas, shipping and logistics and tourism—have been in decline and have shed tens of thousands of jobs since 1980. The BP-DWH oil spill and moratorium negatively impacted related employment and livelihoods. Unemployment in the metro region rose between 2007 and 2012; and foreclosures on mortgages rose between 2008 and 2012. The poverty level in the city of New Orleans remains stubbornly at 29 %, the same level as it was in 1999. Post-Katrina housing is more unaffordable and violent crime is nearly twice the national rate. New Orleans is ranked first for corruption and third in its concentration of poverty in the top 100 US metro areas. Louisiana continues to be

[10] See http://www.nwrc.usgs.gov/releases/pr05_007.htm

[11] See http://www.gnocdc.org/

one of the poorest states in the US—with more than 25 % living below the poverty line and poverty levels have climbed in recent years. The state ties for second in percentage of people living below the poverty line, has the second-highest rate of infant mortality, and ranks fourth in violent crime, 49th in life expectancy and 46th in percentage of people older than 25 with college degrees. Katrina Road Home grants have all but been disbursed and no one still lives in a FEMA trailer. But there are still billions of dollars of government recovery funds yet to be dispersed, including US$ 3.5 billion from FEMA for debris removal and infrastructure repairs.

Why, despite massive recovery investment and effort, has it been so difficult to reduce long-term exposure to coastal storms and reverse social vulnerability; and what does this portend for the future?

Every level of government, and most public leaders, have been harshly criticised for failing to adequately prepare for and respond to the widely predicted disaster; and this failure persisted well into recovery. There were some successes in the government response including intergovernmental cooperation that helped to facilitate pre-landfall evacuation, mobilisation of the National Guard and the search and rescue operations of the US Coast Guard and Louisiana Department of Wildlife and Fisheries (Derthick 2007). But these successes were overshadowed by systemic government failure. Key non-governmental actors, including the Red Cross, were also severely criticised for inadequate preparation and response; and there was ineffective coordination between the public, voluntary and private sectors (Ink 2006; Edwards 2009). Overall, independent studies and reviews show that despite good intentions and the laudable efforts by many, Katrina exposed flaws in virtually every aspect of disaster governance: from failed initiative and leadership (U.S. House of Representatives 2006) to multiple failures in accountability regimes, and systemic flaws in among other things emergency management, engineering, economics, environmental management and inter-governmental cooperation (Cigler 2007). Koliba et al. (2011) argue that Katrina represents one of the most severe breakdowns in governance networks in modern history.

The BP-DWH oil spill reveals congruent insights. It would be naive to think that punishing BP is a panacea for averting future oil spill disasters or that it will halt unsustainable practices in the delta. As the history of the region shows, it will take much more than reforming a single corporation, or even the entire energy sector and associated regulatory framework, to secure the health, safety and sustainability of delta ecosystems and communities. The thinking and practices that drove delta exploitation for at least a century need to be transformed (Bergin 2011; National Commission (2011; Gramling and Freudenburg 2012; Lustgarten 2012). Many analysts argue that future oil spill disasters will be averted only if the underlying drivers of disaster risk are confronted, including, among other things, the pervasive influence of corporate power, corruption and a lack of accountability; a 'timid' congress; the complexity of socio-technological risk in the twenty-first century; government de-regulation; dependence on fossil fuels; and deep societal resistance to embracing a safe and clean energy future (Bergin 2011; Freudenburg and Gramling 2011; Gramling and Freudenburg 2012; Hoffman and Jennings 2011; Juhasz 2011;

**Fig. 15.5** Living on the Edge—Grand Bayou, Louisiana. (Source: photograph by Bruce Glavovic)

Ladd 2012; Lustgarten 2012). But there are scant prospects of such fundamental reform. Two years after the BP-DWH oil spill disaster, one of the two scientists on the National Commission wrote that the US is "largely failing to act on the lessons learned from that experience to ensure that deep-water drilling and production is safe and environmentally compatible" (Boesch 2012). (Fig. 15.5).

The 'big questions' revealed by Katrina and the BP-DWH oil spill therefore go beyond apportioning blame to individuals, firms or agencies. They concern the roles, responsibilities and interactions of key actors from government, civil society and the private sector in creating the conditions that make disasters inevitable; the role of governments in reducing disaster risk and building resilience; and how to institutionalise the harrowing lessons learned (Cigler 2007; Freudenburg and Gramling 2011; Gramling and Freudenburg 2012). Drivers of delta disaster risk are rooted in exposure to an array of perils and the region's history of racism, poverty and inequity and culture of corruption (Cobb 1992; Woods 1998; Campanella 2007; Jurkiewicz 2007a), that has been compounded by the insidious impacts of the 'growth machine' (Freudenburg et al. 2009) and the misguided attempt to separate water from the land (Morris 2012). Many of those living in the delta face waves of adversity that will worsen in coming decades. Is recent 'recovery' experience a troubling harbinger of things to come? Will climate change deepen physical vulnerability and exacerbate inequities between rich and poor in the region (Mutter 2010)? Or will recent experiences galvanise efforts and lead to a sea-change in thinking and practices in the delta? Averting future disasters necessitates protection or relocation of those most 'at risk' and the reversal of social vulnerability, including the drivers of racism, poverty, inequality and social dysfunction (e.g., high levels of violent crime); waning economic opportunities; aging and dilapidated physical infrastructure; and a political legacy of corruption, nepotism and cronyism that casts a dark shadow over delta governance. This exploration of delta history, unsustainable resource practices, disaster risk, vulnerability and resilience reveal four imperatives for the region.

**Fig. 15.6** Shrimp boat near Grand Isle, Louisiana. (Source: photograph by Bruce Glavovic)

## *15.3.2 Delta Imperatives*

Four major lessons or imperatives have been brought to the fore by recent disasters (Fig. 15.6).

*First, stem wetland loss and restore delta ecosystems:* The Katrina and the BP-DWH oil spill disasters underscore the critical importance of arresting wetland loss and restoring delta ecosystems to sustain coastal livelihoods and reduce disaster risk, not to mention the intrinsic value of the wetlands and delta stewardship obligations. Concern about wetland loss and the restoration imperative grew from the 1960s. The upshot of nearly two decades of campaigning led to plans developed by a coalition of all levels of Government called Coast 2050, approved in 1998, that carried an implementation price tag of about US$ 14 billion over 30 years. The White House baulked at this cost and a more modest follow up plan was developed. Despite numerous projects initiated from the 1990s, these efforts remained small-scale and piecemeal. After Katrina, it was recognised that the cost of effective wetland restoration will be significantly higher than previously estimated. The State of Louisiana developed a Coastal Master Plan, initially approved in 2007 and revised in 2012 (LACPRA 2007/2012). The new plan carries a price tag of US$ 50 billion to carry out 109 fully funded projects that will improve flood protection and create a 'sustainable coast.' Recent disasters underscore the need to integrate flood protection and coastal restoration. A 'multiple lines of defence' strategy has been advocated based on complementary measures including restoring barrier islands and shorelines, stabilising banks and shorelines and creating marshes; structural works such as engineered floodwalls, floodgates and levees; non-structural options such as elevating buildings, property acquisition and permanent relocation and land-use zoning and building codes. The Coastal Master Plan recognises this multi-pronged approach. Paradoxically, however, the issue of climate change has been muted in publicly distributed documents because of the contentious nature of this topic in the delta—even though the significance of this issue is well-known to scientists and many others in the region. Rampant coastal erosion and wetland loss is sufficient to galvanise political and public attention without having to invoke climate change risk in support of the imperative to restore delta ecosystems. This imperative has become ever more urgent and compelling given that significant areas of the existing delta will be inundated even if sediment loads in the Mississippi river are restored because projected sea-level rise is at least three times faster than it was during the construction of the delta plain (Blum and Roberts 2009, 2012). Restoration efforts will have to become even more intensive in coming years to offset climate change impacts; and these efforts will need to become less energy-intensive given the anticipated rising cost of energy (Day et al. 2005). But there are extraordinarily complex legal, political, fiscal, institutional and cultural obstacles to translating intentions into practical reality. Securing the necessary funding and implementing the plan will require the coordinated and active engagement of all levels of government in partnership with key stakeholders from the private sector, civil society and the scientific community. The Katrina and BP-DWH oil spill recovery processes underscore the vexed nature of such a complex and contested endeavour.

*Second, overcome the safe development paradox:* Burby (2006) describes the paradoxical consequence of efforts to reduce modest risk associated with high frequency events such as annual flooding by constructing levees that leads to a false sense of security and intensified development within the levees and potentially catastrophic impacts when a low probability event exceeds design standards. Burby (2006, p. 178) argues that the catastrophic flooding of New Orleans "could be viewed as an expected consequence of federal policy rather than an aberration that is unlikely to be repeated." Measures that compel local communities to assume the risk burden of development choices may help to overcome this paradox. But, as revealed by historic practices in the delta, century old practices of ill-advised development—driven by self-interest and short-term thinking—have prioritised financial gain over environmental stewardship, undermining the natural defences of the wetlands and predisposed delta communities and especially New Orleanians to disaster. The only sure way to overcome the safe development paradox is to avoid putting people in harm's way, and where necessary to relocate at risk communities. This overly simplistic prescriptive action is the only sure way to avoid exposure to natural hazards. Once physical development has occurred in at-risk localities, there is understandably deep resistance to relocate or retreat. In the face of escalating climate change driven disaster risk, there is, however, a compelling need to shift from the predominant reliance on 'hard-engineering' protective measures towards approaches that chart sequenced adaptation pathways, including retreat from high risk localities at critical thresholds when the social costs of protection outweigh social benefits. Katrina and the BP oil spill, reinforced by changing perceptions about climate change as a result of extreme weather events, such as Hurricane Sandy in New York state in 2012 (Smith and Jenkins 2013), have brought to the fore alternative options for adapting to climate change, including their incorporation into among other things the Coastal Master Plan. Reducing exposure to storms and other perils is, however, only one dimension of disaster risk reduction (Fig. 15.7).

**Fig. 15.7** Mitigation measures, Grand Isle, Louisiana. (Source: photograph by Bruce Glavovic)

*Third, address the drivers and root causes of social vulnerability that predispose marginalised groups and communities to disaster:* This lesson was cruelly underscored by Katrina (Laska and Morrow 2006). It had long been known that many New Orleanians lacked the financial, human and other resources necessary for coping with a major storm event, let alone catastrophic flooding. A major disaster in New Orleans had been predicted for more than a decade and the location of the worst devastation and who would be impacted had even been anticipated (Laska 2004; van Heerden and Bryan 2006). Katrina unequivocally showed that the drivers and root causes of poverty and inequity need to be confronted if disaster risk is to be reduced. The BP-DWH oil spill reinforced this imperative. Formal and informal institutions and processes shape social choices and mediate access and entitlement to a range of resources and assets necessary for building layers of resilience. Exclusion or marginalisation from political processes and markets institutionalise poverty and inequality and increase social vulnerability and hence disaster risk. Public planning and decision-making processes, including those aiming to reduce disaster risk and build resilient and sustainable communities, are thus imbued with the realities of politics and power which need to be confronted in the design and management of efforts to reverse social vulnerability. The foregoing

**Fig. 15.8** Building community resilience. (Source: photograph by Bruce Glavovic)

three imperatives are dependent on fundamental reforms to the architecture of disaster governance (Fig. 15.8).

*Fourth, reframe governance thinking and praxis that drive disaster risk by causing environmental degradation and social vulnerability:* Disaster governance is an emerging concept in the disaster literature (Tierney 2012) that is qualitatively different from terms like emergency management, disaster management or disaster risk reduction that tend to focus on hands-on actions, often by government, to mitigate impacts before, during and after a major event. Disaster management is a misnomer because by definition disasters are not 'manageable'—they are hazard events that exceed the coping capacity of affected populations. 'Governance' is a well-established but hotly debated term in diverse literatures (e.g., Rhodes 1996, 1997; Stoker 2000; Newman 2001; Bang 2003; Kooiman 2003; Boyte 2005). The term is apt for disaster studies and practice, and natural hazards planning in particular (Glavovic 2010; Glavovic et al. 2010), and especially valuable insights can be gleaned from literatures on environmental (e.g., Lemos and Agrawal 2006; Newell et al. 2012) and risk (e.g., Renn 2008; Aven and Renn 2010; Renn et al. 2011; Klinke and Renn 2012) governance. Drawing from this diverse scholarship, I use the term 'disaster governance' to refer to the steering activities of interacting government, civil society and private sector actors as they seek to diminish the prospect and impacts of disasters. Interactions may be deliberate or inadvertent and take place through a gamut of formal and informal institutions, actor networks and practices. Steering activities include norms, taboos, laws, policies and practices invoked by social groups at different scales (from local to international levels and over time) to collectively address the array of sudden shock and slow-onset perils society faces. Katrina and the BP-DWH oil spill reveal the diversity of actors and institutions that create, bear and share disaster risk and how their interactions shape exposure and vulnerability to hazard events and to building long-term resilience and adaptive capacity or paradoxically entrench pre-event vulnerabilities that predispose communities to repeat disasters. The failure of disaster

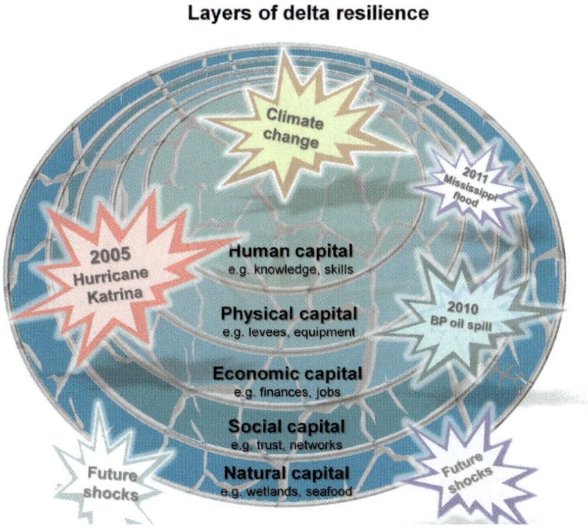


**Fig. 15.9** Waves of Adversity, Layers of Resilience. (Source: Bruce Glavovic, drawn by Paul Schneider)

governance in Katrina and in the BP-DWH oil spill reflects the systemic failure of wider governance networks to effectively advance environmental stewardship and equity. Successive US governments have actively or at best tacitly encouraged rampant and often high-risk exploitation of delta resources with perfunctory regulatory oversight (Gramling and Freudenburg 2012). Building more resilient, adaptive and sustainable delta communities thus requires transformation of prevailing disaster governance thinking and praxis.

Addressing these four imperatives in a coherent manner requires a paradigm shift in thinking about the value of the delta; how to sustainably use deltaic resources and build culturally, socially, economically and ecologically sustainable livelihoods; and in the process reduce exposure and vulnerability and build resilience. Charting development pathways that are adaptive and sustainable necessitates the building of layers of resilience to buffer waves of adversity that are compounded and exacerbated by climate change. What then are the barriers for mainstreaming climate change adaptation into planning and decision-making in the Mississippi delta region; and how can these be converted into opportunities?

## 15.4 Barriers and Opportunities for Building Resilience and Adapting to Climate Change

Figure 15.9 illustrates notional layers of resilience that will help to buffer delta communities against waves of adversity.

Delta communities need to build 'thick' layers of resilience or robust 'critical infrastructure' to cope with sudden shocks and plan for and adapt to change in ways that minimise exposure and sensitivity to climate change impacts. The history of delta disasters reveals the juxtaposition of coupled community resilience and vulnerability in the delta (see Gotham and Campanella 2011). Vulnerability fissures and fractures weaken extant layers of resilience that need to be repaired and 'thickened'. This notion builds upon and extends the 'multiple lines of defence' strategy. But the latter implies a resistance orientation that is not well aligned to the uncertain and turbulent reality of climate change. Building layers of resilience is thus an apt metaphor and necessitates among other things restoring and maintaining the health, productivity and integrity of coastal wetlands; providing robust public infrastructure, including flood protection works; fostering economic well-being whilst acting as stewards of climate-sensitive resources; strengthening social capital; eradicating poverty and inequality; and building inclusive governance institutions that include effective research, monitoring, public awareness and communication systems. One way of thinking about how to translate barriers into opportunities for building layers of resilience is to use the metaphor of 'capital':

*Human capital:* This typically refers to the stock of knowledge, competencies and attributes including education, health, entrepreneurship and skills that enable people to contribute to economic and social life. A broader perspective includes the implicit attitudes and knowledge that confer social advantage; described by Bourdieu (1986) as 'cultural capital'. I use the term human capital to include both the 'technical' and 'cultural' dimensions of knowledge and experience that shape public perceptions and awareness of and attitudes towards disaster risk and sustainability; and to reflect that learning takes place in a social context. There are strong anti-reflexive tendencies in influential circles in the US towards the issue of climate change and sustainability more generally, including "merchants of doubt" (Oreskes and Conway 2010) who deliberately obfuscate the unequivocal scientific evidence of global warming that is very likely driven by human actions (IPCC 2007; see e.g., McCright and Dunlap 2010). Reluctance to make explicit reference to climate change in the Coastal Master Plan is indicative of the knowledge, attitudinal and institutional barriers that need to be overcome. Stimulating reflexivity (self-conscious and self-critical reflection on oneself and society) is crucial for developing the human capital necessary to deepen and extend individual and societal adaptive capacity and resilience. An important starting point is to understand better the nature of risk, the value of ecosystem services and the complexity of coupled social-ecological systems. Integrating such understanding into public planning and decision-making is challenging given the difficulty of bridging the science-policy-practice gap, especially with respect to climate change (Watson 2005). Evolving resilience and risk governance scholarship and praxis underscores the crucial role that social learning plays in helping to bridge this gap (Adger 2003; Folke 2006; Klinke and Renn 2012).

*Physical capital:* Investment in physical capital, including factors of production such as equipment and machinery as well as public infrastructure such as roads and protective works, is vital for economic productivity and public safety. Well established communities seek to maintain social stability and secure existing investments in property and infrastructure and are thus likely to prioritise 'hard engineering' protective works as a defence against natural hazards such as floods and storms over 'soft engineering' or 'design with nature' (McHarg 1971) approaches. There is a compelling but understandable proclivity to resist change and defend societal investment. But, as the 'safe development paradox' indicates, such resistance can prove to be maladaptive in the long-term, especially in the face of climate change and escalating disaster risk. Alternative development pathways that are adaptive, resilient and sustainable therefore need to be explored. Long-term strategic planning provides a distinctive opportunity to chart such pathways. Initiatives such as the Coastal Master Plan demonstrate that physical capital, such as levees, is an integral component of a 'multiple lines of defence' strategy, and that innovative design and management measures geared towards 'living with water' offer practical alternatives to traditional 'defensive' strategies. Reframing this defensive strategy towards one that embraces change and builds 'layers of resilience' is compelling but challenging.

*Economic capital:* The pursuit of private gain and accumulation of financial capital helps to build vibrant communities through job creation and economic development. But relentless prioritisation of short-term profit and economic growth at the expense of ecological sustainability and social equity can among other things escalate disaster risk and reduce livelihood options and community resilience and sustainability. Increasing attention is being focused on developing alternative economic models that take into account the value of ecological services, the nature of risk and the complex inter-connected nature of social-ecological systems (Daly 2005; Costanza et al. 2006; Batker et al. 2010). Such approaches seek to integrate short- and long-term considerations; reconcile private and public interests; and equitably distribute risk, and costs and benefits, to shareholders, stakeholders and citizens. Translating such laudable intentions into practical reality is, however, vexed. But as the recovery experience in the delta demonstrates, there are positive indications that economic interests are being viewed in a wider context of sustainability and resilience.

*Social capital:* This concept is contested but broadly speaking connotes the value of reciprocal relationships or social networks (Coleman 1988; Portes 1998). Notwithstanding evidence of widespread declining social capital (Putnam 1995, 2000), Solnit (2010) shows that, remarkably, in the worst of circumstances, people engage in altruistic and compassionate social actions that help to rebuild shattered communities. In the immediate aftermath of disaster, community members engage in practical social actions that are powerful catalysts for building social capital. There is a strong positive relationship between social capital and community resilience in general (Murphy 2007) and in preparing for, responding to and recovering from disasters (Chamlee-Wright and Storr 2011). Natural hazards planning processes that explore, share and build on these positive community experiences can thus help to build social capital and strengthen resilience. The challenge is to overcome the established patterns of social behaviour and governance institutions and practices that become dominant after disasters and close down the flourish of transient communities of mutual aid that arise organically in adversity (see Solnit 2010). Social capital is thus imbued with power and politics. Political systems and practices that alienate and marginalise particular groups or communities result in inequitable entitlements to the range of assets at the disposal of communities, thus increasing social vulnerability and disaster risk—as Katrina poignantly showed. Deepening and extending democracy through authentic public participation is thus pivotal to natural hazards planning and decision-making processes that build social capital and 'thicken' layers of community resilience.

*Natural capital:* The land, air, water, living organisms and all formations of the biosphere can be thought of as natural capital that provides humanity with ecosystem goods and services essential for survival and well-being (Daly 1990, 2005). Traditional measures of economic performance ignore the value of natural (and social) capital and hence it is neglected relative to human, physical and economic capital. Consequently, natural systems are over-exploited and degraded. The incredible value of the delta's natural capital has been highlighted above and its recognition is integral to securing the requisite investment to restore delta wetlands and build layers of resilience. But to secure the delta's natural capital, and overcome the drivers of unsustainable development stemming from the prevailing hegemony, there is a need to institutionalise approaches that among other things: Maintain environmental diversity and redundancy; manage connectivity; manage slow variables and feedbacks; foster understanding of complex social-ecological systems as adaptive systems; encourage social learning and experimentation; broaden and deepen public participation; and promote multi-scalar and multi-level governance systems (Biggs et al. 2012).

Building human, physical, economic, social and natural capital in the delta in the face of climate change is a 'super wicked problem' for which there are no panaceas. Past coastal innovations (such as deep water drilling for oil) have unwittingly accelerated unsustainable use of coastal resources in general and the delta in particular. Paradoxically, innovation is needed to escape the vulnerability trap that past innovation has set (Glavovic 2013b). A transformative practice of deliberative coastal governance provides a framework for navigating the fraught transition to coastal sustainability (Glavovic 2013c) and 'thickening' the layers of delta resilience.

Importantly, these layers of resilience do not exist in isolation; they are mutually supportive: Delta resilience as a whole is greater than the sum of the individual layers of resilience.

## 15.5 Towards Deliberative Delta Governance

Deliberative delta governance is recommended to address the delta imperatives and overcome the barriers and unlock the opportunities outlined above. The framework proposed by Glavovic (2013c) explains why a deliberative approach is needed to facilitate the transition towards coastal resilience and sustainability, and how this can be achieved. The framework builds upon and extends the Orders of Outcomes framework developed by Olsen and colleagues (Olsen 2002, 2003; Olsen et al. 2009) which recognises that pursuit of coastal sustainability starts with the creation of *enabling conditions for sustained implementation* through agreed goals, supportive and informed constituencies, implementation capacity, and commitments to invest in implementation. Only once the first Order of Outcomes is in place can progress be made towards the second Order of Outcomes, *implementation through changed behaviour*, and then the third, *achieving targeted environmental and societal outcomes*, and the elusive fourth Order of Outcomes, *sustainable coastal development*. Progress to higher Orders of Outcomes is challenging because of maladaptive path dependencies and 'lock-in' due to unsustainable 'business as usual' practices. Glavovic (2013c) argues that the above Orders of Outcomes need to be underpinned by four *deliberative or process outcomes* that provide an essential platform for realising the *coastal outcomes* described by Olsen and colleagues. The foundational pillars for deliberative delta governance are briefly summarised here and readers are referred to Glavovic (2013c) for more detailed explanation.

Deliberation is a non-coercive process of communicative interaction between social actors that stimulates reflection on societal values, preferences and interests in making social choices (Dryzek 2000). It can take place in both formal and informal public settings and involves much more than 'talking'. It includes information sharing, discussion and debate that matures with practice and social learning and enables participants to make well-informed public decisions (Chambers 2003). Deliberation needs to be authentic, inclusive and consequential if it is to yield legitimate social outcomes (Dryzek 2009) and facilitate the transition to sustainable development (Fisher 2000; Baber and Bartlett 2005; Dryzek 2011).

The first pillar of deliberative delta governance is to understand the nature of delta risk and issues, and improve democratic attitudes and skills. 'Safe arenas' for public deliberation need to be created to enable participants to explore and develop a shared understanding of delta concerns. Deliberation can engage and integrate different types of knowledge and knowledge claims, including traditional disciplinary science and local and traditional knowledge. In so doing, coastal concerns, including climate risk, resilience and adaptation options and pathways, can be explored in ways that promote social learning and enhance both the 'technical' and 'cultural' dimensions of human capital referred to above. Deliberation can also help to raise awareness and foster deeper appreciation and tolerance of divergent viewpoints. Democratic attitudes and interpersonal communication skills and judgement can also be improved with deliberative practice that enhances group interactions and decision-making processes.

The second pillar of deliberative delta governance is to catalyse community-oriented action and improve institutional capacity and decision-making. Delta history highlights the difficulty of achieving collective action in the face of strident individualism or where community members are alienated or marginalised from public decision-making processes. Inclusive deliberation helps to build a common purpose and stimulates participation in community activities. It provides a foundation for reconciling contending interests, strengthens community institutional capacity and decision-making competency and enhances the legitimacy of community decisions.

The third pillar of deliberative delta governance is to deepen community problem-solving capacity. Deliberation helps participants' reframe and improve their understanding of delta risk and issues, strengthens institutional capacity and decision-making and engages community members more constructively in community life on a sustained basis. In practice, however, understanding and resolving complex and contested intra-community problems is a difficult and protracted process. Deliberation helps to overcome the barriers and unlocks opportunities explored above to develop community problem-solving capacity over time.

The fourth pillar is facilitating inter-community collaboration through cross-scalar and multi-level processes of authentic and inclusive dialogue, visioning, negotiation and cooperation. Such polycentric deliberation takes place in formal and informal collaborative arenas. Addressing delta risk and adapting to climate change cannot progress meaningfully without being framed in this wider governance milieu. Both intra- and inter-community deliberation is necessary to build layers of resilience and foster delta sustainability.

## 15.6 Priority Practical Actions for Adapting to Climate Change

Three practical priority actions are recommended for building community resilience, adaptive capacity and sustainability based on the notion of deliberative delta governance outlined above.

*First, articulate, share and celebrate community narratives about overcoming adversity and building community resilience and sustainability.* Much can be learned from local 'success stories' and the pathways that offer promise for buffering communities against the waves of adversity that are compounded and exacerbated by climate change. For example, Chamlee-Wright and Storr (2011) demonstrate that collective narratives can shape individual recovery strategies. In their study of St. Bernard parish they found that individual strategies emphasised self-reliance informed by a shared narrative of being a close-knit, family-oriented community made up of hard-working people. "Social capital in the form of collective narratives, we contend, shaped the disaster response and recovery efforts in St. Bernard Parish and helps to explain the surprising signs of resilience in that community" (Chamlee-Wright and Storr 2011, p. 267). Sharing community narratives about overcoming adversity does not imply avoiding self-critical reflection on mistakes made and lessons learned from failed initiatives. Much can be gained by deliberate and reflexive learning from both success and failure. The challenge is then to institutionalise lessons learned.

*Second, design inclusive processes of local community disaster risk reduction and resilience planning.* The deliberative delta governance approach advocated here is predicated on inclusive processes of authentic public engagement, civic science and social learning. The appropriate modalities for engagement and who should participate need to be contingent on the characteristics of the delta risk and issues under consideration. The convoluted and contested New Orleans recovery planning process underscores the complex and inherently political character of such endeavours. A trusted mediator can help to create the necessary 'safe spaces' that enable community voices to be heard, including marginalised voices, and conflicting interests to be resolved.

*Third, region-wide strategic collaborative planning processes are needed to address issues that are beyond the influence of local communities and create the deep-rooted inertia that results in unsustainable and maladaptive outcomes.* Whilst the deliberative delta governance approach outlined above progresses from lower to higher Orders of Outcomes in a generally sequential manner (albeit not in an inexorable linear fashion), building inter-community deliberative capacity is likely to be fostered by first building intra-community deliberative capacity. Delta- and even Gulf-wide initiatives, such as the Gulf of Mexico Alliance, provide encouraging evidence of the emergence of collaborative initiatives that transcend particular interests or geographical communities.

## 15.7 Conclusion

This chapter has shown that the Mississippi delta is made up of social-ecological systems of immense value. Delta communities face a range of perils and have prevailed despite many disasters, including floods, hurricanes and oil spills. Past and prevailing practices are, however, unsustainable and coastal livelihoods are imperilled. To compound matters, the delta is a global hotspot for climate change impacts and sea-level rise in particular. Disaster risk is escalating. Delta communities need to build layers of resilience as a protective barrier against waves of adversity that are likely to intensify with climate change. Recent disasters have exposed the paradox of conjoint resilience and vulnerability; though different groups and communities have varying levels of each over time. There is nonetheless, seemingly inexorable pressure to rebuild 'as before' and 'return to normal' after a disaster. Such a hankering for the physical reconstruction of the past, whilst understandable, is invariably unwise and nigh impossible to realise in practice. There is a post-disaster window of opportunity to chart new pathways that avoid recreating the exposure and vulnerabilities that predisposed communities to disaster in the first place. Recovery choices are, however, complex and contested. Innovative modalities of collaborative planning, and courageous leadership by key actors in government, the private sector civil society, the research community and media, are needed to expedite recovery and make wise choices that foster community resilience and sustainability.

Delta history Katrina and the BP-DWH oil spill reveal four delta imperatives. First, stem wetland loss and restore delta ecosystems to sustain coastal livelihoods and reduce disaster risk in the face of climate change. Second, overcome the 'safe development paradox' created by reliance on structural flood protection measures and explore alternative adaption pathways. Third, address the drivers and root causes of social vulnerability that predispose marginalised groups and communities to disaster. Fourth, reframe governance thinking and praxis that drive disaster risk by causing environmental degradation and social vulnerability. Addressing these four imperatives in a coherent manner requires a paradigm shift in thinking about the value of the wetlands; how to sustainably use deltaic resources and build culturally, socially, economically and ecologically sustainable livelihoods; and reduce disaster risk and vulnerability, and build resilience in the face of climate change.

Barriers and opportunities for mainstreaming climate change adaptation into planning and decision-making in the delta are explored with respect to the different forms of capital (human, physical, economic, social and natural capital) that together construct layers of resilience. Efforts to reduce disaster risk, plan for different hazards, including pre-event planning and post-disaster recovery, and navigate the turbulent waters of future climate change, need to be framed as a practice of deliberative delta governance. Four pillars rooted in deliberation are foundational to such an approach: First, build understanding of risk and coastal issues, and improve democratic attitudes and skills. Second, foster community-oriented action and improve institutional capacity and decision-making. Third, deepen community problem-solving capacity. Fourth, facilitate inter-community collaboration through sustained processes of authentic and inclusive dialogue, visioning, negotiation and cooperation.

Finally, three priority actions are recommended for building community resilience, adaptive capacity and sustainability based on the recommended deliberative delta governance approach. First, articulate, share and celebrate community narratives about disasters, risk and resilience. Much can be learned from local 'success stories' and the pathways that offer promise for buffering communities against the waves of adversity that are compounded and exacerbated by climate change. Second, design inclusive processes of local community disaster risk reduction and resilience planning. A trusted mediator can help to ensure that all voices are heard and conflicting interests resolved. Third, delta- and region-wide strategic collaborative planning processes are needed to address issues that are beyond the influence of local communities and create the deep-rooted inertia that results in unsustainable and maladaptive outcomes.

## References

Adger, W. N. (2003). Social capital, collective action, and adaptation to climate change. *Economic Geography, 79*(4), 387–404.

Allen, B. L. (2003). *Uneasy alchemy: Citizens and experts in Louisiana's chemical corridor disputes*. Cambridge: MIT Press.

Aven, T., & Renn, O. (2010). *Risk management and governance*. Heidelberg: Springer.

Baber, W. F., & Bartlett, R. V. (2005). *Deliberative environmental politics: Democracy and ecological rationality*. Cambridge: MIT Press.

Bang, H. P. (Ed.). (2003). *Governance as social and political communication*. Manchester: Manchester University Press.

Barbier, E. B. (2011). Coastal Wetland restoration and the deepwater horizon oil spill. *Vanderbilt Law Review, 64*(6), 1821–1852.
Barrios, R. E. (2011). If you did not grow up here, you cannot appreciate living here: Neoliberalism, space-time, and affect in post-Katrina recovery planning. *Human Organization, 70*(2), 118–127.
Barry, J. M. (1997). *Rising tide: The great Mississippi flood of 1927 and how it changed America*. New York: Simon and Schuster.
Batker, D. P., de la Torre, I., Costanza, R., Swedeen, P., Jr. Day, J. W., Boumans, R., & Bagstad, K. (2010). *Gaining ground-Wetlands, hurricanes and the economy: The value of restoring the Mississippi river delta*, Tacoma: Earth Economics.
Bergin, T. (2011). *Spills and spin: The inside story of BP*. London: Random House Business Books.
Biggs, R., Schulter, M., Biggs, D., Bohensky, E. L., BurnSilver, S., Cundill, G., Dakos, V., Daw, T. M., Evans, L. S., Kotschy, K., Leitch, A. M., Meek, C., Quinlan, A., Raudsepp-Hearne, C., Robards, M. D., Schoon, M. L., Schultz, L., & West, P. C. (2012). Towards principles for enhancing the resilience of ecosystem services. *Annual Review of Environment and Resources, 37,* 421–448.
Bjorndal, K. A., Bowen, B. W., Chaloupka, M., Crowder, L. B., Heppell, S. S., Jones, C. M., Lutcavage, M. E., Policansky, D., Solow, A. R., & Witherington, B. E. (2011). Better science needed for restoration in the Gulf of Mexico. *Science, 331,* 537–538.
Blum, M. D., & Roberts, H. H. (2009). Drowning of the Mississippi delta due to insufficient sediment supply and global sea-level rise. *Nature Geoscience, 2*(7), 488–491.
Blum, M. D., & Roberts, H. H. (2012). The Mississippi delta region: Past, present, and future. *Annual Review of Earth and Planetary Sciences, 40,* 655–683.
Boesch, D. F. (2012). Deep-water drilling remains a risky business. *Nature, 484,* 289.
Boesch, D. F., Josselyn, M. N., Mehta, A. J., Morris, J. T., Nuttle, W. K., Simenstad, C. A., & Swift, D. J. P. (1994). Scientific assessment of coastal wetland loss, restoration, and management in Louisiana. *Journal of Coastal Research, 20*(Special Issue), 1–103.
Bourdieu, P. (1986). The forms of capital. In J. Richardson (Ed.), *Handbook of theory and research for the sociology of education* (pp. 241–258). New York: Greenwood.
Boyte, H. C. (2005). Reframing democracy: Governance, civic agency, and politics. *Public Administration Review, 65*(5), 536–546.
Burby, R. J. (2006). Hurricane Katrina and the paradoxes of government disaster policy: Bringing about wise governmental decisions for hazardous areas. *Annals of the American Academy of Political and Social Science, 604,* 171–191.
Campanella, R. (2006). *Geographies of New Orleans: Urban fabrics before the storm*. Lafayette: Center for Louisiana Studies, University of Louisiana at Lafayette.
Campanella, R. (2007). An ethnic geography of New Orleans. *Journal of American History, 94,* 704–715.
Chambers, S. (2003). Deliberative democratic theory. *Annual Review of Political Science, 6,* 307–326.
Chamlee-Wright, E., & Storr, V. H. (2011). Social capital as collective narratives and post-disaster community recovery. *The Sociological Review, 59*(2), 266–282.
Cigler, B. A. (2007). The "big questions" of Katrina and the 2005 great flood of New Orleans. *Public Administration Review, 67*(Special Issue), 64–76.
Cobb, J. C. (1992). *The most southern place on earth: The Mississippi delta and the roots of regional identity*. New York: Oxford University Press.
Coleman, J. S. (1988). Social capital in the creation of human capital. *The American Journal of Sociology, 94*: 95–120.
Colten, C. E. (Ed.) (2000). *Transforming New Orleans and its environs: Centuries of changes*. Pittsburgh: University of Pittsburgh Press.
Colten, C. E. (2006a). *An unnatural metropolis: Wresting New Orleans from nature*, Baton Rouge: Louisiana State University Press.
Colten, C. E. (2006b). Vulnerability and place: Flat land and uneven risk in New Orleans. *American Anthropologist, 108*(4), 731–734.

Colten, C. E., & Sumpter, A. R. (2009). Social memory and resilience in New Orleans. *Natural Hazards, 48,* 355–364.

Comfort, L. K., Birkland, T. A., Cigler, B. A., & Nance, E. (2010). Retrospectives and prospectives on hurricane Katrina: Five years and counting. *Public Administration Review, 70*(5), 669–678.

Committee on the Effects of the Deepwater Horizon Mississippi Canyon-252 oil spill on Ecosystem Services in the Gulf of Mexico. (2012). *Approaches for ecosystem services valuation for the Gulf of Mexico after the deepwater horizon oil Spill: Interim report*. Washington, DC: The National Academies Press.

Costanza, R., Batker, D., Day, J.W., Feagin, R.A., Martinez, M. L., & Roman, J. (2010). The perfect spill: Solutions for averting the next deepwater horizon. *Solutions, 1*(5), 17–20.

Costanza, R., Mitsch, W., & Day, J. (2006). Creating a sustainable and desirable New Orleans. (Editorial). *Ecological Engineering, 26,* 317–320.

Daly, H. E. (1990). Toward some operational principles of sustainable development. *Ecological Economics, 2,* 1–6.

Daly, H. E. (2005). Economics in a full world. *Scientific American, 293*(3), 100–107.

Day, J., Barras, J., Clairain, E., Johnston, J., Justix, D., Kemp, P., Ko, J., Lane, R., Mitsch, W., Steyer, G., Templet, P., & Yanez, A. (2005). Implications of global climatic change and energy cost and availability for the restoration of the Mississippi delta. *Ecological Engineering, 24,* 253–265.

Day, J., Boesch, D., Clairain, E., Kemp, P., Laska, S., Mitsch, W., Orth, K., Mashriqui, H., Reed, D., Shabman, L., Simenstad, C., Streever, B., Twilley, R., Watson, C., Wells, J., & Whigham, D. (2007). Restoration of the Mississippi delta: Lessons from hurricanes Katrina and Rita. *Science, 315,* 1679–1684.

Derthick, M. (2007). Where federalism didn't fail. *Public Administration Review, 67*(Special Issue), 36–47.

Dryzek, J. S. (2000). *Deliberative democracy and beyond. Liberalism, critics, contestations.* Oxford: Oxford University Press.

Dryzek, J. S. (2009). Democratization as deliberative capacity building. *Comparative Political Studies, 42*(11), 1379–1402.

Dryzek, J. S. (2011). *Foundations and frontiers of deliberative governance*. Oxford: Oxford University Press.

Dyson, M. E. (2006). *Come hell or high water: Hurricane Katrina and the color of disaster*. New York: Basic, Civitas Books.

Edwards, F. L. (2009). Obstacles to disaster recovery. *The Public Manager, 37*(4), 66–71.

Finch, C., Emrich, C. T., & Cutter, S. L. (2010). Disaster disparities and differential recovery in New Orleans. *Population and Environment, 31,* 179–202.

Fischer, F. (2000). *Citizens, experts and the environment: The politics of local knowledge*. Durham: Duke University Press.

Folke, C. (2006). Resilience: The emergence of a perspective for social-ecological systems analysis. *Global Environmental Change, 16*(3), 253–267.

Freudenburg, W. R., & Gramling, R. (2011). *Blowout in the Gulf: The BP oil spill disaster and the future of energy in America*. Cambridge: The MIT Press.

Freudenburg, W. R., Gramling, R., Laska, S., & Erikson, K. T. (2008). Organizing hazards, engineering disasters? Improving the recognition of political-economic factors in the creation of disasters. *Social Forces, 87*(2), 1015–1038.

Freudenburg, W. R., Gramling, R., Laska, S., & Erikson, K. T. (2009). *Catastrophe in the making: The engineering of Katrina and the disasters of tomorrow*. Washington, DC: Island Press.

Gill, D. A., Picou, J. S., & Ritchie, L. A. (2012). The *Exxon Valdez* and BP oil spills: A comparison of initial social and psychological impacts. *American Behavioral Scientist, 56*(1), 3–23.

Glavovic, B. C., Saunders, W., & Becker, J. (2010). Land-use planning for natural hazards in New Zealand: The setting, barriers, 'burning issues' and priority actions. *Natural Hazards, 54*(3): 679–706.

Glavovic, B. C. (2010). The role of land-use planning in disaster risk reduction: An Australasian perspective, Special Issue of *Australasian Journal of Disaster and Trauma Studies.* http://www.massey.ac.nz/~trauma/issues/2010-1/editorial.htm. Accessed 22 May 2013.

Glavovic, B. C. (2013a). Disasters and the continental shelf: Exploring new frontiers of risk. In M.H. Nordquist, J.N. Moore, A. Chircop, & R. Long (Eds.) *The regulation of continental shelf development* (pp. 225–256). Leiden: Nijhoff.

Glavovic, B. C. (2013b). The coastal innovation paradox. *Sustainability, 5,* 912–933.

Glavovic, B. C. (2013c). The coastal innovation imperative. *Sustainability, 5,* 934–954.

Gotham, K. F., & Campanella, R. (2011). Coupled vulnerability and resilience: The dynamics of cross-scale interactions in post-Katrina New Orleans. *Ecology and Society, 16*(3), 12. http://dx.doi.org/10.5751/ES-04292-160312. Accessed 22 May 2013.

Gramling, R., & Freudenburg, W. R. (2012). A century of Macondo: United States energy policy and the BP blowout catastrophe. *American Behavioral Scientist, 56*(1), 48–75.

Hawkins, R. L., & Maurer, K. (2011). You fix my community, you have fixed my life: The disruption and building of ontological security in New Orleans. *Disasters, 35*(1), 143–159.

Hoffman, A. J., & Jennings, P. D. (2011). The BP Oil Spill as a cultural anomaly? Institutional context, conflict, and change. *Journal of Management Inquiry, 20*(2), 100–112.

Ink, D. (2006). An analysis of the house select committee and White House reports on hurricane Katrina. *Public Administration Review, 66*(6), 800–807.

IPCC, (2007). *The physical science basis.* Cambridge: Cambridge University Press.

Juhasz, A. (2011). *Black tide: The devastating impact of the Gulf oil spill.* Hoboken: Wiley.

Jurkiewicz, C. L. (2007a December). Louisiana's ethical culture and its effect on the administrative failures following Katrina. *Public Administration Review, 2007*(Special Issue), 57–63.

Jurkiewicz, C. L. (2007b December). The responders: Official and grassroots actors in the aftermath of the hurricane. *Public Administration Review, 2007*(Special Issue), 82–83.

Kates, R. W., Colten, C. E., Laska, S., & Leatherman, S. P. 2006. Reconstruction of New Orleans after hurricane Katrina: A research perspective. *Proceedings of the National Academy of Sciences, 103*(40), 14653–14660.

Kessler, R. C., Galea, S., Gruber, M. J., Sampson, N. A., Ursano, R. J., & Wessly, S. (2008). Trends in mental illness and suicidality after hurricane Katrina. *Molecular Psychiatry, 13*(4), 374–384.

King, R. O. (2005). *Hurricane Katrina: Insurance losses and national capacities for financing risk.*Washington, DC: Congressional Research Service, Library of Congress (RL33086. September 15).

Klinke, A., & Renn, O. (2012). Adaptive and integrative governance on risk and uncertainty. *Journal of Risk Research, 15*(3), 273–292.

Ko, J., & Day, J. W. (2004). A review of ecological impacts of oil and gas development on coastal ecosystems in the Mississippi delta. *Ocean and Coastal Management, 47,* 597–623.

Koliba, C. J., Mills, R. M., & Zia, A. (2011). Accountability in governance networks: An assessment of public, private, and nonprofit emergency management practices following hurricane Katrina. *Public Administration Review, 71*(2), 210–220.

Kooiman, J. (2003). *Governing as governance*. London: Sage.

Ladd, A. E. (2012). Pandora's well: Hubris, deregulation, fossil fuels, and the BP oil disaster in the Gulf. *American Behavioral Scientist, 56*(1), 104–127.

Laska, S. (2004). What if hurricane Ivan had not missed New Orleans? *Natural Hazards Observer, 29*(2), 5–6.

Laska, S., & Morrow, B. (2006). Social vulnerabilities and hurricane Katrina: An unnatural disaster in New Orleans. *Journal of the Marine Technology Society, 40*(4), 16–26.

Lee, M. R., & Blanchard, T. C. (2012). Community attachment and negative affective states in the context of the BP Deepwater Horizon disaster. *American Behavioral Scientist, 56*(1), 24–47.

Lemos, M. C., & Agrawal, A. (2006). Environmental governance. *Annual Review of Environmental Resources, 31,* 297–325.

Lerner, S. (2005). *Diamond: A struggle for environmental justice in Louisiana's environmental corridor*. Cambridge: The MIT Press.

Lindahl, C. (2012). Legends of hurricane Katrina: The right to be wrong, survivor-to-survivor storytelling, and healing. *Journal of American Folklore, 125*(496), 139–176.

Louisiana Coastal Protection and Restoration Authority (LACPRA). (2007/2012). *Louisiana comprehensive master plan for a sustainable coast*. Baton Rouge, LA: LACPRA. http://www.coastalmasterplan.louisiana.gov/. Accessed 22 May 2013.

Louisiana Recovery Authority (LRA). (2007). *Louisiana speaks regional plan: Visions and strategies for recovery and growth in South Louisiana*. Baton Rouge, LA: Louisiana Recovery Authority.

Lustgarten, A. (2012). *Run to failure: BP and the making of the Deepwater Horizon disaster*. New York: W.W. Norton.

Malakoff, D. (2012). BP criminal case generates record payout for science and restoration. *Science, 338,* 1137.

Masozera, M., Bailey, M., & Kerchner, C. (2007). Distribution of impacts of natural disasters across income groups: A case study of New Orleans. *Ecological Economics, 63,* 299–306.

McCright, A. M., & Dunlap, R. E. (2010). Anti-reflexvity. The American conservative movement's success in undermining climate science and policy. *Theory, Culture and Society, 27*(2-3), 100–133.

McHarg, I. (1971). *Design with nature*. Doubleday/Natural History Press.

Morris, C. (2012). *The big muddy: An environmental history of the Mississippi and its peoples, from Hernando de Soto to hurricane Katrina*. Oxford: Oxford University Press.

Murphy, B. L. (2007). Locating social capital in resilient community-level emergency management. *Natural Hazards, 41,* 297–315.

Mutter, J. (2010). Disasters widen the rich-poor gap. *Nature, 466,* 1042.

National Commission on the BP Deepwater Horizon Oil Spill and Offshore Drilling. (2011). *Deep water: The Gulf oil disaster and the future of offshore drilling: Report to the President*, http://www.whitehouse.gov/the-press-office/remarks-president-nation-bp-oil-spill. Accessed 17 Dec 2012.

National Research Council. (2006). *Drawing Louisiana's new map: Addressing land loss in coastal Louisiana, Committee on the Restoration and Protection of coastal Louisiana*. Washington, DC: National Research Council.

National Research Council. (2009). *Final report from the NRC committee on the review of the Louisiana coastal protection and restoration (LACPR) program, Committee on the Restoration and Protection of coastal Louisiana*. Washington, DC: National Research Council.

Nelson, M., Ehrenfeucht, R., & Laska, S. (2007). Planning, plans, and people: Professional expertise, local knowledge, and governmental action in post-hurricane Katrina New Orleans. *Cityscape, 9*(3), 23–52.

Newell, P., Pattberg, P., & Schroeder, H. (2012). Mutiactor governance and the environment. *Annual Review of Environment and Resources, 37,* 365–387.

Newman, J. (2001). *Modernising governance. New labour, policy and society*. London: Sage.

Oliver-Smith, A., & Hoffman, S. M. (Eds.). (2002). *Catastrophe and culture: The anthropology of disaster*. Santa Fe: School of American Research Press.

Olsen, S. B. (2002). Assessing progress towards goals of coastal management. *Coastal Management, 30,* 325–345.

Olsen, S. (2003). Frameworks and indicators for assessing progress in integrated coastal management initiatives. *Journal of Ocean and Coastal Management, 46*(3–4), 347–361.

Olsen, S. B., Page, G. G., & Ochoa, E. (2009). *The analysis of governance responses to ecosystem change: A Handbook for Assembling a Baseline, LOICZ R & S Report No. 34.* Geesthacht: GKSS Research Centre.

Olshansky, R. B. (2006). Planning after hurricane Katrina. *Journal of the American Planning Association, 72*(2), 147–153.

Olshansky, R. B., & Johnson, L. A. (2010). *Clear as mud: Planning for the rebuilding of New Orleans*. Chicago: American Planning Association Planners Press.

Olshansky, R. B., Johnson, L. A., Horne, J., & Nee, B. (2008). Planning for the rebuilding of New Orleans. *Journal of the American Planning Association, 74*(3), 273–287.

Oreskes, N., & Conway, E. M. (2010). *Merchants of doubt: How a handful of scientists obscured the truth on issues from tobacco smoke to global warming*. New York: Bloomsbury.

Parent, W. (2004). *Inside the carnival: Unmasking Louisiana politics*. Baton Rouge: Louisiana State University Press.

Portes, A. (1998). Social capital: Its origins and applications in modern sociology. *Annual Review of Sociology, 24,* 1–24.

Putnam, R. D. (1995). Bowling alone: America's declining social capital. *Journal of Democracy, 6,* 65–78.

Putnam, R. D. (2000). *Bowling alone: The collapse and revival of American community*. New York: Simon and Schuster.

Renn, O. (2008). *Risk governance: Coping with uncertainty in a complex world.* London: Earthscan.

Renn, O., Klinke, A., & van Asselt, M. (2011). Coping with complexity, uncertainty and ambiguity in risk governance: A synthesis. *Ambio, 40*: 231–246.

Rhodes, R. A. W. (1996). The new governance: Governing without government. *Political Studies, 44,* 652–667.

Rhodes, R. A. W. (1997). *Understanding governance: Policy networks, governance, reflexivity and accountability*. Buckingham: Open University Press.

Saikku, M. (2005). *This delta, this land: An environmental history of the Yazoo-Mississippi flood-plain*, Athens: University of Georgia Press.

Schneider, S. K. (2005). Administrative breakdowns in the governmental response to hurricane Katrina. *Public Administration Review, 65*(5), 515–516.

Shaffer, G. P., Day, J. W., Mack, S., Kemp, G. P., van Heerden, I., Poirrier, M. A., Westphal, K. A., FitzGerald, D., Milanes, A., Morris, C. A., Bea, R., & Shea Penland, P. (2009). The MRGO navigation project: A massive human-induced environmental, economic, and storm disaster. *Journal of Coastal Research, 54*(Special Issue): 206–224.

Silliman, B. R., van de Koppel, J., McCoy, M. W., Diller, J., Kasozi, G. N., Earl, K., Adams, P. N., & Zimmerman, A. R. (2012). Degradation and resilience in Louisiana Salt Marshes after the BP-Deepwater Horizon oil spill. *Proceedings of the National Academy of Sciences, 109*(28): 11234–11239.

Smith, A., & Jenkins, K. (2013). Climate change and extreme weather in the USA: Discourse analysis and strategies for an emerging 'public', *Journal of Environmental Studies and Sciences*, Published online 28 May, 2013.

Solnit, R. (2010). *A paradise built in hell: The extraordinary communities that arise in disasters.* New York: Viking.

Solomon, G. M., & Janssen, S. J. (2010). Health effects of the Gulf oil spill. *Journal of the American Medical Association, 304*(10), 1118–1119.

Stoker, G. (Ed.). (2000). *The new politics of local governance*. London: Macmillan.

Stokstad, E. (2010). Looking beyond the spill, Obama highlights long-term restoration. *Science, 328,* 1618–1619.

Tierney, K. (2012). Disaster governance: Social, political, and economic dimensions. *Annual Review of Environment and Resources, 37,* 341–363.

Törnqvist, T. E., Kidder, T. R., Autin, W. J., van der Borg, K., de Jong, A. F. M., Klerks, C. J. W., Snijders, E. M. A., Storms, J. E. A., van Dam, R. L., & Wiemann, M. C. (1996). A revised chronology for Mississippi river subdeltas. *Science, 273*(5282), 1693–1696.

U.S. House of Representatives. (2006). *A failure of initiative: Final report of the select bipartisan committee to investigate the preparation for and response to hurricane Katrina*. Washington, DC: U.S. Government Printing Office.

van Heerden, I., & Bryan, M. (2006). *The storm, what went wrong and why during hurricane Katrina -The inside story from one Louisiana scientist*. New York: Viking.

Watson, R. T. (2005). Turning science into policy: Challenges and experiences from the science-policy interface. *Philosophical Transactions of the Royal Society B, 360,* 471–477.

Wisner, B., Blaikie, P., Cannon, T., & Davis, I. (2004). *At risk: Natural hazards, people's vulnerability and disasters*. London: Routledge.

Woods, C. (1998). *Development arrested: The blues and plantation power in the Mississippi delta.* London: Verso.

# Chapter 16
# Conclusions: Integrating Natural Hazards Risk Management and Climate Change Adaptation through Natural Hazards Planning

**Gavin P. Smith and Bruce C. Glavovic**

**Abstract** A significant part of adapting to climate change involves learning from the many lessons found in planning for natural hazards and acting on this information. In the preceding chapters four imperatives have emerged that will be used to summarise our collective findings and structure a set of lessons to help stimulate and inform a still nascent dialogue among local, regional national and international role-players focused on one of the most pressing issues facing communities around the world—adapting to the effects of climate change.

**Keywords** Adaptation imperatives · Natural hazards risk management · Natural hazards planning · Climate resilience · Sustainable development

A significant part of adapting to climate change involves learning from the many lessons found in planning for natural hazards and acting on this information. In the preceding chapters four imperatives have emerged that will be used to summarise our collective findings and structure a set of lessons to help stimulate and inform a still nascent dialogue among local, regional, national and international role-players focused on one of the most pressing issues facing communities around the world—adapting to the effects of climate change.

As explained in Chap. 1 (see Fig. 16.1), and further explored in the preceding chapters, there is an intimate but complex relationship between local community development prospects and risk, including those risks tied to natural hazards as we have historically defined them and now the phenomenon of climate risk. People pursue alternative livelihoods in a vulnerability context that includes natural hazards and climate change. Livelihood alternatives depend upon the availability of assets or resources that can be framed as natural (N), financial (F), physical (P), social (S) and human (H) capital. Access to these resources, which is mediated by prevailing and emergent institutional structures and processes, enable people to

G. P. Smith (✉)
University of North Carolina at Chapel Hill, Chapel Hill, North Carolina, USA
e-mail: gavin_smith@unc.edu

B. C. Glavovic
Massey University, Palmerston North, New Zealand
e-mail: b.glavovic@massey.ac.nz

B. C. Glavovic, G. P. Smith (eds.), *Adapting to Climate Change,* Environmental Hazards, DOI 10.1007/978-94-017-8631-7_16,  

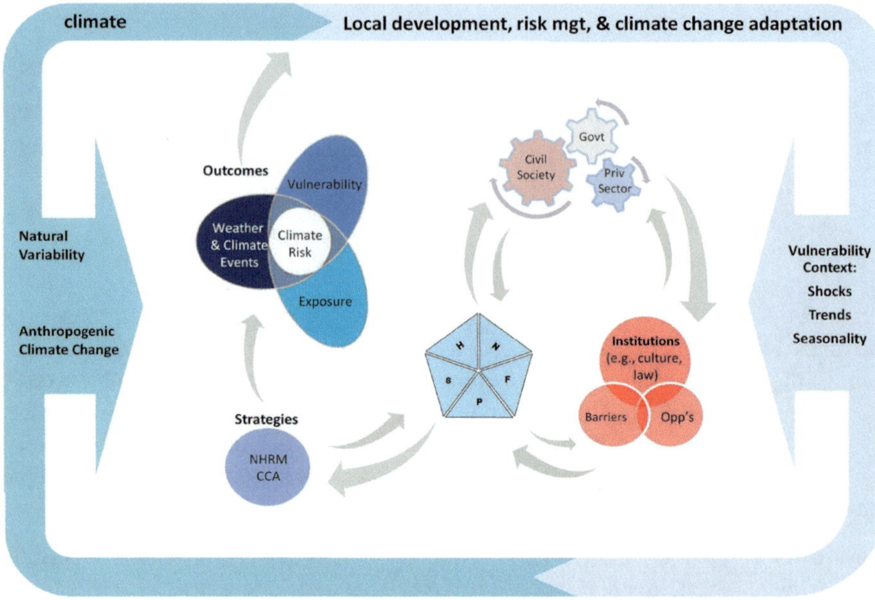


**Fig. 16.1** Local development, risk management and adapting to climate change

pursue livelihood strategies, including those tied to natural hazards risk management (NHRM) and climate change adaptation (CCA), to achieve desired outcomes (such as reduced risk or improved income). Natural hazards planning can be an enabling process to help communities understand their vulnerability context and make more informed governance choices, taking into account institutional barriers and opportunities that integrate natural hazards risk management and adaptation strategies in pursuit of sustainability and resilience.

Sustainability and resilience, which have been used with increasing regularity in the natural hazards risk management and climate change adaptation literatures, serve as powerful integrating themes for describing imperatives and associated lessons drawn from the case studies in the previous chapters. Understood in the context of plan-making for sustainable development, resilience provides an organising principle or vision which when applied to the context of this book might read:

> To enable communities to integrate and effectively implement natural hazards risk management and climate change adaptation strategies that secure sustainable livelihood outcomes by (1) maintaining social-ecological system structures, functions, and feedbacks; (2) where necessary, reorganising and transforming these systems; and (3) learning from experience and adapting to an uncertain and changing future that includes sudden shocks due to extreme events, including those caused or exacerbated by slow-onset climate change (see Sect. 1.2.2 in Chap. 1 and Beatley in Chap. 6).

Next we posit four imperatives for integrating and mainstreaming natural hazards risk management and climate change adaptation (Box 1). These imperatives are

**Imperatives for integrating and mainstreaming natural hazards risk management and climate change adaptation**

*Governance Imperative:* Identify, nurture and sustain collaborative governance networks that span varied spatial and temporal scales.

*Capability Imperative:* Expand the forward-looking use of community assets and pre- and post-disaster resources to enhance risk reduction and adaptation initiatives.

*Planning Imperative:* Invest more in pre-event planning for risk reduction and enhanced collective adaptive capacity.

*Moral Imperative*: Adaptation and risk reduction are ethical issues that require practical, actionable and enduring policies, plans and initiatives to build resilience and sustainability.

typically construed in planning vernacular as goal statements or "an ideal future condition" aspired to by a collective body, such as a local community (Berke et al. 2006, p. 296) each of which are designed to support and realise the vision of resilience and sustainability highlighted above.

We then discuss a series of associated lessons drawn from each of the chapters that help to operationalise these imperatives (see complete list of case study lessons in the Appendix to this chapter). This approach, which draws upon the physical architecture used to construct a plan, is missing a central element of plan-making, which involves an iterative, reflexive and communicative process among those responsible for implementing agreed upon actions as well as those who are ultimately affected by the results of this planning and decision-making process. Engaging in inclusive and authentic deliberation is particularly important when attempting to construct and implement a unifying policy and planning framework that involves large networked governance systems (Innes and Booher 1999, 2003).

Important procedural elements for developing an engaged approach to natural hazards planning described by authors in this book include fostering an improved understanding of the interrelationships among relevant stakeholders; a clarification of terms; the framing of issues; a distribution of responsibilities across institutions, recognising adaptation barriers and opportunities; the identification of clear monitoring and implementation procedures; the identification and rectification of mismatches across scales and between legislative, cultural and behavioral norms; and the need to pursue both incremental and transformative adaptation pathways towards resilience and sustainability in the face of prevailing unsustainable and maladaptive practices. Therefore, the remainder of this chapter is intended to stimulate future discussions, including those held by decision-makers and stakeholders involved in formal and informal governance processes addressing natural hazards risk management and climate change adaptation. It is also intended to inform and strengthen the rapidly growing set of interactions among and between

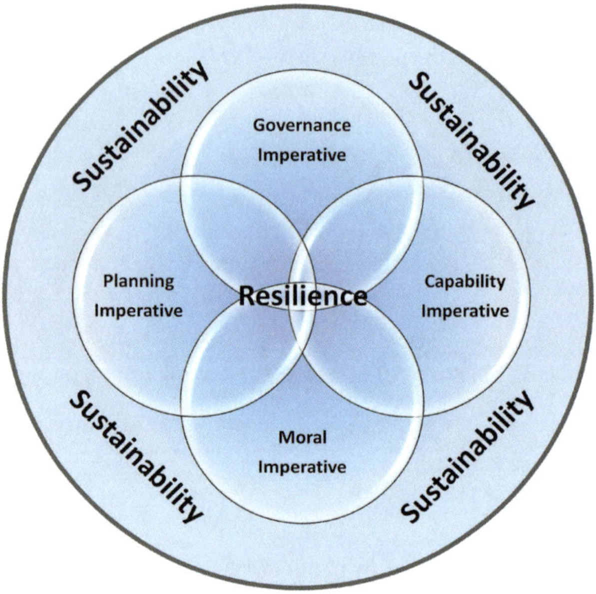


**Fig. 16.2** Imperatives for resilience and sustainability

researchers, policy-makers, planners, other professionals and citizens that seek to better understand and address this complex issue of risk reduction and climate change adaptation.

These imperatives and associated lessons are interconnected and mutually reinforcing as opposed to being linear or sequential steps in a plan- or policy-making process (Fig. 16.2). The case studies in this book show that the pursuit of resilience and sustainability is a normative or ethical stance—a moral imperative—that requires collaborative governance networks to confront the root causes and drivers of vulnerability and exposure to natural hazards. Hence the imperative to build the capabilities of at-risk groups and communities in the face of escalating disaster risk. Poor planning, or planning to maintain the status quo or resist change, is maladaptive and unsustainable in the face of rapidly changing climatic conditions. Natural hazards planning can play a pivotal role in building capability and facilitating more collaborative ways of thinking and working to achieve resilience and sustainability. The processes and outcomes associated with planning are, however, framed by societal values. Realising the promise of planning necessitates that locally specific institutional barriers are understood and overcome so that culturally relevant and appropriate adaptive pathways can be charted in these turbulent times.

## 16.1 Governance Imperative: Identify, Nurture and Sustain Collaborative Governance Networks that Span Varied Spatial and Temporal Scales

An integrative premise that emerges from each of the case studies is the critical importance of collaborative governance networks in facilitating horizontal and vertical integration, dealing with change and integration in policies and plans and addressing the complex interactions across spatial and temporal scales.

### *16.1.1 Horizontal and Vertical Integration*

The concept of horizontal and vertical integration provides an instructive way to describe, bridge and inform governance networks including those that characterise natural hazards risk management and climate change adaptation (Berke et al. 1993; Smith 2011; see also Chaps. 7, 9, 13 and 14). Realising the power and influence of collaborative governance networks requires building vertical connectivity between governance actors, including international, national and sub-national government agencies; the private sector; civil society organisations; and the scientific community. It also means coordinating activities horizontally across sectors and regions and between community organisations, government agencies, businesses, the groups that emerge following extreme events and the individuals that reside in these areas. An expanded horizontally and vertically integrated network connotes the creation of strong and enduring bonds between those that have been involved in attempts to prepare for, respond to, recover from and mitigate against the effects of natural hazard events and disasters as well as those that are now beginning to develop strategies to adapt to a changing climate at the local level. Glavovic (Chap. 10) refers to this transformative process as risk governance in which deliberation and reflexivity are important elements.

Chapter authors found that local governments are most effective when they are supported by higher level (e.g., sub-national and national) institutions and associated assets or resources that are flexible enough so as to nurture locally grounded efforts including those that may have formed organically and irrespective of national assistance programmes. This approach effectively underscores and bolsters the role often played by local governments as the lynchpin of horizontally and vertically integrated partnerships as they have direct contact with members of the community as well as national and sub-national policymakers and those involved in plan-making. Local governments and communities can also serve as advocates for the adoption of climate change adaptation measures regardless of whether national mandates are in effect. In eThekwini (see Chap. 3), South Africa, for instance, the local Environmental Planning and Climate Protection Department assessed local programmes, resource limitations and future business plans in order to determine how these policies and practices could be "mainstreamed" into a larger climate change effort. Lakshmi, Purvaja and Ramesh (Chap. 12), who also note the importance of mainstreaming policies, suggest that disaster risk reduction initiatives must be more closely aligned with development schemes, including those tied to restoring livelihoods among fishing and agricultural communities and the maintenance of protective "coastal shelterbelts" comprised of mangroves and casuarina plantations. In this case, the use of integrated coastal management principles are encouraged and supported by a national programme that is intended to provide a policy framework on which to implement these ideas at the local level.

The Samoa case study (Chap. 13) similarly suggests that a national strategic policy framework should provide broad guidance but rely on local governments (e.g., village councils) and residents to assist in the identification of natural hazards and ways to address them as part of an evolving planning process that respects traditional decision-making methods, recognises the predominance of communally owned land and fosters the adoption of indigenous adaptation measures. Bosomworth, Handmer and Dovers (Chap. 11) also describe the value of drawing lessons from local groups and their long-standing cultural relationship with fire whereas Lakshmi, Purvaja and Ramesh (Chap. 12) argue that fostering a bottom-up approach necessitates conducting a local needs assessment to drive the formation of supporting policies and programmes from higher levels of government that account for unique local conditions and capabilities. Oliver-Smith (Chap. 4) notes in his study of highland populations in Peru that any national policies and investments should be crafted in a way that supports existing local capacities and goals, while Siembieda (Chap. 7) describes how the State of California has developed a broad policy framework tied to achieving higher order goals including resilience and adaptation while allowing the cities of Berkeley and San Francisco to craft collaboratively developed policies and programmes designed to achieve them. In the Manawatu, New Zealand, case (Chap. 10), Glavovic points out that considerable attention has been paid to flood risk since the 2004 flood, including a two year review of existing legislative provisions, and the creation of a New Zealand standard for managing flood risk at the local level, including a discussion of how a changing climate will result in worse flood-related disasters. Yet there remains no legislative directive as to how New Zealand's Territorial Authorities (i.e., city and district councils) and Regional Councils should address flood risk. The effect of the gap in vertical connectivity between government stakeholders merits further study in order to determine how best to fill this void.

Applying the vertical integration typology to the natural hazards risk management-climate change adaptation discussion means expanding the idea beyond the familiar linkage between local, sub-national (e.g., state and region) and national stakeholders to include the international arena and actors at either end of the vertical spectrum (Smith 2011). One of the lessons drawn from the Hurricane Katrina experience is the importance of developing international resource assistance protocols before a major disaster strikes as the United States did not have in place the procedures to accept post-disaster aid from other nations. This not only highlights the need for developed nations to recognise the likelihood that they too may need to create these types of assistance protocols in the age of climate change, there are lessons to be learned from developing nations that already have procedures in place that enable the acceptance of international assistance from a number of sources before and after a disaster. The Samoan case (Chap. 13) demonstrates this expansive vertical connectivity as it includes the involvement of the World Bank (funder of the Samoa Infrastructure Management project), the Planning and Urban Management Agency, the National Disaster Management Office, who led the interaction with local villages, and the consulting firm Beca International, who assisted in capacity building efforts through the training of government staff and local partners. Ziervogel and Parnell (Chap. 3) argue that the sustained provision of resources with flexible eligibility criteria should be provided by more than international organisations in order to ensure collaborative governance and capacity building while local governments, for instance, should ear-mark budgets to ensure a long-term commitment to adaptation.

Oliver-Smith (Chap. 4) describes a process where the Peruvian Ministry for the Environment, working with the Swiss Agency for Development and Cooperation, has developed the Program of Adaptation to Climate Change. The programme, which is implemented through municipal governments, focuses on four areas: an assessment of vulnerability and the identification of possible adaptation strategies, monitoring of those strategies over time, the provision of advisory services for those responsible for the implementation of adaptation projects, and the establishment of operations that integrate climate change adaptation into local development agendas. This process addresses many of the important themes raised by other scholars in this book, including the importance of communicating risk, assessing vulnerability and mainstreaming climate change adaptation initiatives. It also highlights many of the key aspects of planning as described by Berke in Chap. 8.

At the opposite end of a vertically integrated network are individuals. As Smith (2011) points out, individuals play key roles in natural hazards risk management, including the delivery of assistance to others, identifying local needs before and after extreme events, serving as advocates for the disenfranchised and functioning as repositories of rich historical information tied to an understanding of risk. Yet, individuals are often excluded from participating in the formulation of disaster recovery policies and plans that directly affect them. Building enduring networks that span horizontal and vertical axes necessitate investing the time required to understand the underlying interests of varied stakeholder groups, the unique assets or resources they can draw upon and provide to others, striving to identify common goals and maximising the coordinated distribution of these resources over time. Understanding how prevailing institutional structures and processes mediate access to and use of available resources is, however, critical to determining how best to develop and implement effective governance strategies (see Chap. 1). In the Peruvian highlands, for instance, individuals are acutely aware of the effects of a changing climate on their way of life and these concerns have been conveyed to local officials who are actively promoting locally-based adaptation projects (Chap. 4). White (Chap. 5) argues that a key part of moving away from a flood defense paradigm to one that advances a flood risk management approach is expanding the traditional responsibilities of governance networks, giving communities and individuals more power while holding them more accountable for their actions. Berke (Chap. 8) describes the importance of what he calls "accountable implementation" in high quality plans to include a process that should define those responsible for carrying out proposed policies and establish a timeline for completion, including the means by which the policies will be achieved.

### *16.1.2 Change and Integration in Policies and Plans*

The need to foster vertical and horizontal integration does not imply that such actions are easy or happen quickly; indeed political science scholars have suggested that systematic, cross-sectoral changes in policy can take decades to achieve (Sabatier and Jenkins-Smith 1993). Bosomworth, Handmer and Dovers (Chap. 11) similarly note in their case study on Australian bushfires, that change necessitates first recognising the administrative, cultural and political factors that shape the institutional framing of policy issues and then seeking "shared conceptions" as part of larger efforts to ..."expand the range of existing policy perspectives."

The formation of "policy entrepreneurs" capable of coalescing a sufficiently influential group of advocates can affect institutional change. This theory, which is widely recognised as the Advocacy Coalition Framework, has been used to describe changes in disaster policy (Birkland 1997; Olson et al. 1999). Smith, in his chapter on Hurricanes Fran and Floyd in North Carolina, USA, (Chap. 9) notes that the window of opportunity to effectively coordinate the distribution and use of resources across larger networks needed to elicit these changes, while possible, varies significantly over time. Baumgartner and Jones' concept of punctuated equilibrium (1993) describes how long periods of incremental change can be followed by significant policy transformations given appropriate circumstances such as the mobilisation of policy entrepreneurs that redefine the policy issue in a way that attracts new participants and energises others while highlighting areas where policies can be linked across varied policy venues (pp. 238–243). For instance, a number of coastal communities in India noted that after the 2004 tsunami a newfound and improved level of communication with government officials was achieved, leading to an enhanced level of collective capacity (Chap. 12). Smith (Chap. 9) also observes that the transfer of lessons and the institutionalisation of policies and the personnel needed to continue the implementation of these policies over time from one disaster to another is difficult to achieve unless sufficient political, financial, technical and administrative resources are present and sustained over time.

The lessons drawn from the case studies in this book paint a varied portrait of how we might join natural hazards risk management and climate change adaptation policy and practice together in an operationally viable way. For instance, Bosomworth, Handmer and Dovers (Chap. 11) note that the construction of policy frames, including one which recognises defending ourselves from nature (emergency management) or one that assumes humans are part of nature (sustainability) has important policy implications. In the case of the emergency management frame, fire should be suppressed in order to protect human safety whereas the sustainability frame suggests that efforts should be undertaken to balance intertwined social and ecological objectives. Glavovic's case study on the Mississippi Delta (Chap. 15), similarly describes the dire consequences of failing to address the interconnectivity between climate change, environmental degradation and the amplification of social vulnerability in New Orleans and the larger Mississippi river delta.

Iain White shows in his historical overview of flood governance in England (Chap. 5) that changing paradigms from flood defense to flood risk management has led to an expanded network of stakeholders working toward a reframed common goal. White recognises that social inequities are inevitable and their distribution is often geospatially definable, thus requiring new organisational arrangements, moving beyond engineering-dominated flood defense-based methods, to those advocating the adoption of larger, more inclusive flood risk management networks comprised of spatial planners and a devolution of responsibilities from national government to community officials and individuals in order to confront greater levels of uncertainty tied to urbanisation and more extreme climactic events. Glavovic reaches a similar conclusion in his evaluation of the Manawatu floods in New Zealand (Chap. 10), arguing that a protect-based rationale is flawed, particularly in light of new and expanded challenges associated with climate change. Instead, he suggests that a regional river basin catchment planning process is needed that emphasises an avoidance strategy grounded in land use while recognising cross-scale interactions and the unique and changing nature of rural communities due to globalisation and various economic and socio-political factors.

### 16.1.3 Varied Spatial Scales

Developing broad, distributed collaborative governance networks that link individual, community, sub-national, national and international governance actors and actions is highly complex and difficult to achieve in practice. Yet as the globalisation of economies and advancements in communication technologies and risk modeling become increasingly ubiquitous, the successful application of these tools provide important lessons for those who strive to better understand and convey the effects of a changing climate on locally scaled systems such as public infrastructure and human settlements, including unique factors associated with increasing urbanisation (see Chap. 5). In Chap. 2, Birkmann and Pardoe note that the integration of data and models that span differing spatial scales represents an important frontier of research and practice (see also Chap. 8).

Advances in natural hazards risk assessment, including its use to help inform plans and policies at various spatial scales, provide important clues as to how we can downscale data derived from complex meteorological, hydrological and geological phenomena to inform the location, type and density of human settlements in hazardous areas (see Chap. 2). At the same time care must be taken to recognise important historical, cultural, political and economic drivers influencing how vulnerable populations perceive risk. These challenges were evident in North Carolina and Mississippi following several major hurricanes (Chap. 9). For instance, investing in the re-mapping of the State of North Carolina's floodplains in the aftermath of Hurricane Floyd was met with strong state-wide support. The State of Mississippi also adopted updated Flood Insurance Rate Maps (FIRM's) after Hurricane Katrina that significantly increased expected flood depths and the breadth of the floodplain (see Chap. 14). In both states the adoption of new FIRM's utilised the latest technological advances and improved data, the mapping procedures followed the worst disasters in each states respective history and the process involved extensive outreach campaigns and educational efforts targeting local officials and residents. Perhaps most significantly, the adoption of the new maps was required in order for local governments to continue to participate in the National Flood Insurance Program (local government participation in the program is required in order for those owning property in the jurisdiction to purchase a federally-backed flood insurance policy). On the other hand, efforts once supported by the state (and a US$ 5 million Congressional appropriation) to develop scenario-based sea-level rise projections and proposed adaptation strategies in North Carolina have more recently been fiercely opposed by coastal residents, state elected officials, agency leaders and the Governor following a transition from one political party to another. Reconciling this apparently incongruous perception of risk is likely to have significant policy implications across governance networks as the politicisation of risk collides with the creation of inclusive and cooperative groups required to confront a problem of this magnitude and complexity. For instance, as nations and communities develop improved models capable of downscaling results to help inform communities how a changing climate will affect local weather patterns and rising sea-levels, the use of these models, including their integration into scenario-based plans and policies, will benefit from obtaining broad-based support in order for them to be effective and enduring.

In addition to the technical challenges present, conveying the information generated through the assessment of risk in a manner that results in an appropriate and publicly acceptable mix of risk reduction measures taken by policy-makers, planners and members of the larger natural hazards risk management/climate change adaptation network represents perhaps the greatest challenge. Lessons derived from pre- and post-disaster risk communication, disaster agenda setting and the hazard mitigation literature should help to inform the policies and practices adopted. Specific examples can be drawn from the North Carolina and Mississippi case studies (Chaps. 9 and 14). They include: taking advantage of varied post-disaster windows of opportunity, linking re-mapping efforts to compliance with institutionalised hazard mitigation programs (e.g., the National Flood Insurance Program), or new ones to be developed (e.g., sea-level rise), and engaging in extensive outreach efforts that convey the effects of new standards while explaining the cost of non-compliance.

### 16.1.4 Varied Temporal Scales

A temporal dimension that is widely discussed in the literature involves the differential nature of episodic versus slow onset extreme events, including those linked to climate change. In many ways, the manner in which societies have addressed natural hazards risk management over time, including before and after extreme events, is more mature than the similar but still emerging techniques proposed to address a changing climate and much can be learned from many of the successes as well as the repeated mistakes. For instance, post-disaster assistance, including that tied to humanitarian aid, often emphasises a short-term perspective rather than one that provides the support needed to build an enhanced and sustained adaptive capacity through a robust planning strategy coupled with flexible financial assistance. Further hampering long-term adaptive capacity is the inability or unwillingness of those requesting aid to build the "institutional infrastructure" needed to bridge humanitarian assistance and developmental needs that pre-date an extreme event (Chap. 2).

Yet as pointed out by Oliver-Smith (Chap. 4), anthropological studies show that societies have adapted to changing climatic conditions as part of a range of culturally nuanced strategies linked to human well-being and survival as a species and these lessons are equally pertinent. The ability of societies to endure repeated disasters highlights long-standing resilience. The Andean people, for instance, have coped with environmental risks and uncertainty over time through a range of customs and practices (Chap. 4). Over the last 4,000 years, complex societies in Peru have used a number of adaptive strategies, including those that recognise prevalent natural hazards such as drought, flooding, earthquakes and landslides. Specific examples include the …"exploitation of multiple ecological tiers, complex water management systems, dispersed settlement patterns, environmentally appropriate building materials and techniques, communal labor systems, interregional exchange and distribution systems, surplus storage and preparedness, and ideological modes of explanation and meaning formulation for frequent environmental disturbances." Glavovic (Chap. 15) similarly notes that early populations in the Mississippi Delta, some dating back more than 12,000 years, employed a range of adaptive strategies to deal with floods and storms including the location of camps on high ground, building elevated earthen mounds and migrating to and from the area on a seasonal basis.

White (Chap. 5) describes a historical record of flood events in the United Kingdom going back more than 1,000 years, which has shaped the creation and nature of specific approaches, professions and governance systems, including those that are not sufficient to effectively manage flood hazard risk including that influenced by urbanisation and a changing climate. The institutional, cultural and political manner in which societies differentially respond to these types of events, including the factors that enhance an integrated and collaborative effort by governance actors, is proposed to guide a policy and planning framework that includes both natural hazards risk management and climate change adaptation. The connections between natural hazards risk management and adaptation are laden with rich temporal dimensions, each of which plays an important part in the development of policies, plans and day-to-day practices. For example, the efficacy of adaptation measures may take decades to become evident versus the benefits of risk reduction initiatives that may become clear over shorter timeframes, although as Berke (Chap. 8) explains, the immediate benefits of hazard mitigation tend to lag behind the resources needed to implement them.

In order to better reconcile the differing temporal dimensions associated with hazard mitigation, disaster recovery and climate change adaptation, a "collective learning environment" should be created that "recognises the long-term time horizon associated with climate change while taking advantage of 'teachable moments' following extreme events" (Chap. 3). Paradoxically, disasters can create the conditions in which innovation and opportunism can thrive, although these opportunities are seldom equitably distributed across all groups and communities. The variability of recovery outcomes is particularly evident at the local level as communities often struggle to address their needs using external sources of assistance. The ability of national governments to adopt and effectively implement more flexible policies in the aftermath of events, including those exacerbated by a changing climate, are likely to become increasingly important. In one such example noted by Lakshmi et al. (Chap. 12), the government of Tamil Nadu implemented a program that paid laborers who were not eligible for other forms of asset compensation to pick up cyclone-generated debris. This program did not exist prior to the tsunami, but was developed afterwards based on an assessment of local needs. This type of post-event adaptive planning offers important lessons as to the importance of creating an environment in which creative problem solving can thrive, drawing on varied partners to identify and address unexpected issues as they emerge (Smith 2011).

Several authors described the need to coordinate actions in the pre- and post-disaster environment, as pre-event actions, including planning and capacity building initiatives, affect post-disaster trajectories such as the degree to which communities are prepared for disaster-related impacts, the degree to which communities embrace the implementation of risk reduction measures during reconstruction and the degree to which disaster assistance strategies are equitable. For instance, care must be taken to ensure that post-disaster generated "opportunities" do not merely reflect past power imbalances, which have the potential to further disenfranchise socially vulnerable groups or hinder broader public interests through actions such as the forcible post-disaster relocation of the poor to make way for new development, the inequitable distribution of post-disaster assistance or reinvestment strategies that increase hazard exposure (Smith 2011; see also Chap. 15).

As the previous examples suggest, actions taken in the transitional period between the initial response and the early phases of the recovery effort influence the path of long-term recovery policy options as well as overall development. Additional examples of policies affecting the trajectories of recovery include the adoption of a temporary building moratorium and decisions surrounding the siting of emergency housing. Implementing a temporary building moratorium in the aftermath of a disaster can give local officials and the communities they serve the time needed to assess whether it is prudent to rebuild in the same area and/or adopt stronger building codes that account for known and projected hazard threats, including those exacerbated by climate change. Ideally, the moratorium policy is part of a pre-event recovery plan developed over time and informed by a meaningful dialogue among community members and the wider disaster assistance network.

Decisions surrounding the location of emergency housing sites post-disaster also influence future settlement patterns as temporary housing locations can become permanent over time, particularly in locales in which land use policies may be non-existent and squatter settlements are common. In both cases, the intense pressure to rebuild communities quickly can trump more deliberative efforts, including those that involve the adoption of a systemic risk reduction strategy and associated policies and projects. In the age of climate change, risk reduction efforts should also include those policies that consider long-term adaptation measures, integrating them into existing hazard mitigation and disaster recovery policies and programmes. Good pre-event plans and associated policies that account for multiple risk-based scenarios allow for proactive, yet flexible policies that account for uncertainty while building strong and diverse networks capable of blending risk reduction and climate change adaptation (see Chaps. 8 and 10).

## 16.2 Capability Imperative: Expand the Forward-looking use of Community Assets and Pre- and Post-disaster Resources to Enhance Risk Reduction and Adaptation Initiatives

Before and after disasters, resources may be invested in communities that directly and indirectly affect their overall vulnerability to natural hazards, including those hazards that are linked to climate change. Examples include ongoing economic development initiatives, patterns of investment in public works and community infrastructure, the provision of loans and capital to developers, pre- and post-disaster grants, insurance practices before and payouts following extreme events, non-profit advocacy and boundary-spanning efforts and the adoption of government policies and plans that influence current and future human settlement patterns. In the pre-event timeframe, there is potential for resource delivery to be well planned and provided in close coordination with one another, emphasising a commitment to reduce vulnerability and build the collective capacity of the wider governance network to reduce current and future levels of natural hazards risk. In practice, the ability to achieve this objective depends on a commitment to facilitate more equitable access to assets and the direct involvement of governance actors in planning and decision-making processes.

Siembieda's chapter (Chap. 7) shows that partnerships developed to address pre-event vulnerability to earthquakes in the San Francisco Bay Area has led the cities of Berkeley and San Francisco California (USA) to initiate climate change adaptation strategies, in part, because of the dramatic effects of Hurricane Katrina and reflections on the two California cities' exposure to future disasters. Berke (Chap. 8) similarly found in Punta Gorda, Florida (USA) that the development of a working relationship between the city and the Charlotte Harbor National Estuary Program—Climate Ready Pilot initiative was tied to a greater collective recognition of hazards vulnerability. The initial event that precipitated action was Hurricane Charley, which devastated the area in 2004. Additional concerns were raised following studies that showed that Southeast Florida is one of the most vulnerable places in the world to the impacts of climate change, including rising sea levels and more intense hurricanes. Ziervogel and Parnell's case study of eThekwini in South Africa (Chap. 3) notes how coastal storms and flooding precipitated support among elected officials to advocate for a climate change adaptation policy. An important observation noted explicitly in the California case studies and alluded to in other chapters is that it is not important which level of government initiates action. Rather, it is more important that the adaptation strategies are grounded in scientifically derived, easily accessible information that are supported by interconnected governance actors and institutions.

In a number of Samoan villages affected by the 2009 tsunami, locally developed pre-event plans were used to identify "broad development goals"…while "providing a base map of infrastructure (pre-tsunami), coastal hazard zones and a documented list of issues, concerns and possible solutions for each village" that were ultimately used to inform disaster recovery strategies (Chap. 13). Some of these plans had already identified a series of options tied to reducing future impacts of weather related hazards. Options included the relocation of villages away from the coast or a selective movement of villagers' homes while keeping the tourist operations in place. These procedures were implemented after the tsunami struck as a result of pre-event planning and consensus building. Together such processes help to build community capability to integrate and mainstream natural hazards risk management and climate change adaptation strategies in pursuit of resilience and sustainability.

The findings of the Mississippi, USA, case study (Chap. 14), which assessed disaster recovery efforts following hurricanes Camille (1969) and Katrina (2005), suggest that communities exposed to high levels of risk need to gain a better understanding of the nature of this risk and take action following major disasters to alter repetitive losses. Smith states that "The post-disaster environment presents a unique set of conditions that enable adroit governance networks to take advantage of a heightened political saliency of natural hazards and access to resources not available on a regular basis." Not only can these resources be used to physically reconstruct communities in a manner that better accounts for episodic and long-term risks, these resources, including new policies and educational and training initiatives, can serve to build the adaptive capacity of individuals, organisations and networks in advance of the next event, whether that event is a hurricane or rising sea level. While the adoption of new codes and ordinances in the aftermath of Katrina was representative of significant improvements, local governments along the Mississippi coast did not account for a changing climate and associated sea-level rise projections or potential intensification of hurricanes. More recently, however, there is evidence to suggest that coastal communities in Mississippi are beginning to adopt sea-level rise adaptation measures due to educational initiatives and funding provided through the National Oceanic and Atmospheric Administration's Sea Grant Program. But fundamental questions remain about the extent to which marginalised and socially vulnerable groups have been helped or further marginalised in this context; and hence the ongoing imperative to address both exposure and social vulnerability in building overall community capability to reduce risk and build adaptive capacity.

Understanding hazard risks in the face of a changing climate requires engaging in a meaningful discussion about land use and the potential alteration of settlement patterns, including proactive steps that can be taken to reduce exposure and vulnerability to extreme events and the application of post-disaster risk reduction measures. Glavovic points out in the Manawatu, New Zealand flooding case study (Chap. 10) that achieving a robust avoidance strategy is achievable but can be difficult as communities may be resistant to change due, in part, to developmental inertia or strong pro-growth interests. In an effort to address this issue in the State of Mississippi, emphasis was placed on the communication of risk during the reconstruction process as a way to influence the manner in which coastal communities were rebuilt. While Smith argues that the outreach effort was extensive and somewhat successful, the underlying cause of changing settlement patterns has not been a purposeful modification of land use strategies. Rather, the changing face of coastal Mississippi has been driven by the adoption of higher building codes whose additional costs of compliance has led to uneven post-disaster reconstruction, including the replacement of coastal neighborhoods with large-scale condominiums and casinos. Good land use plans that better account for hazard risk were not the driving force behind these changes. The resettlement patterns of Mississippi exemplify recovery outcomes that can be viewed in two different ways. On one hand, the replacement of housing with larger structures financed by investment firms and owners who *may* be better able to bear the costs of future disasters could represent the future of coastal development in the United States. On the other hand, large structures and their associated infrastructure may prove less resilient and adaptive, requiring significant investments in protective measures like levees, seawalls and beach nourishment. Alternatives like the physical relocation of large-scale buildings, including supportive and protective infrastructures, is very costly and difficult to achieve after initial investments have been made. Further, development interests may push for the use of public funding strategies to protect private investments arguing that such actions are necessary to maintain the economic viability of the area.

An oft-neglected question must be confronted: Even if post-disaster reconstruction results in safer and more resilient buildings through what could be called 'disaster gentrification'—the progressive 'improvement' or 'upgrading' of neighbourhoods as wealthier new owners and residents construct more hazard resistant buildings and are better able to bounce-back in future extreme events—what happens to pre-event residents who cannot 'afford' to return and rebuild their community? The post-Katrina diaspora and low return rates by some marginalised and poor communities to neighbourhoods such as the Lower Ninth ward represents a case in point (see Chap. 15).

The initial "success" of reconstruction activities following disasters is typically measured by how quickly rebuilding occurs. At the same time, public officials often suggest that disasters provide an opportunity to build back better than before (Comfort 2005). Yet few communities actually plan for post-disaster recovery. Rather, disaster affected communities and supporting assistance networks tend to engage in a type of adaptive recovery planning, struggling to balance the need for speedy recovery with a more deliberative process that is necessary to secure better long term outcomes (Olshansky 2006). After disasters, large sums of resources are brought into affected communities that have typically failed to plan for this eventuality. As a result, most communities become overwhelmed and struggle to administer the associated funding and coordinate the multitude of entities that provide resources ranging from food and clothing, insurance settlements, crews able to assist in the repair of damaged housing, grants and loans and the development of post-disaster recovery plans intended to guide reconstruction efforts (Smith 2011).

In many of the case studies described throughout this book, new, often novel and expanded governance consortia were created to address the multitude of challenges associated with disaster recovery. These institutional arrangements varied significantly in terms of their commitment to endure beyond the physical reconstruction of communities and the administration of post-disaster funding. In the two chapters written by Smith (Chaps. 9 and 14), the states of North Carolina and Mississippi (USA) have maintained quasi-governmental state-level disaster recovery organisations more than ten and eight years after Hurricanes Floyd and Katrina respectively. An expansion of these types of organisations should include a greater emphasis on injecting risk reduction measures into post-disaster recovery actions and placing a greater emphasis on pre-event planning and capacity building across the larger assistance network, to include climate change adaptation-related activities.

Given the range of institutional, political, cultural, environmental and economic pre-conditions found in this book's case studies, it is not surprising that differing institutional configurations and individual organisation(s) assumed leadership roles in both natural hazards risk management and climate change adaptation-related activities. In some cases, like that described in the Indian state of Tamil Nadu, non-governmental organisations played key boundary-spanning roles linking individual residents, communities and governmental agencies. In Samoa, a national government-led planning framework focused on a coastal infrastructure management strategy operationalised by locally-championed coastal infrastructure plans. The City of Cape Town's adaptation strategy was led by the Environmental Resource Management Department in cooperation with academics, consultants and non-governmental organisations. In California, the state's Natural Resources Agency developed the *Climate Adaptation Strategy* (CAS), which established working groups addressing public health, ocean and coastal resources, water supply and flood protection, agriculture, forestry, biodiversity and habitat and transportation and energy infrastructure. Explicit policies in the CAS targeting natural hazards include avoiding state agency permitting or siting of new development in areas vulnerable to flooding, wildfire or coastal erosion due to climate change; amending the state's environmental review process to include a review of proposed permits in areas susceptible to the effects of climate change; and requiring local governments to identify vulnerable areas when updating land use plans and coastal management plans (Chap. 7).

Several chapter authors noted the importance of local officials and residents working within larger collaborative governance networks to assess risk and develop locally-grounded strategies based on a mutually ascertained collective understanding of and capability to address these threats (e.g., Chaps. 6, 7, 8, 12 and 13). In the past, including events as recent as Hurricane Katrina in Mississippi and Louisiana, risk communication remained focused on episodically occurring extreme events and did not include much public discussion about the additional compounding threats associated with a changing climate. As a result, many of the adopted reconstruction strategies, including coastal housing codes and standards as well as floodplain management ordinances and infrastructure repairs and reinvestments are already outdated.

White refers to this problem in his English case study chapter as one of stationarity, the organising principle of hydrological analysis (Chap. 5). This approach relies on a static representation of risk when it is abundantly clear that we live in a highly dynamic environment subject to change as a result of urban development, the alteration of natural systems and the migration of populations, whose own variable levels of social vulnerability further complicate matters. This probability-based approach has been used to inform England and many other countries' emphasis on flood defenses. Given the additional uncertainties associated with a changing climate, the models we rely upon to assess and communicate risk are in need of significant revision. White goes on to state that it is this uncertainty in our modeling capabilities that has driven many countries to embrace the concept of risk management and more recently resilience, achieved through a more forward-looking approach based on spatial planning and expanded governance.

> Achieving this shift required a much broader engagement with flooding than ever before, within which the management of water now includes professions beyond the traditional spheres, such as engineering to modeling, to incorporate all those with influence over support systems and particularly those connected with where people live and how they act (Chap. 5).

As noted by Beatley (Chap. 6), and described in detail in the next imperative, planning for resilient communities benefits from a collective understanding of and capability to reduce risk as part of an enhanced adaptive capacity that needs to be built and sustained over time, recognising the inevitability of change.

## 16.3 Planning Imperative: Invest More in Pre-event Planning for Risk Reduction and Enhanced Collective Adaptive Capacity

The imperative to plan for multiple possible futures and act on the policies and initiatives derived from this process depends on an enhanced adaptive capacity that can only be achieved through a heightened level of pre-event collective action or collaborative governance. Efforts to build such community-wide capability are often hindered by institutionally or sectorally-focused plans that target governments, communities and businesses as distinct entities rather than as part of a larger collaborative governance network (Chap. 1). The development of broad-based, inclusive governance policies and planning processes focused on pre-event capacity building initiatives that are ultimately ingrained in local communities and wider networks can prove difficult. For instance, many of the institutions that will be tasked with building the adaptive capacity of the network are themselves ill-prepared to adapt to a changing reality in their current form as many of their programs and policies are rigid, inflexible and do not account for local needs. Take for instance, the delivery of post-disaster assistance which is often defined by narrow, inflexible programs that are unable to address variations in local needs (Smith 2011). Compounding this problem is the limited emphasis placed on pre-event capacity building versus an overreliance on the delivery of post-disaster monetary aid, of which, the latter approach tends to disproportionately drive recovery trajectories in a way that is often less sustainable and resilient. While some international relief organisations maintain specific programs that emphasize the provision of pre-event resources as part of a more proactive approach tied to development, the ability to offset weak governmental aid programs in developing nations or the imposition of aid without a good understanding of local needs and appropriate institutional incentives often leads to suboptimal outcomes (Cuny 1983; Harrell-Bond 1986; Gibson et al. 2009).

Berke's (Chap. 8) discussion of planning for multiple climate change scenarios over time is a critical part of any discussion surrounding adaptation. However, the adaptive capacity of those that are directly affected by a changing climate is mixed as noted in several case studies. For instance, this problematic reality is noted by Ziegovel and Parnell (Chap. 3) in their discussion of South African cities while White describes initial efforts to accomplish this aim in his discussion of sea-level rise planning in England (Chap. 5). Oliver-Smith notes that regional governmental organisations (that have the potential to serve an important boundary-spanning role) can be constrained by their narrow technical focus and limited attention placed on "social, cultural and economic concerns" (Chap. 4). Poutasi et al., (Chap. 13) explain how good inclusive planning for climate change adaptation can expand the number and types of options available to stakeholders, including those grounded in an understanding of the barriers and opportunities created by pre-existing institutions and socio-cultural conditions.

Good plans derive their strength from a robust and enduring participatory process that builds "collaborative engagement" and ownership amongst those involved in the development and implementation of plans (Chap. 8). For instance, an important lesson drawn from the post-Katrina recovery process in Mississippi includes the need to more effectively engage the design community in pre-event planning for post-disaster recovery and develop better ways to include community officials and individual residents in the design process in order to more effectively integrate risk reduction into reconstruction efforts. Similar conclusions were reached by Berke et al. in a 2009 study and uncovered in an analysis of planning scenarios in the Punta Gorda, Florida case, which showed that smart growth scenarios placed more property at risk and resulted in more projected property loss due to rising sea levels and increased hurricane intensity than either the traditional urban form associated with sprawling development (policy scenario) or the proposed resilience scenario (Chap. 8).

A number of studies and practice-based results point to the value of inclusive pre-event planning for risk reduction and disaster recovery (Burby and Dalton 1994; Burby 1998; Schwab et al. 1998; Godschalk et al. 1999; Smith 2011). Specific benefits include gaining an improved understanding of risk, injecting hazard mitigation into the recovery and reconstruction process, the development of supportive coalitions advancing an agreed upon set of goals and guiding vision of what needs to be done and by whom and the more efficient and equitable distribution of resources after a disaster occurs (Smith 2011). Berke argues there are important differences in traditional planning practice, including hazard mitigation and planning for climate change and these differences have important policy implications. Yet these differences should not continue to divide those advancing natural hazards risk management and climate change adaptation agendas. Rather, we must strive to expand our collaborative governance networks to devise more inclusive and integrated risk reduction and adaptation strategies. This can be achieved, in part, through scenario-based natural hazards planning, which provides an important bridge between climate change adaptation planning and natural hazards risk management. Siembieda (Chap. 7) shows in his case studies that the "collective acceptance of risk" through planning, including the collection and widespread dissemination of information, helps to foster greater social resilience, including that engendered among individuals and more formal institutions through routine and ongoing interactions. Glavovic's exploration of risk governance posits a decision-making framework undergirded by deliberation and reflexivity across an expanded network of governance actors (see Chap. 15).

According to Berke (Chap. 8), traditional planning practices are more adept at addressing …"human-ecological systems that are relatively stable and predictable over relatively short periods of time". Berke and other planning scholars argue that scenario-based planning offers a new paradigm that can guide collective action related to climate change adaptation (Chakraborty et al. 2011). This process has become recognised as "anticipatory governance" in the literature (Quay 2010, p. 498) and is intimately tied to collaborative planning processes in which technical experts and stakeholders explore a series of potential scenarios or outcomes and develop both "contingent" policies based on one or more of these potential futures, as well as "robust" policies that address multiple scenarios and are often referred to as low regrets or no-regrets policies.

Siembieda describes supportive conditions referred to as "resiliency capacity" as:

> the sum of efforts between government and its citizens expressed through shared responsibilities and an acceptance of the risk conditions they understand at any point in time. It reflects a change in the traditional government paradigm which is 'managing the city' to a new paradigm in which government shares decisions and responsibilities with its citizens and a larger governance network (Chap. 7).

Siembieda's concept is similar to that expressed by White (Chap. 5) who emphasises the need for a greater involvement of individuals in decision-making processes while at the same time holding them more accountable for their actions. This expanded notion of anticipatory and collaborative governance allows for the more effective use of locally held knowledge among individuals that live in hazardous areas, while at the same time ensuring that individuals develop a clearer shared understanding of the ramifications of the choices before them.

The importance and challenges of engendering authentic, inclusive public participation and coordinating multiple plans after a disaster occurs was dramatically evident in New Orleans following Hurricane Katrina (see Chap. 15). Initial post-disaster planning efforts driven by technical experts were strongly opposed by the residents of the city and were ultimately abandoned in favor of a more inclusive neighborhood-level planning process. In a later effort initiated by the City, the Unified New Orleans Plan (UNOP) was created. This plan, which was intended to build on the neighborhood plans, caused confusion among a number of stakeholder groups and threatened the legitimacy of what were referred to as the Lambert Plans. During this time the city quickly developed a hazard mitigation plan in order to ensure their eligibility for post-disaster federal funds. Yet the hazard mitigation plan was not effectively integrated with other plans, nor did it establish clear operational objectives even though it was thought to be a potential avenue to procure the funds needed to implement the UNOP.

The operationalisation of natural hazards planning requires that policies must be linked to clear implementation and monitoring strategies. The implementation of policies requires drawing on a network of governance actors, via their associated institutional structures and processes, who are held accountable for carrying out agreed upon actions in accordance with established timelines. Given the uncertainties of climate change science, the complexity of the adaptation process and the manner in which global change is evident in a designated planning locale, the plan should take into account shifts and/or an expansion of those responsible for taking action, the varied resources needed to accomplish existing and emergent aims and changes in assigned timelines (Chap. 8). Good monitoring practices allow for the tracking of changes in the effects of climate change and the ability to modify policies accordingly.

If we compare the assumptions postulated by Berke with those described in the post-disaster planning environment found in several case studies, a number of important similarities are evident. First, pre-event planning for post-disaster recovery is less common than pre-event hazard mitigation planning, and as a result, much of disaster recovery planning is conducted in a more ad hoc manner. This reality, while problematic for a number of reasons as described in previous cases, requires the disaster recovery assistance network to engage in a form of adaptive planning, albeit one that is more reactive than the scenario-based planning framework described by Berke. Second, good pre- and post-disaster recovery plans should account for the highly dynamic and fluid nature of the post-disaster recovery environment. These conditions include routine changes in policies after disasters, the highly variable and often uncoordinated distribution of assistance among members of assistance networks and the degree to which the resources provided meet differing local needs. Third, good recovery plans embrace the fact that some of the most potent providers of assistance are groups who did not exist prior to a disaster event, but rather emerge following disasters to fill an unmet need that is not being addressed by more formal organisations. Fourth, many hastily developed post-disaster recovery plans fail to utilise scenario-based planning efforts that link the availability of often substantial post-disaster resources with the adoption of forward-looking adaptive strategies.

Accepting the maxim that natural hazards risk reduction is tied to where and how we build in relation to hazards, the advent of climate change and its uncertainty further complicates, but does not diminish the importance of adopting codes and standards or land use strategies as part of the natural hazards planning process. It does, however, require that plans that are designed to merge climate change adaptation and hazard mitigation recognise the implications of differing scenarios based on a changing climate. Good hazard mitigation plans also recognise the dynamism of natural hazards, including the ability to adopt new policies and practices when warranted. Oftentimes this means being prepared to adopt higher codes and standards in the aftermath of a disaster as numerous case studies suggest. It is important to note that adopting higher standards without a similar commitment to new land use policies and resettlement programs can lead to uneven and potentially inequitable redevelopment as evident in the Mississippi, USA, case study. The adoption of new land use measures, which guides that which can be built in certain high hazard areas, benefits from a discourse begun in advance of an extreme event and tied to a well thought out set of options embedded in plans developed by a range of stakeholders, most importantly those who will be affected by and responsible for implementing the policies. In many ways this highlights the multi-dimensional, interconnected nature of resilience, including the importance of addressing social capital and physical infrastructure simultaneously (Alesch and Siembieda 2011; Aldrich 2012). Glavovic's chapter, which addresses the long history of disasters in the Mississippi Delta region (Chap. 15), describes and expands on the interconnectivity of these and other "layers" of resilience, including human, physical, economic, social and natural capital; and underscores the important role that deliberative processes can play in strengthening resilience.

Given the complexity and uncertainties associated with climate change, questions frequently arise as to the appropriate design and land use standards that should be adopted and employed. The effects of a changing climate further compounds and complicates what Burby (2006) refers to as the "safe development paradox" in which design standards associated with the overreliance on protective measures like levees and seawalls have the effect of incentivising development in high hazard areas leading to predictable disasters when these design parameters are exceeded (see Chap. 15). In order to confront this challenge, sound hazard mitigation and climate change adaptation policies should take into account a wider range of options tied to the location, type and density of construction relative to current and projected hazard scenarios that advance broader goals such as sustainability, resilience and adaptive capacity.

When designed as part of an integrated natural hazards planning effort and framed within a set of complementary goals, good plans can provide a significant degree of flexibility as innumerable permutations are possible, thus allowing for changes in policy combinations over time. This is not always the case, however, as some choices result in fewer options or hamper the likelihood of identifying multiple, actionable alternatives. For instance, careful consideration should be made relative to investments in large infrastructure projects, including protective measures, in areas subject to encroaching seas as they are less likely to be relocated or abandoned in the future whereas lower density, smaller infrastructures and those built in an incremental manner are more amenable to change (Chap. 6). At the same time, high risk communities and countries need to develop plans that include the possible relocation of human settlements (i.e., the collective amalgamation of infrastructure, housing and public facilities) away from these areas, while identifying those sites best suited to receive them (Chap. 14). In the Manawatu-Wanganui region of New Zealand, for instance, the One Plan emphasises an avoidance strategy focused on limiting future development in flood prone areas (defined as at least the 200-year flood extent and 500-year case for Palmerston North) while recognising the need for communities to achieve risk reduction and other societal goals (including adapting to a changing climate) in an integrated manner utilising varied approaches.

Notwithstanding the planning imperative, and the promise of natural hazards planning in particular, it is important to recognise and address existing institutional and professional impediments or barriers to integrating and mainstreaming natural hazards risk management and climate change adaptation. For instance, in a national assessment of Hazard Mitigation Plans in the United States, Berke, Smith and Lyles have found that few plans address land use or climate change adaptation (Berke et al. 2012; Lyles et al. 2013). Research conducted by USA scholars Kartez and Faupel (1994), Smith (2009) and Lyles et al. (2013) note a long-standing disconnect between emergency managers (who are often put in charge of developing these plans) and land use planners, many of whom seem to think that disaster risk reduction is the purview of emergency managers. This is problematic as the involvement of land use planners in the creation of hazard mitigation plans has been found to have a strong positive influence on the incorporation of land use measures (Lyles et al. 2013). These observations coincide with similar narrowly defined groups of technical experts noted by Bosomworth, Dovers and Handmer (Chap. 11) as well as White (Chap. 5) and Glavovic (Chap. 10) who describe current emergency management (bushfires) and engineering-dominant (flood) paradigms that hinder efforts to achieve more resilient communities.

An important way to help overcome these impediments and barriers involves targeting collaborative opportunities among those that have all too often remained entrenched in their professional silos. This requires more clearly defining the resources brought to the table by the wider network of risk governance actors, how they are delivered, to whom and how this occurs over time. It also means identifying champions who are able to invest the time and resources needed to marshal collective action. In its current form the resource delivery process is often inefficient, duplicative and sometimes counterproductive. For instance, the delivery of post-disaster reconstruction funding and technical assistance can lead to the rebuilding of at-risk communities to their pre-event condition (see Chap. 15). In the age of climate change this approach is increasingly problematic as it fails to incorporate additional measures or increased standards and practices that recognise the effects of not only natural hazards as we have defined them in the past, but also hazards as understood in the context of a changing climate (e.g., rising sea levels, more intense coastal storms, and more severe droughts, wildfires and floods). A fresh perspective should therefore incorporate this new and still uncertain reality into flexible yet forward-looking design parameters, resettlement patterns, future capital investments and land use choices that account for both threats.

While building the collective capability of an expanded collaborative governance network offers promise, it also presents its own set of challenges. For instance, it requires including actors that have the potential to short-circuit the consensus-building process and support institutional structures and processes that are antithetical to the concepts of resilience and sustainability. In the North Carolina (USA) case study, Smith (Chap. 9) notes the problem of coastal residents and the recent emergence of a state legislature that steadfastly refuses to recognise the realities of a changing climate in a state that is among the most vulnerable in the United States to rising seas and more intense hurricanes. Yet little has been done to bring together the natural hazards risk management community, the proponents of adopting climate change adaptation measures and those who hold that climate change is not occurring. Instead, state agency officials are being told to discontinue these discussions as well as scenario-based planning for sea-level rise. Ironically, this apparent stalemate is occurring in a state that has the most advanced floodplain mapping programme in the country and received a US$ 5 million Congressional appropriation to study the effects of rising seas and develop a series of adaptation options. Furthermore, the state's emergency management office was engaged in one of the largest flood-hazard risk reduction efforts in the nation following Hurricanes Fran and Floyd, developed one of the first state hazard mitigation planning initiatives (prior to the adoption of a similar national programme by the Federal Emergency Management Agency) and invested over US$ 836 million in state funds to develop 22 state-level post-disaster grant programmes advancing sustainable disaster recovery. The North Carolina case study highlights the difficulty of engaging and building collaboration between natural hazards risk management and climate change adaptation stakeholder groups to build a sufficiently strong advocacy coalition capable of influencing the widespread and enduring support of a state-level climate change adaptation strategy that conveys the importance of adapting to an inherently dynamic environment irrespective of rising sea levels or other hazards uniquely tied to a changing climate.

In Cape Town, South Africa (Chap. 3), the city's approach to climate change reflects a unique set of contextual issues including rolling blackouts which helped to foster support for alternative energy strategies. While the results of sea-level rise modeling created strong support for protecting the city's coastline, actions were not framed as adaptation measures. Rather, they were tied to the notion of creating a more disaster resilient city in the face of more frequent storms and rising sea levels. This approach highlights the importance of "acceptable language" and context appropriate strategies that help to integrate climate change adaptation and natural hazards risk management through what has increasingly become known as a "no regrets strategy".

In the Samoa case study, Poutasi et al. (Chap. 13) suggest two approaches to improve the collective base of knowledge surrounding environmental management and natural hazards risk management. These include investing in the training of government officials who can inform other stakeholders (e.g., villagers, infrastructure providers and developers) and the development and implementation of coastal infrastructure management plans that provide a recognised and accepted implementation vehicle. Several development programmes already exist in Samoa that have the potential to serve as further conduits for this complementary information and the efficacy of this approach should be monitored to see how effectively this is achieved over time.

One way to help address many of the challenges cited above is to encourage agencies and organisations in the natural hazards risk management and climate change adaptation community to co-support experimental efforts, pilot programs and educational initiatives in both the pre- and post-disaster environment that clearly demonstrate the co-benefits of natural hazards planning—including hazard mitigation, pre-event planning for post-disaster recovery and climate change adaptation. This approach should be crafted in such a way as to recognise the opportunities that may arise due to the ad hoc methods typically employed after disasters; the identification of practices that work under varied conditions (e.g., across disaster types-including those of varied duration, intensity, speed of onset and scope); the diverse makeup of affected locations (e.g., differing levels of vulnerability, network composition and level of horizontal and vertical integration); and differing levels of political, technical, administrative and economic capacity. These experimental endeavors should span varied spatial and temporal scales including individuals, the community, and regional, sub-national, national and international levels, as well as slow onset and rapid onset events. As lessons accrue it is critical that they are shared with others through a system that is inclusive of at-risk communities and the wider natural hazards risk management and climate change adaptation community. In his chapter describing the condition of the Mississippi Delta, for instance, Glavovic (Chap. 15) suggests that one way to convey this information is through the articulation, sharing and celebration of delta narratives about disasters, risk and resilience. A number of authors, including Glavovic, have described the importance of an approach that is focused on the integration of these lessons through cooperatively developed policies, programmes and agreements.

## 16.4 Moral Imperative: Adaptation and Risk Reduction are Ethical Issues that Require Practical, Actionable and Enduring Policies, Plans and Initiatives to Build Resilience and Sustainability

The case studies have revealed an integrative suite of imperatives and associated lessons from natural hazard experiences that can serve as a means to clarify the linkage between natural hazards risk management and climate change adaptation. The foregoing governance, capability and planning imperatives are interconnected and mutually reinforcing and are underpinned by a fundamental moral imperative that the case studies explored in this book frame as resilience in pursuit of sustainable development.

This moral imperative can help us bridge the natural hazards risk management-climate change adaptation divide. As outlined in Chap. 1, and reiterated and explored in greater depth by Beatley (Chap. 6), this framing of resilience represents a more constructive way of thinking about and practicing natural hazards planning because it includes an emphasis on adaptation to changing conditions and building capabilities to adapt to change over time, including after both sudden shocks and slow-onset events. Citing Godschalk's research on urban hazard mitigation, resilient cities are strong and flexible, comprised of governance networks that are committed to acting in a collaborative manner, anticipatory in nature and willing to learn from past events (2003). Resilience typically connotes "bouncing back" from a shock to the system, disturbance or crisis (Paton and Johnston 2006). Other hazards scholars point to the speed with which communities return to a sense of normalcy. This narrow description fails to account for the importance of seeking a new condition predicated on a more prepared, adaptive, equitable and safer set of societal conditions. Paton, McClure and Burgelt (2006), as cited in Beatley (Chap. 6), recognise the "capacity to draw upon personal and social resources to manage the consequences of disasters". More broadly speaking, resilience means proactively building community capabilities and institutional capacities, achieved through planning and collaborative governance, to chart sustainable recovery outcomes in the face of a changing and uncertain future. It also means learning from past events and amending policies, programmes and plans accordingly. Learning from disaster events and instituting meaningful change means more than the common slogans and rhetoric espoused by elected officials to "build back better". As shown by Vale and Campanella (2005), such claims rarely result in fundamental alterations in pre-event settlement patterns (see Chap. 15).

Beatley (Chap. 6), White (Chap. 5), Glavovic (Chaps. 10 and 15) and others argue that, notwithstanding debate about these concepts, there are compelling reasons and broad support amongst many communities, professional planners and other stakeholders, that building resilient and sustainable communities constitutes a moral imperative for natural hazards planning. This book makes the argument that resilience and sustainability can help communities to frame the integration and mainstreaming of natural hazards risk management and climate change adaptation.

In many ways the operationalisation of this moral imperative through enduring and actionable policies, plans, programmes and initiatives requires enhanced "frame convergence," that in turn, fosters collaborative governance (see Chap. 11). In addition to the overriding importance of governance expressed by all chapter authors, the role of natural hazards planning was seen to be a bridge with the potential to link these broad concepts to actionable policies and practices. Plans were recognised as a practical way to engage and expand vertically and horizontally integrated networks recognising the link between natural hazards risk management and climate change adaptation; draw from existing efforts to build expanded and enduring policy frames; share pertinent information over time, including that which changes due to varied scenarios and new input following extreme events, advances in risk assessment models and locally grounded public engagement initiatives; and build the political support needed to adopt forward-looking policies, including those that advance risk reduction and adaptation strategies.

Berke (Chap. 8) suggests that even though planning has evolved over time to embrace participatory approaches and collaborative decision-making, planning for climate change requires a "fundamental transformation" in order to address high levels of complexity and uncertainty. While seemingly a daunting task to overcome, he also states that the established base of knowledge and policies developed surrounding hazard mitigation offer important insights that can be used to inform climate change adaptation. According to Berke, both include a focus on rapid and slow-onset events, both possess a future orientation and both are focused on anticipating future needs and impacts. The need for transformative natural hazards planning processes to engage innovative and practical approaches that are appropriate for dealing with different risk problems is also underscored by Glavovic in his analysis of the Manawatu flood experience (Chap. 10). There is an urgent and compelling need to avoid treating all risk problems as if they are amenable to traditional 'probability-consequence' analysis and treatment; especially in the light of the complexity, ambiguity and even 'unknowability' about aspects of future climate change. Building and maintaining an enduring commitment to the moral imperatives of resilience and sustainability provide a strong normative basis for critically gauging the appropriateness of actions to integrate and mainstream natural hazards risk management and climate change adaptation; and overcome the institutional barriers that prevail in contemporary society.

According to Berke (Chap. 8) there are a number of major obstacles that have hampered natural hazards planning efforts to date, including: (1) the lack of a public constituency at the local level that places a high priority on risk reduction initiatives; (2) the costs associated with engaging in risk reduction are immediate whereas the benefits may be uncertain or accrue over time (often after the term of elected officials have ended); (3) the creation of safer communities may not be clearly visible; and (4) planning has often failed to include the most vulnerable populations in decision-making activities. These observations often have the effect of limiting proactive, land-use based strategies and influence the prevailing trend among communities to "mitigate" choices made in the past through large-scale buyouts and relocation efforts after a disaster occurs, the construction or repair of damaged protective measures like levees, the installation of urban stormwater management programmes and other flood defenses and efforts to control nature through wildland fire-fighting measures and the draining of wetlands rather than seeking ways to balance social and ecological objectives. Such unsustainable and maladaptive outcomes often take place despite claims and even policy goals that advocate resilience and sustainability.

Conversely, several of the case studies in this book have shown that sound natural hazards planning has led to the more effective engagement of marginalised populations, the adoption of spatial planning practices, the devolution of decision-making and responsibility to the community and individual level, the incorporation of lessons from past extreme events (as well as pre-event measures that reduce the likelihood of disasters occurring in the first place) into sub-national climate change plans and policies and adopting the concepts of resilience and sustainability as a moral imperative. While the case studies are not intended to serve as a representative sample of global practice, they do show that change is occurring in a cross section of nations facing differing threats; policies and plans are being led by differing levels of government (e.g., national, sub-national, regional, city and local) as well as non-governmental actors; and a range of post-disaster reactionary approaches as well as purposeful pre-event initiatives are being undertaken.

A common theme has emerged from these case studies which underscores the need for national or higher order legislative requirements and guidance that can be achieved in multiple ways through the adoption of scenario-based policies and plans implemented at the sub-national and local level. In order to further understand this process, the efficacy of these emerging approaches should be evaluated empirically over long timescales, focusing on the degree to which the adoption and implementation of plans and policies lead to tangible improvements in both natural hazards risk management and climate change adaptation-related outcomes. The identification and institutionalisation of effective integrative approaches needs to be developed collaboratively by the network of relevant governance actors, taking into account, among other things, the lessons from case studies in this book that highlight how to overcome institutional barriers and unlock opportunities for charting adaptation pathways in the light of alternative scenarios.

One of our challenges remains more effectively building a shared understanding of conceptual terms like resilience, sustainability, risk reduction and adaptive capacity and developing policies, plans, programmes and agreements that translate rhetoric into reality. The pragmatic efforts surrounding the development of "no-regrets" policies and initiatives offer some evidence that this can be achieved in practice. The notion of sustainability has gained traction as an underlying part of global development programs and supportive international, national and sub-national policy frameworks have been developed since the 1992 Rio Earth Summit to achieve this end. The case studies in this book reiterate the contested and elusive nature of this ideal. Resilience is similarly gaining widespread recognition and the term has become a ubiquitous catchphrase, potentially deflecting attention from its true value. Operationalising resilience within the larger sphere of sustainability in a way that helps to realise the vision outlined at the start of this chapter, and spans physical, environmental, social and economic dimensions and associated technical and scholarly domains that range from engineering to planning, economics and sociology, remain a real challenge.

The centrality of dealing with uncertainty in building resilience underscores the need to develop policies and plans that are characterised by increased flexibility, reflexivity, deliberation, social learning and systematic monitoring of changing conditions in pursuit of collaborative governance. Given the dire implications of failing to better address the root causes and drivers of vulnerability to climate change in particular, and disaster risk more generally, there is a compelling need to translate this moral imperative into actionable agreements by the range of governance actors in different communities. Yet the degree to which risk reduction measures adopted before and after extreme events accounts for future risk tied to a changing climate remains uncommon worldwide. While this topic has begun to gain support amongst some in both the natural hazards risk management and climate change adaptation communities, the rhetoric has not been systematically integrated into programmes, policies or plans. Nor has the climate change community effectively conveyed the direct implications of a warming climate on site specific areas with the spatial accuracy necessary to regulate development in hazard prone areas. This thorny topic is gaining greater attention as evidenced by the investments in new research and modeling capabilities and a wide range of climate change adaptation initiatives around the world. Translating rhetoric into tangible reality on the ground, as the case studies in this book demonstrate, poses not only a profound practical challenge but a deep moral one.

Escalating exposure to natural hazards, driven in part by climate change, and the depth and extent of social vulnerability, provide us with the moral imperative and integrative rationale for action. One of the key challenges in natural hazards risk management policy-making and practice involves persistent reactive approaches, often in response to an extreme event. Oliver-Smith suggests that linking the concept of disaster to climate change relocates the focus of analysis from an event to a socio-ecological process of vulnerability construction. It also further underscores the important and widely recognised reality that disasters are socially constructed events. According to Oliver-Smith,

> locally specific disasters and climate change take place always in the context of local vulnerability. Any attempt to address the risks of disasters and climate change that does not put that condition at the forefront, will reduce its chances of any durable success (Chap. 4).

Reducing climate risk, and disaster risk more generally, necessitates confronting the root causes and drivers of social vulnerability (Wisner et al. 2004). Much can be learned from long-standing thinking and practices in diverse fields to understand and reduce social vulnerability (see Chap. 2).

Much can also be learned from experience in natural hazards planning, including the problems associated with conducting vulnerability and risk assessments in a vacuum by technical experts; the problems associated with stationarity; and the failure to build a shared understanding of and addressing the underlying causes of vulnerability and rising disaster risk. There is nonetheless value in exploring, and where appropriate, adopting emerging advances in vulnerability and risk assessment, including future conditions risk assessments that account for the dynamism of natural hazards; advances in complex climatological modeling-including greater place-based estimates of risk; and visualisation techniques that convey risk-based data in multi-dimensional arrays. Realising the potential of such advances necessitates that they are embedded in deliberative and transformative processes that engage local communities in developing and implementing locally relevant and integrated vulnerability assessments and risk reduction strategies. Improving the utility of these tools also includes developing new methods to calculate changes in natural hazard return periods that are influenced by climate change and incorporate these findings into new maps, plans, policies and tools. Fundamentally, it means understanding the nature of the dominant risk problem that is pre-eminent and to use appropriate risk assessment and management approaches, some of which involve sophisticated technical and quantitative analysis; others of which involve the design and facilitation of equally sophisticated qualitative processes that focus on authentic deliberation and effective resolution of conflicting interests (see Chap. 10).

Not only must we move beyond the narrow focus of extreme events as something we respond to once they happen, we also need to embrace our reactive nature as humans, albeit in a way that is done in conjunction with a forward-thinking set of activities. Adopting a natural hazards planning approach serves to bind risk reduction and post-disaster recovery as well as natural hazards risk management more broadly construed with adapting to a changing climate. This approach underscores the need for pre-event planning to reduce climate and wider disaster risk, and move beyond a uni-dimensional focus on the large scale expenditure of resources after a disaster, including misplaced efforts that do not account for risk. It means including the additional, albeit uncertain effects of climate change in policies and initiatives intended to proactively address these threats. Not having a definitive answer to how high the seas will rise, how increasing temperatures will affect drought and wildfire conditions or the degree to which warmer oceans will lead to stronger coastal storms is not an adequate excuse for inaction. Berke argues that we can begin to address this uncertainty by crafting a suite of both robust low- and no-regrets policies as well as contingent policies tied to alternative scenarios. In order to better achieve the latter set of policies, Berke suggests that we need to improve our ability to integrate multiple models needed to "assess impacts and reduce uncertainty" and to improve the ability of stakeholders to use the results (Chap. 8).

Natural hazards experience has taught us that failing to invest in more pre-event capacity building initiatives is a foolhardy strategy and yet, communities across the globe do not adequately address this well-known problem. We also know that our vulnerability to disasters, including those created or exacerbated by a changing climate, are a direct reflection of our own actions, including inaction. The alteration of natural systems, building in hazard prone areas and now the physical alteration of our climate requires an expanded understanding of the steps we need to take in order to confront and adapt to this challenge today and over time as conditions merit. Central to this endeavor is the moral imperative to address the root causes and drivers of social vulnerability that contribute to escalating disaster risk, unsustainable practices and maladaptive development trajectories.

The improved practice of natural hazards risk management will require planning for what amounts to an uncertain and turbulent future. It is incumbent on those in diverse multi-disciplinary fields of research and practice to identify ways to work together more effectively and collaboratively to address these uncertainties both proactively and reactively in the face of a changing climate and escalating disaster risk. Ultimately it means modifying the institutional cultures, structures and processes that shape the actions of those involved in natural hazards risk management and climate change adaptation so that lessons learned from past events can be effectively integrated and mainstreamed into practice.

## 16.5 Developing and Operationalising Natural Hazards Planning

The findings of the case studies in this book provide a striking number of common lessons that span developed and developing nations, varied natural hazards types, governance strategies and geographic locales. The case studies in this book also underscore the interdependence between trajectories of local community development and escalating natural hazard risks in an era of climate change. The convergence of common lessons help to bolster our argument for a discrete set of imperatives that strive to better integrate the natural hazards risk management and climate change adaptation communities. Operationalising these imperatives can help to realise the promise of natural hazards planning as developing more integrative risk reduction and adaptive strategies is a relatively recent focus of attention in scholarship, policy and practice.

The promise of natural hazards planning is that by building more deliberative, enabling and collaborative governance networks, communities will better understand the risks they face as well as their own unique contextually-relevant vulnerability and adaptive capacity so that they can chart more resilient and sustainable pathways based on this knowledge. This task is not easy and requires confronting a number of barriers. Prevailing institutional structures and processes often foster inequitable access to available assets and resources, perpetuate maladaptive strategies that resist change and consequently predispose communities to the devastating impacts of extreme events. To compound matters, post-disaster recovery efforts often increase future risk by rebuilding communities in ways that merely reinforce patterns of pre-event exposure and vulnerability. The case studies described in this book reveal valuable lessons for operationalising effective natural hazards planning processes that can enable communities to adapt in the face of significant obstacles.

We posit four imperatives as a way of framing lessons learned for realising the promise of natural hazards planning. Table 16.1 represents a distillation of the lessons derived from the preceding chapters. While many of the lessons span multiple imperatives, they have been situated in the table as found in this chapter's summary of case study findings. The Appendix to this chapter contains a complete list of the lessons, including an attribution to chapter authors who identified them. This approach allows the reader to return to the narrative found in this chapter and associated case studies to gain additional contextual insights. The lessons are intended to help stimulate a dialogue among natural hazards risk management and climate change adaptation scholars, practitioners and stakeholders and facilitate action to build resilience and sustainability through natural hazards planning.

In order to operationalise the **governance imperative** (i.e., identify, nurture and sustain collaborative governance networks that span varied spatial and temporal scales) there is a need to:

1. **Build adaptive governance capacity**: Understand local needs and adaptive capacities across relevant governance networks and develop or expand cultur-

**Table 16.1** Adaptation imperatives and lessons learned from natural hazards experience and scholarship

| |
|---|
| *Governance Imperative: Identify, nurture and sustain collaborative governance networks that span varied spatial and temporal scales* |
| 1. Build adaptive governance capacity |
| 2. Integrate and mainstream development, risk management and adaptation |
| 3. Clarify governance roles and promote inclusivity, collaboration and conflict resolution |
| 4. Bridge varied institutional, spatial and temporal scales |
| 5. Create institutional diversity and flexibility to deal with uncertainty |
| 6. Foster reflexive learning |
| *Capability Imperative: Expand the forward-looking use of community assets and pre- and post-disaster resources to enhance risk reduction and adaptation initiatives* |
| 1. Understand and reduce vulnerability |
| 2. Confront barriers to adaptation |
| 3. Build layers of resilience |
| 4. Develop an expanded network of capacity-builders |
| 5. Deepen and extend knowledge and understanding through dialogue and deliberation |
| 6. Develop enduring and integrative policies and practices |
| *Planning Imperative: Invest more in pre-event planning for risk reduction and enhanced collective adaptive capacity* |
| 1. Institutionalise resilience planning |
| 2. Develop a supportive institutional framework for building local adaptive capacity |
| 3. Adopt a planning approach that focuses on spatial implications of land use decisions in light of risk and employs scenario based planning practices |
| 4. Prioritise pre-event planning and embrace the post-disaster window of opportunity to implement inclusive adaptive strategies and capacity building initiatives |
| 5. Integrate plans and planning processes |
| 6. Encourage experimental and innovative governance strategies |
| *Moral Imperative: Adaptation and risk reduction are ethical issues that require practical, actionable and enduring policies, plans and initiatives to build resilience and sustainability* |
| 1. Build institutional capacity for reaching agreement about a shared vision for the future |
| 2. Operationalise resilience and sustainability through collaborative governance |
| 3. Engage non-specialists, including those with indigenous and/or local knowledge |
| 4. Develop enduring reflexive and deliberative procedures |
| 5. Foster the conditions for collaborative leadership |

ally sensitive, integrative policies and practices that recognise and are tailored for distinctive socio-ecological conditions that shape community well-being.

2. **Integrate and mainstream development, risk management and adaptation**: Natural hazards risk management and climate change adaptation practices need to be integrated and mainstreamed into local community development planning and decision-making. Integrate adaptation measures into recognised natural hazards risk related policy and planning frameworks, document their effectiveness over time, share the findings and modify policies and practices accordingly.
3. **Clarify governance roles and promote inclusivity, collaboration and conflict resolution**: Address gaps, overlaps and contradictions in the roles and responsibilities of governance actors in support of inclusive and collaborative processes

that resolve conflicting interests about local community development, public safety, resilience and sustainability. Deliberately engage marginalised groups by involving historically under-represented groups in the formulation of policies and plans; and respect traditional decision-making processes and indigenous adaptation practices.

4. **Bridge varied institutional, spatial and temporal scales**: Develop policies and practices that recognise that risk reduction in the face of a changing climate and societal change more broadly transcends institutional, spatial and temporal scales. Promote horizontal (e.g., across sectors) and an expanded notion of vertical (e.g., across individuals, spheres of government and international consortia) integration, and reconcile spatial (e.g., local, national and international) and short- and longer term interests. Develop and sustain region-wide strategic collaborative planning processes to address the intractability of climate change that local communities cannot resolve alone.
5. **Create institutional diversity and flexibility to deal with uncertainty**: Rigid institutional structures and processes are ill-suited to deal with the uncertainty and turbulence inherent in a changing climate. Notwithstanding the need to foster predictability and institutional stability, there is a countervailing need to maintain institutional diversity, flexibility and redundancy in the face of change.
6. **Foster reflexive learning**: Critically reflect upon and encourage learning approaches that embrace diversity, disagreement, creativity and frankness in public discourse in an effort to identify and frame alternative governance, policy and practice-based options for adapting to a changing climate.

In order to operationalise the **capability imperative** (i.e., expand the forward-looking use of community assets and pre- and post-disaster resources to enhance risk reduction and adaptation initiatives) there is a need to:

1. **Understand and reduce vulnerability**: Address the drivers and root causes of social vulnerability that predispose marginalised groups and communities to experience disasters. Where appropriate, identify and transform institutional structures and processes that cause vulnerability, and result in maladaptive and unsustainable outcomes.
2. **Confront barriers to adaptation**: Identify and address institutional barriers (both formal and informal) and unlock opportunities that unleash the collective capacity of governance actors and networks. Strategies, policies and investments undertaken at the national level should focus on providing assistance to local communities in developing responses according to local capacities and priorities. Focus particular attention on removing perverse incentives that increase risk, including managerial and financial policies and practices that increase the exposure of human settlements to the negative effects of extreme events and climate change and deepen social vulnerability.
3. **Build layers of resilience**: Adaptive capacity and resilience to withstand sudden shocks and slow onset disasters driven by climate change necessitates, among other things, restoring and maintaining the health, productivity and integrity of ecosystems and natural defences; providing robust public infra-

structure, including, for example, flood protection works in tandem with sound spatial planning; fostering economic well-being whilst acting as stewards of climate-sensitive resources; strengthening social capital; eradicating poverty and inequality; and building inclusive governance institutions that include effective research, monitoring, public awareness and communication systems.

4. **Develop an expanded network of capacity-builders**: Adaptation at the local level is shaped by the capacity building role played by local authorities, the influence of international, national and regional policies and plans as well as partnerships with non-governmental actors including, the private sector, quasi-governmental organisations, non-profits and individuals.
5. **Deepen and extend knowledge and understanding through dialogue and deliberation**: Improve knowledge and understanding of adaptation and risk reduction through the use of ongoing dialogue, deliberation and risk communication strategies. For example, improve the knowledge base of the planning and design community to include the complementary aims of hazard mitigation, disaster recovery and climate change adaptation through curricula development and continuing professional development programmes.
6. **Develop enduring policies and practices**: Adaptation is an evolving process that requires long-term institutional commitment and provisions that reduce risk and facilitate adaptation to a changing climate. This necessitates developing policies and practices that recognise the varied timescales associated with episodic and slow-onset events as well as changes in political party, ideology and priorities.

In order to operationalise the **planning imperative** (i.e., invest more in pre-event planning for risk reduction and enhanced collective adaptive capacity) there is a need to:

1. **Institutionalise resilience planning**: Planning processes should 'build layers of resilience' so that communities can draw upon the range of financial, human, physical, social and natural assets at their disposal in the face of sudden events and slow onset disasters. Such an approach necessitates an understanding of natural hazards, exposure and social vulnerability; and a planning praxis that confronts the root causes and drivers of social vulnerability.
2. **Develop a supportive institutional framework for building local adaptive capacity**: Continually review, and where appropriate, create or modify higher level institutional structures and processes, including national and international strategies and standards, which build the adaptive capacity of local governments to support their communities in charting appropriate adaptation pathways.
3. **Adopt a planning approach that focuses on spatial implications of land use decisions in light of risk and employs scenario based planning practices**: Plans need a sound information base, a set of actionable goals and an implementation, monitoring and evaluation strategy, taking into account changing conditions and uncertainty, including that which is tied to the dynamism of natural hazards and a changing climate as well as shifting public interests. Focus attention on the spatial implications of land use decisions in light of natural hazard risks. Develop natural hazards risk management and climate change strategies tied to land use, includ-

ing the siting of infrastructure and human settlements as well as the relocation of at-risk communities. Employ the use of scenario-based planning techniques that include a flexible collection of robust and contingent provisions that identify thresholds for step-wise adaptation actions as climate change impacts become more apparent. Improve the means by which non-technical experts are involved in the use of modeling outputs and the identification of locally-appropriate risk reduction/adaptive strategies.

4. **Prioritise pre-event planning and embrace the post-disaster window of opportunity to implement inclusive adaptive strategies and capacity building initiatives**: Place greater emphasis on pre-event planning for extreme events, including rapid and slow-onset disasters rather than the traditional emphasis on emergency response and post-disaster monetary aid. In the aftermath of a disaster, harness heightened public awareness and political will to chart pathways that link disaster recovery, risk reduction, sustainable development and climate change adaptation, recognising the need to account for and actively involve socially vulnerable, less powerful populations in policy making.
5. **Integrate plans and planning processes**: Municipal functions (including planning, emergency management, infrastructure and asset management and community development) need to be coordinated and integrated to foster coupled risk reduction and adaptation activities. Among other things, employ risk communication and information strategies using new social media technologies, formulate flexible planning frameworks that emphasise low-cost/low-regrets policies and win/win/no-regrets policies that are integrated into broader day to day decision-making and comprehensive spatial plans and establish risk and vulnerability reduction as integrative themes bridging adaptation and natural hazards risk management.
6. **Encourage experimental and innovative governance strategies**: Incentivise individuals and organisations to take the initiative and adopt experimental and innovative approaches, pilot programmes and educational initiatives in the pre- and post-disaster environment that champion the co-benefits of hazard mitigation, pre-event planning for post-disaster recovery and climate change adaptation. The lessons drawn from these experimental efforts should be captured and shared with others.

In order to operationalise the **moral imperative** (i.e., adaptation and risk reduction are ethical issues that require practical, actionable and enduring policies, plans and initiatives to build resilience and sustainability) there is a need to:

1. **Build institutional capacity for reaching agreement about a shared vision for the future**: Create 'safe spaces' for discussing a vision for the future and explore what resilience and sustainability mean to people. Create the institutional space needed to negotiate agreements within and across governance networks to build a shared understanding and vision of adaptation pathways into the future and improve the distribution of available resources to enable natural hazards risk management and climate change adaptation.

2. **Operationalise resilience and sustainability through collaborative governance**: Develop integrated policy frameworks that can operationalise the tenets of resilience and sustainability. Develop cooperative agreements linking selection criteria and funding for adaptation and risk reduction programmes and projects.
3. **Engage non-specialists, including those with indigenous and/or local knowledge**: Deliberately involve non-specialists in assessing and defining alternative risk reduction and adaptation pathways, including those with indigenous and/or local knowledge.
4. **Develop enduring reflexive and deliberative procedures**: Critically review the effectiveness of prevailing practices. Explore how to frame hazard mitigation and climate change adaptation as developmental issues as opposed to environmental issues. Intentionally reflect upon and deliberate on the underpinning policy frames and formal and informal institutions that shape natural hazards risk management and adaptation practices. Based on this reflection, identify areas of common concern and reframe risk reduction and adaptation approaches.
5. **Foster the conditions for collaborative leadership**: Pursuit of resilience and sustainability is elusive under prevailing business as usual practices. Operationalising these moral imperatives necessitates community wide action stimulated by catalytic visionary leadership that fosters cooperation and collaboration. Such leadership is effectively mobilised when there are high levels of trust.

## Appendix: Adaptation Imperatives and Lessons Learned from Natural Hazards Experience

The lessons that follow are drawn from the preceding chapters. In many cases these lessons are represented as consolidated statements given their similar language and intent. In other cases, they are restated verbatim from the chapter in which they are found. If subtle differences exist, lessons are listed separately. Individual lessons may not be universally applicable, but together, they provide governance actors with a robust framework that can build understanding and enable proactive steps to be taken to collectively address one of the greatest challenges of the twenty first century, building resilience and sustainability in this era of climate change.

**Governance Imperative: Identify, Nurture and Sustain Collaborative Governance Networks that Span Varied Spatial and Temporal Scales.**

**Lesson:** Conduct a global audit of existing international, national and sub-national programmes, policies and funding sources across the natural hazards risk management and climate change adaptation sectors; identify complementary and contradictory policies; and use this information to help develop an integrated natural hazards risk management/climate change adaptation strategy that may help inform others who are in the process of developing a larger national climate change policy framework (Smith, Chap. 9; Ziervogel and Parnell; Glavovic, Chap. 10).

**Lesson:** Conduct a needs assessment in order to inform capacity building and livelihood enhancement initiatives (Lakshmi, Purvaja and Ramesh).

**Lesson:** Involve historically under-represented groups in the formulation of policies and plans (Poutasi, Daly, Kohlhase and Nelson; Glavovic, Chap. 15).

**Lesson:** Clarify role ambiguity across national, provincial and local spheres of government (Ziervogel and Parnell).

**Lesson:** Expand or modify existing policy frameworks to address climate change adaptation and natural hazards risk management (Ziervogel and Parnell; Poutasi, Daly, Kohlhase and Nelson; Lakshmi, Purvaja and Ramesh; Oliver-Smith; Siembieda; Bosomworth, Handmer and Dovers; Glavovic, Chap. 10).

**Lesson:** Mainstream development, natural hazards risk management and climate change adaptation (Lakshmi, Purvaja and Ramesh; Oliver-Smith; Birkmann and Pardoe; Poutasi, Daly, Kohlhase and Nelson).

**Lesson:** Institute a reflexive learning approach that embraces heterogeneity, disagreement, creativity and frankness in an effort to identify different governance, policy and practice-based options (Bosomworth, Handmer and Dovers; Glavovic, Chap. 10; Ziervogel and Parnell).

**Lesson:** Enhance the horizontal and vertical integration of emerging natural hazards risk management and climate change adaptation networks (Siembieda; Lakshmi, Purvaja and Ramesh; Smith, Chaps. 9 and 14; Ziervogel and Parnell).

**Lesson:** Identify sustained sources of financing that advance climate change adaptation measures, including funds tied to municipal budgets (Ziervogel and Parnell; Smith, Chap. 9; Birkmann and Pardoe).

**Lesson:** Develop risk reduction and climate change adaptation policies that recognise the unique conditions found in rural (Glavovic, Chap. 10; Lakshmi, Purvaja and Ramesh; Oliver-Smith) and urban (Ziervogel and Parnell; White; Poutasi, Daly, Kohlhase and Nelson) areas.

**Lesson:** Enhance the role of local government in climate change adaptation and natural hazards risk management initiatives that recognise traditional decision making processes and indigenous adaptation measures (Ziervogel and Parnell; Poutasi, Daly, Kohlhase and Nelson; Lakshmi, Purvaja and Ramesh; Oliver-Smith; Berke; Bosomworth, Handmer and Dovers; Birkmann and Pardoe).

**Lesson:** Develop coalitions that are prepared to advocate for the adoption of climate change adaptation measures following extreme events (Ziervogel and Parnell; Birkman and Pardoe; Beatley).

**Lesson:** Develop agriculturally-focused climate change adaptation programs as part of a larger strategy tied to development (Oliver-Smith).

**Lesson:** Develop policies that recognise varied spatial dimensions of climate change-related hazards (White; Birkmann and Pardoe; Beatley; Berke).

**Lesson:** Develop and sustain region-wide strategic collaborative planning processes to address the intractability of climate change that delta communities cannot resolve alone (Glavovic, Chap. 15).

**Lesson:** Restore livelihoods after disasters in a way that recognises the gender-based roles assumed by men and women in varied cultures (Lakshmi, Purvaja and Ramesh).

**Lesson:** Develop culturally appropriate processes and plans (Poutasi, Daly, Kohlhase and Nelson; Oliver-Smith).

**Lesson:** Integrate measures designed to address specific hazards with programmes to reduce systemic vulnerabilities and social inequality (Oliver-Smith).

**Lesson:** Improve the downscaling of climate change models to help assess local vulnerability (White; Birkmann and Pardoe; Berke).

**Lesson:** Develop adaptation and risk reduction plans that move beyond the notion of stationarity and recognise the changing nature of hazard risk in light of a changing climate (White).

**Lesson:** Develop risk communication and information strategies using new social media technologies (Siembieda; Poutasi, Daly, Kohlhase and Nelson; Lakshmi, Purvaja and Ramesh).

**Lesson:** Develop policies that recognise the varied temporal dimensions associated with climate change adaptation and extreme events (Ziervogel and Parnell; Berke; Birkmann and Pardoe; Beatley).

**Lesson:** Reframe governance thinking and praxis that drive disaster risk through environmental degradation and social vulnerability (Glavovic, Chap. 15).

**Lesson:** Develop pre-existing governance networks capable of accepting and effectively distributing post-disaster resources in a coordinated, equitable and timely manner (Smith, Chap. 14).

**Capability Imperative: Expand the Forward-looking Use of Community Assets and Pre- and Post-disaster Resources to Enhance Risk Reduction and Adaptation Initiatives.**

**Lesson:** Create enduring pre-disaster sub-national, quasi-governmental recovery organisations and expand their roles to include greater emphasis on pre-event planning and capacity building across joint networks while placing a greater emphasis on risk reduction and climate change adaptation (Smith, Chaps. 9 and 14; Siembieda; Oliver-Smith).

**Lesson:** Improve the management capabilities of those organisations and institutions addressing both current slow and fast onset events as well as future climate risks (Oliver-Smith).

**Lesson:** Strategies, policies and investments undertaken at the national level should also focus on providing assistance to local communities in developing responses according to local capacities and priorities (Oliver-Smith).

**Lesson:** Expand the coordinated pre- and post-event blending of public, non-profit, quasi-governmental and private sector investments, risk reduction initiatives and adaptation measures through a clear risk communication strategy that describes the implications of varied choices made by these stakeholders. (Smith, Chap. 14; Siembieda).

**Lesson:** Establish regional and local forums to improve understanding about natural hazard risks and explore opportunities to integrate emergency management, land-use planning, asset management and climate adaptation practice (Glavovic, Chap. 10).

**Lesson:** Embed and align adaptation policies in existing priorities and capacities (Ziervogel and Parnel).

**Lesson:** Increase the involvement and self-reliance of citizens as part of a larger city-wide capacity-building programme focused on resilience and adaptation (Siembieda).

**Lesson:** Limit the placement of structures in areas prone to sea-level rise and other natural hazard areas (Beatley).

**Lesson:** Expand the role of the private sector (e.g., private utility companies) in the development of an enhanced level of local capacity to address risk reduction and climate change adaptation initiatives (Siembieda).

**Lesson:** Move beyond the over-reliance on formal institutional arrangements, recognising the value of informal agreements, organisations and indigenous knowledge (Bosomworth, Handmer and Dovers; Lakshmi, Purvaja and Ramesh; Poutasi, Daly, Kohlhase and Nelson; Oliver-Smith).

**Lesson:** Strategies, policies and investments undertaken at the national level should reflect a greater understanding of local capacities and priorities (Oliver-Smith; Siembieda; Lakshmi, Purvaja and Ramesh; White; Poutasi, Daly, Kohlhase and Nelson; Smith, Chaps. 9 and 14).

**Lesson:** Address the drivers and root causes of social vulnerability that predispose marginalised groups and communities to disaster (Glavovic, Chap. 15; Oliver-Smith).

**Lesson:** Capture lessons and avoid the mistakes associated with large-scale voluntary housing acquisition and relocation programs as part of an eventual national sea-level rise adaptation strategy (Smith Chap. 9).

**Lesson:** Expand pre-existing risk management networks to include those advocating the adoption of climate change adaptation measures (Siembieda).

**Lesson:** Develop an improved means to design and finance large infrastructure projects incrementally (Berke; Beatley).

**Lesson:** Design and build distributed and decentralised infrastructure systems that reduce exposure and vulnerability (Beatley).

**Lesson:** Create incentives through grants, loans or other means to encourage the relocation and resettlement of communities out of flood-prone areas, including those susceptible to sea-level rise (Smith, Chap. 9; Poutasi, Daly, Kohlhase and Nelson).

**Lesson:** Engage in a meaningful dialogue about land use and the potential alteration of pre-event settlement patterns, including those associated with migration (Oliver-Smith) before and after disasters (Smith, Chaps. 9 and 14; White; Poutasi, Daly, Kohlhase and Nelson; Berke; Glavovic, Chap. 10; Beatley).

**Planning Imperative: Invest More in Pre-event Planning for Risk Reduction and Enhanced Collective Adaptive Capacity.**

**Lesson:** Foster the flexibility to achieve higher order goals while giving local governments the ability to choose their own paths to achieving them (Siembieda; Oliver-Smith; Lakshmi, Purvaja and Ramesh; Poutasi, Daly, Kohlhase and Nelson; Berke).

**Lesson:** Develop integrative sub-national natural hazards risk management and climate change adaptation plans in lieu of a clear national climate change strategy (Smith, Chap. 14).

**Lesson:** Improve warning systems to account for traditional modes of communication and varied messaging systems, including new media (Lakshmi, Purvaja and Ramesh).

**Lesson:** The public sector should engage in deliberate reflection on underpinning policy frames and informal institutions (e.g., bushfire management), and based on this reflection, identify areas of common concern with emerging adaptation policy frames (Bosomworth, Handmer and Dovers).

**Lesson:** Improve the knowledge base of the planning and design community to include the complementary aims of hazard mitigation, disaster recovery and climate change adaptation through curricula development and conducting professional association workshops and seminars (Smith, Chap. 14).

**Lesson:** Reduce physical vulnerability through the construction of disaster resilient housing and cyclone shelters, delineation of evacuation routes, development of early warning systems and establishment of better connectivity between roads and bridges (Lakshmi, Purvaja and Ramesh).

**Lesson:** Adaptation and risk reduction strategies should be grounded in sound data tied to the vulnerability and exposure of societies, communities and social-ecological systems (Birkmann and Pardoe).

**Lesson:** Overcome the safe development paradox in which protective measures can incentivise development in known hazard areas leading to future disasters when the design parameters are exceeded (White; Glavovic, Chap. 15; Smith, Chap. 14).

**Lesson:** Link hazard mitigation and preparedness policy frameworks (Poutasi, Daly, Kohlhase and Nelson).

**Lesson:** Invest the time and resources needed to gain the political support of elected officials during the development of a joint hazard risk reduction/climate change adaptation strategy (Siembieda).

**Lesson:** Adopt climate change and risk reduction strategies that place a greater emphasis on spatial planning versus engineering-dominant approaches (White).

**Lesson:** Create well trained and experienced planning and design teams that can be deployed to communities before and after disasters to share their experiences and assist them develop integrated disaster resilience and climate change adaptation plans, policies and projects (Smith, Chap. 14).

**Lesson:** Develop integrative policies that recognise differing norms held by members of the disaster risk reduction and climate change adaptation communities (Birkmann and Pardoe).

**Lesson:** Planning interventions should occur at multiple scales and lead to an integrated, interlocking set of resilience measures (Beatley).

**Lesson:** Improve livelihood diversification through the provision of safe shelter and improved income security (Lakshmi, Purvaja and Ramesh).

**Lesson:** Stem wetland loss and restore delta ecosystems to sustain coastal livelihoods and reduce risk in the face of climate change (Glavovic, Chap. 15; Lakshmi, Purvaja and Ramesh).

**Lesson:** Place a greater emphasis on pre-event planning and adaptive capacity building initiatives, rather than the continued over-emphasis on post-disaster monetary assistance (Smith, Chap. 9; Lakshmi, Purvaja and Ramesh).

**Lesson:** Recognise and act on the consequences of adopting a hazard mitigation strategy focused on codes and standards and less on land use (Smith, Chaps. 9 and 14; Berke).

**Lesson:** Emphasise the role of land use as a key risk reduction and climate change adaptation technique (Lakshmi, Purvaja and Ramesh; White; Berke; Beatley; Smith, Chaps. 9 and 14; Glavovic, Chap. 10).

**Lesson:** Communities and countries should plan for the relocation of human settlements away from high hazard areas and those sites likely to receive them (Poutasi, Daly, Kohlhase and Nelson; Smith, Chaps. 9 and 14; Berke; Oliver-Smith; Glavovic, Chap. 15; Beatley; Lakshmi, Purvaja and Ramesh).

**Lesson:** Apply the lessons drawn from the use of post-disaster temporary housing strategies to inform new resettlement policies tied to climate change adaptation (Smith, Chap. 14; Lakshmi, Purvaja and Ramesh).

**Lesson:** Disaster resilient housing should give greater consideration to the role of site design (including proper drainage) and the physical proximity of housing to areas vulnerable to flooding, including current cyclonic and rainfall events as well as projected rising sea levels (Lakshmi, Purvaja and Ramesh; Beatley).

**Lesson:** Take advantage of the window of opportunity to link disaster recovery, risk reduction, sustainable development and climate change adaptation, recognising the need to account for socially vulnerable, less powerful populations (Ziervogel and Parnell; Smith, Chap. 9; Oliver-Smith; Bosomworth, Handmer and Dovers; Glavovic, Chaps. 10 and 15).

**Lesson:** Provide adequate resources (including staffing and financing) in support of collaborative governance and capacity building (Ziervogel and Parnell; Smith, Chaps. 9 and 14; Glavovic, Chap. 10; Siembieda; Poutasi, Daly, Kohlhase and Nelson).

**Lesson:** Develop pre-event sub-national (state) and local staffing plans to account for increased demands following disasters (Smith, Chap. 9).

**Lesson:** Create plans that establish a compelling vision of the future (Beatley; Berke).

**Lesson:** Improve the knowledge base among key stakeholder groups regarding environmental management, natural resource management and natural hazards risk management across an inclusive network (Poutasi, Daly, Kohlhase and Nelson; Glavovic, Chap. 15; Oliver-Smith).

**Lesson:** Provide institutional "space" for social learning in policy and practice (Ziervogel and Parnell; Siembieda; Bosomworth, Handmer and Dovers).

**Lesson:** Articulate, share and celebrate narratives about disasters, risk and resilience (Glavovic, Chap. 15).

**Lesson:** Create maps, shown at a large scale, that are easily understood by laypeople, include locally-relevant landmarks and are informed by local input and indigenous knowledge (Poutasi, Daly, Kohlhase and Nelson).

**Lesson:** The adoption of new maps that depict hazard risk should employ an extensive, participatory outreach effort targeting multiple stakeholders including local residents and officials, builders, design professionals and others (Smith, Chap. 14).

**Lesson:** Link risk assessment maps and other analytical results to accepted regulatory policies and pre-existing administrative capabilities when possible (Smith, Chap. 14).

**Lesson:** Improve the physical demarcation of natural hazard areas (Lakshmi, Purvaja and Ramesh; Smith, Chap. 14).

**Lesson:** Develop financial and managerial frameworks that reflect the sources (e.g. rivers and sea versus surface water and drains) of risk (White).

**Lesson:** Ensure that risk assessment models account for new and changing understandings of local level risk management (White).

**Lesson:** Encourage agencies and organisations in the natural hazards risk management and climate change communities to co-support experimental efforts, pilot programmes and educational initiatives in both the pre- and post-disaster environment that clearly demonstrate the co-benefits of hazard mitigation, pre-event planning for post-disaster recovery and climate change adaptation (Glavovic, Chap. 15; Smith, Chap. 9; Siembieda).

**Lesson:** Frame climate change adaptation as a development issue not an environmental issue (Ziervogel and Parnell).

**Lesson:** Emphasise the use of scenario-based planning practices to better account for the uncertainties associated with a changing climate while drawing from the lessons observed in natural hazards risk management planning initiatives (Berke; Glavovic, Chap. 10; Birkmann and Pardoe).

**Lesson:** Climate Change Adaptation plans should include a flexible collection of robust and contingent policies (Berke; Bosomworth, Handmer and Dovers).

**Lesson:** Plans should contain a clear, yet flexible, implementation program that holds identified stakeholders accountable for carrying out proposed policies and identifies a timeline for completion, including the means by which this will be achieved, recognising that changes in responsibilities, timelines and the commitment of resources may be necessary based on climate variability (Berke).

**Lesson:** Whenever possible, sanctions and/or information about compliance should be communicated to the public and elected officials to coerce and/or encourage accountability (Berke; Siembieda).

**Lesson:** Develop plans that are easy to read and readily accessible (Poutasi, Daly, Kohlhase and Nelson).

**Lesson:** Plans should contain a clear monitoring and evaluation program that allows for changing conditions and uncertainty, including that which is tied to the dynamism of natural hazards and a changing climate as well as shifting public interests (Berke).

**Lesson:** Plans should include a flexible policy framework to anticipate a range of possible future climate impacts (Berke).

**Lesson:** Develop indicators and protocols for the collection of data used for monitoring at multiple spatial scales and the support of collaborative arrangements among multiple levels of government (Berke; Birkmann and Pardoe).

**Moral Imperative: Adaptation and risk reduction are ethical issues that require practical, actionable and enduring policies, plans and initiatives to build resilience and sustainability.**

**Lesson:** Improve the operationalisation of resilience, sustainability and adaptation (Oliver-Smith; Beatley; Siembieda).

**Lesson:** Develop enhanced levels of institutional adaptability as manifest in the types of policies and governance arrangements adopted in the face of a changing climate (Siembieda).

**Lesson:** Incorporate climate change adaptation measures into recognised hazard mitigation and disaster recovery programmes, policies and associated funding mechanisms; document their effectiveness over time; share the findings with others, including those in the climate change community; and ensure that post-disaster hazards management activities are a key part of any national climate change adaptation strategy (Smith, Chap. 14).

**Lesson:** Conduct further study of the unique and complementary features of hazard mitigation planning and climate change adaptation planning (Berke).

**Lesson:** Encourage FEMA to integrate adaptation measures into the Disaster Mitigation Act and National Disaster Recovery Framework guidance, training materials, policies and funding programs (Smith, Chap. 9).

**Lesson:** Embrace participatory planning processes that emphasise local customs and culture (Poutasi, Daly, Kohlhase and Nelson; Oliver-Smith).

**Lesson:** Improve the collective development and integration of pre- and post-disaster plans (Glavovic, Chap. 15).

**Lesson:** Develop the institutional space to foster post-disaster use of negotiated agreements tied to the improved distribution of resources that meet local needs (Smith, Chap. 9; Bosomworth, Handmer and Dovers; Lakshmi, Purvaja and Ramesh; White).

**Lesson:** Expand the use of internal and external funding to support the planning process rather than the current preference for projects (Ziervogel and Parnel).

**Lesson:** Conduct comparative research on the causes and consequences of plan quality for climate change adaptation (Berke).

**Lesson:** Develop appropriate study parameters, metrics and analytical procedures that can measure the cause and effect relationships between elements of resilience programs and tangible outcomes (Siembieda).

**Lesson:** Develop cooperative agreements where selection criteria and funding for adaptation and disaster risk reduction programmes and projects requires collaboration across disaster risk reduction and climate change adaptation stakeholders (Birkmann and Pardoe).

**Lesson:** Integrate knowledge about collaborative rationality, hazard mitigation and scenario planning into the climate change adaptation process (Berke).

**Lesson:** Develop metrics and methods used to detect how well plans support plan quality principles and assess the degree of success in plan implementation (Berke).

**Lesson:** Conduct research on scenario formulation and testing that can lead to better plans, including the better integration of multiple models to assess impacts and reduce uncertainty (Berke).

**Lesson:** Formulate flexible policy frameworks that emphasise low-cost/low-regrets policies and win/win/no-regrets policies that are integrated into broader day to day decision-making and comprehensive spatial plans (Berke).

**Lesson:** Improve techniques for engaging stakeholders, including "non-technical experts", in the use of modeling outputs and the creation of a range of collaboratively derived adaptive strategies based on multiple futures that are relevant to urban place making (Berke).

**Lesson:** Engage in multi-disciplinary research on climate change and natural hazards risk management to expand on what amounts to narrow disciplinarily-defined policy frames (Bosomworth, Handmer and Dovers).

**Lesson:** Establish risk and vulnerability reduction as integrative themes bridging climate change adaptation and natural hazards risk management (Oliver-Smith; White; Birkmann and Pardoe; Bosomworth, Handmer and Dovers; Glavovic, Chap. 15).

**Lesson:** Apply plan quality principles in the development of climate change adaptation plans (Berke).

**Lesson:** Embed climate change adaptation measures in new and existing natural hazards risk management policies (Poutasi, Daly, Kohlhase and Nelson).

**Lesson:** Adopt a risk management paradigm that expands the responsibility of stakeholders, including the acceptance of blame beyond governmental actors when warranted (White).

**Lesson:** Maximise the use of existing natural hazards risk assessment tools to inform climate change adaptation strategies, while modifying these tools to more readily incorporate climate change-induced/exacerbated hazards (Smith, Chap. 9).

**Lesson:** Develop new methods to calculate changes in natural hazard return periods that are influenced by climate change and incorporate these findings into new maps, plans, policies, building codes and standards, projects and tools (Birkmann and Pardoe; White; Smith, Chaps. 9 and 14).

## References

Alesch, D. J., & Siembieda, W. (2011). The role of the built environment in the recovery of cities and communities from extreme events. *International Journal of Mass Emergencies and Disasters, 30*(2), 197–211.

Aldrich, D. P. (2012). *Building resilience: Social capital in post-disaster recovery*. Chicago: The University of Chicago Press.

Baumgartner, F. R., & Bryan, D. J. (1993). *Agendas and instability in American politics*. Chicago: University of Chicago Press.

Berke, P. R., Kartez, J., & Wenger, D. (1993). Recovery after disasters: Achieving sustainable development, mitigation and equity. *Disasters, 17*(2), 93–109.

Berke, P. R., Godschalk, D., Kaiser, E. J., & Rodrigues, D. A. (2006). *Urban land use planning* (5th ed.). Urbana: Illinois University Press.

Berke, P., Smith, G., & Lyles, W. (2012). Planning for resiliency: Evaluation of state hazard mitigation plans under the disaster mitigation act. *Natural Hazards Review, 13,* 139–150.

Birkland, T. A. (1997). *After disaster, agenda setting, public policy, and focusing events*. Washington, D.C.: Georgetown University Press.

Burby, R. (1998). *Cooperating with nature: Confronting natural hazards with land-use planning for sustainable communities*. Washington, D.C.: Joseph Henry Press.

Burby, R. (2006). Hurricane Katrina and the paradoxes of government disaster policy: Bringing about wise government decisions for hazardous areas. *The Annals of the American Academy of Political and Social Science, 64*(May), 171–191.

Burby, R. J., & Dalton, L. (1994). Plans can matter! The role of land use and state planning mandates on limiting development in hazardous areas. *Public Administration Review, 54*(3), 229–238.

Chakraborty, A., Kaza, N., Knaap, G.-J., & Deal, B. (2011). Robust plans and contingent plans scenario planning for an uncertain world. *Journal of the American Planning Association, 77*(3), 251–266.

Comfort, l. (2005). Risk, security, and disaster management. *Annual Review of Political Science, 8,* 335–356.

Cuny, F. (1983). *Disasters and development*. Oxford: Oxford University Press.

Gibson, C. C., Andersson, K., Ostrom, E., & Shivakumar, S. (2009). *The Samaritan's dilemma: The political economy of development aid*. Oxford: Oxford University Press.

Godschalk, D. R. (2003). Urban hazard mitigation: Creating resilient cities. *Natural Hazards Review, 4*(3), 136–142.

Godschalk, D. R., Beatley, T., Berke, P., Brower, D. J., & Kaiser, E. J. (1999). *Natural hazard mitigation: Recasting disaster policy and planning*. Washington, D.C.: Island Press.

Harrell-Bond, E. (1986). *Imposing aid: Emergency assistance to refugees*. Oxford: Oxford University Press.

Innes, J., & Booher, D. (1999). Consensus building and complex adaptive systems: A framework for evaluating collaborative planning. *Journal of the American Planning Association, 65*(4), 412–423.

Innes, J., & Booher, D. (2003). Collaborative policymaking: Governance through dialogue. In M. A. Hajer & H. Wagenarr (Eds.), *Deliberative policy analysis: Understanding governance in a network society* (pp. 33-59). New York: Cambridge University Press.

Kartez, J., & Faupel, C. (1994). Comprehensive hazard management and the role of cooperation between local planning departments and emergency management offices. Unpublished paper.

Lyles, W., Berke, P., & Smith, G. (2013). Do planners matter? Examining factors driving incorporation of land use approaches into hazard mitigation plans. *Journal of Environmental Planning and Management.* doi:10.1080/09640568.2013.768973. http://dx.doi.org/10.1080/09640568.2013.768973. Accessed 6 Jan 2014.

Olshansky, R. B. (2006). Planning after Hurricane Katrina. *Journal of the American Planning Association, 72*(2), 147–153.

Olson, R. S., Olson, R. A., & Gawronski, V. T. (1999). *Some buildings just can't dance: Politics, life safety, and disaster*. Stamford: JAI Press.

Paton, D., & Johnston, D. (2006). *Disaster resilience: An integrated approach*. Springfield: Charles C. Thomas Publisher.

Paton, D., McClure, J., & Burgelt, P. T. (2006). Natural hazard resilience: The role of individual and household preparedness. In D. Paton & D. Johnston (Eds.), *Disaster resilience: An integrated approach*. (pp. 305–318). Springfield: Charles C. Thomas Publisher.

Quay, R. (2010). Anticipatory governance: A tool for climate change adaptation. *Journal of the American Planning Association, 76*(4), 496–511.

Sabatier, P. A., & Jenkins-Smith, H. C. (1993). *Policy change and learning: An advocacy coalition approach*. Boulder: Westview Press.

Schwab, J., Topping, K. C., Eadie, C. C., Deyle, R. E., & Smith, R. (1998). *Planning for post-disaster recovery and reconstruction*. Chicago: American Planning Association.

Smith, G. (2009). Disaster resilient communities: A new hazards risk management framework. In J. Pine (Ed.), *Hazards analysis* (Chap. 10, pp. 249–267). Washington, D.C.: Taylor Francis.
Smith, G. (2011). *Planning for post-disaster recovery: A review of the United States disaster assistance framework*. Washington, D.C.: Island Press.
Vale, L., & Campanella, T. (2005). *The resilient city: How modern cities recover from disasters*. New York: Oxford University Press.
Wisner, B., Blaikie, P., Cannon, T., & Davis, I. (2004). *At risk. Natural hazards, people's vulnerability and disasters*. London: Routledge.

# Index


**A**
Abel, N., 13
Adams, D., 274
Adams, J., 113
Adaptation, 1–7, 9–13, 15, 17, 20, 21, 25, 42–45, 50, 51, 57–60, 62–70, 79, 85, 94, 132, 133, 151, 152, 167, 171, 172
  aims of, 166
  Anticipatory adaptation, 3
  Autonomous adaptation, 3
  barriers to, 9–11, 407
  definition of, 27
  frameworks of, 64
  measures of, 4, 17, 45, 49, 53, 102, 132, 208, 214, 364, 410, 416, 417, 419, 427, 435
  planning, 13, 67, 175, 178, 187, 326, 423
  plans, 178, 182, 187
  policies on, 11, 58, 178, 181, 182, 187
  programmes on, 47, 53
  projects on, 64
  research on, 58
  strategies on, 44, 50, 52–54, 178, 209
  studies on, 10
Adaptive governance, 15, 59
Adger, W.N., 4, 9, 14, 17, 58, 211, 259, 392
Advisory Base Flood Elevation (ABFE) maps, 350, 352
Agrawal, A., 97
Albala-Bertrand, J.M., 17, 18
Alder, J., 310
Aldrich, D.P., 425
Alesch, D.J., 151, 425
Allen, B.L., 372
Allison, G.T., 275
Ames, D.V., 198
Anderson, L.E., 97
Anderson, M., 17
Anticipatory adaptation, 68
Anticipatory governance, 423
Assmuth, T., 257
Australia, 270, 271
Aven, T., 257, 390
Avoidance, 3

**B**
Baber, W.F., 262, 395
Babu, S., 294, 297, 298, 300, 301
Bachrach, P., 200, 211
Baker, I., 12
Ballinger, R.C., 329
Bang, H.P., 390
Bannock, S., 329, 332
Baratz, M., 200, 211
Barbier, E.B., 383
Barnes, J., 194, 195, 197
Barriers, 9
Barrios, R.E., 377
Barry, J., 364
Barry, J.M., 373
Bartholomew, K., 174
Bartlett, R.V., 239, 262, 395
Bashears, M., 125
Batker, D.P., 370, 393
Bauer, A.J., 11
Bau, M., 298
Baumgartner, F.R., 412
Beatley, T., 17, 18, 20, 24, 25, 124, 132, 133, 344, 345
Beban, J.S., 260
Becker, J.S., 328
Beck, U., 113
Bell, M.L., 124
Beniston, M., 78
Bennett, J., 79, 80
Benseman, M., 243
Benson, C., 46
Bergdoll, B., 132


B. C. Glavovic, G. P. Smith (eds.), *Adapting to Climate Change*, Environmental Hazards,
DOI 10.1007/978-94-017-8631-7, © Springer Science+Business Media Dordrecht 2014

Berghan, T., 250
Bergin, T., 385
Berkeley, 27, 148, 151, 160, 161, 163–167, 417
Berkeley Climate Action Plan, 160
Berkeley's resiliency approach, 160
Berke, P., 175, 178, 185, 201, 205, 219, 220, 326, 347, 348, 426
Berke, P.R., 17–20, 24, 25, 126, 407, 409
Berkes, F., 50
Berrang-Ford, L., 11
Bertram, G., 316
Betsill, M., 59
Bicknell, J., 14, 26
Biesbroek, G.R., 59
Biggs, R., 394
Birkland, T., 223
Birkland, T.A., 16, 223, 254, 412
Birkmann, J., 14, 15, 18, 21, 46–49, 57, 58, 273
Bjorndal, K.A., 383
Blackwood, P., 242
Blaikie, P., 273
Blakely, E., 14
Blanchard, T.C., 382
Blanchi, R., 271
Bloomberg, M.R., 58
Blum, M.D., 372, 374, 388
Blyth, M., 274, 276
Boesch, D.F., 370, 386
Boin, A., 275
Booher, D., 175
Boraine, A., 61
Bosomworth, K., 271
Boura, J., 323
Bourdieu, P., 392
Boyd, E., 1, 4, 9, 10
Boyer, H., 17
Boyte, H.C., 390
BP oil spill, 388
Bradstock, R., 270
Braman, D., 274, 285
Brinkley, D., 342
Brody, S., 175
Brooks, L., 179
Brooks, N., 6, 11
Brower, D., 133
Brower, D.J., 344, 345
Brown, K., 17
Brundritt, G., 61
Bryan, M., 389
Buckle, P., 127, 129
Bührs, T., 239
Building Codes, 4, 23, 199, 212, 215, 224, 347, 351, 354, 381, 387, 417, 419
Bulkeley, H., 25, 59
Burby, R., 173, 174, 221, 222
Burby, R.J., 17, 18, 23–25, 104, 248, 377, 388, 423
Burch, S., 4, 9, 12
Burgelt, P.T., 126, 429
Burris, S.C., 58
Burton, C.B., 155
Burton, I., 6, 24, 42
Bushfires, 270, 272, 281, 284, 285, 426
Butler, C., 282
Button, G.V., 16
Bye, P., 105

**C**

Campanella, R., 374, 376, 377, 386, 392
Campanella, T., 429
Campanella, T.J., 17, 19, 126, 134
Cannon, T., 17
Capability imperative, 436
Cape Town, 27, 58, 60, 61, 427
Carbonell, A., 14
Carey, M., 89, 94
Carlyon, G., 245
Carmin, J., 14
Carpenter, S.R., 19
Cartwright, A., 58, 61, 64
Carvalho, P., 16
Cary, G., 270, 272
Caviedes, C., 82
Chakos, A., 163
Chakraborty, A., 176, 177
Chambers, R., 7, 328
Chambers, S., 262, 395
Chamlee-Wright, E., 393, 396
Chapin III, F., 172
Chapin, T., 183
Cicin-Sain, B., 310
Cigler, B.A., 385, 386
Clark, C., 18
Climate adaptation, 171–173, 175, 176, 179, 182, 187, 188, 334
Climate change, 77–80, 82, 88–90, 92, 93, 95, 96
  impact of, 80, 90
  issues of, 93
Climate change adaptation, 2, 3, 7, 9, 11, 13, 17, 20, 22, 23, 25, 26, 28, 42, 50, 58, 64, 65, 67–69, 78, 80, 93–95, 97, 146, 168, 183, 193, 194, 199, 200, 205, 208, 209, 211–213, 217, 219–224, 251, 256–259, 262–264, 273, 308, 318, 333, 334, 343, 358, 359, 391, 398, 406, 407, 409–412, 414–418, 422, 423, 425–430, 432–435, 437, 438
  aspects of, 47

assessment of, 7
challenges for, 44, 145
fiscal aspect of, 22
initiatives of, 21, 326
issues of, 17, 226
literature on, 406
plans for, 12
policies on, 14, 209, 223, 316, 418, 425
projects on, 93
social barriers to, 10
strategies on, 9, 26, 27, 219, 225, 339, 364, 406
studies on, 17, 210
Climate change impacts, 3, 4, 12, 19, 65, 66, 77, 174, 185, 186, 238, 250, 334, 335, 370, 388, 392, 397, 438
Climate change issue, 1, 26, 48, 50, 51
Climate-related hazards, 2, 315, 319, 321, 322
Climate risk, 2, 5–8, 25, 70, 95, 178, 233, 238, 239, 251, 254, 259, 260, 263, 395, 405, 432
Coastal communities, 27, 63, 123, 125, 129, 130, 132, 133, 139, 141, 197, 198, 214, 216, 305, 310, 341, 343, 348, 351, 352, 354, 361, 363, 365, 412, 418, 419
Coastal hazards, 131, 233, 251, 310, 315, 321, 328
Coastal infrastructure, 137, 304, 319–321, 326, 333, 428
Coastal planning, 125, 130
Coastal resilience, 130, 131, 139, 140, 142, 394
principles of, 130
Cobb, J.C., 372, 386
Coffman, M., 282
Coleman, J.S., 393
Coleman, T., 271
Collaborative governance, 408, 409, 411, 439
Colten, C.E., 371, 373, 376, 377, 383
Comfort, l., 420
Comfort, L.K., 377
Community Action Programme for Seismic Safety (CAPSS), 158
Community behavior, 156
Community effectiveness, 155
Community efficacy, 146, 159
Community resilience, 2, 50
Concepts, 2, 14, 17, 24, 27, 45, 117, 129, 172, 256, 260, 273, 322, 324, 326, 427, 429, 431
Connor, R., 275
Considine, M., 274, 275
Conway, E.M., 1, 392
Conway, G.R., 7
Cook, A., 244
Cooke, B., 328
Cook, N.D., 85, 86
Cooney, C.M., 48
Coping, 80
concept of, 80
mechanism of, 324
strategies on, 80
Coping capacity, 6
Corburn, J., 57
Corfee-Morlot, J., 58
Costanza, R., 382, 393
Cox, P.A., 326, 331
Creed, W.E.D., 273
Cromwell III, J.E., 213
Cullen, H., 356
Culver, S.J., 198
Cumming, G.S., 50
Cunningham, L., 363
Cuny, F., 17, 18, 422
Cutter, S., 155

**D**

Dalton, L., 423
Daly, H.E., 393, 394
Daly, M., 321, 329, 332
Danielsen, F., 302
Dash, N., 363
Das, S., 302
David, B., 407
Davoudi, S., 20, 25, 179
Day J., 372, 374, 388
Day J.W., 374
Day, T.J., 238, 250
de Boer, J., 273, 276
Delia, A.A., 194
Deliberative delta governance, 394–398
Department of Housing and Urban Development, 204
Derthick, M., 385
Developing country, 63
Development, 318
Dewulf, A., 273, 277, 285
Deyle, R., 183
Deyle, R.E., 24, 175
Diamond, J., 9, 19
Dietrich, K., 59
DiMaggio, P.J., 274, 275
Disaster, 2, 6
Disaster governance, 390
Disaster management, 14, 45, 119, 302, 308, 319, 323, 333, 390
Disaster Mitigation Act of 2000 (DMA), 204, 220

Disaster recovery, 2, 14, 16, 17, 21, 24, 28, 133, 194, 199, 205, 207–209, 211, 213, 214, 217, 219–222, 224, 226, 301, 346, 348–351, 353, 358, 360, 364, 365, 416, 418, 420, 423, 424, 437
  definition of, 16
  strategies on, 15
Disaster risk, 6
Disaster risk escalation, 4
Disaster Risk Reduction (DRR), 2, 6, 7, 11, 15, 22, 27, 42, 43, 45–49, 51–53, 78, 80, 93, 95, 97, 272, 273, 279, 280, 281, 284, 285, 303, 308, 310, 388, 390, 396, 398, 410, 419, 426
  goals of, 50
  perspectives of, 284
  planning of, 274
Dixon, K.L., 198
Dodman, D., 14, 26
Dombeck, M.P., 284
Douglas, C.H., 322
Douglas, I., 109
Dovers, S., 17, 19, 272, 273, 275, 279
Dronkers, J., 3
Druckman, J.N., 273
Dryzek, J.S., 262, 395
Duany, A., 347
Dunbar, K.W., 89
Dunlap, R.E., 392
Dupre, J., 281
Durack, P., 271
Dyer, R., 272
Dymond, J., 244
Dyson, M.E., 373, 377

E
Economic resilience, 129
Eder, K., 279
Edwards, F.L., 375, 385
Eikeland, P.O., 59
Eisner, R., 155
Ekstrom, J.A., 9, 59
Elander, I., 58
Elmqvist, T., 326, 331
Emanuel, K., 340
Emergency housing, 201, 349, 352, 359, 416, 417
Emrich, C.T., 155
Enarson, E., 324
Environmental planning, 176
Environmental Protection Agency, 182, 225
Epstein, C., 274
Ericksen, N.J., 233
Ericksen, P., 59
Esnard, A.-M., 24, 356, 359
Espinoza Villar, J.C., 89
eThekwini municipality, 60
Evans, E., 105, 114
Exposure, 2, 3, 7, 16, 19, 43, 44, 52, 66, 87, 95, 106, 107, 117, 124, 137, 205, 233, 240, 241, 247, 248, 259, 261, 273, 279, 321, 334, 343, 363, 375, 377, 386, 388, 390, 392, 418, 419, 434, 437
  causes of, 4

F
Fairhurst, L., 64
Faupel, C., 426
Federal Emergency Management Agency (FEMA), 340, 427
Fenna, A., 275
Fernando, N., 47
Few, R., 25, 42, 67
Finch, C., 377
Fire management, 270–272, 276, 277, 280–285
Fischer, F., 262, 273
Flatt, V., 220
Fletcher, A.L., 277
Flooding, 15, 65, 82, 85, 88, 89, 102–104, 106, 107, 109, 110, 112, 113, 118, 120
Flood Insurance Rate Maps, 205, 207, 212, 348, 351, 414
Flood risk, 27, 46, 104–107, 111, 114, 115, 118, 232–234, 242, 244, 246, 247, 250, 255, 259, 263, 358, 410
  assessment of, 106, 111
  consequences of, 119
  evaluation of, 215
  management, 10
  management of, 102, 107, 109–111, 113, 115, 116–119, 238, 239, 241, 250, 251, 413
  source of, 102, 107, 118
Folke, C., 19, 59, 275, 392
Fordham, M., 18
Ford, J.D., 12
Freudenberg, W., 200
Freudenburg, W.R., 364, 374, 375, 385, 386, 391
Fuller, I., 235

G
Gaillard, J.C., 6, 17, 21, 80
Gailus, J., 221
Gailus, J.L., 18
Gallardo, M., 94
Gallopin, G.C., 19

Gamson, W.A., 277
Garnett, J.D., 26
Gawronski, V., 16
Gawronski, V.T., 16, 216
Geipel, R., 24
Geis, D., 18
George, A., 306, 308, 309
Gibb, J., 315
Gibson, C.C., 18, 422
Giesecke, A., 82
Gill, A.M., 270, 284
Gill, D.A., 382
Gillen, M., 270
Girard, C., 363
Glavovic, B.C., 14, 17–19, 23, 24, 238, 241, 256, 260, 262, 390, 394, 395
Gobel, B., 87
Godschalk, D., 172, 173, 175, 407
Godschalk, D.R., 17–19, 23–25, 126, 127, 344, 345, 423
Goffman, E., 273
Goldammer, J.G., 272
Gomathy, N.B., 294, 299
Goodwin, E., 237, 250
Gotham, K.F., 392
Governance, 5, 11, 12, 59, 60, 102, 107, 262, 272, 273, 276, 277, 309, 386, 390, 394, 398, 412
  arrangements, 186
  Collaborative governance, 6, 17, 21, 26, 28, 413, 421–424, 427, 429, 431, 434
  frameworks of, 102, 103, 109, 119
  issues of, 52
  networks, 417, 418, 429, 438
  systems, 415
Governance imperative, 434
Governor's Commission on Recovery, Rebuilding and Renewal, 346, 352, 361
Governor's Office of Recovery and Renewal, 346, 348–350, 352, 353, 361
Gramling, R., 385, 386, 391
Granberg, M., 58
Grist, N., 17
Gunderson, L.H., 19
Gupta, J., 9

**H**
Hagemann, M., 14
Hajer, M.A., 277
Halperin, M.H., 275
Hancox, G.T., 235
Handmer, J., 17, 19, 24, 259, 271–273, 279
Hansen, L.J., 353
Hanson, S., 64
Hansson, S.O., 113
Harrell-Bond, E., 18, 422
Hart, P.t., 276
Harvey, D., 104
Hauknes, J., 274
Hawkins, R.L., 377
Haynes, K., 325
Hazard mitigation, 14, 17, 23–25, 27, 28, 127, 172–175, 183, 187, 188, 194, 199–205, 207, 208, 210, 211, 213–216, 218, 219, 221, 222, 224–226, 328, 329, 340, 344–347, 351, 353, 354, 358, 360, 362–365, 379, 381, 414, 416, 417, 423–427, 429, 430, 437
  measures of, 22
  strategies on, 24
Hazard Mitigation Grant Programme (HMGP), 202, 355
Hearn, P.D., 341, 342, 344, 345
Hedao, P., 302
Heinrichs, D., 58
Helmke, G., 274
Hennessy, K., 233, 270
Higgins, H., 183
High, C., 274
Hoffman, A.J., 375
Hoffman, J.R., 353
Hoffman, S.M., 18
Holling, C.S., 18, 19, 79
Hooper, A., 324
Horner, M., 105
Horton, R., 57
Howe, J., 114
Huq, S., 49, 50
Hurlimann, A.C., 25
Hurricane Camille, 341, 343–346, 351, 361, 362, 373
Hurricane Floyd, 194, 195
Hurricane Fran, 195, 197, 200–203, 218
Hurricane Katrina, 16, 125, 136, 216, 220, 340, 341, 346, 348, 350, 352, 358, 362–365, 374, 376, 417, 421, 424

**I**
Implementation, 5, 6, 13, 22, 42, 109, 175, 177, 178, 182, 185, 187, 188, 194, 217, 219, 220, 224, 250, 276, 319, 323, 326, 329, 334, 336, 359, 381
  tools, 131, 332
Incoll, R., 270
Inderberg, T.H., 59
Ink, D., 385
Innes, J., 175, 407
Institutional analysis, 280, 281

Institutional barriers, 7, 10, 25, 258, 259, 263, 392, 406, 408, 430, 431, 436
Institutions, 88, 97, 141, 273, 274, 276, 277, 281, 282, 340, 393, 418, 422
Integrated Hazard Risk Management (IHRM), 207, 209, 210, 212
Isendahl, N., 277

**J**
Janssen, S.J., 382
Jasanoff, S., 282
Jenkins, K., 388
Jenkins-Smith, H.C., 278, 412
Jennings, P.D., 375
Jenwick, J., 316
Johnson, C.L., 119
Johnson, L.A., 147
Johnston, D., 17, 19, 146, 429
Jones, B.D., 412
Jones, L., 9, 10
Jones, P., 319
Jones, R., 271
Jones, S.A., 260
Jordan, A., 58
Juhasz, A., 386
Juhola, S., 9
Jurkiewicz, C.L., 375, 376, 386

**K**
Kahan, D.M., 274, 285
Kaiser, E., 173
Kaiser, E.J., 407
Kartez J., 426
Kartez, J., 24, 409
Kates, R.W., 17, 377
Kates, W., 24
Kathiresan, K., 302
Kay, R., 310
Keay, J., 256
Kelman, I., 6, 80
Kemp, R., 59
Kessler, R.C., 377
Khademian, A.M., 275
Kingdon, J., 119
King, R.O., 376
Klinenberg, E., 16
Klinke, A., 6, 257, 261, 390, 392
Knecht, R.W., 310
Knight, F., 257
Knight, F.H., 6
Koch, P., 274
Koenig, T., 277, 278
Ko, J., 374
Koliba, C.J., 385
Kooiman, J., 5, 258, 262, 390
Kothari, K., 328
Kraft, K., 85
Kroon, M.B.R., 276
Kumar, S., 328
Kunreuther, H., 24
Kurian, V., 305
Kurien, J., 294
Kutzmark, T., 18

**L**
Ladd, A.E., 386
Lakshmi, A., 298, 300, 301, 306, 308–310
Landscape resilience, 135, 140
Land use, 9, 20, 21, 23, 24, 198–200, 205, 208, 217, 244, 246, 248, 253, 319, 345, 362, 421, 425, 426, 437
Laska, S., 376, 389
Lauth, H.-J., 274
Lavell, A., 79, 80
Lavorel, S., 284
Law, F.M., 103
Lawrence, J., 241, 252
Lawson, N., 11
Lay, T., 291
Lea, J.P., 319
Leavell, D.N., 82
Lebel, L., 9, 10
Lee, B.-Y., 239
Lee, M.R., 382
Lees, L., 280
Leichenko, R.M., 271
Lemos, M., 6, 14, 17
Lemos, M.C., 390
Lenhart, S.S., 97
Lenschow, A., 274
Lerner, S., 372
Lesnikowski, A.C., 11
Lester, M., 16
Levin, K., 10, 257
Levitsky, S., 274
Lewis, J., 322
Lindahl, C., 373
Lipton, J., 85, 87, 89, 90, 96
Livelihoods, 1, 9, 79, 87, 90, 96, 232, 234, 235, 299–301, 371, 397, 405
Living with water, 393
Local capacity, 69, 348
Local community development, 2, 4, 405, 434–436
Local government, 12–14, 23, 24, 63, 66, 68–70, 124, 173, 185, 187, 207, 210, 213, 218, 220, 226, 238, 239, 241, 250, 354, 362, 409, 410, 421

Loss and damage, 45, 51, 53
Lowe, I., 245, 247
Lowndes, V., 274
Lucas, C., 270
Lustgarten, A., 385, 386
Lyles, L.W., 24, 25
Lyles, W., 178, 201, 205, 219, 220, 426
Lynn, K., 271

**M**
Macchi, M., 77, 78
Maceda, E., 325
MacGregor, D.G., 275
Mackay, M., 246
Mackinson, S., 326
Mainster, J., 363
Mainstreaming, 12, 28, 70, 257, 258, 275, 302, 333, 398, 411, 426
Malakoff, D., 383
Mallinson, D.J., 198
Manawatu, 233–235, 410
Manuel-Navarrete, D., 11, 20
March, A.P., 25
March, J.G., 274, 276
Martens, P., 59
Masozera, M., 377
Mather, A., 66
Maurer, K., 377
May, P., 24
May, P.J., 17
McAnany, P.A., 9, 19
McBean, G., 1, 2, 4
McClure, J., 126, 252, 429
McCright, A.M., 392
McEvoy, D., 20
McGranahan, G.D., 295
McHarg, I., 393
McPherson, M., 125
Measham, T.G., 12
Mehmood, A., 179
Meinzen-Dick, R., 282
Mendelsohn, R., 271
Mercer, J., 321
Merz, B., 117
Methodology, 110–113, 118, 324, 330
Migration, 44, 87, 96, 97, 197, 198, 352, 421
Mileti, D., 197, 221, 222
Mileti, D.S., 16–18, 23, 24, 173
Miller, L.L., 277
Mismatches, 47–51, 53, 54, 407
Mississippi Alternative Housing Program, 348, 359
Mississippi delta, 28, 370, 391, 397, 413, 415, 425, 428
Mississippi Renewal Forum, 347–349, 354
Mitchell, T., 94
Mitigation planning, 160, 166, 173
Moench, M., 42
Molotch, H., 16
Moore, M., 26
Moral imperative, 408, 429, 431–433, 438, 439
Morris, C., 370, 375, 386
Morrow, B., 376, 389
Morrow, B.H., 324, 363
Moseley, M., 81, 85
Moser, S.C., 9, 59
Mouillot, F., 272
Mukheibir, P., 58
Munasinghe, M., 18
Murphy, B.L., 393
Murra, J.V., 85
Murthy, K.S.R., 295
Mutter, J., 386

**N**
Nadkarni, N., 14
Naidu, S., 63
Nakagawa, Y., 18
Nakićenović, N., 179
National Adaptation Programmes of Action (NAPAs), 5, 45, 49
National Disaster Recovery Framework (NDRF), 218, 220
National Flood Insurance Program, 207, 216, 218, 225, 343, 348, 351, 353, 362, 414
National Oceanic and Atmospheric Administration (NOAA), 225, 419
Natural hazards planning, 2, 18, 23, 25, 241, 248, 253, 255, 257, 258, 260, 262, 264, 275, 390, 393, 394, 406–408, 423–426, 428–430, 432–434
Natural hazards risk management, 2, 6, 7, 9, 11, 13–15, 17, 18, 20–24, 26, 205, 208, 212, 219, 221, 222, 224, 226, 240, 245, 246, 253, 254, 261, 406, 407, 409, 411, 412, 415, 418, 420, 423, 426–435, 438, 439
  aspects of, 21
  evolution of, 16
  fiscal aspect of, 22
  frameworks of, 21
  policy framework, 225
  strategies on, 9, 22
Needs assessment, 306, 311, 410
Neighbourhood Empowerment Network (NEN), 156
Nelson, D.R., 17, 20, 51

Nelson, L., 59
Nelson, M., 377
Neumann, J.E., 213
Newell, P., 390
Newman, J., 390
Newman, P., 17
New Orleans, 137, 341, 342, 371–373, 375–381, 384, 389, 424
New Orleans principles, 134, 136
New Urbanists, 348, 350, 358, 359, 363
New Zealand, 14, 27, 232, 233, 241, 250, 318, 324, 410, 413, 426
Norms, 4, 5, 25, 50–52, 54, 119, 262, 274, 407
Norris, F.H., 146
North Carolina Floodplain Mapping Programme, 207
North Carolina Sea Level Rise Study (NCSLRS), 207, 209, 210, 212
Norton, R., 175
Nottestad, L., 326

**O**

O'Brien, G., 17
O'Brien, K., 42, 51
O'Brien, K.L., 271
O'Keefe, P., 17
Oliver-Smith, A., 18, 24, 85, 375
Olsen, J.P., 274
Olsen, S., 394
Olsen, S.B., 394
Olshansky, R.B., 24, 377, 420
Olson, R.A., 16, 216, 412
Olson, R.S., 16, 216, 412
Oreskes, N., 1, 392
Orlove, B., 79, 90, 91
Ostrom, E., 274, 282
Ouroussoff, N., 132

**P**

Paleo, U.F., 14, 24
Pall, P., 107
Paradigm, 331, 391, 411, 413, 423
Parent, W., 372, 375
Parker, D.J., 104
Parreño, J.C., 64
Parsons, W., 275
Participatory approach, 310, 430
Passive survivability, 136, 137, 140
  concept of, 141
Paton and Johnston disaster resiliency model, 152, 156
Paton, D., 17, 19, 126, 127, 146, 321, 326, 328, 429
Patterson, R.G., 208
Peacock, W.G., 172, 363
Pearce, D.G., 318
Peduzzi, P., 44
Pelling, M., 11, 16–18, 20, 26, 44, 51, 272
Peñalba, L.M., 9
Penuelas, J., 272
Perrow, C., 119
Peru, 77, 78, 81, 85, 86, 89, 92, 94, 97, 415
Picketts, I.M., 25
Pidgeon, N., 277, 282
Pielke, R.A., 57
Pielke, R.A. Jr., 345
Pilkey, O.H., 197, 198
Piper, J., 25
Pitt, M., 106
Planned adaptation, 3, 65
Planning imperative, 426, 429, 437
Plater-Zyberk, E., 347
Platt, R., 24
Policy, 78, 79, 106
Policy change, 68, 119, 194, 199, 200, 206, 210, 213
Policy frames, 272–274, 276, 277, 283, 285, 430
Porter, L., 25
Portes, A., 393
Post-Katrina Emergency Management Reform Act (PKEMRA), 220
Powell, W.W., 274, 275
Pre-event planning, 16, 26, 200, 212, 214, 221, 224, 225, 359, 365, 398, 418, 420, 422–424, 428, 433, 437, 438
Preston, B.L., 12
Price, C., 272
Price, J.C., 213
Private adaptation, 3, 5
Public adaptation, 3
Public policy, 258, 261, 263, 276
Puszkin-Chevlin, A., 18, 23
Putnam, R.D., 141, 393

**Q**

Quade, D., 241, 252
Quay, R., 174–177, 423

**R**

Rabinovici, S.J.M., 164
Rajendran, N., 302
Ramesh Kumar, M.R., 304
Reactive adaptation, 3
Ready, S., 316
Rearic, D.M., 317
Reflexive learning, 25, 26, 276, 280, 282–285, 396

Reice, S.R., 197
Reid, P., 237, 238
Rein, M., 273
Reisinger, A., 233, 252
Reith, G., 118
Renaud, F., 89
Renn, O., 6, 257, 261, 390, 392
Residential Seismic Retrofitting Incentive Programme, 163
Resilience, 17, 19
Resilience planning, 396, 398, 437
Resiliency, 149, 154, 162, 166, 167
  indicators of, 155
  issues of, 160
  planning for, 159
  programmes on, 167
  strategies of, 157
Rhie, C., 14
Rhodes, R.A.W., 5, 390
Richards, J., 114
Riggs, S.R., 195, 198
Risk governance, 14
Risk management, 14
Risk reduction, 8, 14–17, 24, 45, 48, 49, 53, 54, 159, 163, 166, 205, 215, 225, 252, 261, 271, 353, 358, 362, 416, 422, 426, 430, 431, 434, 436–438
  approaches to, 145
  measures for, 158
Rittel, H.W.J., 257
Riverine flooding, 123, 371
Roberts, D., 58, 62, 63, 66
Roberts, G., 332
Roberts, H.H., 372, 374, 388
Rodgers, C., 1, 2, 4
Rodrigues, D.A., 407
Romero-Lankao, P., 57
Romieu, E., 8
Rose, J., 17
Roths, R.J., 24
Rouse, H., 233, 250, 251, 260
Rubin, C., 16, 223

S
Sabatier, P.A., 412
Saikku, M., 370
Salagrama, V., 300
Salt, D., 17
Samoa, 28, 315–320, 322, 324, 326, 328, 335, 336, 418, 420
San Francisco, 27, 145–148, 152, 156–159, 166, 167, 410
San Francisco Bay Conservation Development Commission (BCDC), 148
San Francisco Climate Action Plan, 168
San Francisco Mayor's office, 153
San Francisco Planning and Urban Research Association (SPUR), 154
San Francisco resiliency effort, 166
San Francisco resiliency movement, 153
Sankar, S., 304
Sardans, J., 272
Satterthwaite, D., 14, 26, 57, 58
Saunders, W.S.A., 260
Scales, 2, 11, 12, 14, 17, 21, 23, 25–27, 48, 49, 51, 53, 128, 131, 141, 174, 263, 409, 413, 428, 436
Scharpf, F.W., 274
Schipper, E.L.F., 17, 171
Schipper, L., 17, 18, 21, 26, 42, 44, 272
Schmidt, V.A., 275, 277
Schneider, R.O., 18
Schneider, S.K., 376
Schon, D., 273
Schultz, P., 163
Schwab, A., 133
Schwab, J., 24, 423
Scott, W.R., 274
Sea level rise, 61, 64, 123, 124, 128, 130, 132, 134, 147, 148, 161, 165, 171, 173, 179, 182, 183, 185, 186
Shaffer, G.P., 374, 380
Shawm, R., 326
Shaw, R., 18
Shea, E., 322
Shepherd, J., 244
Siembieda, W., 151, 425
Silgado, E., 82
Silliman, B.R., 382
Smit, B., 7, 16, 18, 24, 25
Smit, G., 16, 23, 24
Smith, A., 388
Smith, G., 24, 25, 146, 152, 153, 166, 178, 199–202, 205, 208, 211, 216–221, 224, 340, 347, 358, 360, 363–365, 409–411, 416, 420, 422, 423, 426
Smith, J.B., 97, 213
Smith, L., 64
Smith-Lovin, L., 125
Smith, R.A., 175
Smith, W., 235, 241
Social capital, 128, 140, 141, 152, 154, 157, 166, 392–394, 396, 437
Social exclusion
  causes of, 10
Solnit, R., 393
Solomon, G.M., 382
Song, Y., 185, 347, 348

South Africa, 58, 60–62, 67, 427
Spangle, W., 24
Spatial planning, 46, 109, 115, 119, 182
Speck, J., 347
Spence, A., 277
Sperling, F., 81, 89, 90, 95, 97
Stakeholder responsibilities, 110, 115
State Acquisition and Relocation Fund (SARF), 206, 215, 218
State Disaster Management Authorities (SDMAs), 309
State disaster recovery organization, 216, 360
State disaster recovery programmes, 200
State recovery plan, 217, 218
State recovery programmes, 203, 206, 217, 218
Stern, N., 1
Stern, N.H., 1
Stevens, M., 24, 185, 347, 348
Stirling, A., 6, 257
Stoker, G., 390
Stokstad, E., 383
Storbjörk, S., 9
Storch, H., 58
Storr, V.H., 393, 396
Strategic National Action Plans (SNAP), 46
Subramanian, B.R., 294
Surface water flooding, 106, 108, 111, 112, 114
Sustainable development, 6, 17, 18, 24–26, 308, 318, 322, 358, 395, 429, 438
  literature on, 18
Sustainable disaster recovery, 14, 201, 206, 220, 427
Sustainable livelihoods, 7, 256, 398
Sutter, C., 64
Swart, R., 179
Sweetman, C., 324
Sylves, R., 224

T
Tadross, M., 64
Tam, L., 147
Tanali, T.R., 16
Tang, Z., 13
Taule'alo, T., 318
Technical Assistance, 205, 213, 348, 426
Temporary Housing, 340, 359, 360, 417
Termeer, C., 9, 10
Thiele, L.P., 272
Thomalla, F., 17
Thomas, C.D., 272
Thomas, R.B., 81
Thomsen, D.C., 13
Tibbits, A., 271
Tierney, K., 14, 390
Tobin, G.A., 20
Tobin, L.T., 163
Tompkins, E.L., 7, 11, 12, 14
Topping, K., 160
Törnqvist, T.E., 372
Toulmin, C., 49, 50
Trigoso Rubio, E., 78, 90, 94
Tryhorn, L., 271
Tschakert, P., 59
Tsunami, 2, 6, 28, 123, 241, 291, 292, 294, 295, 297–299, 301, 302, 305–310, 315, 317, 320, 334, 416, 418
Turner, B.L., 256, 273
Twining-Ward, L., 318

U
United Nations Framework Convention on Climate Change (UNFCCC), 2, 12, 42, 45, 53, 272
Unnikrishnan, A.S., 304
Urban city-scale adaptation, 27
Urban planning, 173, 174, 319
Urwin, K., 58

V
Vale, L., 429
Valency, R.A., 329
van den Belt, M., 262
van Heerden, I., 389
van Zeijl-Rozema, A., 59
Venhaus, H., 135
Village consultation, 319, 324, 325, 328
Vincent, J.R., 302
Vogel, J.M., 213
von Teichman, K., 46, 47, 48
Von Teichman, K., 15, 18
Vulnerability, 2, 3, 6, 7, 9, 16, 19, 20, 23, 26, 42–44, 48, 49, 52, 77, 78, 80, 82, 85–88, 93, 94, 97, 102, 117, 125, 127, 137, 139, 178, 181, 183, 185, 187, 197–199, 209, 210, 213, 216, 222, 256, 259, 264, 272, 273, 277, 278, 284, 285, 300–302, 305, 310, 311, 319, 334, 348, 363, 372, 375–377, 384, 386, 389, 390–392, 397, 398, 408, 411, 417–419, 421, 431–434, 436, 437
  assessment of, 7
  causes of, 4, 10, 11, 19
  social, 18
  studies of, 10

**W**

Wagenaar, H., 277
Walker, B.H., 17, 18
Walker, B.S., 126
Wamsler, C., 11
Wandel, J., 7, 16, 18, 24, 25
Ward, C., 235
Watson, B.E., 324
Watson, J., 247, 249
Watson, R.T., 392
Watters, R.F., 316
Waugh, J., 233, 238
Weber, E.P., 275
Weber, M.M., 257
Welsh, M.G., 24, 356, 359
Wenger, D., 200, 208, 221, 409
Westerhoff, L., 9
Westlake, S., 250
Wheeler, S.M., 13, 171, 173
White, G., 197
White, G.F., 6
White, G.R., 24
White, I., 103, 104, 106, 114
Wilbanks, T., 57
Wilbanks, T.J., 17
Wilkinson, C., 25
Williams, W., 24
Wilson, A., 136, 137
Wilson, E., 25
Window of Opportunity, 68, 194, 200, 210, 254, 276, 302, 397, 412
Winsvold, M., 59, 60
Winterhalder, B., 81
Wisner, B., 17, 19, 146, 151, 167, 375, 432
Woodcock, F., 316
Woodrow, P., 17
Woods, C., 372, 386
Wright, K., 235
Wuerthner, G., 279
Wynne, B., 113, 282

**Y**

Yoffee, N., 9, 19
Young, K., 85, 87, 89, 90, 96

**Z**

Zevenbergen, C., 112
Ziervogel, G., 58, 59, 64, 66
Zito, A.R., 274